Masters of Theory

Masters of Theory
Cambridge and the Rise of Mathematical Physics

ANDREW WARWICK

The University of Chicago Press
Chicago and London

Andrew Warwick is senior lecturer in the history of science at Imperial College, London. He is coeditor of *Teaching the History of Science* and *Histories of the Electron: The Birth of Microphysics*.

The University of Chicago Press, Chicago 60637
The University of Chicago Press, Ltd., London
© 2003 by The University of Chicago
All rights reserved. Published 2003
Printed in the United States of America

12 11 10 09 08 07 06 05 04 03   1 2 3 4 5

ISBN: 0-226-87374-9 (cloth)
ISBN: 0-226-87375-7 (paper)

Library of Congress Cataloging-in-Publication Data

Warwick, Andrew.
    Masters of theory : Cambridge and the rise of mathematical physics / Andrew Warwick.
       p.   cm.
    Includes bibliographical references and index.
    ISBN 0-226-87374-9 (cloth : alk. paper) — ISBN 0-226-87375-7 (pbk : alk. paper)
    1. Mathematical physics—History—19th century. 2. University of Cambridge—History—19th century. I. Title.
    QC19.6 .W37   2003
    530.15′09426′59′09034—dc21

                                                        2002153732

∞ The paper used in this publication meets the minimum requirements of the American National Standard for Information Sciences—Permanence of Paper for Printed Library Materials, ANSI Z 39.48-1992.

Contents

List of Illustrations ～ *vii*
Preface and Acknowledgments ～ *ix*
Note on Conventions and Sources ～ *xiii*

1. **Writing a Pedagogical History of Mathematical Physics** ～ 1

2. **The Reform Coach** Teaching Mixed Mathematics in Georgian and Victorian Cambridge ～ 49

3. **A Mathematical World on Paper** The Material Culture and Practice-Ladenness of Mixed Mathematics ～ 114

4. **Exercising the Student Body** Mathematics, Manliness, and Athleticism ～ 176

5. **Routh's Men** Coaching, Research, and the Reform of Public Teaching ～ 227

6. **Making Sense of Maxwell's** *Treatise on Electricity and Magnetism* **in Mid-Victorian Cambridge** ～ 286

7. **Joseph Larmor, the Electronic Theory of Matter, and the Principle of Relativity** ～ 357

8. **Transforming the Field** The Cambridge Reception of Einstein's Special Theory of Relativity ～ 399

9. **Through the Convex Looking Glass** A. S. Eddington and the Cambridge Reception of Einstein's General Theory of Relativity ～ 443

Epilogue: Training, Continuity, and Change ～ 501

Appendix A: Coaching Success, 1865–1909 ～ *512*
Appendix B: Coaching Lineage, 1865–1909 ～ *524*
Bibliography ～ *527*
Index ～ *549*

Illustrations

1.1. Derivation of the Poynting Vector   5
1.2. Poynting's 1876 Examination Script (investigation)   19
1.3. Poynting's 1876 Examination Script (problem)   20
1.4. Pages from Newton's copy of Barrow's 1655 Latin edition of Euclid's *Elements*   31
2.1. Cambridge University Senate House   54
2.2. William Hopkins   81
2.3. Title Page of J. M. F. Wright's *The Private Tutor*   83
2.4. The "Star of Cambridge" Stagecoach   91
2.5. A College Lecture   93
3.1. Smith's Prize Questions   121
3.2. A Disputation   123
3.3. Written Examination in Trinity College   124
3.4. Examination Script by William Garnett   165
4.1. "Keeping an Exercise"   193
4.2. Boat Races on the river Cam, 1837   195
4.3. "Training"   196
4.4. St. John's College Rugby Team, 1881   199
4.5. The Order of Merit in the *Times* newspaper, 1909   203
4.6. Reading the Order of Merit   204
4.7. Presentation of the Senior Wrangler   207
4.8. Ceremony of the Wooden Spoon   209

4.9. Parading the Wooden Spoon outside the Senate House   210
4.10. The Trinity First Boat   220
5.1. Edward John Routh   232
5.2. Syllabus of the Mathematical Tripos, 1873   239
5.3. The Senior Wrangler in his Rooms   245
5.4. Robert Rumsey Webb   248
5.5. Robert Webb's Coaching Notes   250
5.6. Francis S. Macaulay   259
5.7. Title Page of Wolstenholme's *Mathematical Problems*   262
5.8. Professorial and Intercollegiate Lectures in Mathematics, 1880–1881   268
5.9. Robert Herman   283
6.1. Pedagogical Geography of Mid-Victorian Cambridge   290
6.2. Routh's Lecture Notes on Electrostatics   308
6.3. Examination Script by George Pitt   312
6.4. Examination Script by Joseph Ward   314
6.5. Routh's Team for the Mathematical Tripos of 1880   316
6.6. W. D. Niven   318
6.7. Examination Question, 1886   341
7.1. A Torchlight Procession in Belfast   365
7.2. Tripos Questions, 1901 and 1902   378
7.3. Jeans's *Mathematical Theory*, Articles 257, 258   383
8.1. Ebenezer Cunningham   410
8.2. Harry Bateman, P. E. Marrack, and Hilda Hudson   417
8.3. Jeans's *Mathematical Theory*, Articles 226, 227   419
8.4. Jeans's *Mathematical Theory*, Articles 231, 232   420
8.5. Lecture notes by L. H. Thomas   438
9.1. A. S. Eddington (1904)   450
9.2. Alice through the Convex Looking Glass   466
9.3. Lecture notes by L. H. Thomas   481
9.4. Tripos Problems, 1927 and 1929   488

Preface and Acknowledgments

This study offers a new account of the rise of modern mathematical physics. The history of the mathematical sciences has until recently played a relatively minor role in the development of social and cultural studies of science. As an eminent historian in this field remarked to me a few years ago, the project of writing a cultural history of mathematical physics sounded to many like a contradiction in terms. Taking Cambridge University as my example, I contend in this study that mathematical physics has a necessarily rich culture which is most effectively made visible through the history of training, the mechanism by which that culture has been reproduced and expanded for more than two hundred years. We tend to think of "theories" as collections of propositions or fundamental mathematical equations, but this is how they appear to those who have already mastered the physicist's craft. It takes special aptitude and more than a decade's education to make human beings who can see a theory in a jumble of mathematical squiggles, and I am interested in how that shared vision is produced.

My approach is naturalistic in that it explores theory not in terms of method or logic but through the experience of the student struggling to become its master. Nor do I treat theorizing as a uniquely cerebral process. Historians of experiment now speak routinely of machines, technicians, and embodied practical skills, but what of the material technology and skilled practices of theoreticians? Cambridge undergraduates in the late eighteenth century found it very odd that they were being required to learn by sketching and writing rather than by reading and talking, and to display their knowledge by solving mathematical problems on paper instead of disputing in Latin. Their testimony highlights the peculiarity of mathematical knowledge made using hand, brain, paper, and pen. It is part of my task to understand how these new skills were taught and learned, and to explore the relationship between drawing, writing, and knowing. I also contend that *how* and *by whom* students were taught is as important as *what* they were taught. The mathematical coaches who trained and motivated the Cambridge elite placed great emphasis on pedagogical technique, and required their students to spend long hours rehearsing mathematical methods in the solitude of their college rooms. The mathematician's body, gender, and sexuality were also implicated in the making of the modern theoretician. Those who would master celestial mechanics in Victorian Cambridge believed that competition on the playing field was as important as competition in the classroom, that irregular sexual activity sapped a man's strength, and that women's minds and bodies could not withstand these trials. I want to understand how these embodied values contributed to what contemporaries characterized as the industrialization of the learning process.

Above all, my study is a historical one. A good deal has been written about the rise of big, mainly experimental science in the twentieth century, but mathematical theory got big, at least in relative terms, a century earlier. It did so by successfully colonizing undergraduate studies in the expanding universities. My interest is focused not so much on why that colonization occurred as on its consequences for the emergence of mathematical physics as a discipline. That said, the first half of this study is as much a contribution to institutional and educational history as it is to the history of science. In my opinion historians have yet to take on board the profound changes wrought in the kind of knowledge prized in the universities as essentially medieval practices of teaching, learning, and examining were swept away by a new pedagogical economy of model problem-solutions, private tutoring, and paper-based learning and examination. I am also concerned with the way novel theories are made teachable, the power of training to shape research, and the several and contradictory ways in which training regimes are conservative. On the one hand, training provides a mechanism by which the esoteric culture of mathematical physics is preserved and replicated at new sites; on the other, the peculiarities of each site produce distinct and local research cultures. Attempts to exchange knowledge across these boundaries are especially interesting as they highlight tacit forms of expertise not carried in published works. The second half of my study explores these issues by looking at the establishment and operation of Cambridge's first research school in mathematical physics (electromagnetic theory) and the Cambridge reception of the special and general theories of relativity. These examples illustrate my more general thesis that the cultural history of mathematical physics is not an alternative to its technical history but an explanation of how the latter was made possible.

I began this study as a doctoral student in the Department of History and Philosophy of Science at Cambridge University during the mid-1980s. It was a rare privilege to study in that exciting and stimulating intellectual environment, and I would like to thank all the teaching staff and research students with whom I then worked. In particular I should mention my supervisor Mary Hesse, Simon Schaffer, whose inspiring intellectual leadership awakened my interest in the social history of science, and two especially like-minded doctoral students, Rob Iliffe and Iwan Morus. It has been my good fortune in subsequent years to meet many scholars who offered friendly encouragement and freely shared with me their own insights and expertise. In most cases the studies I have found useful are cited at appropriate points in the book, but I would like here to mention the following people who were kind enough to read and comment on one or more draft chapters: Ron Anderson, Jed Buchwald, Tony Crilly, Olivier Darrigol, David Edgerton, Moti Feingold, Ivor Grattan-Guinness, Bruce Hunt, Myles Jackson, Chris Lawrence, John Norton, Joan Richards, John Stachel, Chris Stray, and Alison Winter. Special thanks are due to Rob Iliffe and Yves Gingras, both of whom read the entire typescript and offered penetrating criticism and helpful advice in equal measure. I doubt that any of those mentioned above would agree entirely with the arguments and claims made in the following pages, and any errors are my own. This study could not have been completed without the generous institutional and financial support provided by the British Academy, the Dibner Institute (MIT), Imperial College (London), St John's College (Cambridge), and the Verbund fur Wissenschaftsgeschichte (Berlin). I am much indebted to my commissioning editor, Susan Abrams, for rescuing me from the publishing doldrums and for sage advice on how to turn a typescript into a book. The latter process was greatly assisted by my copy editor, Richard Allen, who also prompted me to clarify and sharpen a number of important points in the text. My most heartfelt thanks are reserved for my wife Suzanne and for our two sons James and Michael. Without their love, patience, and indulgence this project would not have come to fruition.

Note on Conventions and Sources

Almost all of the historical characters referred to in this study are graduates of the Cambridge Mathematical Tripos. Since it is often useful to inform or remind the reader of a character's position in the order of merit and year of graduation, I shall provide this information in abbreviated form after the first, and occasionally subsequent, mentions of the character's name. Thus James Clerk Maxwell (2W 1854) indicates that Maxwell was "second wrangler" in the Mathematical Tripos of 1854. The letters "SW," "SO" and "JO" used in the same convention refer respectively to the senior wrangler, the senior optimes, and the junior optimes (these terms are explained in chapter 2).

References to unpublished letters and manuscripts will be followed by a letter code in parentheses indicating the collection and archive or library in which the materials are held. Thus "Box 5 Add 7655/vk/8, verso (JCM-CUL)" refers to a numbered manuscript sheet in the James Clerk Maxwell collection held at Cambridge University Library. A complete list of these codes and the collections to which they refer is provided at the beginning of the bibliography.

When citing published scientific and mathematical papers, I shall whenever possible refer to edited collections of the author's work. In these cases the collection is identified by the author's name and an abbreviated form of the title. In the case of *The Collected Papers of Albert Einstein* (cited as: Einstein, *Collected Papers*) I shall whenever practicable refer to sections within the numbered documents. This will enable the reader to refer with equal ease to the original German language edition or English language translation of the papers. All other citations of published sources refer by author surname and year of publication to the bibliography.

Writing a Pedagogical History of Mathematical Physics

> Late to bed and early rising,
> Ever luxury despising,
> Ever training, never "sizing,"
> I have suffered with the rest.
> Yellow cheek and forehead ruddy,
> Memory confused and muddy,
> These are the effects of study
> Of a subject so unblest.
>
> JAMES CLERK MAXWELL, NOVEMBER 1852 [1]

1.1. Learning and Knowing

At 9 A.M. on 3 January 1876, J. H. Poynting (3W 1876) and around a hundred of his undergraduate peers concluded their mathematical studies at Cambridge by embarking on nine days of gruelling examination in the University Senate House. They were anxious men. After ten terms of intensive study, they would be ranked by their examiners according to the number of marks they could accrue answering increasingly difficult mathematical problems against the clock. At approximately half past ten on the morning of the fourth day, the most able candidates encountered the first problem set by Lord Rayleigh, the examiner with special responsibility for mathematical physics:

vi. Investigate the equations of equilibrium of a flexible string acted upon by any tangential and normal forces.

An uniform steel wire in the form of a circular ring is made to revolve in its own plane about its centre of figure. Show that the greatest possible linear velocity is independent both of the section of the wire and of the radius of the ring, and find roughly this velocity, the breaking strength of the wire

---

1. Campbell and Garnett 1882, 613. From Maxwell's poem, "A Vision," written during his seventh term of undergraduate mathematical study at Cambridge. "Sizings" were extra orders of food in the college hall.

being taken as 90,000 lbs per square inch, and the weight of a cubic foot as 490 lbs.[2]

Rayleigh's question was considered difficult for a paper set for the first four days of examination, but Poynting knew he could take no more than eighteen minutes to complete the investigation and solve the problem if he were to remain among the top half dozen men.[3] He would almost certainly have committed a form of the required investigation to memory and begun at once to reproduce the relevant sections as quickly as possible. The next part of the question was more difficult. He had to apply the general theory just derived to solve a problem he had never seen before. What remains of Poynting's efforts that January morning is reproduced below in figures 1.2 and 1.3.

On 14 December 1883, J. H. Poynting, Professor of Physics at Mason College Birmingham, sent a paper to Lord Rayleigh asking in a covering letter whether he would communicate it to the Royal Society for publication in the prestigious *Philosophical Transactions*. Poynting's work would turn out to be one of the most important ever published in electromagnetic theory. In the paper he derived a simple mathematical expression relating the flow of energy in an electromagnetic field to the electrical and magnetic forces of which the field was composed. The importance of the expression lay in the fact that it helped to clarify a number of conceptual difficulties that had troubled students of James Clerk Maxwell's new field theory of electricity and magnetism for almost a decade. The "Poynting vector" (as the expression became known) showed, for example, how the energy of a battery was transmitted during conduction around an electrical circuit, even though, as Maxwell's theory seemed to require, nothing flowed along the conducting wires. The final steps in the derivation and Poynting's explanation of the result are shown in figure 1.1.

1.2. Physics in the Learning

At the close of the nineteenth century, Cambridge University had for some seventy-five years been one of Europe's foremost training grounds in math-

---

2. Rayleigh 1899–1920, 1:280, question vi.
3. Joseph Ward (SW 1876) noted in his diary that evening that Rayleigh's questions were "uncommonly high for the fourth day. What he will give us in the [final] five days I cannot think" (6 January 1896, JTW-SJCC).

ematical physics. Among those who had learned their craft within its walls were William Hopkins, George Airy, George Stokes, John Couch Adams, P. G. Tait, William Thomson, James Clerk Maxwell, Edward Routh, J. W. Strutt, Osborne Reynolds, W. K. Clifford, G. H. Darwin, Horace Lamb, J. H. Poynting, J. J. Thomson, Joseph Larmor, W. H. Bragg, E. T. Whittaker, and James Jeans. These men, together with a much larger number of only slightly less well-known figures, had helped to develop what in the twentieth century would be known as "classical physics."[4] From the cosmic sciences of celestial mechanics, thermodynamics, and electromagnetism to the humbler dynamics of the billiard ball, the boomerang, and the bicycle, they considered themselves the mathematical masters of every known phenomenon of the physical universe.[5] Joseph Larmor, shortly to become Lucasian Professor of Mathematics, was about to publish a remarkable treatise, *Aether and Matter* (1900), in which he would attempt to explain the whole material universe in terms of tiny charged particles moving in a ubiquitous dynamical ether. This was an ambitious project, yet one befitting a man working at the intellectual hub of the largest empire the world had ever known. Cambridge University, situated in a small market town in the East Anglian fens, was home to more expertise in mathematical physics than any other place in Britain or her empire—perhaps more than any other place on earth.

This study aims to deepen our understanding of the nature and historical origins of that expertise by exploring it from the perspective of pedagogy or training. The subject of scientific education has received considerable attention from historians over recent years, yet few studies have made any sustained attempt to use the educational process as a means of investigating scientific knowledge.[6] Likewise, little attempt has been made to provide a historiography of the rise of modern mathematical physics in terms of the formation and interaction of communities of trained practitioners. This failure to explore the relationship between learning and knowing is surprising, moreover, as it is now several decades since philosophers of science such as Thomas Kuhn and Michel Foucault drew attention to the importance of training both in the production of knowing individuals and in the formation of the scientific disciplines. Kuhn's most important claim in this regard was that scientific knowledge consisted in a shared collection of craft-like

---

4. In the period 1825–1900, some 8,000 students were awarded honors in mathematics, around a third of whom were placed in the first class (Tanner 1917).

5. For dynamical studies of billiards, the boomerang, and the bicycle, see respectively Hemming 1899, Routh 1892, 138, and Whipple 1899.

6. The few exceptions to this generalization are discussed later in this study.

skills leaned through the mastery of canonical problem solutions. Foucault's was that regimes of institutionalized training introduced in the decades around 1800 found new and productive capacities in those subjected to its rigors, and imposed a new pedagogical order on scientific knowledge.[7] But, despite numerous references to their works, Kuhn and Foucault's suggestive comments have received little critical discussion, especially from a historical perspective.[8] Historians of the mathematical sciences have in the main focused on the history of theoretical innovation while simply assuming that their key characters are able to innovate and to communicate with their peers via a taken-for-granted collection of shared technical skills and competencies.[9] As an initial point of departure, therefore, I would like to explore the proposition that the history of training provides a new and largely unexplored route to understanding the mathematical physicist's *way of knowing*.[10]

Consider the second of the episodes in J. H. Poynting's early career outlined above. His 1883 paper on the flow of energy in an electromagnetic field is a classic example of the kind of published source commonly used by historians of mathematical physics to write the discipline's history (see fig. 1.1). What are its key characteristics? First and foremost it presents a neat and novel result in mathematical form. The so-called Poynting vector was recognized in the mid 1880s as a major contribution to electromagnetic field theory and remains to this day an important calculating device in physics and electrical engineering. Second, the paper opens with a clear statement of the origin and nature of the problem to be tackled. According to Poynting, anyone studying Maxwell's field theory would be "naturally led to consider" the problem "how does the energy about an electric current pass from point to point"? Third, the solution to the problem offered by Poynting seems to emerge naturally from Maxwell's theory. Starting with Maxwell's field equations and expression for the total energy stored in an electromagnetic field, Poynting deduces the new vector by the skilful manipulation of second-

---

7. See Kuhn 1970, especially the "Postscript," and Foucault 1977, Part 3.

8. Important exceptions to this are Dreyfus and Rabinow 1983, and Rouse 1987. Goldstein (1984) discusses the relevance of Foucault's work to the history of the professions and disciplines, but has almost nothing to say on the history of training.

9. This focus on novelty is not confined to the history of science. Histories of technology, for example, similarly focus on the *invention* rather than the *manufacture* or *use* of new devices. See Edgerton 1999.

10. I use the phrase "way of knowing" to remind the reader that knowledge is always embedded in those practical skills and technologies employed by communities of knowing individuals, from which it ought not to be separated. The term is used in Pickstone 1993. Kuhn uses the phrase "manner of knowing" in a similar fashion; see Kuhn 1970, 192.

But from the values of P′, Q′, R′ in (5) we see that

$$\frac{dQ'}{dz} - \frac{dR'}{dy} = -\frac{d^2G}{dtdz} - \frac{d^2\Psi}{dxdz} + \frac{d^2H}{dtdy} + \frac{d^2\Psi}{dxdy}$$

$$= \frac{d}{dt}\left(\frac{dH}{dy} - \frac{dG}{dz}\right)$$

$$= \frac{da}{dt} = \mu\frac{d\alpha}{dt} \quad (\text{Maxwell, vol. ii., p. 216})$$

similarly

$$\frac{dR'}{dx} - \frac{dP'}{dz} = \frac{db}{dt} = \mu\frac{d\beta}{dt}$$

$$\frac{dP'}{dz} - \frac{dQ'}{dx} = \frac{dc}{dt} = \mu\frac{d\gamma}{dt}$$

Whence the triple integral in (6) becomes

$$-\frac{\mu}{4\pi}\iiint\left(\alpha\frac{d\alpha}{dt} + \beta\frac{d\beta}{dt} + \gamma\frac{d\gamma}{dt}\right)dxdydz$$

Transposing it to the other side we obtain

$$\frac{K}{4\pi}\iiint\left(P\frac{dP}{dt} + Q\frac{dQ}{dt} + R\frac{dR}{dt}\right)dxdydz + \frac{\mu}{4\pi}\iiint\left(\alpha\frac{d\alpha}{dt} + \beta\frac{d\beta}{dt} + \gamma\frac{d\gamma}{dt}\right)dxdydz$$
$$+ \iiint(X\dot{x} + Y\dot{y} + Z\dot{z})dxdydz + \iiint(Pp + Qq + Rr)dxdydz$$
$$= \frac{1}{4\pi}\iint\{l(R'\beta - Q'\gamma) + m(P'\gamma - R'\alpha) + n(Q'\alpha - P'\beta)\}dS \quad . \quad . \quad . \quad (7)$$

The first two terms of this express the gain per second in electric and magnetic energies as in (2). The third term expresses the work done per second by the electromagnetic forces, that is, the energy transformed by the motion of the matter in which currents exist. The fourth term expresses the energy transformed by the conductor into heat, chemical energy, and so on; for P, Q, R are by definition the components of the force acting at a point per unit of positive electricity, so that $Ppdxdydz$ or $Pdxpdydz$ is the work done per second by the current flowing parallel to the axis of x through the element of volume $dxdydz$. So for the other two components. This is in general transformed into other forms of energy, heat due to resistance, thermal effects at thermoelectric surfaces, and so on.

The left side of (7) thus expresses the total gain in energy per second within the closed surface, and the equation asserts that this energy comes through the bounding surface, each element contributing the amount expressed by the right side.

This may be put in another form, for if G' be the resultant of P′, Q′, R′, and θ the

FIGURE 1.1. This page, showing the final steps in Poynting's 1884 derivation of the eponymous vector (equation 7), is fairly typical of the way mathematical physicists presented their work in late Victorian Britain. Notice the reference to Maxwell's *Treatise on Electricity and Magnetism*, the explanation of some (but by no means all) steps in the mathematical argument, and the explanation of the physical meaning of equation 7. Poynting, *Collected Scientific Papers*, 180. (By permission of the Syndics of Cambridge University Library.)

order partial differential equations and volume and surface integrals. Finally, the paper concludes with the application of the vector to seven well-known electrical systems. Poynting shows in each case that the vector makes it possible not only to calculate the magnitude and direction of energy flow at any point in the space surrounding each system but also to trace the macroscopic path of the energy as it moves from one part of the system to another.[11]

Papers of this kind have long appealed to historians of physics, partly because they are easily accessible in research libraries around the world but also because they express the ideals and aspirations of physicists themselves. Poynting's publication was intended to assert his priority in deriving a new and significant result, thereby building his fledgling reputation as an original researcher. This route to establishing one's name in the mathematical sciences has existed since at least the late seventeenth century, the annual volumes of such prestigious journals as the Royal Society's *Philosophical Transactions* providing an internationally recognized and seemingly unproblematic record of the accretion of new knowledge in mathematical physics. It is important to note, however, that sources of this kind tend to impart a particular and partisan view of the discipline's history. Notice, for example, the stress laid on novelty and priority. Poynting's paper is memorable because it tells us when and by whom a neat and important equation was first published. It tells us nothing of why the equation was found at that particular time nor why it was Poynting rather than another student of Maxwell's work who found the new result. The paper also offers an implicit account of how the vector was derived, and, by broader implication, of how work of this kind in mathematical physics generally proceeds. Taking the published derivation at face value one might imagine that mathematical physicists generally start with a well-recognized theoretical problem, solve the problem by the application of logical mathematical analysis, and conclude by applying the result to a number of previously troublesome cases. Perhaps most significantly of all, Poynting's narrative makes Maxwell's electromagnetic theory the central player of the piece. We are led to believe that this theory not only prompted his initial investigation but lent meaning to his result. On this showing, the history of mathematical physics is essentially a history of physical theories, mathematics being the mere language used by individuals in making more or less original contributions to the theories' development.

---

11. I provide references here only to material not discussed and cited in chapter 6. The technical meaning of the vector is discussed in chapter 6.

But to what extent does Poynting's paper provide a faithful account either of his motivations for undertaking research in electromagnetic theory or of the route by which he obtained the new result? Contrary to his assertion, one is not led "naturally" to the energy-flow problem through the study of Maxwell's work. Only one other student of Maxwell's theory, Oliver Heaviside, attempted to investigate a special case of this problem, and he had reasons for doing so that were not shared by Poynting.[12] Poynting himself almost certainly took little if any research interest in Maxwell's theory until the autumn of 1883 and began to do so only after he had hit upon the new result by a different route. Up until that time his research had concentrated mainly on precision measurement and the theory of sound. Moreover, the neat derivation of the vector published in the *Philosophical Transactions* was not even entirely his own work. Both of the referees appointed by the Royal Society to review the paper recommended that a number of revisions be made before it appeared in print.[13] Finally, the problem that had actually led Poynting to the new vector was tucked away at the end of the paper as if it were a mere application of the more general expression. In summary, the structure and implicit rationale of the published work is a skilful fabrication bearing little relation to its actual origin and offering an artificial view of how research is undertaken.[14]

My purpose in introducing Poynting's work is not, at least at this stage, to provide a detailed analysis of its origins or content, but rather to highlight the ways in which publications of its kind display some aspects of the mathematical physicist's craft while concealing others. First and foremost it is an exemplary piece of ready-made science, carefully tailored for the consumption of the wider expert community.[15] A glance at the page reproduced in figure 1.1 will immediately convince most readers, especially those without a training in advanced mathematics and physics, that this is a specialized form of technical communication aimed at a small and select readership. For most people it will seem a dizzying sequence of incomprehensible jargon and symbols, an unwelcome reminder perhaps of the impenetrability of modern

---

12. Heaviside was particularly interested in the way energy was conveyed along a telegraph wire. See Hunt 1991, 119.

13. The referees were William Thomson (2W 1845) and William Grylls Adams (12W 1859). See Royal Society, RR.9.234 (11 April 1884) and RR.9.235 (10 June 1884) respectively.

14. Medawar (1996, 38) famously noted that the "totally misleading narrative" of articles in the biological sciences reflects an inductive ideal of how scientific research *ought* to proceed.

15. The term "ready made" is used by Bruno Latour (1987, 13–15, 258) to designate completed bits of science that efface the resources, trials, and disputes from which they were formed.

physics or of frustrating hours spent in mathematics classes at school. Poynting clearly assumes that his reader is familiar with the rudiments of Maxwell's electromagnetic field theory and with the operations of higher differential and integral calculus. There are nevertheless limitations to what he expects even his specialist audience to be able to follow. Notice for example that he provides precise page references to equations borrowed from Maxwell's *Treatise on Electricity and Magnetism*. Poynting does not expect the reader to share his own intimate knowledge of this difficult work. Notice too that he sometimes offers a few words in explanation of a step in the argument and often refers back to equations numbered earlier in the derivation. These remarks and references indicate places where Poynting thought the reader might require additional information or clarification before assenting to the current step in the argument. He effectively anticipates and manages the reader's progress through the technical narrative. Notice lastly the confident and succinct way in which Poynting explains the physical meaning of the final and forbidding equation (7). If the reader agrees with Poynting's derivation and interpretation of this equation, the rest of the paper follows straightforwardly.

What this work reveals, then, is how a brilliant young mathematical physicist chose to present part of his research to a broad audience in the mid 1880s. What it conceals is how he produced the work in the first place, why he chose to present it in the form described above, and how his peers responded to the paper. We might ask, for example, how long it took to write. Was it a few days, a few weeks, or perhaps a few months? Unfortunately the paper offers no clues to the temporality of its own composition. Why did Poynting present his work as a logical succession of problems, solutions, and applications when, as I have already noted, it was produced by a quite different route? Were there conventions regarding the way work of this kind was structured, and, if there were, how did Poynting learn them? How did he recognize the importance of his work to Maxwell's theory so quickly given that he had previously shown no research interest in this area? Did all mathematical physicists of Poynting's generation have an unspoken knowledge of Maxwell's *Treatise on Electricity and Magnetism* and the tricky theoretical questions it raised but left unanswered?

We might in a similar vein ask how long it took Poynting's contemporaries to read and to make sense of the paper. Was its meaning immediately self-evident to them or did they struggle to follow its argument and complex mathematical manipulations? The referees appointed by the Royal Society focused on the computational power of the new vector and seem not to have appreciated the conceptual clarity it brought to Maxwell's theory.[16] By contrast, one of Poynting's peers, J. J. Thomson, clearly grasped the latter point but noted that the author's interpretation of the new vector was open to serious objection on both mathematical and physical grounds.[17] How was consensus reached concerning the work's technical reliability and proper interpretation? Broadening our question from one of time to one of scale and geography, we might also inquire how many people were capable of following a work of this kind in the mid 1880s, how many actually mastered the paper and in which universities they worked.[18] How did Poynting know what level of technical competence he could reasonably expect of his readers? Were the assumptions he made in this regard reasonable, and, if they were, how was the expert community for which he wrote generated and sustained?

These are not issues that are usually raised in such explicit and systematic terms by historians of science, largely because they call into question technical skills that are for the most part taken for granted in historical writing. Material intended for a general readership is normally couched in nontechnical language, while that written for those with a training in physics simply assumes the reader's technical competence. In the one case the reader remains firmly outside the expert community, in the other he or she has already been admitted. This study is in part an attempt to understand what it means to cross this *Great Divide* that separates the knowing expert from the unknowing outsider. The Great Divide is a term coined to designate the gulf believed by some anthropologists to separate primitive from advanced societies, those that are understood to live by superstition, magic, and myth from those that have made the leap to rationality, science, and history.[19] But the term might equally be applied to the gulf separating the lay person from the technical expert. Few people in the industrialized world could give a technically informed account of quantum mechanics or the theory of relativity, yet members of physics departments in universities in many parts of the developing world not only understand these theories but contribute to their growth. Insofar as a Great Divide exists today, it does not separate different

---

16. Neither referee even commented explicitly on this point. See Royal Society, RR.9.234 (11 April 1884), and RR.9.235 (10 June 1884).

17. Thomson 1886, 151–52.

18. Barely any physical scientists in the United States, for example, would have been capable of following Poynting's work at this time. Those who visited Europe circa 1900 were astonished at the mathematical knowledge possessed by their European peers and found themselves unable to follow lectures on mathematical physics. See Servos 1986.

19. Goody 1977, 3; Latour 1987, 215–32.

peoples or races so much as subcultures of experts from everyone else. As Steven Weinberg has noted, "since the time of Newton, the style of physical science has spread among the world's cultures [and has] not been changed by the increasing participation of physicists from East Asia nor ... by the increasing number of women in physics" (2000, 8). Whether working in Tokyo or Texas, Moscow or Beijing, today's culturally diverse practitioners of a discipline founded by a relatively small number of Western European men pursue similar goals using similar tools.

But the emergence of this global enterprise is surely a remarkable phenomenon and one that needs to be explained rather than taken for granted. As I noted above, scientific work of the kind published by Poynting presupposes the existence of a community of readers capable of filling in numerous tacit steps in the technical narrative and of recognizing new results as meaningful contributions to an ongoing research project. The published paper is but one medium of exchange in a much larger economy of technical skills and competencies, most of which are taken for granted not just by historical actors like Poynting and his peers, but also by those historians and philosophers of science who have studied their work. The expertise in question would, moreover, be extremely difficult to reconstruct from published papers alone. Imagine a historian with no technical training at all in mathematics or physics (and no access to anyone with such expertise) trying to understand the meaning of Poynting's mathematical argument in fig 1.1. Today's historians of physics find arguments of this kind at least partially accessible because a physics training still includes some mathematical methods and physical concepts that were familiar in Poynting's day.[20] Insofar as his work remains accessible to the present, global generation of physicists, it does so through the preservation of those taken-for-granted skills referred to above. But, as Weinberg's remarks remind us, in the late seventeenth century this kind of technical expertise was confined to a handful of individuals in Europe. How, then, has such a bounded and esoteric culture of technical skill in the mathematical sciences been developed and propagated in the centuries since Newton, and why has the nature of this skill and the mechanism of its propagation apparently been of such little interest to most historians and philosophers of science?

As a first step towards answering these questions, we can usefully turn to recent work in the social and cultural history of the experimental sciences.

---

20. This is not to say that a late twentieth-century training in physics allows one easily to grasp the many subtleties of Poynting's work as understood in late Victorian Britain. The latter, as we shall see in chapter 6, requires a good deal of historical reconstruction.

During the last twenty years, historians have sought to explain the rise of these sciences by relating them to a range of contemporary developments in European culture. In order to do so they have emphasized science's important links with politics, craft skills, and material resources, gender, travel, commerce, imperialism, and war, as well its collective study in institutions such as scientific societies, museums, academies, universities, and laboratories.[21] One important and productive characteristic of much of this scholarship has been a deep concern with the local and contingent resources by which new scientific work is accomplished and with the labor and additional resources required to make such work reproducible at new sites. It is now widely appreciated that experimental knowledge is closely bound to the instruments, machines, and embodied skills by which it is produced, and that the transmission of that knowledge depends to a considerable degree upon the successful reproduction of these skills and material resources. But, despite the profound effect these studies have had on the way historians explain the scientific accomplishments of the past, it is fair to say that they have been far more successful in enriching our understanding of the experimental and field sciences than of work carried out inside the theoretician's study. It is now commonplace for historians concerned with material culture to display the historical and geographical contingency of experimental work by referring to the "site-specific instruments, techniques and materials" employed by experimenters both individually and in groups, but when the tools of the trade are what have been described as *"readily transportable* mathematical techniques and abstract theoretical concepts," the focus on craft skill, localism, and collective production is almost invariably abandoned.[22]

This difference in approach to mathematical and experimental work in science has many causes, but two are of particular relevance to the present discussion. The first concerns the way science studies has developed over recent years. The turn to social and cultural studies of science was to some extent a reaction against an earlier tradition of writing science's history as a succession of theories, a tradition rightly associated with intellectual history and the history of ideas.[23] Coupled with a roughly contemporary wane in the popularity of philosophical accounts of science dominated by attention to a universal scientific method and to the logical structure of

---

21. There is an enormous literature in this area, but of particular relevance to my present concerns are McCormmach 1969, Gooding et al. 1989, Schaffer 1992, Geison and Holmes 1993, Galison 1997, and Galison and Warwick 1998.

22. Geison and Holmes 1993, 232; my italics.

23. McCormmach 1969, vii–viii.

theories (especially in physics), this turn helped to propel the shift towards social studies of science concerned with locality, machines, instruments, technicians, and the possession and transfer of craft skill in experimentation.[24] Despite this shift, however, mathematically formulated theories have continued, largely by default, to be regarded as the province of the history of ideas and therefore to have little or no place in social histories of science concerned with labor and material culture.

A second and closely related reason for the difference derives from the deeply enshrined distinction in Western society between the terms "theory" and "practice." Where the latter is associated with embodied skills, the tools of the trade, and specific locations, the former, especially in the case of the mathematical sciences, conjures a vision of contemplative solitude and cultural transcendence. As Steven Shapin has argued, the place of the purest forms of knowledge in Western society is "nowhere in particular and anywhere at all" (1990, 191). This deeply ingrained sense that theorizing and experimenting are essentially different sorts of activity helps to explain why theoretical innovations are still more likely to be attributed to acts of isolated contemplation or the insights of genius than to the deployment of site-specific skills that have been collectively developed and painstakingly learned. Furthermore, the depiction of new theories as revelations of genius seems to support an implicit belief that the truth of a new theory ought to be apparent to any other open-minded scholar. By attributing, say, Isaac Newton's theory of universal gravitation to nature itself (rather than to Newton's artifice) we make the theory a transcendent truth which, once stated, should have been self-evident to those of his peers interested in mathematics and natural philosophy.[25] Unlike its experimental counterpart, new knowledge made in the theoretician's study is not understood to be bound to specific sites and individuals by material culture and embodied skill but to be as easily transportable as the printed pages on which it is inscribed.

I shall argue that this distinction between theoretical and experimental work is in many ways artificial and that by acknowledging a number of similarities between these activities we can begin to build a more symmetrical account of theory and experiment. The first noteworthy similarity is that novel theoretical work travels no more easily to new sites or practitioners

---

24. On the shift from theory-led accounts of scientific development see Galison 1988, esp. 207–8.

25. Since we now believe Newton's mechanics and theory of gravitation to be false, we ought not to evoke their truth as an explanation of their appeal to other scholars.

than does its experimental counterpart. In order to illustrate this point one has only to consider the difficulties experienced by Newton's contemporaries in mastering the technical details of his *Principia Mathematica* (1687).[26] Most of Newton's Cambridge peers found the more technical sections of the book impenetrable, while even the greatest mathematicians in England, Scotland, and continental Europe struggled to master some of the book's more abstruse theorems and problem solutions.[27] Christiaan Huygens, for example, who was not only a leading natural philosopher and geometrician of the day but one of a tiny handful of scholars believed by Newton to be capable of appreciating his work, found some of the book's mathematics and physics difficult to follow. Despite paying several visits to Newton in 1689 to discuss their common interest in the science of mechanics, Huygens subsequently expressed the hope that a second edition of the *Principia* would explain what he continued to regard as the "many obscurities" contained in the first.[28]

The reason the vast majority of scholars found much of the *Principia* impenetrable is highlighted by the experience of Richard Bentley, a Cambridge classicist and one of Newton's early disciples. Daunted by the *Principia*'s reputation as a brilliant but technically demanding work, Bentley turned first to the mathematician John Craig and then to Newton himself for advice on how best to approach the book.[29] Both men advised Bentley to begin not with the *Principia* itself, but by studying lengthy lists of works covering the method of tangents, the quadrature of figures, geometry, algebra, mechanics, optics, hydrostatics, and astronomy. Newton claimed that mastery of the works on his list would be "sufficient" to understand the *Principia*'s basic doctrines, but added that "the perusal" of Huygens's *Horologium Oscillatorium* would make Bentley "much more ready" for the task.[30] What this advice reveals is the remarkable breadth of mathematical and natural-philosophical knowledge that Newton assumed of a reader capable of following his work. He was well aware, moreover, that even knowledge of the forbidding list of readings recommended to Bentley would not guarantee insight into the *Principia*'s more technical sections. In practice, Newton did not expect him

---

26. For an overview of the early reception of the *Principia*, see Bertoloni Meli 1993, chap. 9.

27. The attempts of the great Scottish mathematician, David Gregory, to master the *Principia*'s contents are discussed in Guicciardini 1999, 179–84.

28. Guicciardini 1999, 124.

29. Although a very accomplished mathematician, Craig himself had great difficulty understanding the more technical sections of the *Principia* (Iliffe 2003, 51).

30. Iliffe 2003, 52–53. Newton had partly followed the model of Huygens's *Horologium* when writing the *Principia*; see Guicciardini 1999, 118.

to get beyond the first sixty pages of Book I.[31] Only scholars of rare ability who had devoted much their lives to the study of mathematics and natural philosophy could realistically expect to be in command of the knowledge assumed by Newton, and, as with Huygens, even these men invariably found much of the *Principia* very heavy going.[32] Newton had sometimes employed his new and unpublished method of fluxions (known today as the differential and integral calculus) in obtaining important results quoted in the book, so that scholars to whom these methods had not been revealed were unlikely to be able to prove the results for themselves.[33]

This example both illustrates the point that novel theoretical work does not travel easily to new sites and casts light on why this form of translation is difficult. Just as the replication of a novel experimental result is heavily dependent on the reproduction of the skills, instruments, and material resources by which the result was first obtained, so the understanding of a new mathematical theory is likewise dependent on a common mastery of the technical skills assumed in the theory's original statement and application.[34] These skills can moreover be divided into two different sorts which are illustrated respectively by the experiences of Bentley and Huygens. Bentley's difficulties in comprehending even the first few sections of the *Principia* sprang from his ignorance of technical expertise that was commonplace among the leading mathematicians and natural philosophers of Europe. Cases of this sort are useful in revealing the kind of knowledge that is so taken for granted by the leaders in a given field that it is seldom explicitly discussed. Huygens's difficulties with the *Principia* sprang from a different source. In his case it was the more abstruse sections of the book that proved troublesome, partly because Newton had sometimes adopted a novel and therefore unfamiliar approach to tackling problems in dynamics, but also because he had not revealed the methods by which some of the problems had originally been solved.[35] Cases of this kind are informative because they

---

31. Iliffe 2003, 53. It is unlikely that Newton really expected Bentley to master the forbidding list of preparatory readings.

32. The other European scholar whose knowledge of mathematics and mechanics made him an ideal reader of Newton's work was Gottfried Leibniz. For an account of Leibniz early struggles to master the *Principia*'s technicalities, see Bertoloni Meli 1993, chap. 5.

33. The extent to which Newton used unpublished mathematical techniques to discover results quoted or proved by other methods in the *Principia* is a complicated historical issue. For an overview appropriate to the present study, see Guicciardini 1999, 3–6, 95–98.

34. The classic study of the problems of experimental replication is Collins 1985.

35. Huygens did not master the differential and integral calculus developed independently by Newton and Leibniz until the mid 1690s. Newton revealed his version of the calculus only to a small group of close acquaintances in England. See Guicciardini 1999, 104, 120–21, 183.

highlight the fact that technical arguments in books like the *Principia* make implicit reference to skills and assumptions that are taken for granted by the author yet remain unfamiliar and sometimes impenetrable even to the author's most able peers. It should also be noted that both kinds of skill can be considered *tacit* in the sense that they are neither explicitly stated nor otherwise easily communicated in published accounts of novel work.

This brings us to a second way in which historical explanations of the origins of theoretical and experimental work can usefully be made more symmetrical. I noted above that whereas new experimental results are now routinely discussed in terms of the local skills and resources by which they were first produced, new theoretical work is still frequently attributed, albeit implicitly, to the insights of an isolated genius.[36] Once again, Newton provides an appropriate example. Until recently, scholars have shown little interest in relating his expertise in mathematics and natural philosophy to the context of his student studies in Grantham and Cambridge. As historian Mordechai Feingold has noted, the received image of Newton has long been that of the brilliant autodidact who "in defiance of the traditional curriculum and without supervision . . . made the great leap into realms undreamed of (let alone shared) by his contemporaries at Cambridge" (1993, 310). This depiction of Newton exemplifies my earlier remarks that theoretical work tends to be characterized in terms of solitary contemplation and deemed worthy of investigation only insofar as it is understood to be wholly original. It also goes some way towards explaining why the issue of education figures so little in discussions of Newton's early work. If our primary concern is to celebrate the truths revealed by Newton's genius in the privacy of his rooms in Trinity College or in the intellectual isolation of Woolsthorpe Manor, then it appears superfluous, perhaps even disrespectful, to pry into what well-known philosophical principles and mathematical methods he might have learned from lesser scholars. Even to suggest that Newton required such education and struggled as a student to master the works of his predecessors will seem to some to call his unique brilliance into question. Yet, as Feingold has also noted, it is through an unwillingness to explore the resources (beyond private study) on which Newton might have drawn in acquiring his technical expertise that historians have effectively isolated him from his intellectual milieu.[37] Only occasionally and relatively recently have scholars recognized

36. For recent attempts to localize innovation in theoretical physics, see Galison 1998 and 2000a.

37. Feingold 1993, 310. Feingold traces this unwillingness in part to the historical aftermath of Newton's priority disputes with Hooke and Leibniz.

and begun to exploit Newton's educational biography as a potentially important factor in explaining his remarkable achievements in mathematics and natural philosophy in the mid 1660s.[38]

This brief discussion of Newton's work not only illustrates my point that histories of theory and experiment can be made more symmetrical, but points to a general strategy for doing so. The key difference in our understanding of the two activities is that where experimental work is now generally regarded as just that, *work,* what is sometimes called "theoretical work" continues to be treated as an essentially cerebral, contemplative, and introspective accomplishment of isolated and outstanding individuals.[39] In order to break down this distinction it is necessary to approach theorizing not in terms of inaccessible cerebration but as the deployment of practical and embodied skills that have to be learned, developed, and actively communicated. It is perhaps not immediately obvious, however, where a theoretical counterpart to the material history of experiment is to be found. When, as in figure 5.3, we catch a rare glimpse of a Victorian mathematical physicist posed "at work," we see no material culture in the form of machines, instruments, or support staff, just a solitary man writing at a desk, a few shelves of books, and a large wastepaper basket. Yet paper, ink, and wickerwork, mundane as they are, can provide an important clue to a practical strategy for analyzing the theoretician's craft in terms of material culture.

In order to appreciate this point, we can usefully reflect for a moment on the way anthropologist Jack Goody has sought to understand the relative intellectual accomplishments of different cultures. Goody has suggested seeking explanations in terms of the representational forms employed by cultures, especially the written form. Goody insists that such "material concomitants" of what he refers to as "mental domestication" are extremely important because "they are not only the manifestations of thought, invention, creativity, they also shape its future forms" (1977, 9). This attempt to understand the Great Divide in terms of representational techniques suggests a cultural rather than cerebral analysis of theoretical production. Although Goody's

---

38. On Newton's mathematical studies at King's School, Grantham, and his private mathematical studies at Cambridge, see Whiteside 1982. On Newton's Cambridge education and his relationship with the first Lucasian Professor of Mathematics, Isaac Barrow, see Feingold 1993. On Newton's early original work in Cambridge and Woolsthorpe, see Westfall 1980. Westfall notes that, contrary to common mythology, the pace of Newton's early mathematical and natural philosophical achievements was slowed rather than accelerated by his intellectual isolation in Woolsthorpe in 1665–66.

39. The notion of "science as craftsman's work" is discussed by Ravetz (1996, chap. 3), though his discussion focuses mainly on experiment and says little about training or its history.

remarks were intended to undermine explanations of changes in human thought couched in terms of the essential mental qualities of different ethnic groups, they might equally apply to those special mental abilities sometimes attributed to great scientists. As I noted above, the Divide that separates literate from nonliterate cultures can also stand for the division between subcultures of technical experts and those who do not share their expertise. The sorts of "division" that I have in mind are nevertheless flexible and manifold. Every group of practitioners possessing technical expertise is to a greater or lesser extent divided from the rest of society, while even within a group there may be serious problems of technical communication between subgroups.[40] The point I want to emphasize is that these are not simply divisions of mentality or intellect, but of representational technology, training, skill, and experience.

Returning to the case of Newton, we can say that what separated him from Bentley was not just that the former had greater cognitive ability or powers of abstraction, but that Newton, by virtue of his educational biography, private study, interests, and abilities, had mastered and was able to deploy a range of technical skills that Bentley did not possess.[41] Moreover, this mastery of notations, proofs, theorems, and problem solutions was neither a static nor a purely cerebral accomplishment but one gradually embodied through protracted routines of learning using a material culture of books, paper, pens, ink, and mathematical instruments. I would suggest that the very manuscript record of Newton's labors—from which historians have painstakingly reconstructed his working practices and route to important new results—is in fact a testimony to the material technology of paper and pen by which he learned and practiced his craft.[42] It is also important to appreciate that many of these skills were the common possession of the leading mathematicians in Europe. This was the knowledge that enabled them to participate in a shared discipline and to find meaning in new and highly technical works such as the *Principia*. In so far as new mathematical results do move relatively easily from one site to another, it is these networks of shared competence that enable them to do so. Indeed, in the *Principia*'s case

---

40. For example, on the problems of communication between experimenters and theoreticians in high energy physics, see Galison 1997, chap. 9.

41. This is not to deny that Newton was a mathematician of rare ability, but to note that such ability is a necessary but not sufficient explanation of his unique accomplishments.

42. The manuscript record of Newton's "crawling progress" towards mastering Descartes's geometry is discussed in Whiteside 1982, 113. Westfall notes how the manuscript record of Newton's early original investigations reveals a "human process which is comprehensible in a way that bursts of insight are not" (1980, 116).

it was extremely important that a handful of the finest mathematicians felt able—after some initial difficulties—to vouch for the reliability and brilliance of most of Newton's proofs and problem solutions. Thanks to this endorsement, the incomprehensibility of Newton's masterpiece to the majority of scholars became a hallmark of his genius.

This approach to theoretical knowledge in terms of learned and embodied skills promises a more naturalistic understanding of what I referred to above as the mathematical physicist's *way of knowing*. By *naturalistic* I mean an understanding that starts not, as it were, from above, with the certainty of completed and disembodied theories or with the contributions of genius to theories in the making, but from below, with the uncertainty of the student struggling to enter the expert's world. It is an approach that seeks to illuminate the mathematical physicist's knowledge by exploring how that knowledge is acquired through education. Following the educational process can provide insight into those sensibilities, skills, and competencies that all aspiring theoreticians master only slowly and painstakingly but then so take for granted that they are seldom mentioned. Like the five-finger exercises of a concert pianist or brush strokes of a great artist, these are hard-learned skills that are invisible in the execution of a work yet bar any nonexpert from replicating the performance. Goody's comments also remind us that these skills, together with the written trace left by the paper-and-pen performance, are not merely the outward manifestations of otherwise inaccessible thoughts, but an integral part of theoretical production. Understanding what kinds of skill are held together through the material culture of theory can provide insight into the resources that, on the one hand, enable some theoreticians to innovate and, on the other, enable the wider community to follow and to criticize their work.

1.3. Entering the House of Theory

We can put some of the above remarks into more concrete form by comparing J. H. Poynting's paper on energy flow in the electromagnetic field with his attempt to answer Rayleigh's examination question of 1876 (figs. 1.2 and 1.3). The two documents are in several respects polar opposites. Whereas we cannot tell how long Poynting spent redrafting and revising the former, we know exactly where and when the latter was written and (within a matter of minutes) how long it took to write. Indeed, examination scripts are almost certainly the closest thing we have to real-time records of mathematical physics in the making. They are further distinguished by the fact that, unlike

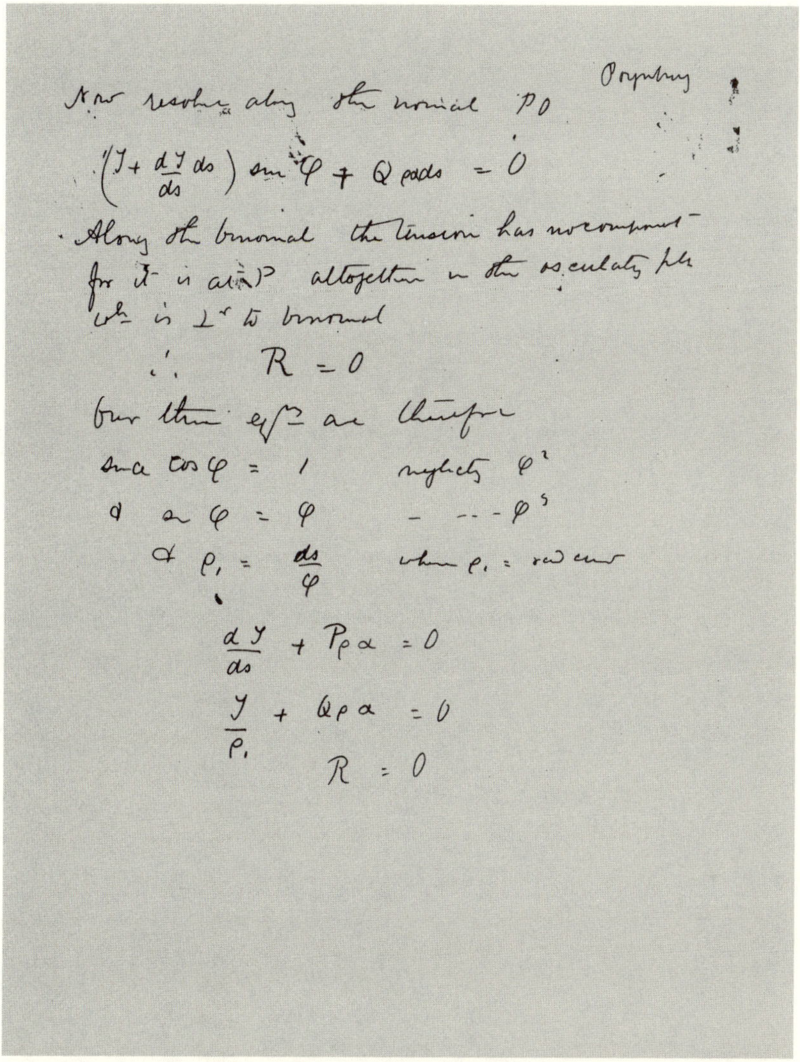

Poynting

Now resolving along the normal PO

$$\left(T + \frac{dT}{ds}\,ds\right)\sin\varphi + Q\rho\alpha\,ds = 0$$

Along the binormal the tension has no component for it is axial altogether in the osculating plane wh. is $\perp^r$ to binormal

$$\therefore \quad R = 0$$

But these eq^{ns} are changed

since $\cos\varphi = 1$ neglecting φ^2

& $\sin\varphi = \varphi$ — — φ^3

& $\rho_1 = \dfrac{ds}{\varphi}$ when $\rho_1 =$ rad curv

$$\frac{dT}{ds} + P\rho\alpha = 0$$

$$\frac{T}{\rho_1} + Q\rho\alpha = 0$$

$$R = 0$$

FIGURE 1.2. The final sheet (half full size) of Poynting's investigation of the "equations of equilibrium" as required by Rayleigh's question. The equations are given in the last three lines. P, Q, R are respectively the tangential, normal, and perpendicular forces per unit mass acting at a point on the string, and T, s, ρ, ρ_1, α are respectively the tension per unit cross section, distance along the string's length, its density, radius of curvature at a point and area of cross section. The examples of student scripts reproduced in this book are taken from the manuscripts of Lord Rayleigh and James Clerk Maxwell. Both men acted as examiners in Cambridge and subsequently drafted scientific papers on the back of examination scripts they had marked. From RC-PLRL, unpublished calculations, verso. (By permission of the Phillips Laboratory Research Library.)

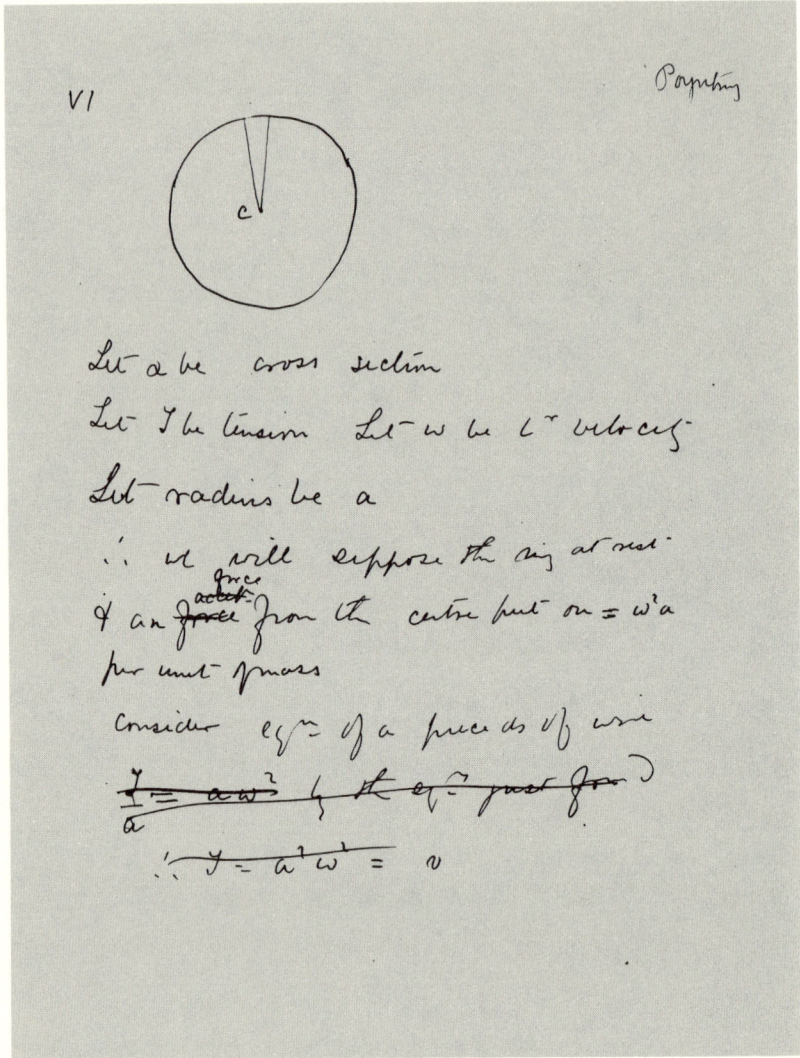

FIGURE 1.3. Poynting's attempt (two sheets) to solve the problem in Rayleigh's question. Note the corrections made as the calculation unfolds and the failed attempt to explain the meaning of the result

draft or published works, we know to what questions they were supposed to constitute answers and that they were written without the immediate aid of reference works or other individuals. These peculiarities of work done under examination conditions enable us to broach issues that would be very difficult to investigate using published material. We can for example assess

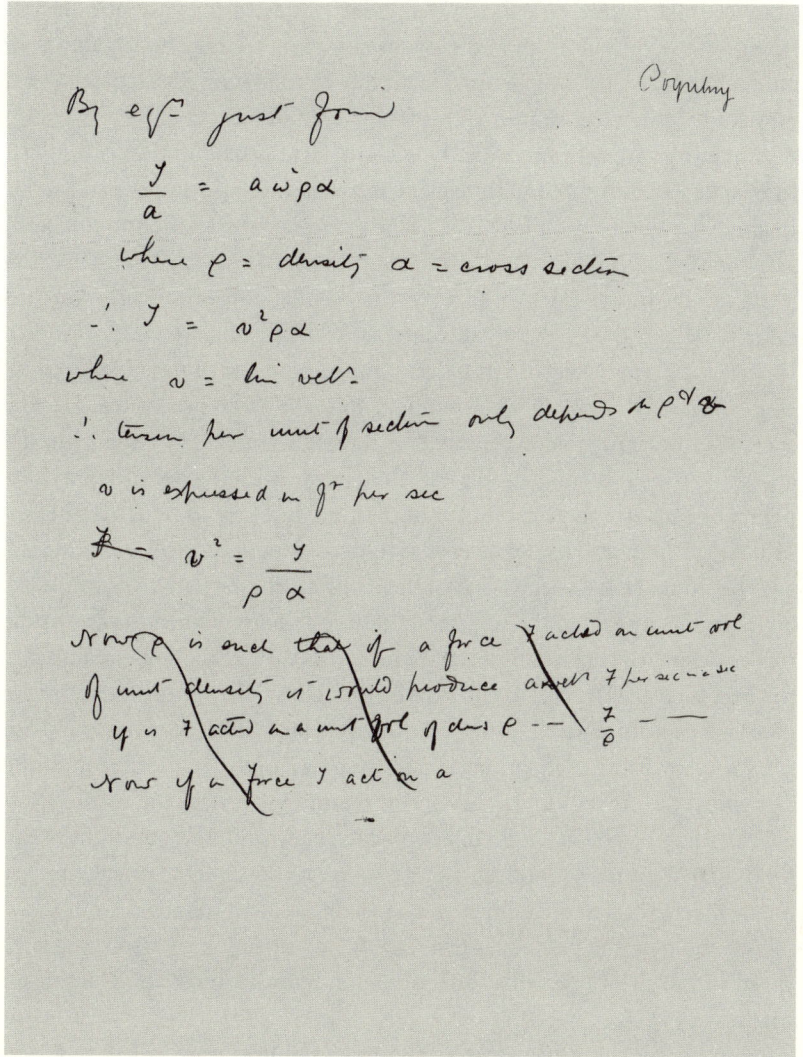

obtained. From RC-PLRL, unpublished calculations, verso. (By permission of the Phillips Laboratory Research Library.)

the breadth of knowledge in mathematical physics present at the fingertips of a student like Poynting about to embark on a physics career in the mid 1870s. The neat and fluent derivation concluded on the first page of figure 1.2 shows that he knew how to resolve component forces using appropriate limit and approximation methods from the differential calculus and algebraic geometry, and how to apply these methods to the case of a flexible string acted on by normal and tangential forces. We may reasonably conclude therefore that he was capable rapidly of following the steps of an argument in which techniques of this kind were implicit.

Poynting's attempt to tackle the second part of the question is similarly informative. Confronted by a problem he had not seen before, he had quickly to apply the theoretical principles of the previous investigation to find a solution. Working under examination conditions, he had time neither to explore a range of strategies for tackling the problem nor to conceal the errors he made along the way; he had rather to work through a solution in real time, leaving a paper record of his reasoning for the examiner, and for us, to assess later. Let us pause for a moment to see how he proceeded. He first sketched the arrangement described by Rayleigh as an aid to visualizing and analyzing the problem. The analysis then began with the definition and labeling of each of the variables Poynting thought likely to figure in the imminent calculation. His next step was to conceptualize the problem in such a way as to make it a special case of the investigation he had just completed. To do this he imagined the ring brought to rest and redefined the *centripetal* force due to the ring's rotation as a static *centrifugal* force acting radially from the ring's centre. The ring could now be thought of as a continuous string of constant curvature acted on by a uniform normal force. Now he could apply the second of the equations already derived to obtain the algebraic and numerical results required. This line of reasoning displays the sort of strategy Poynting had been taught to adopt when confronted by questions of this kind. The single most important skill it reveals is the ability to conceptualize a new mechanical arrangement as one example of a broader class of problems soluble by the principles given in the investigation. It also reveals the speed with which these skills could be deployed by the most able Cambridge mathematicians.

Another noteworthy feature of Poynting's solution, especially when compared with the published work, is the error and uncertainty it displays. He is briefly unsure whether to redefine the mechanical effect of the ring's rotation as a force or an acceleration and then makes a serious error in applying the equation from his investigation.[43] Perhaps most strikingly of all, he is unable at the end of the calculation to explain why the relationship he has (correctly) found between the tension produced in the ring and its velocity

---

43. On the second sheet Poynting crosses out "force" in favor of "accel[eration]" only to revert again to "force." At the bottom of the page he also crosses out an erroneous application of the second equation from the first sheet.

of rotation constitutes an answer to the first part of the problem. After three-and-a-half lines of unconvincing explanation (bottom of third sheet) he realizes he is getting nowhere and crosses his remarks out.[44] These twists and turns offer an alternative to the usual, purely cerebral view of theoretical production. They not only draw our attention to the material apparatus and symbolic language of theoretical work, but illustrate the embodied interaction between sketching, writing, and thinking that constitutes the creative process of solving a novel problem. Poynting's errors and corrections remind us of the uncertainty with which a theoretician works when tackling a new problem, the written trace revealing what might be regarded as a series of experimental attempts at finding an acceptable solution. Such errors also remind us of the time, effort, and skill that go into producing the robust derivations, explanations, and applications that appear eventually in the public domain. We shall see in chapter 5 that there are very significant differences between solving examination problems and the process of research, but examination scripts nevertheless provide a rare glimpse into the kind of activities that routinely take place behind the closed door of the theoretician's study.

We should also note that the examination was a communal as well as a personal experience. Poynting entered the Senate House with around a hundred other undergraduates, the best of whom struggled with the same questions over the nine days of examination. The scripts of other able candidates show that they tackled proofs and problems in similar but by no means identical ways to Poynting. There was clearly scope for difference in the strategy and style of reasoning adopted in answering questions, provided that the written record led eventually to the correct result. What was important was that the steps in the recorded argument were deemed legitimate by the examiners. A key point to grasp in this context is that the examination provided a clear standard for the few top men who went on to become professional mathematicians or physicists. The experience of the final examination informed them of what they might reasonably expect their colleagues to be able to follow and agree with. We should note, for example, that both of the men appointed by the Royal Society to referee Poynting's paper, William Thomson (2W 1845) and William Grylls Adams (12W 1859), had in their day sat a similar examination in the same room. All five of Poynting's examiners as well as his strongest advocate and critic, J. J. Thomson (2W 1880), had not only sat the examination in the same room but had been trained by the same

---

44. Whether Poynting tried another explanation on a fresh sheet or just moved to the next question we cannot now tell.

great mathematics teacher, Edward Routh (SW 1854). Moreover, Poynting and Thomson had both learned Maxwell's theory in the Trinity College classroom of W. D. Niven (3W 1866), yet another graduate of the Mathematical Tripos and student of Routh's. The kind of script reproduced in figs. 1.2 and 1.3 is thus indicative not just of the skills possessed by an individual like Poynting but of those held communally by a much broader group of practitioners with whom he could readily communicate.

The form of Rayleigh's question also provides an intriguing clue to the structure of Poynting's published work. The sequence of general problem, general solution, and application that is built into the question is more or less the one followed by Poynting in announcing the new result in electromagnetic theory. This might at first sight seem a somewhat farfetched analogy, but, as we shall see in chapter 5, it was commonplace for Cambridge mathematicians both to announce new mathematical results in the form of examination problems and to publish solutions to examination questions in research journals. Several senior Cambridge mathematicians in fact complained that the research publications of recent Cambridge graduates were far too much like model examination problems. The gap between the most difficult problems and actual research was also much narrower than one might today expect. The elementary example discussed here was chosen for its accessibility, but some of the more difficult problems Rayleigh set towards the end of the examination were based on the research he was undertaking in the autumn of 1875. The most able students were thus required to recognize when an as yet unpublished research problem had been expressed in soluble form and to demonstrate extremely rapidly how the solution was effected. Here too we can see how examination problems served to generate a community in which many of the steps in the solution of a new problem could be left as implicit exercises to the implied reader. We can also see that a young mathematical physicist like Poynting, lacking any other model, would instinctively present his research as a sequence of problems, solutions, and applications, a presentation that would have been readily recognized by his Cambridge-trained peers.

This example illustrates some of the connections I want to draw between learning and knowing. Old examination scripts are not a source to which historians and philosophers have typically turned when pondering the nature of scientific knowledge, but, as we have just seen, both in their formulation and solution they can provide a window onto that large collection of generally taken-for-granted skills that constitute the mathematical physicist's way of knowing. The skills exhibited on Poynting's script are but a small subset of those that not only enabled him to conceptualize, tackle, and solve new problems, but enabled a wider community of his peers to recognize the solutions as legitimate and to fill in tacit steps in the argument where necessary. Starting with theory-in-the-learning rather than theory-in-the-making also offers a different perspective on the production of new knowledge in mathematical physics. As we shall see in chapter 6, the historical analysis of Poynting's route to the new vector reveals direct links between training, examination, and research. The question of how energy flowed in an electromagnetic field was one that emerged implicitly from Poynting's undergraduate training, as did the mathematical tools and physical concepts with which he eventually solved the problem. In this case, at least, the creative act through which new theory was produced cannot be properly understood unless it is placed in the pedagogical context of its production.

The examination script provides a form of real-time access to the technical performance of which the most able Cambridge-trained mathematical physicists were capable during the mid-Victorian era. What it does not reveal is how the performance was learned in the first place. Far from tracing a journey across the Great Divide, this source only measures the extent to which a student had mastered a performance for which he had been rehearsing for several years. In order to inquire into the nature of the journey itself we must turn to the techniques by which undergraduates learned and by which they were taught. Here again Poynting's undergraduate experience provides a useful starting point for discussion. His studies at Cambridge in the mid 1870s were managed by Edward Routh, the most successful mathematics coach of the Victorian era. It was in Routh's classroom that Poynting was taken through a carefully contrived course of progressive mathematical studies, each lecture being supplemented with textbook readings, graded examples, past examination problems, and manuscripts covering the most advanced topics. Poynting himself spent long hours in his college rooms attempting to reproduce the skills displayed by Routh, mainly by "writing out" the proofs and theorems he was required to commit to memory and by struggling with numerous exercises and problems from past examination papers. Routh for his part not only offered model answers and detailed corrections to Poynting's efforts, but performed the most difficult problem solutions on a blackboard at the beginning of each lecture. The body of Routh's pupils was also broken down into a number of separate classes according to mathematical ability. This enabled him to teach each class as rapidly as possible and to preserve a strong competitive spirit through regular written examinations. Poynting would have been aware from week to week of his relative standing in the elite class and worked at an exhausting pace to preserve his position. He also shared with his peers the belief that intellectual endeavour had to be balanced against physical exercise. As we shall see in chapter 4, this combination was thought to preserve mental and physical health and to aid the development of the competitive spirit on the playing field as well as in the classroom. A keen and able oarsman, Poynting trained regularly on the river Cam and rowed competitively for his college boat club.[45]

These were the main components of the training system through which undergraduates acquired their technical skill in mid-Victorian Cambridge. An important aim of the present study is to investigate how each component contributed to the production of a student's mathematical knowledge and how the components functioned together to generate a large community of knowing individuals. Is it the case, for example, that an isolated individual could acquire this knowledge from a textbook or book of problems alone, or is personal interaction with fellow students and an able teacher a necessary part of the training process? One difficulty with an investigation of this kind is that most of the pedagogical devices referred to above are now so commonplace in elite technical education that they seem barely to warrant comment—how else would one train mathematical physicists other than by a progressive syllabus, lectures, tutorials, textbooks, graded examples, written examinations, and so on? A powerful way of surmounting this difficulty and revealing the productive capacity of training is to highlight the latter's relatively recent historical origins. Despite their present-day familiarity, most of these devices were virtually unknown even in Europe until the late eighteenth and early nineteenth centuries. A similar issue arises in connection with the sites at which mathematical education takes place. Today we associate advanced technical training and research respectively with undergraduate and postgraduate studies in the universities, but this too is largely a development of the last two hundred years. By studying the transformation in undergraduate studies that made the mathematical sciences central to university education, we can explore the new institutional home of these sciences, the productive power of the new pedagogical regimes, and the nature of the knowledge that such regimes produced and propagated.

45. Poynting 1920, xvi.

1.4. Pedagogical Histories

The education in mathematics and mathematically related disciplines offered by the universities until at least the late eighteenth century was necessarily elementary. From their foundation in the medieval period, the universities had primarily been theological seminaries whose job it was to provide a broad education for undergraduates and to prepare the most able scholars for higher studies in medicine, law, and theology. The study of mathematics beyond elementary arithmetic, geometry, and algebra was far too specialized for these purposes and was in any case unsuited to the oral methods of teaching and examining used in most universities.[46] Furthermore, the discipline of mathematical physics, which emerged during the seventeenth and eighteenth centuries, was in some respects even more alien to undergraduate studies than were those established subjects of mathematics and physics from which it emerged. At the beginning of the seventeenth century there existed only a handful of mathematically expressed laws concerning the physical behavior of the terrestrial world. These relationships, such as the law of the lever, the law of flotation, and the law of reflection, were expressed as simple geometrical relationships, most having been handed down as part of a mixed-mathematical tradition from the ancient and Islamic worlds.[47] The most technically demanding of the mathematical sciences at this time was astronomy, but even the technicalities of planetary motion, although too advanced for detailed undergraduate study, remained kinematic rather than dynamical.[48]

In the early seventeenth century, then, most of what was known at the time as "mixed mathematics" was not overly demanding from a technical perspective. However, it was considered by the majority of scholars to be of only secondary importance to the study of "physics," the search for the true causes of natural phenomena. Thus Galileo's famous law of falling, published in 1638, was not necessarily recognized at the time as part of physics because it told one *how* but not *why* a heavy body moved spontaneously towards the

---

46. For a detailed discussion of the place of the mathematical sciences in the English universities in the seventeenth century, see Feingold 1997. At this time the mixed-mathematical sciences were at least as likely to be cultivated and taught in Jesuit colleges and by mathematical practitioners in towns like London as in the universities. See Bennett 1986, 8–9, Johnston 1991, 342–43, Lattis 1994, 30–38, and Dear 1995, 32–33.

47. Shapin 1994, 232. On the transmission of the tradition of mixed mathematics to Europe, see Kuhn 1976, 35–41, and Heilbron 1993, 107–12.

48. What I wish to emphasize here is that Ptolemaic and Copernican astronomy were easier to grasp conceptually than was Newton's dynamical account of planetary motion.

center of the earth.[49] This is not to say that mathematics itself was not held in high regard. It had been recognized since ancient times that mathematical proofs, especially in the form of Euclidian geometry, provided the most reliable form of demonstration. The discipline also acquired considerable status from its long association with astronomy, the study of the noble heavens. Mathematics nevertheless remained subservient to physics, partly because of its associations with practical and commercial calculation but mainly because its demonstrations were logical rather than causal and because the entities with which it dealt (numbers, points, lines, and idealized geometrical figures) were not physically real.[50] Within this hierarchy, physico-mathematics occupied a complicated and contested position. Although it dealt mathematically with physical phenomena, it was neither based on self-evident axioms (as was geometry) nor offered the causal explanations of physics.[51]

There were two lines of development during the seventeenth century that led to the emergence of mathematical physics as a discipline. The first was the gradual abandonment of the search for the causes of physical phenomena in favor of the mathematical certainty of what was known from the early seventeenth century as "physico-mathematics." As Peter Dear has argued, the work in mechanics of men like Galileo, René Descartes, Christiaan Huygens, and Isaac Newton constituted "a process of disciplinary imperialism, whereby subject matter usually regarded as part of physics, was taken over by mathematics [thereby] upgrading the status and explanatory power of the mathematical sciences" (1995, 172). One branch of this tradition reached an important high point with the publication of Newton's *Principia*. Newton unified geometry, motion, mechanics, and astronomy in order to derive the movement of the heavens from a series of mathematically expressed laws and principles. Some scholars criticized him for failing to offer a mechanical explanation of the law of gravitational attraction, but most were more enthralled by the technical brilliance of his accomplishment than repelled by his refusal to explain how gravity worked. The second and related line of development was the invention of the mathematical methods that would eventually lie at the heart of mathematical physics. Foremost among these were the algebra of François Vièta, the algebraic geometry of Descartes and Pierre Fermat, and the differential and integral calculus de-

49. Dear 1995, 36–37, 44.
50. Shapin 1994, 315–21; Dear 1995, 35–46, 161–68.
51. On the disputed status of mathematics in natural philosophy in the late sixteenth century, see Lattis 1994, 32–38. On the emergence of "physico-mathematics" as a recognized discipline in the seventeenth century, see Dear 1995, 168–79.

veloped independently by Newton and Leibniz. These mathematical tools formed the basis of what in the eighteenth century would be known as "analysis."[52] It was with the gradual translation of Newton's mechanics and theory of gravitation into this new language in the early years of the eighteenth century that the fundamental techniques of mathematical physics, indeed the discipline itself, began slowly to come into existence.[53]

Despite its beginnings in the decades around 1700, it was only during the eighteenth century that many of the mathematical methods and physical principles now associated with classical mechanics began to emerge. The reworking of Newton's *Principia* in the language of the Leibnizian calculus is a case in point. This process was begun in the late 1690s by Pierre Varignon, Johann Bernoulli, and Jacob Hermann, but it was not until the mid 1730s that Bernoulli's former student, Leonhard Euler, was able to show how most of the mechanical problems raised by Newton could be reliably solved in the form of systematic solutions to a small number of differential equations.[54] Throughout the mid-eighteenth century, the development of the calculus remained closely tied to the solution of physical problems. Many mathematicians in this period made little distinction between the physical problems they were trying to solve and the mathematical techniques they employed. Novel problems often prompted the introduction of new mathematical methods while the physical plausibility of the solutions obtained acted as a check on the reliability of the methods themselves.[55] The most important of these problems involved the curves traced by bodies moving under certain constraints, the motion of vibrating strings, the shape of the earth, the trajectory of comets, and the motion of the moon. The mathematical techniques introduced or developed in association with these problems included the interpretation of equations as multivariate functions (rather than algebraic curves), the solution of partial differential equations, D'Alembert's Principle, the Principle of Virtual Velocities, the calculus of variations, and early formulations of the Principle of Least Action. These methods were formalized, further developed, and systematically expounded in the decades around 1800 in such great treatises as Joseph Louis Lagrange's *Mécanique analytique* (1788) and Pierre-Simon Laplace's *Traité de mécanique céleste* (1798–1827). In the early decades of the nineteenth century a clearly recog-

---

52. Bos 1980, 327.
53. Ibid., 334.
54. Guicciardini 1999, chap. 8.
55. Garber (1999, chap. 2) goes so far as to argue that mathematical physics in the eighteenth century should be regarded essentially as mathematics.

nizable discipline of mathematical physics began to emerge as these methods were applied by men like Joseph Fourier, Augustin Fresnel, and Georg Ohm to such terrestrial phenomena as heat, optics, and the electrical circuit.[56]

The aspect of these developments of most immediate relevance to our present concerns is that the increasingly technical nature of physicomathematics from the mid-seventeenth century made it ever more alien to undergraduate studies. As Feingold has noted, subjects of this kind required "a degree of aptitude, dedication and specialization not yet common in other disciplines" (1997, 446). Mathematics beyond the elementary level had in fact long been recognized as a very difficult subject that was not accessible to all scholars. In the sixteenth century the discipline was regarded by some as a cabalistic practice more akin to the art of the magician or the "mystery" of a practical trade than to ordinary learning. The subject acquired this aura because it was mastered only through long periods of private study under the guidance of an adept, and was comprehensible only to a small number of individuals of apparently supernatural ability.[57] In the seventeenth century concerns of this kind assumed a different form. Although mathematics was increasingly prized for the certainty with which it enabled consequences to be drawn from mathematically expressed physical principles, exponents of the experimental method, most notably Robert Boyle, were concerned that an overly mathematical natural philosophy would severely restrict the size of the scientific community. Experimental demonstrations offered a *public* form of natural knowledge that could be witnessed and judged by most educated members of civic society. By contrast, the esoteric demonstrations of advanced mathematics were directly accessible only to a tiny and closed community of experts.[58]

Furthermore, the oral lectures, catechetical tutorial sessions, guided readings, and Latin disputations, through which most university students were taught and examined, were ill suited to imparting the skills of advanced mixed-mathematics. Unlike the more elementary parts of Euclid's *Elements*, in which the propositions and demonstrations were expressed in verbal form, the new analysis relied on the mastery and application of several new and highly specialized symbolic languages (fig 1.4).[59] The latter skills not

---

56. Greenberg 1995, 620–26; Garber 1999, chap. 4.
57. Watson 1909, 279; Johnston 1991, 320.
58. Shapin 1994, 335–38.
59. On the rapid displacement of verbal by symbolic reasoning in mathematics during the seventeenth century, and on Thomas Hobbes's complaint that algebraic geometry excluded philosophers from mathematical debate, see Cajori 1993, 1: 426–27.

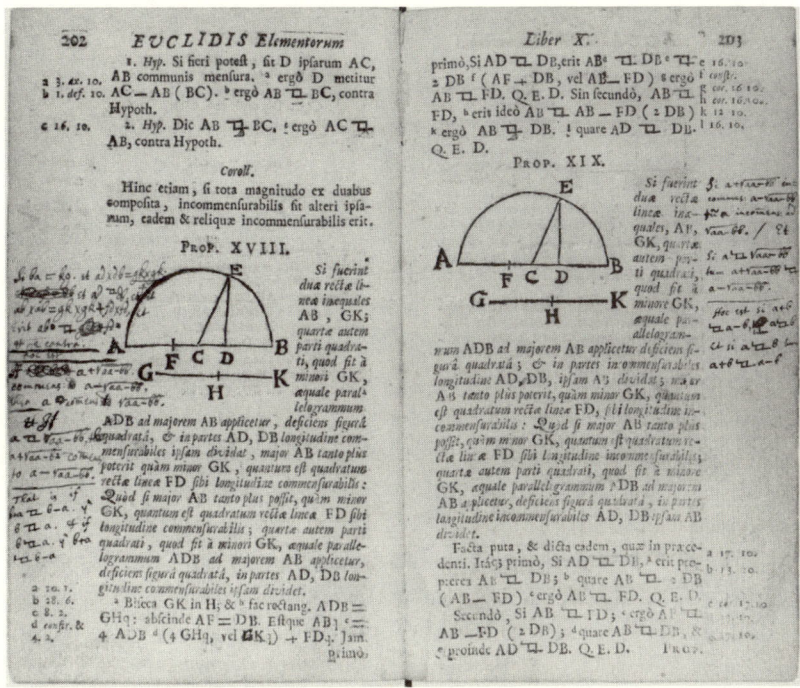

FIGURE 1.4. Pages from Newton's copy of Barrow's 1655 Latin edition of Euclid's *Elements*. Newton has translated the lengthy verbal proposition (Book 10, Proposition 18) into the language of mid-seventeenth-century algebra. Expressed algebraically, the proof demonstrates that, x being determined from the equation $4x(a - x) = b^2$, if x is commensurable with $(a - x)$, then $(a^2 - b^2)^{1/2}$ is commensurable with a (and conversely). To those who had mastered the paper-based manipulation of algebraic symbols the truth of this and similar propositions was quickly and easily proved. To those without these skills the symbolism was incomprehensible. (By permission of the Master and Fellows of Trinity College Cambridge.)

only demanded aptitude on the student's part, but were best learned through a combination of direct guidance using paper-and-pen demonstrations and long hours of solitary practice in the manipulation of symbols and the solution of problems. In the universities it was the new professors of mathematical subjects who possessed these technical skills, but they continued to teach in the traditional way through formal lectures. As Thomas Hobbes noted in this context, even those "who had most advanced the mathematics, attained their knowledge by other means than that of public lectures." Furthermore, professorial lectures were supposed to contain some account of the latest developments in a field, an aim that was almost impossible to combine with the needs of undergraduate beginners. Thus Edmond Halley remarked in the introduction to a series of lectures on the geometrical construction of algebraic equations, that whereas those already "learned" in mathematics came to hear that which was "new and curious" and were satisfied only with "elegant demonstrations, made concise by art and pains," ordinary students "demand[ed] explanations drawn out in words, at length."[60] In practice, the handful of students who possessed sufficient ability and enthusiasm to try to master advanced mathematics did so through a mixture of private study and the informal encouragement and instruction of a willing lecturer, college tutor, or professor.[61] It is for this reason that it is now so difficult to ascertain precisely how Newton mastered advanced mathematics so rapidly in the early 1660s. Manuscripts from the period reveal that he certainly learned a great deal through private study, but the extent to which he also relied upon the informal encouragement, advice, and instruction of his teachers, especially Isaac Barrow, must remain largely a matter of conjecture.[62]

The rapid development of mathematical physics over the eighteenth century exacerbated the pedagogical problems outlined above. This was the period during which very long and complex chains of paper-based mathematical reasoning began to be employed to tackle extremely difficult physical problems using a wide range of advanced analytical techniques and physical principles.[63] These techniques took years of progressive study to master and were best learned in the manner of a trade or craft, through a close master-pupil relationship. At the beginning of the century, for example, the Leibnizian calculus and its applications were understood by only a tiny handful scholars, and even otherwise able mathematicians found the new mathematics extremely difficult to learn without direct contact with those who were already its masters. By far the most important center for the initial transmission of what Bernard Fontenelle referred to as "a kind of mystery, and, so to speak, a Cabalistic Science" of infinitesimals, was Basel. Johann and Jacob Bernoulli not only gave lectures in Basel University on the Leibnizian calculus and its applications to problems in mechanics but offered private lessons on these subjects at home to students and paying guests.

---

60. Hobbes and Halley quoted in Feingold 1997, 385. Newton's successor in the Lucasian Chair, William Whiston, taught undergraduates from the *Principia* by "rendering Newton's calculations into paragraphs of connected prose" (Snobelen 2000, 41).

61. Feingold 1997, 370–71, 384–87.

62. Newton's possible relationship with Barrow is discussed in Feingold 1993.

63. For an analysis of a calculation of this kind concerning tautochronous motion, see Greenberg 1995, 289–308.

It was through this kind of regular and informal contact with skilled individuals, together with supervised practice at solving problems on paper, that the techniques and applications of the new analysis were propagated to a new generation of mathematicians.[64]

But despite these rare centers of training run by talented enthusiasts, the methods of teaching employed by most universities in the eighteenth century remained unsuited to propagating advanced mathematics to large numbers of undergraduates. As with Christoph Clavius and Newton, it remains unclear by exactly what route many outstanding analysts of the latter eighteenth and early nineteenth centuries learned their craft.[65] For example, Laplace first displayed his mathematical abilities in the mid 1760s as a student in the Faculty of Arts at the University of Caen and subsequently went to Paris with a letter of recommendation to the academician Jean LeRond D'Alembert.[66] It is unclear precisely how Laplace's talent for mathematics was recognized and nurtured in the Faculty of Arts, but it is probably significant that one of the philosophy professors at Caen, Christophe Gabled, was among the first to teach mathematics above an elementary level (including the calculus) in a French university. The way in which Laplace learned mathematics as an undergraduate nevertheless remains obscure, as does the extent to which D'Alembert guided his early studies in Paris.[67] Likewise at the beginning of the nineteenth century, two of the founders of German mathematical physics, Wilhelm Weber and Georg Ohm, owed their rare knowledge of mathematical physics far more to family connections and private study than to formal university training.[68] As in the seventeenth century, those who mastered higher mathematics in this period appear still to have done so through a mixture of personal inclination, exceptional talent, private study, and the good fortune to find a teacher who was willing and able to tutor them pri-

64. Guicciardini 1999, 196–99. Fontenelle judged that only five or six people understood the Leibnizian calculus circa 1696.

65. Despite his important role in introducing mathematics to the curriculum of the Jesuit schools, we know virtually nothing of how Christoph Clavius learned his subject. See Lattis 1994, 13–15.

66. Gillispie 1997, 3–4. The origins of D'Alembert's own knowledge of higher mathematics are similarly diverse and obscure. See Hankins 1970, 19–20. For an overview of the development of mathematical physics in eighteenth-century France, see Greenberg 1986.

67. Brockliss 1987, 384–85. On the introduction of higher mathematics to French engineering schools in the mid-eighteenth century, see Alder 1999. According to Brockliss, Gabled "tutored" Laplace, though it is unclear whether this refers to formal lecturing or individual tuition.

68. Jungnikel and McCormmach 1986, 45–46, 51–52.

vately.⁶⁹ This form of passing knowledge from one generation to the next could only produce a small number of able analysts and made their de facto mathematical training extremely reliant on those individuals and books that they chanced to find.

Mixed mathematics in Britain developed somewhat differently over the eighteenth century from its Continental counterpart. British mathematicians used Newton's fluxional calculus (rather than the d-notation due to Leibniz) and stuck with the mechanical principles used by Newton in the *Principia*. Despite these differences, mathematicians on both sides of the Channel initially took a keen interest in each other's work, those in Britain developing fluxional techniques that were roughly comparable to the Leibnizian ones developed by their Continental peers. From the middle decades of the century, however, this contact diminished and mixed mathematics in Britain did not undergo the profound transformation wrought in Continental analysis by the introduction of those new mechanical principles and mathematical methods outlined above.⁷⁰ Indeed, by the 1770s one of Cambridge's most brilliant mathematicians, Edward Waring (SW 1757), found it virtually impossible even to follow Leonhard Euler's careful exposition of the use of partial differential equations and the calculus of variations to tackle problems in mechanics.⁷¹ Where the novelty of Newton's methods had made the *Principia* very hard going for Europeans in the late seventeenth century, Continental analytical mechanics posed similar difficulties for the British in the late eighteenth and early nineteenth centuries. The development of higher mathematics in Britain during the latter period was characterized by a growing appreciation of the superiority of French analysis and its application to celestial mechanics and mathematical physics, and by the gradual adoption of analytical methods in the British universities and military schools.⁷²

Like their Continental counterparts, the leading British mathematicians of the eighteenth century came to practice their craft through a mixture of happenstance, enthusiasm, and ability. This point is exemplified by Nicholas Saunderson who, despite having no university education or degree, became the leading teacher of Newtonian natural philosophy and fluxional calculus

---

69. In order to master the "immense colossus" of French analysis in the mid 1820s, the young Gustav Dirichlet (who had been a student of Ohm's at the gymnasium in Cologne) left his native Germany to study for five years with French mathematicians in Paris. There he was befriended by Joseph Fourier, who nurtured his interest in mathematical physics and Fourier series (Jungnikel and McCormmach 1986, 55, 171–72).

70. Guicciardini 1989, 139–42.

71. Ibid., 89–91.

72. For an overview of the reform, see Guicciardini 1989, chaps. 7–9.

at Cambridge in the early eighteenth century. As far as can now be ascertained, Saunderson taught himself mathematics 'at home' and made a living in Cambridge first as a private tutor and then, from 1711, as the fourth Lucasian Professor of Mathematics in the university.[73] There was also a great deal of public interest in Newton's natural philosophy and fluxional calculus in Britain in the middle decades of the century, an interest that was nurtured through public lectures in coffee houses, the establishment of mathematical societies, the publication of introductory books on the "doctrine of fluxions," and popular magazines such as the *Ladies' Diary* that posed mathematical questions to be answered in the following issue.[74] The level of mathematical discussion among these so-called philomaths was relatively elementary, but it was from within this public culture that some of the most able British mathematicians emerged.[75] Consider, for example, the careers of the three most active and original "fluxionists" of the mid-eighteenth century, Thomas Simpson, John Landen, and Edward Waring. Both Simpson and Landen were largely self-educated, and initially established their mathematical reputations as contributors to the *Ladies' Diary* (which Simpson subsequently edited from 1754–60). Landen remained an amateur mathematician who earned a living as a land surveyor, while Simpson made his way as an astrologer, weaver (associated with the Spitalfields Mathematical Society), and, finally, mathematics master at the Royal Military Academy at Woolwich. Waring was the only university-educated member of the trio, but had been taught mathematics only to a relatively elementary level as a Cambridge undergraduate in the mid 1750s. It is unclear precisely how these men mastered higher mathematics and chose their research problems, and none passed on his knowledge directly to a new generation of students.[76]

It is also worth noting that higher mathematics in Britain was not regarded with the same reverence accorded to it by the leading philosophers of the European Enlightenment, especially in France.[77] Where analysts such as D'Alembert, Euler, Laplace, and Lagrange played leading roles in the great

---

73. Saunderson was blinded by smallpox at the age of twelve months and appears to have taught himself mathematics by recording formulae and figures with pins on a wooden board. For an outline of his career, see the *DNB* and Guicciardini 1989, 23–24.

74. Guicciardini 1989, chap. 4; Stewart 1992, chap. 4.

75. Guicciardini 1989, 27, 85, 104, 108–23.

76. Simpson taught only elementary mathematics at Woolwich while Waring, Lucasian Professor of Mathematics at Cambridge from 1760 to 1798, declined to lecture at all. A contemporary noted that Waring's "profound researches" were "not adapted to any form of communication by lectures." See Guicciardini 1989, 89, 110; and "Edward Waring," *DNB*.

77. On the high status of "analysis" in the French Academy in the eighteenth century, see Terrall 1999.

academies in Paris, Berlin, and St. Petersburg, Britain's premier scientific institution, the Royal Society of London, took a more equivocal view of mathematicians. An informative manifestation of this state of affairs occurred in 1784 when a small group of mathematicians threatened to secede from the society following what they regarded as political finagling on the part of the President, Sir Joseph Banks, to marginalize their influence.[78] The relevance of this episode to our current concerns is that it highlights the difficulty of accommodating extremely mathematical studies in a society whose broad membership communicated mainly through the pages of its journal, the *Philosophical Transactions*, and sociable gatherings at which papers were read to the assembled members. At the heart of the dispute was a heated debate over whether the society's membership should be constituted of groups of experts, each capable of adjudicating matters of great technical complexity, or men of broader learning who could assess the credibility of any claim to new scientific knowledge. The mathematicians asserted that a well-rounded intelligence was not sufficient to assess the quality of mathematical papers submitted for publication in the *Philosophical Transactions* nor to judge the "minutiae . . . and finesse of those dry studies which mostly occupy us in the times of our meetings." Banks's supporters responded—in terms redolent of Boyle's worries a century earlier—that "real philosophers" were likely to succumb to their own arcane enthusiasms and that only "men of general literature" were capable of offering "a right opinion concerning the general value of the philosophical observations and experiments which are produced at the Society's meetings."[79]

An important question raised by this debate concerns the kind of community within which knowledge in the form of higher mathematics can circulate and develop. The problem from the Royal Society's perspective is easily appreciated by leafing through a volume of the *Philosophical Transactions*, say the one for 1779. Alongside fairly accessible papers ranging in topic from the manufacture of new airs and a description of a total eclipse of the sun to a cure for St. Vitus's Dance by electricity and why orang outangs are unable to speak, a paper by Waring entitled "On the General Resolution of Algebraical Curves" stands out immediately as one that would have been utterly incomprehensible to all by a tiny handful of mathematicians. As a Fellow and ex-Vice President of the Society, Samuel Goodenough observed in 1820, the "algebraical pursuits . . . cannot but be in *the reading* very uninter-

---

78. This episode is discussed in Heilbron 1993.
79. Quoted in Heilbron 1993, 86, 88.

esting, not to say tiresome."[80] This antipathy to original mathematical studies is further explained by recalling that even in the late eighteenth century higher mixed-mathematics was comprised almost entirely of celestial mechanics, geodesy, and a limited range of terrestrial problems such as the dynamics of vibrating strings. Just prior to the discipline's expansion to embrace branches of physics such as heat, light, electricity, and magnetism, many fellows of the society would have considered the tedium and incomprehensibility of advanced mathematics too high a price to pay for its limited utility.[81] A culture in which "dry studies" of this kind could flourish would need to be one whose members possessed the technical skills to appreciate the "minutiae" and "finesse" of the problems and solutions under discussion.[82]

Even as the debate between Banks and the mathematicians was raging in London, Cambridge University was in the process of developing an undergraduate culture of just this kind. From the mid-eighteenth century, mixed mathematics had begun to displace more traditional subjects such as theology and moral philosophy, while competitive written examinations replaced the Latin disputation as the dominant form of student assessment. By the end of the eighteenth century, mathematics dominated undergraduate studies to the almost complete exclusion of all other subjects, a situation that continued until the mid-nineteenth century. An important aspect of this enormous expansion in mixed-mathematical studies is that it represented a major change not just in *what* was taught but also in *how* and *by whom* it was taught. I noted above that mathematics could not easily be learned from formal lectures, oral debate, and reading alone. In order rapidly to master complicated mathematical methods and their application to difficult physical problems, the vast majority of students needed an ordered and progressive plan of study, long periods of private rehearsal tackling graded examples on paper, and regular face-to-face interaction with a tutor prepared to correct their written work, explain difficult concepts and techniques, and demonstrate good practice. In Cambridge the latter function was fulfilled not by college lecturers or university professors, but by private tutors (usually college fellows) who were prepared to manage a student's studies for a fee. During the first third of the nineteenth century, these tu-

---

80. Samuel Goodenough to James Smith, 6 September 1820, quoted in Miller 1983, 29.
81. Heilbron 1993, 101–2.
82. Even the Spitalfields Mathematical Society did little to cultivate higher mathematics, concentrating instead on experimental demonstrations and natural philosophy. See Stewart and Weindling 1995.

tors assumed almost complete responsibility for preparing the most able undergraduates for increasingly difficult and competitive written examinations in mixed mathematics. Their success as teachers was based on the power of the pedagogical methods they developed to impart to their pupils a very high level of technical skill over a relatively short period of time. These methods were the forerunners of those employed by tutors such as Routh and have subsequently become commonplace in technical education.

This pedagogical revolution marked a major watershed both in the nature of undergraduate studies and in the development of the mathematical sciences. From the early nineteenth century the skills of higher mathematics and rational mechanics became the focus of intense and competitive undergraduate study, while the most efficient methods of teaching them became the subject of extensive pedagogical experiment. These developments radically altered the forms of sociability peculiar to undergraduate studies. The open and oratorical spaces of the lecture hall, disputation, and tutorial conversation gave way to the closed and private worlds of the coaching room, the written examination, and solitary study. Moreover, as the once individual experience of mastering the higher mathematical sciences became increasingly routinized and communal (albeit in small groups), the skills on which successful mathematical practice relied, as well as the peculiarities of the techniques by which they were best taught, learned, and tested, became more accessible to historical analysis. By tracing the development and impact of such seemingly mundane pedagogical devices as face-to-face training in problem solving on paper, written examinations, educationally orientated treatises, and end-of-chapter exercises, and by listening to the comments of tutors and students as the development occurred, I shall make visible a remarkable revolution in training that brought the modern mathematical physicist into being.

Having identified the historical specificity and productive power of these pedagogical devices, I look more closely at their role in generating a local community of knowing individuals at one or more sites. Training clearly offers a promising route for tackling the problem I raised above concerning the spread of physics to the world's cultures during the late nineteenth and twentieth centuries. Once a well-defined regime has been devised to generate a particular form of technical expertise at a given site, that expertise ought to be as reproducible at new sites as is the training regime itself. It is nevertheless important to preserve a distinction between *what* it taught and *how* it is taught. Although there will generally be a close relationship between the detailed process and content of training, broadly similar pedagogical techniques can be used to generate quite different skills and sensibilities. One line of enquiry in the present study will be to investigate the extent to which the pedagogical methods specific to Cambridge produced a local community of mathematical physicists with comparably specific technical skills and research interests. I noted above that Poynting's training enables us to account for the kind of skills he routinely deployed, the research problems he believed to be important, what he assumed his peers would be able to understand, and how he structured his publications. Does this imply, however, that physicists trained in a different tradition, say in Germany, would find his results difficult or even impossible to understand?

A similar question arises in connection with the dissemination of knowledge in the form of novel theories. If the physicist's way of knowing is closely bound to his training, in what circumstances can he recognize a new theory as superior to one with which he is already familiar, and how easy will it be for him to master the new theory's canonical techniques? Morever, once a theory has become widely accepted, how is it accommodated within an extant training system so that it will become the orthodoxy for a new generation of students? In order to address questions of this kind I shall consider two general issues regarding the relationship between training and knowing. First, to what extent does training in a new physical theory inculcate skills and sensibilities that are invisible when the theory is formally stated in a published paper or textbook presentation? If grasping an author's meaning requires tacit knowledge of this kind, then effective communication within an expert community must be more closely tied to patterns of training than is generally supposed. Second, in what ways do specific kinds of pedagogical regime have the power to produce or respecify the content and operation of a technical discipline? This power can refer to the kind of proof, theorem, or problem solution deemed legitimate within the community, but it can also refer to the location and development of important new capacities in the discipline's practitioners.

A pedagogically orientated historiography of the kind outlined also prompts a reappraisal of Cambridge's relative standing with respect to other European institutions of higher education at the end of the eighteenth century. Georgian Cambridge has sometimes been depicted as a moribund university that squandered Newton's intellectual legacy by failing to remain at the forefront of research in the development of the calculus and celestial mechanics. But, as I have already noted, few if any European universities taught mathematics beyond the elementary level until the late 1760s, most Continental research in analysis and its applications taking place in the academies. Moreover, from a pedagogical perspective, Georgian Cambridge was as successful as any contemporary European institution of higher education in producing students with high levels of technical expertise in mathematics.[83] Thus where Edward Waring's mathematical *researches* were certainly of less significance than those of some of his French contemporaries, the problems he set as an examiner for the most able Cambridge undergraduates from the 1770s were regarded by some as "the most severe test of mathematical skill in Europe."[84] While it is certainly the case, therefore, that academicians such as Laplace and Lagrange worked in an environment that encouraged and enabled the production of original works to a far greater extent than did Cambridge, this did not preclude the emergence of a powerful pedagogical tradition in the latter institution. Traditions of research and training developed independently in the late eighteenth century, and both must be taken into account in explaining the expansion of mathematical physics over the subsequent two hundred years.[85]

1.5. Purpose and Plan

These questions and issues are explored in the following chapters through a study of the rise of mathematical physics in Cambridge University from the 1760s to the 1930s. The study begins in chapters 2 and 3 by tracing the dramatic changes in undergraduate training that took place from the mid-eighteenth to the late nineteenth centuries. The first of these focuses on the rise of mathematical studies in Georgian and Victorian Cambridge with special reference to the gradual emergence of the private tutor as the most important teacher of undergraduate mathematics. These tutors developed a range of new training methods capable of producing ever higher levels of technical competence in their most able pupils. I describe the historical origins of these methods, contrast them with more traditional forms of university teaching, and discuss the heated controversy they generated over the very nature, content, and purpose of undergraduate studies in the university. As we shall see, the cultural significance of this revolution was neatly captured circa 1830 in the term "coach," a word coined to describe not just the private tutor himself but also the experience of working under his tute-

---

83. The uses of mathematics in Georgian Cambridge are discussed in detail in chapters 2 and 3.
84. "Edward Waring," *DNB*.
85. On the development of a pedagogical tradition in analytical mechanics in France in the eighteenth century, see Alder 1999.

lage. The broader pedagogical apparatus of the revolution is further explored in chapter 3, where I turn to what I describe as the *material culture* of mathematical studies. I noted above that an important characteristic of mathematical physics as it developed over the eighteenth century was the use of paper and pen as a material medium for tackling long and difficult technical problems. This apparatus of mathematical work is so commonplace that it is seldom even referred to by historians or philosophers of mathematics, but it was made starkly visible around 1800 when mathematical problem solving was introduced to an undergraduate culture long based on oral methods of teaching and examination. By studying how student sensibilities were gradually refashioned to fit a new culture of problem solving with paper and pen, I lend visibility to mathematical skills that are usually tacit, and reveal the pedagogical devices invented to teach them as rapidly as possible to large numbers of undergraduates. I shall also show that pedagogical devices such as mathematical textbooks, examination questions, and end-of-chapter exercises have a rich and closely interrelated history. That history includes the construction of an archive of canonical exemplars, and I begin in this chapter to explore ways in which learning through exemplars shapes the mathematical physicist's knowledge and research style.

The next two chapters look more closely at the student's experience of mathematical study in Victorian Cambridge through what might be described as a historical ethnography of undergraduate life. In chapter 4 I investigate the ideals of hard work, competitive study, and physical exercise that emerged coextensively with private teaching and written examination in the early nineteenth century. Those students who aspired to score highly in the annual mathematical examinations adopted an increasingly ascetic lifestyle and devised complementary regimes of physical and intellectual exercise. This chapter explores the relationship between the daily experience of student life, the numerous incitements to hard study deployed by the university, and the emergence of the elite mathematics student as a manly ideal in whom the rational mind and body were perfectly combined. Although some aspects of this ideal faded in the latter nineteenth century, many of the sensibilities it engendered have remained at the heart of elite technical education. In chapter 5 I turn to a detailed analysis of mathematical coaching in action. By describing the training techniques and devices used by Edward Routh and his contemporaries in the mid and late Victorian periods, I reconstruct the rich pedagogical economy of face-to-face teaching, guided problem solving, regular examination, textbooks, coaching manuscripts, graded exercises, and private study, through which the subtle skills of the Cambridge mathematical physicist were produced. This chapter also develops the questions first raised in chapter 3 concerning the extent to which specific forms of training generate similarly specific technical skills, and ways in which those skills are in turn generative of specific forms of research. The former question is tackled indirectly by looking at attempts to train Cambridge-like mathematicians at other sites, the latter more directly through the testimony and work of several successful Cambridge graduates. Chapter 5 concludes with a discussion of the relationship between educational reform and the rise of a new research ethos from the late 1860s, and the resultant appropriation of the coach's teaching methods by mathematics professors, lecturers, and supervisors.

In chapters 6 and 7 the focus shifts from the introduction of new pedagogical methods to the interaction between training and mathematical theory. In 1873 James Clerk Maxwell published his *Treatise on Electricity and Magnetism*, a book that contained the most complete account of his new field theory of electromagnetic phenomena. But, despite the fact that Maxwell wrote the *Treatise* in large measure as a textbook for Cambridge undergraduates, it proved extremely difficult even for senior members of the university to understand its case-by-case solutions to difficult electrical problems and new physical concepts. In chapter 6 I discuss the process by which a consistent theoretical interpretation of the book was simultaneously produced and propagated in the university through a mixture of private and collective study, coaching, college lecturing, and the manufacture of canonical examination questions. The process is particularly effective in revealing those aspects of this major new theory that remained tacit and incommunicable in Maxwell's written account of his work. The skills involved in the Cambridge appropriation of the theory were preserved and propagated within the training system, and the chapter concludes by discussing ways in which this system shaped the research of such famous Cambridge "Maxwellians" as J. J. Thomson and J. H. Poynting. In chapter 7 I undertake a more detailed account of Joseph Larmor's work on electromagnetic theory. A Cambridge contemporary of Thomson and Poynting's, Larmor established his reputation in the 1890s by developing a new electrodynamics of moving bodies that refounded Maxwell's field theory on the concept of discrete units of electric charge, or "electrons." This chapter shows how Larmor's work grew out of his Cambridge training and was passed on through the same medium to the next generation of Cambridge mathematical physicists. One of the most striking aspects of Cambridge electromagnetic theory in the first decade of the twentieth century is that it contained many theoretical concepts and computational techniques that were shortly to be rendered false or illegitimate by the theory of relativity. I shall show how such now discredited notions as the electromagnetic ether, the electronic theory of matter, and the Lorentz-FitzGerald contraction hypothesis were not only made familiar and legitimate entities to a generation of Cambridge mathematical physicists but routinely deployed by them in a number of highly technical and ongoing research projects.

In the final two chapters I investigate the relationship between scientific knowledge and pedagogical tradition from a new perspective by discussing the Cambridge response to Einstein's special and general theories of relativity. Up to this point my analysis has been concerned primarily with the production and replication of knowledge at one site, but I now turn directly to questions of pedagogical geography and to the claim that a radically new physical theory is "incommensurable" with its predecessor. Thomas Kuhn made the important and controversial claim that successive physical theories are "incommensurable" in the sense that physicists cannot move by clearly stated rules or logical steps from one to the other. In order to master the new theory they have to immerse themselves in its principles and canonical problem solutions until they undergo a kind of conversion experience or gestalt switch.[86] My purpose is not to embark on a philosophical analysis of this claim but to note that it can usefully be construed in geographical rather than temporal form. Instead of debating the relationship between successive theories, I shall focus on the different meanings ascribed to published work by readers from different traditions. The problem then becomes one of understanding how different meanings are generated around the same group of texts, and under what conditions the authorial meaning can be preserved and successfully propagated. What is *incommensurable* on this showing is not the essential meaning of theories themselves but the particular skills and assumptions that go into generating a working interpretation of them at different sites. I shall argue that these skills and assumptions are to a large extent the products of local regimes of training and that the problem of incommensurability is therefore best construed by the historian not so much as one of time in the form of sequential theories as one of space in the form of pedagogical geography.

The respective Cambridge responses to the special and general theories provide an informative historical illustration of the practical consequences and limitations of incommensurability. In the former case Einstein, then a

---

86. The classic discussion of revolutions as changes in world view is Kuhn 1970, chap. 10.

little known patent office clerk in Bern, suggested a novel method for tackling a series of problems in electrodynamics that Cambridge physicists believed they could solve perfectly well using techniques introduced by Larmor. Why, then, did a tiny handful of Cambridge men take an interest in the papers of a little-known Swiss, given that they did not even share his account of the paper's meaning? This question is addressed in chapter 8, where I show how some of Larmor's students generated new meanings for Einstein's work as they appropriated it within a local research tradition. At the heart of this discussion lies the problem of the fragility of knowledge when circulated in purely printed form and the difficulties this raises for the transmission of novel work to new sites. In chapter 9 I investigate the circumstances in which these difficulties can be overcome by exploring the successful transmission of the general theory of relativity to Cambridge during the second and third decades of the twentieth century. This case is particularly relevant to the issue of text-based transmission because the outbreak of World War I effectively barred any face-to-face meetings between Cambridge mathematical physicists and the Continental masters of relativity. In this chapter I trace the course of events by which the Cambridge astronomer A. S. Eddington not only became the first person in Britain to master the conceptual subtleties and forbidding mathematical apparatus of the general theory, but developed the pedagogical devices by which it could be introduced to undergraduate studies in Cambridge. During the latter process, the pedagogical system that forms the main focus of the present study switched from acting as a conservative barrier to the arrival of the general theory in the university to become instead one of the most important mechanisms for its preservation and propagation in the interwar period.

It is also worth clarifying at this point the sense in which the present study should be regarded as a *social* or *cultural* history of the mathematical sciences. Early studies in the sociology of scientific knowledge were often aimed at undermining the belief that the historical development of the natural sciences was explicable solely in terms of their technical content or the application of a special scientific method. To establish this point it was necessary to show only that cognitive judgments were not wholly determined by an internal logic of scientific enquiry. For those who sought to go further and to provide a social *explanation* of cognitive judgments, the usual procedure, as Jon Harwood has noted, was to try to identify "which social structural features of the society in question were responsible" for the judgments (1993, 12). Explanations of this kind usually proceeded by seeking a strongly causal relationship between the theoretical commitments peculiar to a national or otherwise well-defined group of scientists, and recognizably similar commitments held by the same group in other cultural sectors such as politics, religion, and the arts.[87] To my mind social studies of this kind make claims about science that are by turns too strong and too weak. They are too strong in implying that much of the technical practice of a discipline like mathematical physics can be understood by reference to broader, extrascientific social factors, and too weak in failing to investigate the role of forms of sociability that are peculiar to the technical practice itself. As Bruno Latour has pointed out, if providing "social explanations" of modern physical theory requires finding "features of society mirrored by mathematics in some distorted way" (1987, 246), then we ought to conclude that such theory is barely social at all. But if by "social" we are referring to those shared competencies that enable individuals to participate in an activity and which in some cases are only acquired through protracted periods of training, then we might conclude that it is the most esoteric and technical disciplines that are actually the *most* social.

My study certainly places scientific knowledge in a broader social setting insofar as it seems appropriate to do so. Thus we shall see in chapter 2 that *what* and *how* Cambridge students were taught in the mid-Victorian period was partly a matter of social policy and political strategy regarding the purpose of undergraduate education. However, the primary aim of my study is not to show how some scientific judgements were shaped by homologous "styles of thought" in politics or religion, but how the very existence of a large community of individuals capable of making extremely technical judgements of a particular kind was built on a new apparatus of training.[88] In this sense my approach can be regarded as "social" (if the term is meaningful when used this way) and has something in common with studies of science that emphasise the *tacit* or *personal* dimension to scientific knowledge. Like the "styles" genre just described, these studies also deny that scientific work can be explained in terms of a metascientific method, but they do so in a different way. Instead of explaining cognitive judgments in terms of broader cultural ideals, they explain them by reference to skills and sensibilities which pass invisibly from master to pupil during training and which are tacit in published scientific work.[89]

87. Classic studies of this kind in the history of physics are Forman 1971 and Wynne 1982.
88. The aims of the "styles of thought" genre are clearly discussed in Harwood 1993, chap. 1.
89. Classic studies in this field are Polanyi 1958, Collins 1985, and Ravetz 1996, chap. 3. The use of the term "tacit knowledge" in the literature on research schools is discussed in Olesko 1993.

To date these insights have been most effectively exploited by historians in the growing literature on research schools. This literature is nevertheless concerned almost entirely with schools of experimenters rather than theoreticians, and, insofar as it deals with knowledge production, has focused mainly on each school's leader as a role model for his students. My study complements this literature in several ways. First, I am concerned with the introduction of a series of new pedagogical devices in the mathematical sciences which, although initially confined to a small number of institutions, have since been adopted on an international scale. Second, my analysis focuses closely on the everyday role of this pedagogy in generating technical skill in students. Although many studies have alluded to the importance of a tacit or taken-for-granted dimension to scientific work, few have investigated beyond the most elementary level how this dimension is transmitted, let alone how the means of its transmission have changed with historical time. Third, my study differentiates between the production of technical skill for its own sake and the role of such skill in the formation of research schools. It is noteworthy, for example, that although Cambridge was producing students with very high levels of technical competence in mathematical physics from the 1830s, no discernable "research school" in the subject emerged until the 1870s. By investigating how Maxwell's *Treatise* became a focus of collective research in the university during the last quarter of the nineteenth century, the present study shows how one kind of research school was gradually built on a wider and more individualistic culture of mathematical problem solving.

The sense in which the present study can be considered a contribution to *cultural* history is that it is intended in part as a historical ethnography of a little studied and esoteric form of life. Major sections of the next four chapters are written, insofar as is possible, from the "natives'" point of view, drawing on a wide range of contemporary sources to build a detailed picture of everyday undergraduate experience. What is unusual about my study is that it attempts to understand the production of technical expertise as an integral part of this cultural world. This point is usefully emphasized by comparison with another ethnography of Georgian Cambridge. Historical anthropologist Greg Dening uses a comparative study of Cambridge University and the Hawaiian archipelago in the 1780s to accentuate a number of cultural differences between these contemporary but geographically distant settings. It is a notable feature of Dening's admirable study that while his analysis of the costume, ritual, and ceremony of the university brilliantly depicts Cambridge as a theatrical microcosm of the social order in Georgian England, it has almost nothing to say about the mathematical knowledge that was so highly prized in the university.[90] Where what counted as knowledge for the Hawaiians is described and explained as an integral part of their local cultural life, the same apparently cannot or, at least, need not be done for what the Cantabrigians thought they knew. My purpose is precisely to understand how the rituals, ceremonies, and forms of sociability (including the practices of doing and examining the mathematical sciences) peculiar to Cambridge constituted a way of knowing that must be counted extraordinary by almost any comparison.

This brings me to one last feature of the study that should be of special interest from the perspective of social anthropology. The cultural turn in the history of science over recent years has prompted studies that, like the ethnographies on which they are modeled, explore science in a rich social context but over relatively short periods of time. What has not been studied from a cultural perspective is science's stability through successive generations over much longer periods. The importance of this issue is appreciated by recalling that disciplines such as mathematical physics are sometimes regarded as forms of knowledge that are so technically difficult, cognitively counterintuitive, and fragile in transmission, that they might one day be lost to humanity.[91] My study is perhaps the first directly to address this issue not only by describing the kind of culture that can preserve esoteric knowledge in the form of mathematical or theoretical physics, but by showing how that culture has successfully developed and reproduced itself for more than two hundred years. Throughout this period, knowledge in the form of specific mathematical methods and theoretical concepts came and went, while most of the underlying mechanisms of cultural reproduction remained intact. As far as I am aware, the present study is unique in providing such a detailed and diachronic historical ethnography of mathematical life at one institution, and I offer it as a model of how a cultural history of mathematical physics might be written.

Finally, a brief word of explanation is in order regarding the changing level of technical expertise required of the reader as the study unfolds. I emphasized above that my main purpose is to explain rather than to assume the kind of technical competence possessed by mathematical physicists. This

---

90. Dening 1995, esp. 77–125.
91. See for example McCauley (1998), who compares the fragility of science with the "naturalness" of forms of religious belief.

will indeed be the case for roughly the first five chapters of the book, but from chapter 6 onward I begin increasingly to assume that the reader possesses at least some expertise in mathematical physics. My reasons for doing this are discussed at the beginning of chapter 7, but I would like here to reassure the general reader that most of the narrative together with the gist of the main arguments should remain broadly comprehensible to those without such training.

∼ 2 ∼
The Reform Coach
Teaching Mixed Mathematics in Georgian and Victorian Cambridge

> "Are you stopping at Baymouth, Foker?" Pendennis asked.
> "I'm coaching there," said the other with a nod.
> "What?" said Pen, and in a tone of such wonder, that Foker burst out laughing, and said, "He was blowed if he didn't think that Pen was such a flat as not to know what coaching meant."
> "I'm come down with a coach from Oxbridge. A tutor, don't you see old boy?"
> PENDENNIS, 1848 [1]

2.1. The Private Tutor and Cambridge Mathematics

In his classic study of academic life in Victorian Cambridge, *The Revolution of the Dons,* Sheldon Rothblatt made the novel suggestion that the system of private teaching (or "coaching") which emerged to dominate undergraduate training from the late eighteenth century should be seen as a *positive* development which "made a major contribution to the reform of teaching in the second half of the nineteenth century" (1968, 208). Rothblatt's remark is striking because it challenges a widely held view that private teaching was an illicit and iniquitous trade which undermined liberal educational values and the authority of officially appointed teachers, fostering instead the worst excesses of competitive examination and mindless cram.[2] But, as Rothblatt pointed out, it was the leading private tutors, rather than donnish college lecturers or distant university professors, who provided the prototypical model of the professional academic who would emerge in the latter decades of the nineteenth century. In this chapter I argue that private tutors did indeed play

---

1. Thackeray 1994, 37. First published (as a serial) in 1848. Thackeray coined the terms "Oxbridge" and "Camford" (from Oxford and Cambridge) to refer indirectly to the two ancient English universities. The former is now commonly used to refer collectively to Oxford and Cambridge.
2. For an influential assessment of coaching in Victorian Cambridge by one of the reformers who eventually brought about its demise, see Forsyth 1935.

a major role in the reform of undergraduate teaching after 1850, but that both their rise to eminence during the late eighteenth and early nineteenth centuries, as well as their subsequent half century of pedagogical rule, were of much deeper and more far-reaching historical significance than even Rothblatt supposed.

The emergence of private teaching has rightly been linked, albeit implicitly, to the combined effect of the introduction of written examinations in mathematics coupled with the lack of willingness on the part of the public and college teachers in the university adequately to prepare the most ambitious students to compete in such examinations.[3] W. W. Rouse Ball, for example, the first, and very influential, historian of the Cambridge school of mathematics, saw private teaching as the natural companion of written examination to such an extent that he felt any further explanation of the marginalization of other forms of teaching to be superfluous.[4] On Rouse Ball's showing, attempts by the university to restrict the practice around 1780 were entirely futile, a view he believed to be substantiated by the systematic repeal of such restrictions during the first three decades of the nineteenth century. But a more detailed study of the historical evidence suggests a quite different course of events. I do not deny that written examinations provided a prime motivation for able undergraduates to employ private tutors, but I shall claim that the displacement of public and college teachers by private tutors was a much more managed, protracted, and controversial process than Rouse Ball allowed. If we are to grasp both the significance of the pedagogical revolution initiated by these tutors and the historical events which made the revolution possible, it is necessary to disaggregate their rise to prominence into several distinct periods and to look carefully at what, how, and who they actually taught.

In the first half of this chapter, I argue that private teaching *was* strictly controlled within the university until at least the second decade of the nineteenth century, and that the gradual relaxation of the rules thereafter was a controlled process, calculated both to curtail the influence of private teachers beyond Cambridge and to bolster the technical competence of the most able students. Once this point is established, moreover, it reveals a new and previously unexplored relationship between private teaching and the so-called analytical revolution of the 1810s and 1820s. The introduction of Con-

---

3. The term "public teacher" refers to a paid teacher (usually a professor) whose lectures are open free of charge to all members of the university. Each college had its own staff of lecturers and tutors (paid by the college) whose teaching was restricted to members of the college.

4. Rouse Ball 1889, 160–63.

tinental analysis to Cambridge has been widely studied by historians of mathematics, but very little attention has been paid to the mechanisms by which the new techniques were actually popularized and propagated within the university. The translation of analysis from Paris to Cambridge was not effected simply by overcoming local prejudice and getting a few appropriate questions onto the examination papers; it required, rather, new textbooks, new teaching methods, and years of hard work on the part of tutors and students. A careful study of the people, techniques, and strategies by which the revolution was practically accomplished highlights two very significant points: first, that it was private tutors who actually mastered and taught the d-notation calculus and other analytical methods in Cambridge; and, second, that it was only during this transitional period (and not before) that private tutors displaced the officially appointed teachers of the university to become the guardians of the intellectual elite. The tough, progressive training methods developed by private tutors were much better suited to teaching advanced analysis and its application in the solution of difficult problems than were those of the college lecturer, and it seems very likely that the final relaxation of the rules governing private tuition in the mid 1820s was intended to encourage the rapid uptake of analytical methods. The profound change in undergraduate studies through the 1820s and 1830s must therefore be viewed as much a revolution in pedagogy and personnel as it was in mathematical content.

Despite the almost complete dominance of private teaching from the 1830s, however, the methods and values of the private tutor remained controversial and but one of a number of competing ideals of undergraduate education. This was especially true when it came to debates over the purpose of undergraduate mathematical studies in the university. On this contentious issue, of which more is said in subsequent chapters, there was little consensus among influential members of the university until the 1850s. For example, William Whewell, Master of Trinity College from 1841, believed that mathematics was primarily a tool for cultivating and testing the rational mind and that it should form only part of a broader moral and spiritual education. As far as he was concerned, it was not the purpose of undergraduate teaching to prepare students for research after graduation, and any desire on their part to do so ought to be regarded as a worrying sign that they had failed to be convinced by the explanations of their tutors. George Peacock, by contrast, Professor of Astronomy and Geometry from 1836, thought that undergraduate teaching *should* equip the best students for research. He insisted nonetheless that theology and moral philosophy should also form part of undergraduate studies and that learning should proceed at a leisurely pace consonant with private study and the development of independent habits of thought. William Hopkins, the first mathematics coach in Cambridge to derive a permanent living from private teaching, offered yet another view. He thought that elite undergraduates should be systematically trained in mathematical physics to the highest level and that they should be encouraged as much as possible to engage in research after graduation. Hopkins argued that only the teaching methods employed by private tutors could produce researchers of the appropriate caliber and that hard study and submission to mathematical rigor were themselves sufficient indications of moral virtue.

In the second half of the chapter, I examine the teaching methods employed by private tutors, discuss the origin and cultural resonances of the term "coach," and explore the way coaches were portrayed to the early Victorian public in serialized novels. I also discuss the resolution of the pedagogical differences between Whewell, Peacock, and Hopkins during the 1840s and 1850s. Hopkins's reputation as a pioneer of the new discipline of physical geology, together with his unique status as an outstanding mathematics teacher, made his an influential voice in debates over the role and purpose of mathematical studies both nationally and in Cambridge. From the early 1840s he not only defended private teaching in the face of vitriolic attacks from men like Whewell and Peacock, but argued forcefully that the skills of the mathematical physicist should be turned to the commercial and industrial needs of Britain and her Empire. His pleas finally found favor with the Parliamentary Commissioners who investigated teaching at Oxford and Cambridge in the early 1850s. In the politically liberal climate of mid-Victorian Britain, Hopkins's utilitarian view of mathematical studies, and his free-market approach to preserving standards in undergraduate teaching, offered an appropriate model for the reforms that would take place in Cambridge mathematical education over the next thirty years. From the mid 1850s the key question was no longer *whether* the most able students should be drilled and examined in mathematics using the techniques developed by the leading coaches, but rather *how* these methods could most effectively be brought within the formal teaching structure of the university.

2.2. Mixed Mathematics and the Senate House Examination

Before discussing the role of private teaching in Georgian and early Victorian Cambridge, it will be helpful briefly to outline the structure of undergraduate studies and student assessment in the university during the eighteenth century. These remarks will provide the necessary historical background to the events discussed in the sections below and introduce a number of contemporary customs, ceremonies, and expressions peculiar to Cambridge. In the eighteenth century, the university (as an educational collective) consisted of a largely informal affiliation of some sixteen colleges, a few of which—most notably Trinity and St. John's—were disproportionately powerful by virtue of their immense wealth.[5] The colleges were primarily religious seminaries whose main responsibilities were to offer a broad education to the sons of wealthy families and to provide the clergy for the Church of England. Each college was an autonomous body governed by its own fellowship (most fellows having taken holy orders) and completely responsible for the education of its own students. The *university* (as a formal body), by contrast, was a small administrative institution whose main function was to award degrees in accordance with the statutes.[6] Through its official head, the Chancellor, the university was ultimately answerable to the Crown and Parliament. Despite its responsibility for the maintenance of academic standards, however, the university's only formal contribution to teaching was the appointment of a number of professors. But since the students were educated by their colleges, attendance at professorial lectures was optional and frequently extremely low. The only route by which the university could exercise any effective control over what was actually taught in the colleges was through the statutory examination requirements.

The statutes of the university, dating back to the Elizabethan era, decreed that an undergraduate became eligible for a Bachelor of Arts degree by keeping a certain number of public disputations, or "Acts," intended to test his knowledge of oral Latin, philosophy, theology, logic, and mathematics, as well as his general intellectual acuity.[7] These studies constituted the Cambridge ideal of a "liberal education" and were also intended to prepare the most able men for higher studies in medicine, law, and theology. During the early eighteenth century, the long-established procedure of assessing students by Acts was severely disrupted by the building of a new University Senate House. This work rendered the university buildings in which the disputations had been held temporarily inoperable, and it became practically

---

5. These colleges derived income from the land given to them by the Crown at their foundation. For an excellent overview of the administrative structure of the university and the colleges, see Searby 1997, chaps. 3 and 4.

6. On the functions specified for the university and the colleges by the Elizabethan statutes, see Rouse Ball 1889, 245–46.

7. For a discussion of public disputations in Cambridge from the sixteenth to the eighteenth centuries, see Rouse Ball 1889, chap. 9.

FIGURE 2.1. A late Victorian illustration of the inside of the Senate House during the degree-awarding ceremony. Written accounts suggest that the scene during the annual Tripos examination would have resembled that depicted in figure 3.3 (see chapter 3). The exterior facade of the front of the Senate House is shown in figure 4.8. (Courtesy of the Cambridgeshire Collection, Cambridgeshire Libraries.)

impossible for the authorities to compel degree candidates to fulfil all of the required Acts. When this twenty-year interregnum was finally ended by the opening of the new Senate House (fig. 2.1) in 1730, a new method of assessing students was introduced and gradually formalized.[8] Candidates for the degree of Bachelor of Arts were henceforth required both to keep a certain number of Latin Acts and (at the beginning of the January following their tenth term of residence) to present themselves for oral examination, in English, in the new Senate House.

The formal Acts had always been presided over by one of two "moderators" (appointed by the university) whose job it was to suggest and approve topics for debate, see fair play during the disputations, and assess each participant's performance. In the new "Senate House examination" (or "Tripos" examination as it was also known), the moderators assumed the additional responsibility of questioning the degree candidates to determine where they should appear in the "order of merit" (a continuous ranking of all candidates

---

8. Here I am drawing upon Rouse Ball 1889, Rothblatt 1974, and Gascoigne 1984 and 1989.

according to how well they had acquitted themselves overall).⁹ From 1747 the order of merit was printed and circulated within the university, the annual Senate House examination becoming shortly thereafter a formally recognized event in the examination process. The importance of the Latin Acts was gradually reduced by the rising prestige of the Tripos examination, their primary use from the mid 1760s being to preclassify students into one of eight divisions so that they could be questioned at an appropriate level in the Senate House. The bottom two divisions, the so called "non-reading" men, were generally destined to obtain only non-honors degrees, whereupon they were known as the "hoi polloi" or "poll men."¹⁰ Those in the four intermediate divisions usually became "senior" or "junior optime," the holders of second or third class degrees respectively. Candidates in the top two divisions, the so-called reading men, were in the main destined to become the elite "wranglers," the holders of first class degrees.¹¹ During their tenth and final term, immediately prior to examination in the Senate House, all of the students were known as "questionists."

In 1779 the university decreed that the Senate House examination would be formally extended to last four days and that several new "examiners" would be appointed to assist the moderators. It was decided that the examiners would normally include the moderators of the previous year, a method of appointment that lent continuity to the examination (moderators and examiners being solely responsible for the content of the questions) and which was followed throughout the nineteenth century. The only difference in the work undertaken by the moderators and examiners respectively is that the former gradually assumed responsibility for the difficult problems set to the most able students (I shall generally refer collectively to moderators and examiners as "examiners").¹² The four-day duration of the examination was also divided into two stages. After three days of examination the relative performance of each candidate was ranked by the examiners, students of roughly equal ability being "bracketed" together. During the fourth day of examination, the students within each bracket were ranked by further examination,

---

9. For details of the structure of undergraduate studies around 1800, see Becher 1980a, 4–8. The origin of the term "tripos" is discussed in Rouse Ball 1889, 217–19.

10. The term "hoi polloi" is Greek for "the many," from which came the term "poll men."

11. In the late eighteenth century roughly 120 students graduated each year with a Bachelor of Arts degree, around 45 obtaining an honors pass. For an analysis of student numbers in Cambridge 1750–1880, see Searby 1997, 61. On the origin of the terms "wrangler" and "optime," see Rouse Ball 1889, 168–70.

12. Walton 1863, xix.

after which the moderators posted the complete order of merit.[13] This continuous ranking of students then stretched from the highest man, the "senior wrangler," to the last of the junior optimes (known as the "wooden spoon"), each man carrying his position into later life as a lasting mark of his intellectual ability.[14] Two weeks after the Senate House examination, the top wranglers had an opportunity to obtain further distinction by competing in a final examination for the first and second Smith's Prizes. This most demanding of examinations, first set in 1769, was intended to select the "best proficients in Mathematics and Natural Philosophy" and, in practice, offered the top two or three wranglers a final contest before they became candidates for valuable college fellowships.[15] The acquisition of a fellowship was one of the few recognized routes by which a young man from a relatively poor background, but with academic ability, could make his way up the social scale in Georgian Britain.[16] A fellowship offered the leisure and status to begin preparation to take holy orders and the possibility of eventually acquiring a permanent college post or clerical living.[17]

During the first few decades of its existence, the Senate House examination, like the Latin Acts, was an oral test which covered the broad range of studies taught in the colleges. However, in the 1760s it began to change in three important ways: first, the emphasis on mathematics and related natural philosophical subjects began to increase dramatically; second, the candidates for honors degrees began to be required to submit *written* answers to questions; and, finally, the examination became increasingly competitive. Scholars of eighteenth-century university history have suggested a wide

13. A candidate could move up within a bracket day by day, or into a higher bracket, by challenging the student immediately above him. A detailed account of the examination process in this period is given in Glaisher 1886–87, 10–13. For a historical ethnography of Cambridge student life in the 1790s, see Dening 1995, 75–111.

14. The last of the junior optimes was presented with a malting shovel, known as the "wooden spoon," indicating perhaps a preference for beer over reading. For a reference to the possible origin of this custom, see Wright 1827, 2:28.

15. It was almost always the senior and second wranglers who won the first and second Smith's Prizes respectively. The prize was judged by the Chancellor or Vice Chancellor, the Master of Trinity College, the Lucasian Professor of Mathematics, the Plumian Professor of Astronomy and Experimental Philosophy, and the Lowndean Professor Astronomy and Geometry. For details of the foundation of the prize and a list of recipients, see Barrow-Green 1999.

16. On the social origins of the Cambridge elite for the period 1830–60, see Becher 1984. Virtually none of these men came from working class backgrounds, while the most successful were from the middle classes.

17. There is some evidence that college fellowships began to be awarded in the latter eighteenth century on the strength of performance in the Senate House examination. This would have provided an obvious mechanism for the increase in competition.

range of possible explanations for these changes, and there is a consensus that the changes should be seen as part of a complex interaction between local institutional and broader sociopolitical factors.[18] The rise of Parliamentary rule and the incipient Industrial Revolution were undermining the established authority of church and aristocracy upon which the ancient universities relied. In an age of commerce and religious fragmentation, these institutions found it increasingly difficult to justify their immense wealth and privilege solely on the grounds of their role as Anglican seminaries.[19] Both Oxford and Cambridge sought new methods for ranking student performance in the late eighteenth century, and both ultimately employed written examinations for this purpose. It is nevertheless upon the uniqueness of some aspects of the Cambridge system that our interest must focus. Where Oxford retained formal logic, classics, and theology as the mainstays of its undergraduate teaching, Cambridge, more surprisingly, turned almost exclusively to mathematics and natural philosophy.

An important factor in the development of undergraduate studies in eighteenth-century Cambridge is the role played by Isaac Newton's natural-philosophical works. Although Newton did not found a school in Cambridge before leaving the university in 1696, some of his clerical disciples used his work in the early decades of the eighteenth century to build an alliance between Newtonian natural philosophy and Anglicanism. In his monumental works, the *Principia* and the *Optics,* Newton was believed to have shown how mathematics and experiment could be used to unlock the secrets of God's creation.[20] These books, together with Newton's almost holy status in Cambridge, provided a powerful pedagogical device for linking theology, natural philosophy, and mathematics in undergraduate studies. Initially it was the more qualitative, natural theological aspects of Newton's work that were emphasized, but with the rise of competitive, written examination and the strict ranking of students by marks, the emphasis began to shift towards the technical content of the *Principia*.[21] The discipline of mathematics was especially well suited to an examination system that sought to discriminate be-

---

18. Opinion differs widely over the extent to which the rise of competitive examination should be seen as an attempt to make student assessment more "meritocratic." See Rothblatt 1974, 281–84, Gascoigne 1984, and Searby 1997, 103–6.

19. For discussion of some institutional and wider social factors in the emergence of written examination in Georgian Cambridge, see Gascoigne 1984 and Rothblatt 1974.

20. Gascoigne 1989, 174–84.

21. Gascoigne 1984, 550–53, 555. On Newton's status and the natural-theological role of his works in Georgian and early Victorian Britain, see Gascoigne 1988 and Yeo 1988.

tween the performances of large numbers of well-prepared students. Unlike questions in theology and moral philosophy, those in mathematics could be made more and more difficult as the examination proceeded and be straightforwardly assessed according to a scheme of marks. By the end of the eighteenth century, Book I of Newton's *Principia* had become the absolute pinnacle of elite undergraduate studies and "mixed mathematics" (arithmetic, algebra, geometry, fluxional calculus, astronomy, mechanics, optics, and hydrostatics) had come to dominate the Senate House examination to the virtual exclusion of all other subjects.[22]

Part of the explanation for the rapid rise of mixed-mathematical studies in Cambridge seems to be the almost self-perpetuating relationship that was established between the study of mathematics, the increasing use of competitive, written examinations, and the introduction of a new form of teaching. The rising prestige of the Senate House examination led the colleges to introduce similar preparatory examinations so that the awarding not just of fellowships but also of scholarships and college prizes gradually came to depend on a student's ability to perform successfully in competitive written examinations.[23] This raised the standard of technical performance and prompted ever fiercer competition among the students, which, in turn, prompted the examiners to employ more mathematical problems in order to rank the students effectively. The profound implications of the switch from oral to paper-based examination in mathematics are explored in the next chapter, but there is one aspect of this switch that is highly relevant to our present concerns. Once it became clear to students that success at Cambridge depended upon their ability to reproduce mathematical knowledge on paper and to solve difficult mathematical problems, they began to turn in increasing numbers to private tutors who were willing and able to explain the technical content of such difficult books as Newton's *Principia* and to drill them individually in the problem-solving techniques they had to master. As we shall see, this kind of teaching played a major role not just in raising technical standards in the university, but in the complete reform of undergraduate teaching.

2.3 The Rise of Private Teaching in Georgian Cambridge

One reason that the rise of private teaching initiated a major transition in mathematical pedagogy is that although it was originally employed to sup-

---

22. On possible reasons why mathematics became so dominant in Cambridge undergraduate studies, see Gascoigne 1984, 566–73.

23. Gascoigne 1984, 651; Searby 1997, 103–6.

plement traditional teaching methods in Cambridge, it actually represented a new approach to university education and imparted new kinds of knowledge and skill. As we shall see in the next chapter, the majority of university teaching in the eighteenth century continued to be based on reading and oral discourse, making very little use of writing or technical problem-solving on paper. In order to clarify this point let us consider for a moment the various ways in which students learned in Cambridge in the latter eighteenth century. The main form of undergraduate teaching was large thrice-weekly college lectures in which students were taught the basics required for a pass or mediocre honors degree. Each lecture lasted an hour, the lecturer generally being a young college fellow preparing to take holy orders or waiting to assume a higher college, clerical, or other professional appointment. The main job of the mathematics lecturer was to go through required sections of such important texts as Euclid's *Elements*, ensuring that the majority of students had a least a minimal grasp of geometry, arithmetic, algebra, mechanics, hydrostatics, optics, and astronomy. Lectures were run at the pace of the average student and appear, as we shall see in the next chapter, to have consisted mainly in the lecturer asking students in turn to state proofs and theorems or to solve simple problems orally.

For the more ambitious students there were several other ways to acquire knowledge in preparation for the Latin disputations. First, the college lecturer could provide manuscripts or recommend books to supplement his lectures. Second, students were given varying amounts of help individually by their personal college tutors. In the case of the most able students, the tutor might discuss conceptually difficult problems, recommend further reading, and supply privately circulated manuscripts designed to explain opaque passages in advanced books such as the *Principia*.[24] Third, students could pay a member of the university to tutor them privately, a form of teaching that was especially common among the tiny handful of students annually who wished to learn the mathematical sciences to an advanced level. Beyond its more elementary operations and applications, mathematics was a difficult and highly specialized discipline that few college or university teachers had the inclination or ability to teach.[25] Fourth, students could simultaneously acquire additional knowledge and gain familiarity with the process of oral examination by attending the Acts kept by their peers and elders. This was a

---

24. Each student was assigned a personal college tutor who looked after his moral, spiritual, and domestic welfare, and kept an eye on his academic progress. See Searby 1997, 120–29.

25. This form of private instruction in the mathematical science had been common since the seventeenth century. See Feingold 1997, 368–89.

very important form of learning for students in the eighteenth century as it showed them precisely how their knowledge would be tested and how to conduct themselves in a public disputation. Fifth, students were entitled to attend the public lectures given by the university professors.[26] This was a relatively unimportant form of teaching since many mathematical professors of this period chose not to lecture at all, and even those who did lecture seldom chose a topic or level of delivery suitable for undergraduates.[27]

The main point to take from this brief survey of Cambridge pedagogy is that professorial and college lectures, tutorial sessions, public disputations, and private study were all forms of learning based in the first instance on reading or oral debate. With the gradual introduction of written examinations the preferred form of teaching began to change to suit the new form of assessment. Success in the Senate House examination depended increasingly on the ability to write out proofs and theorems and to solve difficult problems on paper. Ambitious students accordingly turned to private tutors, usually young college fellows seeking to supplement their meager stipends. These tutors taught students individually or in small groups and concentrated on imparting technical proficiency on paper. College fellows sometimes fulfilled more than one of the above teaching roles, and, in every case, would have been a senior or very high wrangler a few years previously. In the following three sections I trace the development of private teaching from the late eighteenth century until the early 1850s, paying special attention to the efficacy of the legislation intended to control the practice and to the changing division of labor between private tutors and college lecturers. Since there is virtually no official documentation concerning the contemporary development of undergraduate teaching in Georgian Cambridge, I have pieced together my account mainly from the diaries, correspondence, and autobiographical reminiscences of those students and more senior members of the university who experienced the events in question. This approach has the disadvantage of requiring the presentation and assessment of a good deal of detailed and occasionally contradictory evidence, but it is compensated by the production of rich insights into the undergraduate's experience of changing pedagogical practice.

26. There were three professorships in mathematical subjects in Cambridge in the eighteenth century: the Lucasian Professorship of Mathematics, the Plumian Professorship of Astronomy and Experimental Philosophy, and the Lowndean Professorship of Astronomy and Geometry. For a list of the incumbents of these chairs, see Rouse Ball 1912, 322–23.

27. Rouse Ball 1889, 107, 158.

Although sources on the emergence of private teaching in the latter eighteenth century are sparse, there is little doubt that it was the rise of the competitive Senate House examination in the mid-eighteenth century that prompted the proliferation of the practice. The first direct references to private teaching occur in the 1760s at exactly the time the technical content of the Senate House examination was beginning to rise, and, by the 1780s, following the introduction of written examination, the use of private tutors was commonplace. Senior members of the university evidently disapproved of students engaging private tutors, as it undermined the notion that the order of merit provided a reliable index of intellectual ability.[28] Students who could not afford a private tutor were clearly placed at a comparative disadvantage, while poor but able undergraduates sometimes compromised their chances in the Senate House by taking students themselves to provide an income. Most troubling of all were cases in which tutors (private or college) were also moderators or examiners, a conflict of interest that laid them open to accusations of partiality and threatened to undermine the integrity of the whole examination process.[29]

In 1777 the university senate passed Graces barring tutors from the office of moderator and forbidding questionists to take private pupils. A further Grace in 1781 forbade all undergraduates from studying with private tutors in the two years prior to their sitting the Senate House examination.[30] It is difficult to assess how effectively these regulations were enforced, but there is very little evidence to support Rouse Ball's influential assertion that "all such legislation broke down in practice."[31] Rouse Ball cites Whewell as an authority on this matter, but Whewell actually makes no such claim and, in fact, acknowledges that the legislation was "for a time effectual." The latter claim is also substantiated by Christopher Wordsworth, a contemporary of Rouse Ball, who argued that the legislation was effective circa 1800, and that private teaching only became more prevalent during the early nineteenth

---

28. According to Rouse Ball (1889, 162), the first documented case of a private tutor being engaged explicitly to prepare a pupil for a high wranglership occurred in 1763. On the suppression private teaching in the late 1770s, see Atkinson 1825, 502, Whewell 1850, 220, Wordsworth 1877, 260, Rouse Ball 1889, 162, Gascoigne 1984, 555, and Searby 1997, 130.

29. Gunning 1854, xx; Gascoigne 1984, 555; Searby 1997, 130.

30. Winstanley 1935, 389. The "Senate" was the university legislature. It was made up (with one or two minor restrictions) of all resident and nonresident doctors and Masters of Arts of the university. Decisions made by the Senate were known as "Graces." See Searby 1997, 52–54.

31. Rouse Ball 1889, 162. Rouse Ball's assertion is followed, for example, by Winstanley 1935, 332–33, Gascoigne 1984, 555, and Searby 1997, 130.

century when the legislation was in any case relaxed.[32] Strangely, both Rouse Ball and more recent studies following his lead cite the case of John Dawson in support of the claim that private tuition was common in Cambridge beyond the fourth term despite the legislation banning it.[33] Dawson was a retired medical practitioner and largely self-taught mathematician living in the Lake District who offered his services as a mathematics teacher to students both before they went to the university and during vacations throughout their undergraduate careers.[34] Although without any Cambridge connection, Dawson was responsible for training at least eleven senior wranglers between 1781 and 1800, the period when the legislation against private tuition was most severe.[35] This evidence surely suggests, if anything, that private tuition was indeed an effective method of preparing students for competitive examinations in mathematics, but that such tuition beyond the fourth term was not readily available in Cambridge.[36] While, then, there is no reason to doubt the word of a contemporary such as William Frend who claimed, in 1787, that private tutors were "universally sought after" and that they were "absolutely necessary to everyone who wished to make a tolerable figure in the Senate," there is equally little reason to doubt that such private tuition generally took place either during the first four terms (as prescribed by the Graces) or else outside of Cambridge during vacations.[37]

These conclusions are further supported by evidence from the first decade of the nineteenth century. One of Dawson's latter successes, George Pryme (6W 1803), provides a rare contemporary account of undergraduate teaching practice in this period, and he claimed explicitly that private tuition was not common in the university in his day.[38] Pryme studied regularly with Dawson both before coming into residence at Cambridge and during the long vacations while preparing for his degree. Lodging in Dawson's house,

32. Whewell 1850, 220; Wordsworth 1874, 114–15. Elsewhere, Whewell (1837, 74–75) does actually suggest, vaguely and indirectly, that private tuition was hard to suppress, but he seems to be referring to the 1820s.
33. Rouse Ball 1889, 162; Gascoigne 1984, 555.
34. For Dawson's life and career, see the *DNB*.
35. Dawson attracted pupils from all over England, but a connection between Sedbergh Grammar School and St. John's College made him especially popular with Cambridge men.
36. Had private tuition been freely available in Cambridge, it seems unlikely that Dawson would have dominated the training of senior wranglers so thoroughly. See Rouse Ball 1889, 162.
37. Frend 1787, 15.
38. Pryme 1870, 29. Rouse Ball concludes that Pryme must have been an exception to the general rule that "nearly every mathematical student read with a private tutor" (1889, 163). It is more likely, however, that Pryme was drawing a contrast with the later period when all students read with private tutors throughout their undergraduate careers.

Pryme was taught six hours a day, the tutor working at the appropriate pace for his pupil, pausing to explain difficulties as they arose.[39] But despite the success of Dawson's pupils, it is clear that even in the 1800s the services of a private tutor were not a prerequisite to Tripos success. Frederick Pollock (SW 1806), for example, made no explicit reference to working with a private tutor in an account of his undergraduate studies.[40] Having read basic geometry and algebra (up to quadratic equations) before arriving in Cambridge—perhaps more than the average freshman would have known—he made excellent use of college lectures and supplemented his knowledge through private study using manuscripts in general circulation.[41] Pollock ascribed his subsequent examination success to an excellent memory—he claimed to know Book I of Euclid's *Elements* "word by word, letter by letter"—and "great rapidity and perfect accuracy" of writing.[42] As we shall see in the next section, it remained possible to become senior wrangler without the services of a private tutor until at least the early 1820s.

The assertion by Rouse Ball and others that attempts to limit private tuition were ineffectual appears to be based on two dubious assumptions: first, that the complete domination of undergraduate teaching exercised by private tutors from the 1840s—and thus familiar to a wrangler of the 1870s such as Rouse Ball—was the natural and inevitable accompaniment of competitive written examinations and must, therefore, always have coexisted with such examinations; and, second, that the gradual repeal of the rules governing private tuition between 1807 and 1824 was simply an acknowledgement of the futility of such legislation.[43] There is however little reason to doubt that attempts to control private teaching within the university from the late 1770s *were* successful, at least until the first decade of the nineteenth century, and that the gradual repeal of the rules governing the period which had to elapse between working with a private tutor and sitting the Senate

---

39. Dawson charged three shillings and sixpence a week for lodgings and five shillings a week for tuition. See Pryme 1870, 29 and Dawson's entry in the *DNB*.

40. Pollock recounted his undergraduate years in response to a letter from DeMorgan in 1869 requesting a "trustworthy account" of the mathematical reading habits of Cambridge men in the early nineteenth century. See Rouse Ball 1889, 111–14.

41. Rouse Ball claims that William Dealtry (2W 1896) was Pollock's "coach," but Pollock states only that "there were certain mss floating about which I copied—which belonged to Dealtry" (Rouse Ball 1889, 111).

42. Ibid., 112. Students could build up a large number of marks in the early stages of the examination by reproducing standard proofs and theorems rapidly and accurately.

43. Rouse Ball did not assess the relative prevalence of private teaching at different points in the century, and might well have been using his history to undermine attempts to curtail private teaching in the 1880s (see chapter 5). Rouse Ball 1889, 162–63.

House examination (the period was reduced to 18 months, 12 months, and 6 months in 1807, 1815 and 1824 respectively) occurred for other reasons.

The incentive for all prospective honors students to engage private tutors beyond the fourth term would certainly have increased after the turn of the century as the emphasis on mathematics in the examination continued to grow. In 1799, for example, the enormous importance of mixed mathematics in the Senate House examination was finally acknowledged by the university. The senior moderator that year took the unusual step of commenting publicly on the mathematical "insufficiency" of the questionists from some colleges and, even more significantly, took the unprecedented step of announcing that henceforth no student would receive a degree unless he displayed a "competent knowledge of the *first* book of *Euclid, Arithmetic, Vulgar* and *Decimal Fractions, Simple* and *Quadratic Equations,* and Locke and Paley."[44] This important announcement marked the first step in the emergence of a formal university-wide undergraduate syllabus, each college being thereafter responsible for ensuring that its students met these minimum requirements. By the end of the eighteenth century, therefore, the prestige and emergent syllabus of the competitive Senate House examination provided a common standard for college teaching throughout the university, and the largest and wealthiest colleges—especially Trinity and St. John's—fought fiercely to dominate the order of merit.[45] At exactly the same time that the rules on private tuition were being relaxed, moreover, the structure of the Senate House examination was being formally altered by the Senate to confine questions on logic to a new fifth day of examination. The original three days would now be devoted entirely to mathematical examination, which, in practice, would dictate a student's place in the order or merit.[46] The relative performances of the most able students during these three days were also being judged increasingly on written answers to problems rather than on oral disputations. Indeed, when the Parliamentary Commissioners reported in the early 1850s on the development of undergraduate studies in Cambridge, they identified the first decade of the nineteenth century as the period when mathematics, which had become "more and more difficult to express orally by reason of its symbolic form," had caused "paper Examination" to overtake the disputation as the prime means of assessing student ability.[47]

44. Wordsworth 1877, 56.
45. Gascoigne 1984, 556–57.
46. Rouse Ball 1889, 209.
47. PP 1852–53. HC xliv [1559]: *Report of the Commissioners Appointed to Inquire into the State, Discipline, Studies, and Revenues of the University and Colleges of Cambridge: Together with the Ev-*

This emphasis on the skilled reproduction of mathematical knowledge on paper doubtless encouraged students to engage private tutors, and keen intercollegiate competition meant that they would have done so with the blessing of their colleges. The evidence nevertheless suggests that private teaching *was* practiced in accordance with the regulations, and that some students continued to achieve high placing in the Senate House without the benefit of any private tuition at all. This being the case, the relaxation of the rules in 1807, which enabled students to work with a private tutor until the end of their sixth term, quite possibly occurred to enable, or even to encourage, ambitious students to exploit the resources of the private tutor while discouraging the use of non-Cambridge tutors such as Dawson. It might have been a minor embarrassment to the university that parents who had already paid substantial college and university fees were then expected to pay a college fellow privately to help prepare their sons for the Senate House, but once it was clear that mathematics had become the primary discipline through which the minds of the Cambridge elite were developed and tested, it would have been unacceptable for teaching to have lain substantially in the hands of non-Cambridge tutors.[48] With the proviso that it was sufficiently controlled to suppress accusations of rank unfairness, therefore, private teaching had become a stable and important component of undergraduate instruction: it provided an efficient mechanism for organizing the first two years of undergraduate study, thereby raising the level of elite performance in what was emerging as Cambridge's primary discipline, mathematics; it prevented undergraduates from turning to outsiders for tuition; it provided an income for young fellows; and it spared college lecturers and tutors the grind of either drilling students in basic mathematical technique or making special provision for the most able pupils. In a primarily undergraduate institution, the first concern of most fellows was, as we have seen, to further their own careers, which, in most cases, lay neither in Cambridge nor in mathematics. But it was not merely the methods of teaching and examining undergraduates that were changing in Cambridge during the first two decades of the eighteenth century. This was also the period when new mathematical methods from Continental Europe were gradually entering the university. The introduction of this new mathematics not only accelerated several of the

---

idence, and an Appendix (Graham Commission Report and Evidence), 109. Hereafter, *Commissioners' Report*.

48. Whewell (1837, 76 and 1850, 218) claimed that private teaching encouraged students to study harder in the summer vacation rather than in the formal terms and that this was unacceptable to the university.

trends discussed above but also had a profound effect on the role of the private tutor in training the undergraduate elite.

2.4. The Analytical Revolution from Below

During the eighteenth century, Cambridge tutors had taken little interest in Continental developments in mathematics. The acrimonious priority dispute between Newton, Leibniz, and their respective followers over the invention of the calculus had soured relations, and, from the late 1780s, the dramatic political events in France and the subsequent Napoleonic wars produced a serious obstacle to intellectual exchange between England and France. Continental methods would, in any case, have seemed irrelevant to many Cambridge tutors, as the study of mathematics in the university was intended not to produce professional mathematicians but to educate the student's mind through mastery of Newton's mathematical methods and natural philosophy. As I noted in chapter 1, however, during the eighteenth century such mathematicians as Euler, D'Alembert, Clairaut, and Daniel Bernoulli had developed a major new field of mathematical study, "analysis," by the systematic application of the Leibnzian calculus to a wide range of problems in algebra, mechanics, and algebraic geometry.[49] With Newton's laws expressed in differential form—an approach first published by Euler in 1736—dynamical problems could be tackled systematically using the power of the calculus. By the end of the eighteenth century, Lagrange had recast mechanics and dynamics in fully analytical form using the principle of virtual work and the calculus of variations, while in his monumental *Traité de mécanique céleste,* Laplace had brought Newton's laws of motion and theory of gravitation to a new level of mathematical sophistication.[50] This was surely a book with which Newton's Cambridge followers ought to have been familiar, but, as a sympathetic British reviewer of Lacroix's great French textbook on the calculus observed in 1800, the average English mathematician would be "stopped at the first pages of Euler or D'Alembert . . . from want of knowing the principles and the methods which they take for granted as known to every mathematical reader."[51]

There was nevertheless some interest in Continental mathematics in Britain, and around the turn of the century there were calls both within and out-

---

49. On the development of analysis in the eighteenth and early nineteenth centuries, see Grattan-Guiness 1970 and 1990.

50. Lagrange 1788; Laplace 1798–1827.

51. Quoted in Becher 1980a, 8.

side the university for a reform of undergraduate teaching which would take some account of analytical methods.[52] In 1803, Robert Woodhouse (SW 1795) attempted to introduce the d-notation calculus and other analytical methods into Cambridge by publishing *The Principles of Analytical Calculation*, but the book had very little immediate effect on undergraduate studies. As we have seen, both the ideology and structure of teaching in Cambridge were extremely conservative and would not be easy to change. The lecturers, college and private tutors, university professors, textbook writers, and examiners had all been high wranglers in their day, and many played several of these roles. Most would have been hostile to learning difficult new mathematical methods which would be of no use in furthering their careers and which, for many, carried unwelcome associations with the atheism and radical politics of revolutionary Paris.[53] Abandoning Newton's methods would have appeared to many as a move calculated to undermine both the authority of the Anglican Church and the reputation of England's greatest natural philosopher. Furthermore, with teaching taking place in the colleges and no formal university body to dictate what was taught, it would be very difficult to reform the system by edict without major Parliamentary intervention. Official changes to university procedure or policy could only be passed by a majority vote of the Senate, and it is extremely unlikely that any proposal formally to replace Newton's methods on the examination papers with those of French mathematicians would have been carried.

As Becher and Enros have shown, the movement that eventually accomplished the so-called analytical revolution in Cambridge began with undergraduates working inside the system.[54] The most famous manifestation of the growing undergraduate enthusiasm for French analytical mathematics was the formation of the short-lived Analytical Society in 1812. The aim of the society was to invigorate the professional study of mathematics in England by introducing the techniques of French analysis into the potential power centers of the discipline. The society did not survive long enough to fulfil these ambitious goals, but three of its more prominent members did go on to achieve partial success. Charles Babbage, John Herschel (SW 1813), and George Peacock (2W 1813) realized that in order to bring about change it would be important both to produce textbooks from which analysis could be learned and to have analytical questions set in the Senate House so that

52. Becher 1980a, 8–14, and 1980b; Enros 1983, 25–26; Guicciardini 1989, 126–31.
53. The politics of the analytical revolution is discussed in Becher 1995.
54. On early attempts to reform Cambridge mathematics see Becher 1980a, Enros 1983, Panteki 1987, and Guicciardini 1989, chap. 9.

such studies would pay in the examination.[55] In 1816 they accordingly completed a translation of Lacroix's introductory textbook on differential and integral calculus, and in 1820 they published a further volume containing practical examples.[56] In 1817, one of their number, George Peacock, was appointed a moderator for the Senate House examination, and he infuriated many of his more conservative colleagues by setting problems in analytical mathematics using the Continental d-notation.[57] Two years later both of the moderators—Peacock and Richard Gwatkin (SW 1814)—were ex-members of the Analytical Society, and both made free use of analytical techniques and notation. Over the next decade, more, and more advanced, Continental mathematics found its way into the Senate House examination so that, by the end of the 1820s, ambitious undergraduates had to be familiar with the new methods if they hoped to become high wranglers.

This familiar chronology of the introduction of analytical methods into Cambridge rightly points to the importance of new textbooks and the infiltration of the Senate House examination in establishing the new mathematics. It reveals nothing, however, of the process by which the analytical revolution was practically accomplished at the level of undergraduate teaching. As one of the founders of the Analytical Society, Charles Babbage, later recalled, even after the translation of Lacroix's textbook and the introduction of questions in the d-notation calculus in the Senate House examination, the "progress of the notation of Leibnitz at Cambridge was slow" (1864, 39) and taught only by young enthusiasts such as Peacock at Trinity and Gwatkin at St. John's. At what point and through which teachers, then, did the new mathematics become commonplace in undergraduate studies throughout the university? These are important questions because, as we shall see, it was private tutors who were primarily responsible for spreading the new mathematics and, in doing so, they not only altered the content, purpose, and structure of undergraduate studies, but also assumed a control over the training of the mathematical elite which would remain unchallenged until the last quarter of the nineteenth century.

We know very little about undergraduate teaching methods in Cambridge in the late eighteenth and early nineteenth centuries, but for the pe-

---

55. On the move by men like Peacock to a more pragmatic approach towards the introduction of analysis, see Becher 1995.

56. In translating Lacroix's textbook, Babbage and Herschel altered and supplemented the mathematical content to suit their own purposes. See Richards 1991, 299.

57. Questions in d-notation calculus had already been set in examinations at St. John's College by John Herschel. See Becher 1995, 414.

riod during which the analytical revolution took place—roughly 1815–25—we have three fairly detailed accounts of undergraduate experience. The earliest and most detailed account is that of J. M. F. Wright, an able student from King's Lynn Grammar School who arrived in Cambridge in October 1815.[58] Having learned the basics of arithmetic, algebra, geometry, and trigonometry before arriving at the university, Wright consolidated his knowledge of these and the other required mathematical topics in his first year through college lectures given by John Brown (2W 1799).[59] At the beginning of his second year Wright was able, apparently unexpectedly, to engage the services of a private tutor when a "friend from the country" sent some money expressly for the purpose.[60] Private teaching—or "pupilizing" as it was commonly known—was certainly commonplace at this time, as Wright notes that most high wranglers who obtained fellowships were now prepared to "receive" private pupils both during term time and the long summer vacation. Wright also claimed that the university had "fixed" the fees payable for these periods but that "fellows of colleges, and others, who are in more request" were able to change substantially more.[61] Apart from revealing the very substantial income a young fellow could now derive from pupilizing, these remarks also point to an emergent distinction between those whose only recommendation as a private tutor was their own success in the Senate House, and those who had an established reputation as an experienced and effective teacher. This was an important distinction which became more pronounced over the next two decades, culminating in the emergence of the mathematical "coach" who made his living by taking pupils.

Wright engaged John Brass (6W 1811), a private tutor of great reputation who, since his normal hours pupilizing were already full, agreed to see Wright for an hour before morning chapel. Wright describes how, arriving freezing cold at six o'clock at Brass's rooms in Trinity Great Court, the latter would jump out of bed and begin teaching in his night clothes. Wright would be set writing out an "elegant demonstration of some important proposition in algebra, such as Newton's Binomial Theorem" and, having made sure his student understood what he was copying, Brass would retire to the bed-

58. Wright registered at the university in 1814 but began his studies in 1815 (Searby 1997, 118).
59. Wright 1827, 1:5. Brown was also one of two Trinity tutors, the other being James Hustler (3W 1802).
60. Wright 1827, 1:171. Wright claimed that the majority of students were unable to afford the services of a private tutor (1:76).
61. The rates for a term and the summer vacation were £14 and £30 respectively (those in greater demand charged £20 and £50 respectively). These rates were probably "fixed" only in an informal sense, as is suggested by the variable rates. See Wright 1827, 1:171.

room to dress, occasionally peering round the door to check his pupil was not "stuck" (1827, 1:172). Wright's account of his studies with Brass provides the earliest written confirmation of the special benefits conferred by a private tutor. The most notable of these were the provision of teaching manuscripts which developed difficult topics in a way designed to facilitate student understanding, and the opportunity for the student to study at exactly his own pace. Working "together at the table" Brass would push Wright along at what the latter described as "'the speed of thought'—that is as fast as my understanding would carry me," a process which contrasted sharply with the plodding progress of a college lecture delivered to up to a hundred students at the pace of the mediocre (1827, 1:172–73).[62] The close personal interaction between tutor and pupil was extremely important in imparting the subtle skills of problem solving, an aspect of private teaching that is explored further in chapter 5. Wright's private tuition, we should also note, was carried out well within the limits prescribed by the regulations which, in 1815, had been altered again to enable students to work with a private tutor until the end of their seventh term.[63]

One aspect of Wright's undergraduate aspirations that separates him sharply from students like Pryme and Pollock of a decade earlier is his eagerness to learn the new Continental mathematics, especially analytical mechanics. By the end of his second year (the early summer of 1817) Wright was keen to begin studying not just Newton's *Principia*, a required book in the third year, but also Monge's *Géométrie descriptive*, Lagrange's *Méchanique analytique*, and Laplace's *Traité de mécanique céleste* (Wright 1827, 2:2). The first Senate House examination containing questions on Continental mathematics had taken place just six months earlier, and it is clear that Peacock had succeeded in generating considerable enthusiasm for the new mathematics even among students who were not directly under his tutorial control.[64] What also emerges from Wright's account of his first two years of undergraduate study is that although the most ambitious students studied the advanced parts of traditional subjects from privately circulated manuscripts or with private tutors, the latter did not, circa 1817, teach the analytical methods which now provided an alternative to traditional geometrical and fluxional techniques. Over the summer of 1817, Wright began to study the above books

---

62. For a strikingly similar firsthand account of private teaching in the late 1820s, see Prichard 1897, 38–39.
63. Searby 1997, 130.
64. Peacock was at this point an assistant tutor working under James Hustler.

and although he gradually made progress with the *Principia*—the mathematical methods of which he had been taught in his earlier studies—he found himself unable to follow the French works.[65] He managed only the first seven pages of Laplace's *Mécanique céleste*, for example, before being stumped by "the doctrine of Partial Differentials, which had not yet found its way into any work on the subject of Fluxions, in the English language" (1827, 2:2).[66] The "uncommon fame" of Laplace's work in Cambridge at this time was nevertheless so great that Wright was determined to master it somehow. For the time being he returned to further study of Newton's *Principia*, "as far as the Eighth section," and then made further study of the fluxional calculus in which he already had a good grounding.[67] Wright at last made some progress when, possibly at Peacock's suggestion, he began to study the latter's recently completed English translation of Lacroix's introductory textbook on the calculus. But even after working through this text he still found Laplace in the original "too much" for him (1827, 2:3).

At the start of his third year, Wright began again to attend Brown's lectures, the whole of the seventh term being devoted to the first book of Newton's *Principia*. Brown, who was also Wright's personal tutor, was well aware that his lectures would not be sufficient for the most able men, and he provided manuscripts to get them through the more advanced parts of Newton's work. According to Wright, it was impossible for even the best student to "make his way through the Principia" without such manuscripts, the majority of which were "in the hands" of private tutors.[68] The relaxation of the regulations on pupilizing in 1815 had made it permissible to work with a private tutor until the end of the first term of the third year, and, Wright reveals (1827, 2:24), the significance of this concession was that it enabled the top men to get a solid foundation in this most important of texts. Unable to afford a tutor in this crucial term, Wright began to skip Brown's lectures in order to push ahead with the first, second, and parts of the third book of the *Principia* using manuscripts he had "scraped together" over the summer. He also continued his study of fluxions, making such rapid progress on all fronts that, towards the end of the term, he found himself so far ahead of his

65. Wright did not appear to have difficulty reading French, but he gives no indication where he learned the language.

66. Methods equivalent to partial differentiation were employed in the English fluxional calculus, but they were neither identified by a special notation nor subject to systematic development. See Guicciardini 1989, 140.

67. Wright learned fluxional calculus from Dealtry 1810.

68. Several of these manuscripts later formed the basis of introductory textbooks on the *Principia*. See the prefatory comments in Carr 1821, Wright 1828, and Evans 1834.

peers that, with Brown's collusion, he ceased to attend lectures altogether (1827, 2:24–25).[69]

It was in the latter part of this term and over the Christmas vacation that Wright finally began to make substantial progress with the French works. He made a systematic study of Lagrange's *Méchanique analytique* and several other minor French works and, during the second term of his third year, worked with great difficulty through Lacroix's three-volume treatise on the calculus. In the last term of the third year the lectures were on optics and astronomy, and, feeling he had little to learn from Brown, Wright at last worked his way through much of Laplace's *Mécanique céleste* and studied Jean Baptiste Biot's recently published *Traité de physique* (1816).[70] It was also the custom this term to start working through problems from past examination papers in preparation for the Senate House, and Wright felt it advantageous to learn the physical topics using analytical methods because of their "more ready and convenient application in the resolution of problems" (1827, 2:46). Wright would at this point have been able to draw on Peacock's examination papers of the previous year which included problems in Continental mathematics, and, with Peacock and Gwatkin the moderators for 1819, he would have anticipated more of the same in his examination.[71]

Wright's undergraduate experience reveals the extent to which even the most enthusiastic student had to struggle by himself to master the new mathematics in the years immediately after 1815. The majority of college lecturers and private tutors would have been unable or unwilling to help their pupils in this respect. As Babbage's comments above imply, we must assume that, apart from rare cases such as Wright, the few students who obtained a good grounding in the new mathematics circa 1820 did so because they had the good fortune to have an enthusiast—such as Gwatkin at St. John's or Peacock at Trinity—as a lecturer and personal tutor.[72] An excellent example of the latter case is that of George Airy, who arrived at Trinity in 1819, and whose undergraduate experiences make an interesting contrast with Wright's. Unlike Wright, Airy was extremely well prepared in mathematics

---

69. Of the several courses of professorial lectures attended by Wright (including Vince's on experimental natural philosophy), none was on mathematics (1827, 2:27–35).

70. On the British reception of Biot's work, see Crosland and Smith 1976, 7–8, 36–41.

71. Wright should have been a high wrangler but achieved only a pass degree. Due to a bizarre series of oversights and accidents, he failed to keep a required Act in his ninth term and could not compete fairly in the Senate House examination. See Wright 1827, 2:35–60. For further biographical details of Wright's subsequent career, see Searby 1997, 120.

72. Peacock stated in 1817 that he intended to use his position as a mathematics lecturer at Trinity to further the analytical revolution (Rouse Ball 1889, 121).

by the time he arrived at the university. Having received a very sound grounding in arithmetic and algebra by the age of twelve, he went on to grammar school in Colchester where he studied geometry and fluxional calculus.[73] Commencing in the spring of 1817 he also went twice a week to be tutored by an ex-Cambridge mathematician, Mr. Rogers, who had come as a mathematical master to the school.[74] Acting as a private tutor, Rogers taught Airy geometry, algebra, mechanics, hydrostatics, optics, trigonometry, fluxional calculus, and Newton's *Principia* to the end of the ninth section—the standard topics required for an honors degree at the university. In a little over a year Airy had outstripped his teacher—who thereafter declined to teach him—and, although Airy was somewhat disparaging of Rogers's powers as a mathematician, he did not underestimate the benefit he received "for its training [him] both in Cambridge subjects and in the accurate Cambridge methods of treating them" (1896, 20). In the summer of 1819, Airy was examined in mathematics by a fellow of Trinity College who was so impressed with his knowledge and ability that he sent Airy's problem solutions to Trinity to be seen by other fellows including his future tutor, Hustler, and Hustler's assistant tutor, George Peacock.

Airy was admitted to Trinity on the strength of his mathematical ability, and Peacock, who was working to consolidate the place of the new mathematics in undergraduate studies, quickly took him under his wing. Despite his head start in mathematics, Airy attended college lectures to make sure that he could reproduce elementary mathematics in exactly the form required in college examinations and the Senate House. Realizing that this would hardly tax Airy's ability, Peacock gave him a copy of the translation of Lacroix's *Differential Calculus* as well as a copy of the recently completed *Examples* in differential and integral calculus.[75] Airy "betook" himself to these with "great industry" together with William Whewell's new d-notation treatise on mechanics (1819), and, quickly perceiving that he would need to read French in order to "read modern mathematics," he began the study of the language with the help of his sister during the Christmas vacation. By the long vacation at the end of his first year, Airy was reading Poisson in the original "struggling with French words." As he followed college lectures and studied advanced French mathematics in his own time, Airy recalled that Peacock

---

73. Airy 1896, 19. Airy, like Wright, learned the fluxional calculus from Dealtry 1810.

74. If the biographical fragments given by Airy are correct, "Mr. Rogers" must have been Thomas Rogers (10SO 1811).

75. Comparison with Wright's undergraduate accomplishments suggests that Airy's knowledge was roughly that of a good second year man when he entered Trinity.

"always had some private problems of a higher class for me, and saw me I believe every day" (1896, 24–26, 29).[76] Airy makes no reference to having worked with a private tutor as an undergraduate, and it is not difficult to see why he would have had no need of one. The majority of such tutors would not have been able to teach him the analytical mathematics he desired to learn, and he had already mastered most of the traditional mathematical topics required in the Senate House. It is in any case clear that Peacock willingly played the part of an unpaid private tutor, setting him problems, helping him with difficulties, and giving Airy free access to his own library of mathematical books when absent from Cambridge during the vacations (1896, 28, 30).[77] Airy's sound mathematical training and confidence also enabled him to make substantial progress on his own, even to the extent of making original contributions to mathematical physics. When he had great difficulty following the standard physical explanation of precession in Samuel Vince's (SW 1775) *Astronomy* (1797–1808), Airy wrote out an explanation of his own in analytical form which later became part of his extremely influential *Mathematical Tracts* (1826). Likewise in trying to work through Robert Woodhouse's (SW 1795) *Physical Astronomy* (1818), Airy was compelled to master the technique of "changing the independent variable," a learning experience that required him to "examine severely the logic of the Differential Calculus" (1896, 29–30). Through his second year, Airy attended Peacock's college lectures while, in his own time, obtaining a thorough mastery of the differential and integral calculus together with its physical applications, and solving problems set privately by Peacock.

At the beginning of his third year, Airy was told by Hustler, who gave the third-year lectures, that there was no point in him attending formal classes any further. It was arranged instead that Airy and his two most able peers would join the questionists of the year above in their final term's examination preparation. Unfortunately the questionists appear to have been offended by the presence of the precocious younger men—referring to them scoffingly as "the impudent year" and the "annus mirabilis"—and, to avoid further bad feeling, Airy and his contemporaries went instead three times a week to Peacock's rooms where he set them questions. Airy completed his third year in the summer of 1822 with further studies of advanced mathematics, privately tutoring less able undergraduates, designing a calculating

---

76. It is not clear which of Poisson's works Airy was reading, but it was probably the *Traité de méchanique* of 1811.

77. Airy moved in his second term to rooms on Peacock's staircase in Neville's Court.

machine, undertaking simple experiments in optics and mechanics, and completing his first original mathematical manuscripts. In October 1822 he became a questionist himself, attending thrice weekly problem classes, and, in the Tripos of 1823, he was senior wrangler and first Smith's prizeman (1896, 33–40). Airy was clearly as well prepared in French mathematics, and especially French mathematical physics, as any Cambridge undergraduate could have been in the early 1820s. After his examinations, he began taking private pupils to generate some income, was elected to a fellowship and mathematical lectureship at Trinity, and continued his original investigations in mathematical physics. In the mid 1820s, he also helped to consolidate the position of the new mathematics in the university by writing his *Mathematical Tracts* (which provided an introduction to lunar theory, the figure of the earth, precession and nutation, and the calculus of variations) and, as Lucasian professor, by giving well-attended lectures in 1827 and 1828 on mechanics, optics, pneumatics, and hydrostatics (1896, 49–66; Crosland and Smith 1978, 14).

The reminiscences of Wright and Airy tell us a good deal about the undergraduate training system around the time of the analytical revolution and about the means by which the new mathematics was first introduced. Wright learned most of his mathematics as an undergraduate but, even in the years 1815–19, was not taught Continental calculus either by his college lecturer or by his private tutor.[78] Even in mastering advanced English mathematics he relied heavily on privately circulated manuscripts and the training offered by his private tutor. That Airy's experience a few years later was very different was due not to any major reform in the content or style of undergraduate teaching, but to his exceptional knowledge of mathematics before entering the university and to his good fortune in having Peacock as a lecturer and tutor. Peacock must have realized that the analytical revolution would be accomplished only if he and his fellow enthusiasts produced a generation of students who could answer the analytical questions set in the Senate House and then go on themselves to teach the new methods to other undergraduates. In the case of Airy, Peacock's efforts were well rewarded.

But despite the accomplishments of students such as Wright and Airy, it remained a mark of the uphill struggle faced by reformers in the early 1820s that, even in 1822, it was still possible to become senior wrangler with little

---

78. Both Brown and Brass graduated before the formation of the Analytical Society in 1812, and neither became a member of the society. On the formation of the society and its members, see Enros 1983.

or no knowledge of the new mathematics. Here the case of Solomon Atkinson is informative. The son of an impoverished Cumbrian farmer, this extremely ambitious and largely self-taught student—who as an undergraduate had "much discussion" about mathematics with Airy—was unable to afford the services of a private tutor. Used to working on his own, Atkinson cleverly used college lectures and standard treatises to guide his own private study. Like Pollock before him, it seems that hard work, an excellent memory, and a thorough grasp of the relatively elementary mathematics which still dominated the examination papers enabled Atkinson to outpoint his more affluent and knowledgeable peers in the Senate House.[79] Atkinson recalled with satisfaction that he had beaten Gwatkin's "little coterie" of St. John's men who had enjoyed "all the best advantage that could be derived from the ablest instructions" (1825, 510), a comment which suggests that Gwatkin's enthusiasm for the new mathematics might actually have prejudiced his students' chances in the examination.[80] It is almost certainly the case, however, that Atkinson was one of the last undergraduates to win the coveted senior wranglership without a substantial knowledge of advanced analytical methods and without the help of a private tutor. According to Babbage (1864, 40), it was only "a very few years" after the publication of the book of examples in Leibnizian calculus in 1820 that the d-notation was widely adopted within the university, and this adoption was a prelude to the systematic introduction of more advanced analytical methods to undergraduate studies. In order to see how and why this occurred we must look more closely at several changes in student life taking place in the 1820s and early 1830s.

First, during the early 1820s both the place of mathematics in undergraduate studies and the apparatus of mathematical examination were further formalized. In 1823 the system of appointing just two moderators and two examiners to take complete responsibility for the examination was instituted. It was also confirmed that, in order to preserve continuity, the moderators of one year would become the examiners the next year, a system which further entrenched the power of a small number of moderators and examiners to control the content of the examination.[81] In 1824 a new Classical Tripos

79. Atkinson 1825. Atkinson published his article anonymously and concealed his identity by implying that he had been an undergraduate circa 1805. He was actually the senior wrangler of 1822, as is confirmed by Venn 1940–54. On his interaction with Airy, see Airy 1896, 30.

80. It should also be noted that although Atkinson was very critical of the practice of private tuition, claiming that, by the early 1820s, a private tutor was a virtual necessity for honors students, he made no suggestion that the regulations were being abused.

81. Glaisher 1886–87, 20. Prior to 1823 the number of examiners assisting the moderators had varied from year to year.

was founded in order to satisfy the complaints of those who felt that the dramatic growth in mathematical studies was marginalizing classical studies to an unacceptable degree.[82] From this point, the Senate House examination was often referred to as the "Mathematical Tripos" in order to differentiate it from the new "Classical Tripos." Second, the appearance during the 1820s of new books giving accounts of various physical topics in the Continental d-notation calculus made it easier for enthusiastic college lecturers and private tutors to teach the new methods, and, once such methods were being widely taught, they could be used to set more, and more advanced, problems in the Tripos examination. In addition to the translation of Lacroix's textbook and the book of examples produced by Peacock, Babbage, and Herschel, Whewell published textbooks giving analytical treatments of mechanics and dynamics in 1819 and 1823 respectively, Henry Coddington (SW 1820) and Woodhouse published accounts of optics and astronomy respectively in 1823, and, in 1826, Henry Parr Hamilton (9W 1816) published an analytical geometry and Airy published his *Mathematical Tracts*. To accommodate the new topics the number of days of mathematical examination in the Mathematical Tripos was extended to four in 1828.[83] The new regulations also made the examination more progressive: the first two days, sat by all candidates, were devoted to easy questions (solvable without the aid of the calculus) while, on the last two days, the candidates were given questions on the "higher and more difficult parts of mathematics" according to a preclassification into four groups by Acts. It was also decided that all the examination questions would henceforth be given in printed form—rather than given out orally—so that future candidates could study the papers to see what kinds of topic and problem they were likely to be set.[84] Finally, the number of students obtaining honors in the Senate House examination increased dramatically throughout the period in question. Between 1813 and 1827 the numbers almost doubled, an increase which further swelled the numbers attending college lectures and heightened the pervasive atmosphere of competition.[85]

All of the above developments served, either directly or indirectly, to enhance the role of the private tutor as an undergraduate teacher. Atkinson's

---

82. Until 1850 students wishing to sit the Classical Tripos had first to obtain honors in the Mathematical Tripos (Searby 1997, 166–67).

83. Hopkins's *Evidence to Commissioners* (1852–53, 239).

84. Glaisher 1986–87, 14.

85. Cambridge admissions more than trebled between 1800 and the early 1820s from around 120 to over 400 (Searby 1997, 61).

account of his undergraduate studies makes it clear that, despite the peculiarity of his own case, by the early 1820s the majority of ambitious students found the services of a private tutor extremely valuable. And as the most able wranglers trained by tutors such as Peacock and Gwatkin began themselves to become moderators and set Tripos questions in the new mathematics, so their peers, in the role of private tutors, began to prepare the most able students to answer such questions.[86] Since there were no official regulations prescribing which mathematical topics (beyond the most elementary) might or might not be examined in the examination, young moderators and examiners were free to set whatever problems they thought the best students might reasonably be able to tackle.[87] Through this mechanism, both the range of subjects making up the unofficial mathematical syllabus as well as the level of technical performance expected of a high wrangler came to be defined and continually inflated by the best private tutors and their most able pupils. The effect of this inflation was to make the private tutor a useful aid to any aspiring honors student and a virtual necessity to those seeking one of the fiercely contested places among the upper ten wranglers.

The very content of analytical mathematics also lent itself to this system of tough competitive learning and examining. As Becher has noted, the teaching of analysis, especially to an advanced level, required "specialised and cumulative learning over a period of years" (1980a, 2). This was a process ideally suited to an educational regime intended to push students to their intellectual limits and then to differentiate minutely between their relative performances in written examinations.[88] Such cumulative, competitive learning was also accomplished more effectively by private tutors using individual tuition, specially prepared manuscripts, and graded examples and problems, than it was by college lecturers teaching large classes at the pace of the mediocre. Even George Peacock, a college lecturer who did not as a rule take private pupils, found it expedient to train Airy in the new mathematics using

---

86. Whewell claimed that the rapid rise of competitive examinations had been aided by the fact that "Examiners, for the most part, alternate the employment of examining with that of private tuition" (*Commissioners' Report*, 74).

87. Glaisher 1886–87, 19; Hopkins 1854, 7. The Mathematical Tripos lists in Tanner 1917 show that during the period 1820 to 1840, more than 75 percent of the examiners were between just four and eight years from graduation. A considerable majority of those setting the examination papers would therefore have been under thirty years of age.

88. Cambridge private tutors seem to have been more successful in teaching analytical methods to large numbers of students than were their contemporaries, including Lacroix, in Paris. See Richards 1991, 302–3.

the methods of the private tutor. During the 1820s, as success in the Senate House examination came to depend upon the ability to solve difficult problems using advanced analysis, it was young college fellows who had just sat the Tripos examination and who wished to derive an income by pupilizing, who were best able to drill ambitious undergraduates in the required techniques. It is noteworthy in this context that, by the mid 1820s, even the established college lecturer and reformer, Richard Gwatkin, was reputed to be the finest private tutor (rather than lecturer) in the university, a reputation that bespoke Gwatkin's realization that pupilizing had become the most effective (and most financially rewarding) means of furthering the analytical revolution.[89]

I suggested above that the gradual relaxation of the regulations on pupilizing should not be understood, as Rouse Ball suggested, as a straightforward acknowledgement that the practice was impossible to suppress. As we have seen, the relaxation in 1807 served a range of purposes, while that of 1815 enabled the elite students to advance their knowledge of the *Principia* beyond the level reached in college lectures. The final relaxation of the rules in 1824, which enabled undergraduates to work with a private tutor until the end of their ninth term, should also be understood in the above context. Even circa 1820 college lectures were of little value to ambitious students like Wright and Airy beyond the seventh term, and, following the adoption of the d-notation calculus and the rapid proliferation of analytical topics during early 1820s, it would have been clear that, without major reform, the traditional structures of college and university teaching could offer little to aspiring wranglers in their final year. It seems likely that, by the mid 1820s, many members of the university realized that a further influx of analytical methods was inevitable, perhaps even desirable, and that it was neither practicable nor in their own interests to leave the intense preparation required in these subjects to college lecturers or university professors. Faced with the invidious choice of either completely reforming undergraduate teaching—which would have been very difficult to get through the Senate—or relaxing the rules on private teaching, the university adopted the easier and, in the short term probably more effective, latter option.[90] A further advantage

---

89. Atkinson 1825, 505.

90. It would not have been necessary to lift the restrictions on private tuition entirely, as the last summer vacation and tenth term were traditionally used for revision and practice at problem solving. Private tutors soon came to oversee this work, however, and there is no doubt that, in this sense, the rule came to be flagrantly ignored.

to legalizing the use of private tutors throughout the nine undergraduate terms was that it made the most effective teaching in Cambridge accessible to students from all colleges (assuming they could afford the fees).[91] The rise of competitive examinations as the measure of the intellectual elite had created severe problems for the smaller and poorer colleges in the university who could not afford to keep a large staff of able mathematical lecturers. The rapid increase in the content and standard of the Mathematical Tripos during the 1820s would have exacerbated this problem and given further advantage to the large and wealthy colleges.[92] By offering their services to any able student who could pay, the best private tutors eased this problem and, as we shall see, provided an important model for a university-wide teaching staff.

A useful contemporary account of the rise of the private tutor during the mid 1820s is provided by William Hopkins (7W 1827), who was not only an undergraduate during this period but went on to become the first private tutor to make a life-long career of pupilizing (fig. 2.2). It was in 1822, following an unsuccessful career in agriculture and the death of his first wife, that Hopkins, already thirty years old, enrolled himself as a student at Peterhouse College. Being ineligible for a normal college fellowship—having remarried while an undergraduate—Hopkins settled in Cambridge and made his living as a private tutor.[93] Drawing upon his firsthand experience of undergraduate life in the mid 1820s, Hopkins recalled that the training of a rapidly expanding number of ambitious students and the introduction of Continental mathematics had not been accomplished by college lecturers who, he noted, had singularly failed to rise to the challenge, but by private tutors.[94] Hopkins believed, moreover, that it was precisely the rapid and simultaneous expansion in both student numbers and the content and difficulty of analytical mathematics in undergraduate studies (led by young private tutors and examiners) which had virtually forced the university to increase the number of written papers in the Senate House examination, and dramatically enhanced the status of private tutors.[95] By the end of the 1820s, he esti-

91. Private tutors employed a discretionary scale of fees. Pupils from aristocratic or wealthy families were expected to pay above the normal rates while bright but poor students paid "about one half" the normal rates. See Bristed 1852, 1:150.

92. *Commissioners' Report*, 79.

93. Hopkins also held several of the university posts for which he was eligible. For biographical details see the *DNB*.

94. Hopkins also recalled that Airy's professorial lectures of 1827–28 had been useful to those learning the new mathematics.

95. Charles Prichard (4W 1830) recalled that his college, St. John's, not only encouraged him to work with a private tutor but was prepared, if necessary, to help cover the expense (Prichard 1897, 39).

FIGURE 2.2. William Hopkins, the first private mathematics tutor in Cambridge to derive a permanent living from "coaching." Hopkins was renowned for "encouraging in his pupils a disinterested love of their studies," and for his "researches on the application of mathematics to physics and geology" *(DNB)*. This likeness is from a portrait that hangs in the Hall at Peterhouse College, Cambridge. (By Permission of the Master and Fellows of Peterhouse College, Cambridge.)

mated three decades later, "the effective mathematical teaching of the university was almost as completely, though not perhaps so systematically, in the hands of Private Tutors, as [in 1854]" (1854, 6–7).[96]

The enormously increased importance of the private tutor by the end of the 1820s is made visible in a different way by the work of J. M. F. Wright. Having failed to obtain an honors degree and a fellowship, Wright earned his living as a mathematics teacher in Cambridge and by producing a series of self-help aids for undergraduate mathematicians. In 1829 he published books on "self-instruction" in pure arithmetic and "self-examination" in

---

96. Charles Prichard also claimed that by the late 1820s "the most important parts of the tuition were carried on by private tutors" (1897, 38).

Euclid's geometry, and, in 1830, he published an edition of Newton's *Principia* which included "notes, examples, and deductions" covering required aspects of this difficult text (Wright 1829a, 1829b, 1830). Lastly, and most remarkably, in 1830 he began a weekly journal, *The Private Tutor and Cambridge Mathematical Repository* (1830–31, see fig. 2.3) which was intended to make "the usual aids" of the eponymous private tutor directly available to all undergraduates, especially those unable to afford the fees of an actual tutor. Appearing every Saturday, Wright's journal contained expository essays on the new mathematics, problems for the reader to solve (solutions being given the following week), and detailed accounts of how examinations, both written and oral, were conducted.[97] Students were informed what they might usefully "write out" to gain speed and accuracy, which mathematical formulae ought to be committed to memory, which treatises contained serious errors (corrections being given), and the questions most commonly sent in by readers were answered in catechetical style. Each issue of the journal was divided into "four principal compartments," one for each year of undergraduate study, the first three compartments "comprising every branch of mathematics in order of lectures at Trinity College." After two years the *Private Tutor* had grown into a complete self-tutor for the undergraduate mathematician, and Wright announced he could now bring his "labours to a satisfactory conclusion."[98]

That Wright never updated these notes, despite the continued growth of analysis in undergraduate studies, suggests that the majority of students chose to work with a real private tutor, but the two volumes do highlight the importance of private tuition by 1830 and provide some insight into the relationship between private and college teaching circa 1830. The sectional divisions of the *Private Tutor* indicate that college lectures continued to provided a basic framework for ordinary undergraduate study and now included the new analytical subjects. In the second year both the d-notation calculus and analytical geometry were major topics of study, while, in the third year, students were expected to acquire some knowledge of the calculus of variations, finite differences, and celestial mechanics according to Laplace. By this time, however, the training of the most ambitious students was firmly in the hands of increasingly professional private tutors, and,

97. Wright 1830–31, 1:145–50.
98. Wright 1830–31, 1:1; 2:406. A typical example of a reader's question is "What is an imaginary angle, and is its cosine imaginary?" (156). The fourth "department" was concerned with the solution of problems.

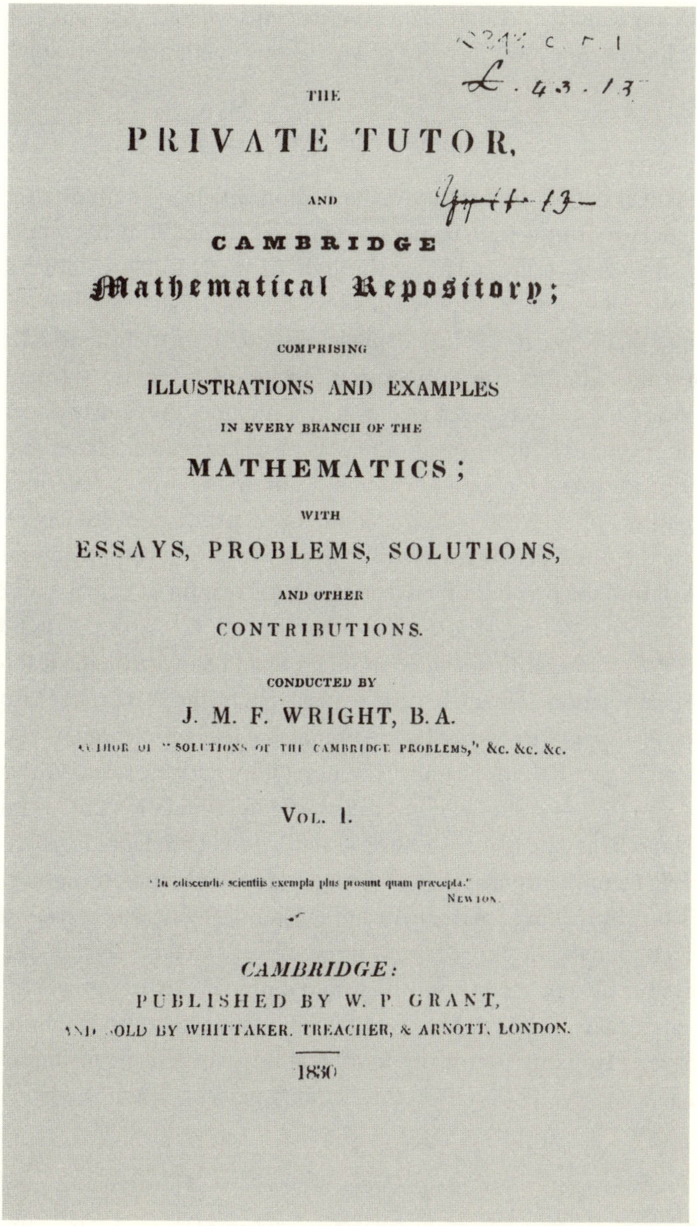

THE
PRIVATE TUTOR,
AND
CAMBRIDGE
𝕸𝖆𝖙𝖍𝖊𝖒𝖆𝖙𝖎𝖈𝖆𝖑 𝕽𝖊𝖕𝖔𝖘𝖎𝖙𝖔𝖗𝖞;

COMPRISING

ILLUSTRATIONS AND EXAMPLES

IN EVERY BRANCH OF THE

MATHEMATICS;

WITH

ESSAYS, PROBLEMS, SOLUTIONS,

AND OTHER

CONTRIBUTIONS.

CONDUCTED BY

J. M. F. WRIGHT, B.A.

AUTHOR OF "SOLUTIONS OF THE CAMBRIDGE PROBLEMS," &c. &c. &c.

VOL. I.

"In discendis scientiis exempla plus prosunt quam præcepta."
NEWTON.

CAMBRIDGE:
PUBLISHED BY W. P. GRANT,
AND SOLD BY WHITTAKER, TREACHER, & ARNOTT, LONDON.

1830

FIGURE 2.3. The titlepage of J. M. F. Wright's weekly magazine, *The Private Tutor*. The list of contents illustrates the stress laid on building technical skill through examples and problem solving. (By permission of the Syndics of Cambridge University Library.)

Hopkins reveals, by the mid 1830s even students who aspired only to a pass degree had begun to supplement college lectures with private tuition.[99]

2.5. From *Pupilizing* to *Coaching*

The changes in the Senate House examination which had supported the introduction of Continental mathematics and intensified student competition during the 1820s continued apace through the 1830s. In 1833 another day of mathematical examination was added to accommodate more new topics, and, in 1836, it was decreed that all papers were to be formally marked, moderators and examiners being thereafter forbidden to rely on a general impression of the answers given as they had sometimes previously done. New regulations in 1839 added yet another day of mathematical examination for yet more advanced topics, and made position in the order of merit dependent entirely upon written examination papers. All honors students would now sit the same progressively difficult papers, while preclassification by oral Acts and the practice of bracketing in anticipation of a final day's competition in the Senate House were discontinued. As in the 1820s, young moderators and examiners had continued to add new mathematical topics to the examinations, which, in turn, were taught by the best private tutors.[100]

The 1830s also witnessed a further important development in undergraduate teaching with the appearance of a new kind of private tutor. We have already seen that, even in the 1810s, tutors of good reputation could generate a substantial income from private teaching, but William Hopkins, barred by marriage from the traditional Cambridge career path, took this a stage further in 1827 by making a permanent and substantial living from private teaching.[101] Hopkins soon showed outstanding ability as a teacher, becoming the first of four men in succession who would dominate the training of the elite mathematicians. In the twenty-two Tripos examinations from 1828 to 1849 inclusively, Hopkins personally trained almost 50 percent of the top ten wranglers, 67 percent of the top three, and 77 percent of senior wranglers—

---

99. Hopkins (1854, 7) reveals that prospective non-honors students continued to rely on college lectures until the mid 1830s.

100. Glaisher 1885, 14–15; Rouse Ball 1889, 183, 212–14.

101. The "regular fee" for private teaching in the early Victorian period was £7 a term (three lessons per week). Students paid similar rates to be taught over the long summer vacation. A full-time private tutor of high reputation earned between £800 and £900 per year, a good professional salary. See Bristed 1852, 1:148, and Hopkins 1854, 25–26.

a remarkable record which soon earned him the title "the senior wrangler maker" in the university.[102] Hopkins success derived from his teaching methods, his own ability and enthusiasm for mathematics, and his reputation as a tutor. Unlike the majority of private tutors, he taught students in small classes—between ten and fifteen pupils—composed of men of roughly equal ability. This meant that the class could move ahead at the fastest possible pace, the students learning from and competing against each other. Hopkins considered it an "immense advantage in Class Teaching when there is a sufficient equality in the ability and acquirements of each member of the class" that every student would: "hear the explanations which the difficulties of others might require, and thus be led to view every part of the subjects of his studies, through the medium of other minds, and under a far greater variety of aspects than those under which they would probably present themselves to his own mind, or would be presented by any Tutor teaching a single pupil" (1854, 19–20). Hopkins's teaching methods were thus designed to optimize the benefits of intensive, progressive, and competitive learning. He also developed and exploited an avuncular intimacy with his pupils which would have been quite alien to most college lecturers and university professors. As Francis Galton wrote in 1841:

> Hopkins, to use a Cantab expression, is a regular brick; tells funny stories connected with different problems and is in no way Donnish; he rattles us on at a splendid pace and makes mathematics anything but a dry subject by entering thoroughly into its metaphysics. I never enjoyed anything so much before. (Pearson 1914, 163)

In the friendly atmosphere of his teaching room, Hopkins combined the admiration of his students with his own infectious enthusiasm for mathematics to promote the competitive ethos and a dedication to hard work. As his reputation grew, Hopkins could also afford to select the students he taught. He would invite promising freshmen to informal parties when they arrived in Cambridge in order to assess their potential and, in the case of the most outstanding men, would keep a watchful eye over their development during the first year.[103] He would only begin teaching students personally in their

---

102. The actual figures, provided by Hopkins in response to an enquiry by Henry Gunning in 1849, are 108 in the first ten, 44 in the first three, and 17 senior. See Gunning 1854, 2:359, and Rouse Ball 1889, 163.

103. S. P. Thompson 1910, 32. On Hopkins's guidance of William Thomson's studies during his first year, see Smith and Wise 1989, 69.

second year, whereupon he "formed a select class of those who had shown in their first year promise of becoming high wranglers."[104]

More typical of the part-time private tutor of this period was Hopkins's chief rival in Cambridge, John Hymers (2W 1826). An undergraduate at St. John's—where he was very likely tutored by Gwatkin—Hymers combined private teaching with a successful college career: he became a fellow of St. John's in 1827, an assistant tutor in 1829, a tutor in 1832, a senior fellow in 1838, and president in 1848. In 1852 he left Cambridge to take up a college living at the rectory of Brandesburton in Holderness.[105] Even as a young man, Hymers was renowned for his "acquaintance with the progress of mathematics on the continent," and he wrote several important treatises which helped to introduce Cambridge students to the new methods. When invited to act as a moderator for the Mathematical Triposes of 1833 and 1834, he was just thirty years old, had already completed a treatise on analytical geometry in three dimensions, and was writing a treatise on integral calculus which introduced elliptic integrals to English students for the first time. His early life provides an excellent illustration of the way a young enthusiast could simultaneously pursue a traditional college career and promote the new mathematics in the university. Although unable to rival Hopkins's unique record as a private tutor, Hymers's work exemplifies the mixture of private teaching, textbook writing, and examining through which the new mathematics was established in undergraduate studies.[106]

The role of the private tutor continued to expand through the 1830s until, by the end of the decade, it was virtually impossible for any undergraduate to contemplate high placing in the order of merit without private tuition.[107] Leslie Ellis, the senior wrangler of 1840, is probably the last man for whom any independence from private teaching was claimed, but careful inspection reveals that he was only a partial exception. Like Airy before him, Ellis mastered a great deal of mathematics while still at school, was taught privately by a Cambridge mathematician before entering the university, and was given personal tutoring by Peacock at Trinity.[108] Ellis's private journal reveals that, also like Airy, having stolen a substantial lead on his peers, he

---

104. See Hopkins's entry in the *DNB*.

105. Thompson 1910, 95. Hymers remained Hopkins's chief competitor until the 1840s. For further biographical details, including publications, see Hymers's entry in the *DNB*.

106. Hymers's best-known student was John Couch Adams (SW 1843). This was one of only five occasions in which Hopkins's best man was beaten.

107. Peacock 1841, 153; Hopkins 1854, 7.

108. Walton 1863, xiii–xiv.

was able to read a great deal of advanced mathematics by himself as an undergraduate. It is important to note that by this time Ellis was considered remarkable precisely *because* he did not read with a private tutor in his second year and that, notwithstanding his remarkable mathematical ability, he did go to Hopkins in his final year in order that his "reading should be arranged, and put in a form suitable for the Cambridge examinations."[109] By 1840, even the most talented and well-prepared undergraduates could not afford to bypass more than a year of expert coaching if they aspired to a place among the top wranglers. Typical of the course of mathematical studies pursued by such students from the early 1840s is that undertaken by George Stokes (SW 1841). Stokes spent the two years preceding his entry to Cambridge at Bristol College, a school at which the Principal, J. H. Jerrard (27SO 1828), was a Cambridge graduate and at which the most able students were expected to "become acquainted" with differential and integral calculus, statics, dynamics, hydrostatics, optics, conic sections, spherical trigonometry, physical astronomy, and the first book of Newton's *Principia*.[110] Whether Stokes was privately tutored during his first undergraduate year at Cambridge is unclear, but he displayed sufficient promise to come entirely under Hopkins's tutelage from his fourth term. Over the next fifteen years Hopkins had an almost complete monopoly of the training of the elite mathematicians including Arthur Cayley (SW 1842), William Thomson (SW 1845), Isaac Todhunter (SW 1848), Peter Guthrie Tait (SW 1852), and James Clerk Maxwell (2W 1854).[111]

One of the earliest accounts of this training system provided by an overseas visitor to Cambridge is that of the American, Charles Bristed, who studied for the Mathematical Tripos between 1841 and 1845. In his book on student life in Cambridge, Bristed devoted a whole chapter to private teaching and gave a succinct summary of the skills possessed by a good tutor. He should be expert: "In working up a clever man whose previous training has been neglected, in cramming a man of good memory but no great brilliancy, in putting the last polish to a crack man and quickening his pace, so as to

---

109. Walton 1863, xv. Ellis's journal reveals that on 15 May 1839 he paid Hopkins £42 for a year's coaching, a sum that indicates he was coached six days a week (RLE-CUL).

110. Stokes recalled in his reminiscences (written more than sixty years after the event) that he had not studied calculus when he entered Cambridge, but contemporary evidence suggests that he underestimated his mathematical knowledge at this time (Stokes 1907, 6–8; Wilson 1987, 30–31).

111. Bristed claimed that, by the mid 1840s, to appear among the top ten wranglers a man had to be "remarkably clever," show "considerable industry," and to "read mathematics professionally" (1852, 1:308).

give him a place or two among the highest. . . . In such feats a skilful tutor will exhibit consummate jockeyship" (1852, 1:148–49). By the early 1840s, the average private tutor had to possess a range of teaching skills tailored to the needs of pupils of different abilities. Those taught in pairs were generally matched for their ability so that, like Hopkins's elite class, they could proceed as rapidly as possible, learning from each other's errors. It was also an important job of the tutor to organize an only tacitly understood syllabus, and to coordinate learning beyond his classroom by providing manuscripts on the most difficult topics, recommending sections from the best textbooks and setting appropriate problems from past Tripos papers. Bristed was astonished at the pace at which his own knowledge developed under his private tutor, noting that he "certainly put more into me in seven months than I could have acquired by my own unassisted labours in two years" (1852, 1:94). It was a also a major responsibility of the private tutor to ensure that his pupils worked consistently and hard. The changes in student sensibilities engendered by tough regimes of competitive learning will be explored in depth in chapter 4, and it is sufficient to note here that students of all abilities relied on their private tutors to stop them slacking. According to Bristed, it was an "acknowledged requisite" of a good tutor that he "blow up any of his team who [gave] signs of laziness. . . . 'I'm afraid of going to T———,' you may hear it said, '*he don't slang his men enough*'" (1852, 1:148–49).

Despite the severity with which a private tutor might occasionally admonish a lazy student, the relationship between tutor and pupil was generally much warmer and more personal than was the case with college lecturers or university professors. Echoing Galton's comments on Hopkins, Bristed observed that the "intercourse" between teacher and student "was of a most familiar kind, the former seldom attempting to come Don over the other." Both parties had a strong interest in the student's satisfactory progress, and the relationship which developed, "blending amusement with instruction" whenever possible, was the most conducive to learning. The best tutors would regularly set their pupils short examinations—known as "fights"—to provide practice in problem solving under pressure and to establish a competitive ethos through a local order of merit. And, unlike a college lecturer, the private tutor would carefully mark all of the work submitted by his students, showing them, in the case of problems, the most elegant way of obtaining the required result as well as any tricks and shortcuts which could legitimately save precious seconds in the Senate House. It was, in short, a major part of the private tutor's craft to learn how to impart his own mastery of problem solving to his students; a good tutor, Bristed mused, "seems to throw a part of himself into his pupil and work through him" (1852, 1:149–50).

These outstanding features of the new regimes of tough progressiv9e training pioneered by Hopkins and others during the 1820s and 30s were succinctly captured in student slang by the invention of the word "coach" to describe the private tutor. The first printed use of the term I have found occurs in a satirical student pamphlet published in 1836 by an Oxford undergraduate in which a mock examination question invites the student to "trace analogically the application of the word coach, when it is said by a man that he has 'just taken such a coach to help him through his small.'"[112] For how long and how widely the term had been in use before this is hard to determine, as such student slang is rarely recorded in print, but, at least as far as Cambridge is concerned, it seems unlikely that it was in use before the early 1830s.[113] J. M. F. Wright's book of 1827 discusses everyday undergraduate life in some detail (including student slang) but makes no mention of the term, while Charles Prichard (4W 1830), who remained in Cambridge until 1833, recalled that in his day "private tutors were not designated by the expressively equivocal name of 'coaches.'"[114] It therefore seems fairly safe to assume that the term was coined around 1830, probably in Oxford.[115] Regardless of its town of origin, the term "coach" soon became most closely associated with the Cambridge private mathematical tutor.

In order to grasp the contemporary meaning captured by the term "coach," it is important to understand that in the decade immediately preceding the widespread building of railways, the latest stage and mailcoaches, running on newly Macadamized road surfaces, represented the most exciting and rapid form of modern transport. These were the years when the coach, rather than the train, was the icon of progressive transport, and the generation of young men growing up in the 1820s and early 30s were highly "coach conscious," associating coaches and coachmen with speed, adven-

---

112. Caswall 1836, 28. The "small" was a preliminary examination at Oxford.
113. The first written use of the term "coaching" by a Cambridge mathematics student of which I am aware occurs in a letter from David Foggo to William Thomson in January 1845. See Smith and Wise 1989, 82.
114. Prichard 1897, 38. Even student satires, which generally make much of undergraduate slang, do not use the term until the latter 1830s. See for example Gooch 1836, which always uses the term "private tutor" in humorous references to undergraduate teaching.
115. The term is more likely to have been coined in Oxford, as it was a major coaching station on the way to the Midlands and northern industrial towns. More than seventy coaches passed through the Oxford each day, and coachmen were part of everyday life. See Hanson 1983, 55.

ture, and—unlike the trains that replaced them for long-distance travel in the late 1830s and 40s—privilege and wealth.[116] George Pryme recorded in his reminiscences that during the 1800s people expressed "disbelief" that a coach could travel from Cambridge to London in just seven and a half hours, while, by the latter 1820s, another Cambridge graduate marveled of the Times coach that "its rate of travelling is so extremely great" that it could reach London in just five-and-a-half hours.[117] By the mid 1830s, the operators of the Star Coach (fig. 2.4) claimed to be able to make the journey to London in a remarkable four-and-a-half hours.[118]

These perceptions of coaches and coach travel help us to appreciate the qualities that mathematics students sought to make visible and gently ridicule in their tutors. The tutor, like the coachman, *drove* his pupils along a predetermined course even when they felt inclined to stray or slack. This quality was complemented by the use of the term "team"—from team of horses—to describe the students working with a particular coach.[119] Another, more subtle yet equally important quality possessed by coachman and tutor was the ability convey a team from A to B unnaturally fast by skilful driving and the careful design of the path traversed. This quality points to the pedagogical significance of the teaching techniques developed by the mathematical coaches. As we have seen, private tutors carefully designed a whole apparatus of progressive tuition which, like stagecoach and road, enabled a student to proceed much more rapidly than he could by his own unaided efforts or even with the help of traditional lectures. These analogues are also consistent with the supposition that the term "coach" was coined, or at least became descriptive of the Cambridge private tutor, in the mid 1830s, since this, as we have also seen, was the period when private tutors were most visibly displacing college teachers as masters of the new mathematics and educators of the undergraduate elite.

The term "coach" as undergraduate slang began to enter the consciousness of the educated Victorian public through several serialized novels

---

116. Faith 1990, 38. The low cost of train travel was initially as much a factor in its popularity as was speed. Venn (1913, 255) recorded that when trains began to run from Cambridge to London in 1845 the cost of a first class fare (12 shillings) was the same as the cost of an outside seat on a coach.

117. Pryme 1870, 62; Appleyard 1828, 2.

118. The accounts of contemporary travelers indicate that advertised travel times were seldom approached in practice. See, for example, Bury 1967. For a timetable of coaches to and from Cambridge, see *The Cambridge Guide*, 1830, 304–7.

119. Bristed (1852, 1:54) records "team" in his list of Cambridge student slang in use in the early 1840s.

FIGURE 2.4. The "Star of Cambridge" leaving The Belle Sauvage Inn on Ludgate Hill (London) for Cambridge. Running on newly Macadamised road surfaces, these stagecoaches were icons of rapid and romantic travel in the 1830s. The advertisement below the illustration boasts that the coach can "perform the distance between London and Cambridge in four hours and a half." (Reproduced by kind permission of Cambridge and County Folk Museum.)

around 1850. In this new literary genre, pioneered by Dickens in the late 1830s, the topic of undergraduate life, especially at Oxbridge, became very popular toward the end of the 1840s.[120] The first reference to coaching in this context, as far as I am aware, occurs in Thackeray's *Vanity Fair*. In this case the dull and clumsy William Dobbin, treated with contempt by his schoolfellows because his father is a grocer, reveals unexpected academic talents, especially in mathematics, when he is coached by the "brilliant" Reginald Cuff.[121] On this occasion Thackeray places the word "coached" in quotation marks, suggesting that in 1847 it would still have been unfamiliar or considered disrespectful by some of the reading public. Thereafter Thackeray and others generally used the term as a literary device to distinguish between those who had been to Oxbridge and those who had not, and

120. Sutherland 1995, 101.
121. Thackeray 1967, 45.

between the naive freshman and his older or more worldly-wise contemporaries. In the epigraph to this chapter, Thackeray both exploits and explains the pun on the word coach while using it to establish the altered relationship between Pendennis and an old school friend, Foker, now in his first year at "Oxbridge."

Another popular novel chronicling the experiences of an Oxford freshman, Cuthbert Bede's *Verdant Green,* uses the term in a similar way.[122] Expressing astonishment at hearing an older student describe a "Mr Cram" as "the first coach in Oxford," the aptly named Green is told in superior tone "Oh I forgot you didn't know college slang. I suppose a royal mail is the only gentleman coach *you* know of. Why, in Oxford a coach means a private tutor, you must know" (Bede 1853, 63). Yet another very popular monthly of the early 1850s, F. E. Smedley's *Frank Fairlegh,* was based entirely on "scenes from the life of a Private Pupil."[123] As with Thackeray's Dobbin, the eponymous Fairlegh made remarkable academic progress at Cambridge once under the guidance of his coach—defined by Smedley as a "vehicle for the conveyance of heavy learning"—quickly becoming the "formidable rival" of men of "infinitely superior" intellect and ending his undergraduate career fourth wrangler in the Mathematical Tripos.[124] Both Thackeray and Bede also employ appropriate forms of the word "coach" as proper names for their characters: in *Pendennis* "Mr Coacher," a widower and private tutor, is too engrossed in his work and books to notice that his daughter, Martha Coacher, is being wooed and jilted by a succession of his pupils; in *Verdant Green* a college lecturer is named "Mr Slowcoach" in recognition of the relative value of college and private teachers (fig. 2.5).[125]

References to coaching in the work of Thackeray and Smedley are the more intriguing as both wrote from personal experience of undergraduate life in Cambridge. Pendennis's unsuccessful career at "Oxbridge" is loosely based on Thackeray's own experiences at Trinity College in 1829–30, while Smedley, crippled as a child and never formally an undergraduate at the university, was coached in Cambridge around 1836 by the chaplain of Trinity

---

122. Bede 1853. Cuthbert Bede was a pseudonym used by the surgeon and humorist Edward Bentley, who contributed popular writings to *Bentley's Magazine* and the *Illustrated London News,* where part of *Verdant Green* appeared. Verdant Green's adventures were so popular that Bentley published two more novels based on his later life. See Boase 1908.

123. Smedley 1850. The early chapters of *Frank Fairlegh* appeared originally in *Sharpe's Magazine,* which was edited by Smedley. See Sutherland 1995, 101.

124. Smedley 1850, 240–41, 259.

125. Thackeray 1994, 94–95; Bede 1853, 34.

FIGURE 2.5. This sketch by J. L. Roget (2SO 1850) illustrates the relative unimportance of college lectures (as compared with private tuition). The sketch also provides a rare illustration of a blackboard, which was a fairly recent pedagogical innovation at this time (chapter 5). Roget, *Cambridge Customs and Costumes* (1851). (By permission the Syndics of Cambridge University Library.)

College.[126] Through the fictional characters popularized by these men, the undergraduate terms "coach" and "coaching" transcended Oxbridge slang to became part of everyday speech.[127] By the 1850s, any tutor or pupil engaged in small group teaching might be said to be "coaching" while, in the 1860s, the verb was sufficiently well recognized and respectable for one leading London school, Wren's Coaching Establishment, to adopt it as a proprietary name.[128] Nor was it only the name of the Cambridge private tutor that this and similar institutions exploited. As we shall see in chapter 5, Wren's em-

---

126. Thackeray 1994, vii–xxxi. Smedley (1850, 240) implies that the term "coach" was actually in use when he was in Cambridge in 1836. For biographical details of F. E. Smedley see the *DNB*.

127. It is unclear how and when the term coach acquired its more recent meaning in competitive sport. According to the OED the first such usage occurs in the popular press in connection with the Boat Race in 1867. It is likely that the term moved from the coaching room to the river bank around mid-century and then into other competitive sports.

128. The school was located at 7 Powis Square, Westbourne Park. On the founder and principal, Walter Wren, see *Who Was Who*? On Joseph Edwards (4W 1878), who coached in mathematics at Wren's until 1915, see Venn 1940–54.

ployed high wranglers who not only carried the teaching methods of the leading coaches beyond Cambridge but specialized in pre-preparing the next generation of would-be wranglers.

2.6. Public Teaching, Private Teaching, and Research

We have seen that private mathematical tutors played a major role in the analytical revolution, accomplishing in the process a parallel pedagogical revolution from which they emerged as the most important teachers of the undergraduate elite. Enthusiastic young reformers such as Peacock and Gwatkin had gone along with these developments in the 1820s and early 1830s believing, no doubt, that the diminution in authority of public and college teachers in the university was a temporary phenomenon, and one which had to be tolerated if the new mathematics was to be quickly and firmly established in undergraduate studies.[129] But by the end of the 1830s, several senior members of the university, including Peacock, were becoming concerned by what they saw as the uncontrolled introduction of new mathematical topics, an excessive emphasis on competitive examination, and the almost total collapse of influence of college lecturers and university professors. The former stability of an undergraduate syllabus tacitly based on such established texts as Euclid's *Elements* and Newton's *Principia,* and including a strong moral and theological component, was rapidly being swept away in favor of tough private training for competitive examinations in advanced analysis. These developments were especially troubling to more conservative members of the university who feared that the study of higher mathematics was undermining the broader educational goal of developing all the mental faculties and moral sensibilities of undergraduates. The most influential spokesman for this faction was William Whewell (2W 1816). From the mid 1830s until the 1850s, Whewell struggled to preserve liberal educational values in undergraduate studies and to reinstate the authority of public teachers. Whewell's differences with Hopkins and, to a lesser extent, Peacock, highlight the struggle that took place during this period between those who saw undergraduate mathematics primarily as a pedagogical vehicle and those who saw it as a practically useful and morally worthwhile end in itself. And in both cases, the methods by which mathematics was to be taught were as much at stake as was the content of undergraduate studies.

[129]. The fact that the university passed legislation extending the number of days of the Senate House examination in 1828, 1833, and 1839 indicates that a majority of senior members must during this period have been in favor of the continued expansion of new mathematical topics.

Whewell began his career as a private mathematics tutor and proceeded, via a fellowship at Trinity and the professorships of mineralogy and moral philosophy, to the mastership of Trinity in 1841. As a student and young mathematics tutor, Whewell had recognized the importance of French analysis, especially in establishing new results in mechanics, and initially participated in the introduction of the new methods. His textbooks on mechanics and dynamics (discussed further in chapter 3) introduced students to the d-notation calculus, but Whewell remained close to traditional geometrical methods whenever possible and always presented analysis as a useful tool in solving physical problems. As a staunch Anglican and political conservative, Whewell did not share the wider objectives of the Analytical Society and disagreed with the implicit political, anticlerical, and professionalizing agenda of the more radical reformers.[130] For Whewell, the prime purpose of the study of mathematics in the university was not to train professional mathematicians and foster research in pure analysis, but to form the basis of a liberal education for the intellectual elite. The purpose of a liberal education, he argued, was to "develop the whole mental system of man" such that his "speculative inferences coincide with his practical convictions," a coincidence best nurtured through practical examples which taught students how to "proceed with certainty and facility from fundamental principles to their consequences."[131] According to Whewell, both the most reliable fundamental principles and the best method of proceeding through strict reasoning to certain conclusions were to be found in mathematics, especially Euclidian geometry and Newtonian mechanics, and it was in the mastery and application of these disciplines that students most assuredly exercized their powers of reasoning on the ground of certain knowledge.

Whewell's belief in the pedagogical role of mathematics as a reliable means of exercising and sharpening the student intellect made him antipathetic to the wholesale introduction of Continental analysis into undergraduate teaching. Analytical methods encouraged students to circumvent the intuitive geometrical or mechanical steps which ought properly to be followed in solving difficult problems. While, then, he had little objection to advanced analysis being taught to a few of the most outstanding undergraduates—though by professorial lectures rather than private tuition—he insisted that such teaching must not become the primary purpose of public

130. On Whewell's career and views on the role of mathematics in a liberal education see Becher 1980a and 1991, and Williams 1991.
131. Whewell 1835, 5. For a comparison of the pedagogical uses of analysis in France and England during this period, see Richards 1991.

instruction at Cambridge. In his opinion, it was imperative to preserve the Mathematical Tripos as the elite course of undergraduate study which nurtured the nation's finest intellects and passed on fundamental, permanent, and divine ideas.[132] This vision of mathematical studies made Whewell wary of both competitive examinations and private teaching. He was especially worried (see chapter 3) by the potential of competitive examinations to undermine his notion of a liberal education, but, in the mid 1830s, he preserved a reasonably tolerant attitude towards private teaching. As we have just seen, this was the period when private tutors were beginning completely to displace college lecturers as teachers of the undergraduate elite, and this was a trend that Whewell certainly regarded as undesirable. Officially appointed college and university teachers had taken, or were preparing to take, holy orders, and had been selected by their elders as suitable moral, spiritual, and intellectual guardians of the nation's future leaders. The public nature of the lectures and sermons delivered by such men also preserved the dignity and decorum of their office. Private tutors, by contrast, taught behind closed doors, in whatever manner most effectively got their pupils through examinations and without the formal sanction of the university.

But despite the fact that Whewell felt private tuition to lack the broader "culture" conveyed by public teaching, he appears to have viewed the trend towards private tutoring as representing only a potential threat to public teaching, and one which, in the mid 1830s, remained manageable and reversible.[133] Private tutors played a useful role, he argued, especially in helping students who through "slowness of intellect, or previous defects of education" (1837, 73) could not otherwise receive the full benefit of college lectures. And if any parents wished to hire a coach to improve their son's knowledge of mathematics, it would, in Whewell's opinion, be "absurd to legislate against their sacrificing a pecuniary to an intellectual advantage" (1837, 77). He nevertheless thought it very important that college teachers be responsible for recommending students to private tutors as and when necessary, and that the former should keep a watchful eye on what was being taught. The best method of instituting and preserving this subservient role for the coach, Whewell thought, would be to ensure that the Tripos examination reflected what college teachers taught and did not become the sole focus of undergraduate studies. His hopes of reinstating the authority of public teachers were not realized. As we have seen, there existed no formal

132. Williams 1991, 147.
133. Whewell 1837, 60.

mechanism through which senior members of the university could dictate what mathematical topics were set in the Senate House examination, and the close informal relationship between coaches and their elite students who went on to become moderators and examiners had generated an inflationary cycle that would be very difficult to break.

Whewell was not the only senior member of the university troubled by the dominance of private teaching. By 1841 Peacock, a political Whig, was prepared to take an even stronger stand against coaching than had Whewell four years earlier. To understand why Peacock and the more conservative Whewell found common cause on this issue we need to return for a moment to the analytical revolution and its aftermath. Peacock's introduction of Continental mathematics to the Tripos of 1817 had prompted such outrage among conservatives that a counterproductive backlash seemed inevitable. In order to preserve broad support for the new mathematics in the university, Peacock found it expedient thereafter to distance himself from the more radical reformers and to compromise with men like Whewell by focusing on the physical applications of analysis.[134] Peacock's strategy proved highly successful through the 1820s and 1830s, but the enormous expansion in the content of undergraduate mathematics over that period together with the introduction of a common written examination for all questionists in 1839 had some deleterious effects on the Tripos. The examination had lost its former stability, and students of all abilities felt their best interests were served by ignoring the elementary subjects and cramming the higher ones which carried more marks.[135] Both of these shortcomings could, with some justice, be blamed on the prevalence of private tuition. The coaches had played a major role in the informal expansion of the syllabus, and the singular importance of the written Senate House examination encouraged dull or lazy students to remain idle for one or two years and then to cram with a private tutor for a few months in order to learn a handful of carefully chosen proofs and theorems by heart. It was in attempting to address these problems that Whewell, Peacock, and even Hopkins found common ground, but, as we shall shortly see, these three men actually held rather different views of the purpose of undergraduate study and the value of the methods employed by professional private tutors.

---

134. The events surrounding the introduction of analytical methods are discussed in detail in Becher 1995. Airy's *Mathematical Tracts* (1826) are an excellent example of the presentation of analytical methods in a broadly acceptable form.

135. On the analytical topics introduced in the 1830s, see Becher 1980a, 23.

In 1841, Peacock, now middle aged and firmly established as Dean of Ely and Lowndean Professor of Astronomy and Geometry in Cambridge, published a wide-ranging critique of undergraduate studies which included a lengthy and very strongly worded attack on private tuition. This "evil of the most alarming magnitude," he claimed, had substantially increased the cost of undergraduate study and, even more seriously, all but destroyed the influence of public teachers in the university.[136] Typical private tutors were, Peacock claimed, "young men of very limited attainments" who were "incompetent to convey to their pupils any correct or enlarged views of the subjects which they teach." Only through the intimacy of their relationship with their pupils were these men able to ensure that it was they, rather than the college lecturers, who would "generally be obeyed" in the regulation of studies. It is important to note that Peacock's criticism of private teaching was not merely that some young coaches were incompetent teachers—which might well have been the case—but that such teachers employed unnatural teaching methods and undermined the proper intellectual development of the most able students. He was prepared to acknowledge that William Hopkins had shown "great skill and pre-eminent success as an instructor of youth," but insisted, nonetheless, that "even in its most favourable form, the [coaching] system is one of forced culture, which, though it may accelerate the maturity of the fruit, is inconsistent with the healthy and permanent productiveness of the tree." Carefully contrived regimes of hard study deprived undergraduates of "leisure and freedom," thereby "interfering with the formation of habits of independent study and original research amongst the more distinguished students." The only remedy, Peacock concluded, was the "strict and preemptory prohibition" of all private teaching.[137] He also recommended that the university limit the number of examinable mathematical topics and that the Senate House examination be sat at the end of the eighth term, measures intended to reduce the pressure of competition and open the syllabus to more nonmathematical studies. It is a mark of the extent to which Peacock attributed the expansion of undergraduate mathematics to the influence of private tutors that he be-

---

136. Peacock 1841, 153–54. Peacock calculated that the total sum paid to mathematical coaches was three times that paid to all official college and university teachers. Note also that Peacock saw the rapid rise of private tuition in the late 1820s and 1830s as a "consequence" of the repeal of the regulations in 1824, and gave no indication that the regulations had previously been flouted.

137. Ibid., 154–56. Peacock conceded that the top coaches should somehow be absorbed into formal university teaching.

lieved these restrictions alone (barring a complete ban) would break their monopoly.[138]

These remarks reveal that although Peacock shared Whewell's worries that private teaching tended to undermine the moral dimension of public teaching and make undergraduate studies overly technical, he was even more troubled that private teaching was now undermining the research style of the most able students. To appreciate just how sharply Whewell and Peacock differed on this issue, it is important to remember that, in Whewell's scheme, undergraduate mathematics was supposed to be a closed and uncontentious discipline which instilled correct and logical habits of thought but otherwise left men's minds "passive and inert." [139] The fact that some young wranglers had begun to publish research papers and treatises on analysis in the early 1830s was therefore interpreted by Whewell as a very worrying sign because it indicated that they had completed their studies dissatisfied with their knowledge of mathematics. The effect of teaching advanced analysis — rather than the foundations of established mathematical principles — to undergraduates was, Whewell claimed, that it "rouse[d] them to speculate for themselves"; it was not therefore surprising (though highly undesirable), he concluded, that a young college fellow in his "season of leisure" and "with some skill in calculation," would try to "settle his own views" on various topics, and come to the belief that his understanding was "more clear" than were those of his "masters." [140] Peacock's ideal of undergraduate study, by contrast, was best illustrated by men like Airy and Ellis, students who attended college lectures but who had the knowledge and ability to pursue their own interests at their leisure with some guidance from a tutor or professor. Yet, as a professor himself, Peacock would have become acutely aware by 1840 that few among even the brightest students bothered to attend the professors' public lectures — they trusted entirely to their coaches.[141]

William Hopkins was not prepared to let these attacks on the teaching methods he had successfully pioneered pass unchallenged. In a pamphlet published in 1841 he revealed his own ideal of undergraduate studies by

138. Ibid., 151, 155. Peacock thought that it was the continued introduction of new advanced topics which prevented college teachers from competing with coaches.

139. Whewell 1835, 41.

140. Ibid. The effect of advanced analysis on the undergraduate mind was, Whewell claimed, that the "vessel not being ballasted by any definite principle, rolls into a new position of transient equilibrium with every new wave."

141. Whereas the lectures of young professors, such as Airy, aided the analytical revolution in the 1820s, attendance declined rapidly in the latter 1830s; see Becher 1980a, 37.

defending private teaching as incomparably superior to the public lectures offered in the university and colleges. Hopkins tacitly agreed with Whewell that analysis was of greatest pedagogical value when taught through its applications to the most mature branches of mathematical physics (especially optics and celestial mechanics). Only when a student approached such "great theories, as Pure Astronomy and Pure Optics" did he "fully appreciate the real importance and value of pure mathematical science, as the only instrument of investigation by which man could possibly have obtained to a knowledge of so much of what is perfect and beautiful." Hopkins also thought that only in the course of these advanced studies did the student "form an adequate conception of the genius which has been developed in the framing of those theories, and [felt] himself under those salutary influences which must ever be exercised on the mind of youth by the contemplation of the workings of lofty genius." Reducing either the range of advanced subjects or the duration of undergraduate studies would, Hopkins contended, undermine these positive influences. It was not until the final undergraduate year that the most able students really mastered advanced analysis and, he reminded the reader, Cambridge offered little incentive for students to pursue such studies after graduation.[142]

Hopkins differences with Peacock were brought most sharply into focus when he set out his view of the role of training as a preparation for research. Hopkins rejected the claim that private tutors had undermined public teaching in the university, claiming instead that they had supplied a need generated by the "absence of any continuous and well arranged system of Professorial tuition, and the inefficiency of the collegiate system" (Hopkins 1841, 7). He believed that the very system of training employed by the best private tutors had not only generated the high standard of mathematics in the university but had become a prerequisite to successful research. As he noted a decade later in a plan to reform mathematical education in Cambridge, far from cramming students so that they were saved the trouble of having to think for themselves, the coaches had developed a form of teaching that "presents to the student a logical exposition of the subject, combined with such detailed explanations and varied elucidations as shall meet his real individual wants, while it leads him to conquer his difficulties by the effort of his own mind" (Hopkins 1854, 33). Hopkins was loth in 1841 to claim outright that such systems were required by men of genius—he would certainly have been accused

---

142. Hopkins 1841, 10, 14. Hopkins estimated that there were scarcely a dozen public appointments where "higher mathematical attainment" was an "essential recommendation."

of underestimating the unique abilities of Newton—but he was nevertheless prepared to make the more cautious observation that it was "extremely difficult to define the extent to which even the highest order of genius may be aided by such systems." And when it came to that "scarcely less essential class of men" who explored the new fields of mathematical enquiry opened up by the genius, Hopkins felt sure that an efficient training system was now essential. Searching for the right words to describe the process by which "the most active students" were conveyed from elementary mathematics to "the last improvements, contained in memoirs, journals, and pamphlets" in just ten terms, Hopkins wrote that the "more distinguished of our students have been conducted, as it were, to an eminence from which they might contemplate the fields of original research before them; and some have been prepared to enter therein immediately after the completion of their undergraduate course" (1841, 7, 14–15). He felt it inappropriate to "particularise recent and sterling instances" of such researchers, but he no doubt had in mind such men as Duncan Gregory (5W 1837), who became the founding editor of the *Cambridge Mathematical Journal* just a year after he graduated, and several of his contemporaries who contributed to this and other research journals.[143] As far as Hopkins was concerned, carefully organized courses of hard and competitive study ought not to be seen as stifling the free spirit of the leisured and curious gentleman, but the proper and efficient way to train first-class researchers.

2.7. The Reform Coach and the Politics of Pedagogy

Whewell and Peacock's efforts to halt the proliferation of new analytical topics through reform came to fruition in the late 1840s. In order to encourage the majority of students to master elementary mathematics while preserving a place for some advanced topics, Whewell proposed a course of undergraduate study which separated what he termed the "permanent" curriculum (which would include geometry, mechanics, algebra, optics, and elementary aspects of the calculus) from the "progressive" curriculum (which would include higher algebra and geometry, analysis, analytical mechanics, and celestial mechanics). In 1846 these proposals were effectively implemented when the university Senate accepted the recommendation of a specially formed

---

143. On the research publications of wranglers between 1815 and the 1840s, see Becher 1980a and Grattan-Guinness 1985. Some notable researchers who graduated in this period were J. J. Sylvester, Duncan Gregory, George Stokes, Arthur Cayley, John Couch Adams, and William Thomson.

syndicate that the examination regulations be further reformed.[144] By the new regulations, which came into effect in 1848, the examination was extended from six to eight days and divided into two parts: the first three days were now devoted to examination of the elementary branches of mathematics, only students who achieved honors in this part being allowed, eight days later, to compete in the second five days devoted to the "higher parts of mathematics."[145] The Senate also adopted a recommendation by Peacock that a Board of Mathematical Studies be instituted which would assume responsibility for the content of the Tripos and prevent young moderators and examiners from introducing new topics at will. One of the first acts undertaken by the newly constituted board was to remove some advanced analytical topics as well as some physical topics—including electricity, magnetism, and heat—which were widely regarded as lacking "any axiomatic principles whatsoever."[146]

The reforms of 1848, which were broadly acceptable to all parties, restored stability to the Mathematical Tripos and finally introduced a formal and revisable syllabus of undergraduate mathematical studies. But, as Hopkins had predicted (1841, 7), they did nothing to resolve the tension between private, college, and public teaching. Indeed, by the mid 1840s, private teaching had so displaced college and professorial lectures in the university that Whewell had come to share Peacock's abhorrence of the practice. In 1845 Whewell published a further defence of a liberal education in which he once again highlighted the superior pedagogical value of classical geometrical constructions over what he saw as the mere symbolic manipulations of analysis. This time he pointedly contrasted geometrical *reasoning*, in which "we tread the ground ourselves, at every step feeling ourselves firm, and directing our steps to the end aimed at," with analytical *calculation*, in which "we are carried along as in a railroad carriage, entering in at one station and coming out of it at another, without having any choice in our progress in the intermediate space." The latter method, he concluded, "may be the best way for *men of business* to travel, but it cannot fitly be made a part of the gymnastics of education" (1845, 40–41). This barely metaphorical analogy between the algorithms of analysis and the contrived ease of railway travel throws further light on Whewell's now profound antipathy to coaching. Although there were signifi-

---

144. The political finagling which culminated in these reforms is discussed in Becher 1980a, 32–37 and 1995, 425.
145. Glaisher 1886–87, 16–17.
146. On the formation and early work of the Board of Mathematical Studies, see Wilson 1982, 337. On the status of electricity and magnetism circa 1850, see *Commissioners' Report*, 113. For a detailed list of advanced topics removed at this point, see Glaisher 1886–67, 20.

cant differences between the cultural meanings of the stagecoach and the railway carriage in the 1830s and 1840s respectively, there were senses in which they were similar. Coaches and trains carried their passengers ever more rapidly along carefully constructed pathways, and both were associated with the expansion of industrial and commercial values. In this sense, Whewell's comments on the railway carriage in the 1840s applied equally to the stagecoach in the 1830s. During the 1830s, moreover, the stagecoach had been adopted as a metaphor for the vehicle of Whig reform, a political cause associated explicitly with the mathematical reformers and professionalizers in Cambridge in the same period.[147] The mathematics coach—with his carefully contrived course of study, graded examples sheets, competitive exercises, and friendly, intimate style—was precisely the vehicle of the amoral, irreligious, and technically meritocratic training system that threatened to undermine Whewell's ideal of a liberal education. His analogy between mechanized transport and the nature of analytical calculation thus drew an appropriately disparaging parallel between the pedagogical shortcomings of higher analysis and the means by which it was being taught by men like Hopkins.

In the Whig-appointed Parliamentary Commission set up in the early 1850s to investigate undergraduate studies at Cambridge, Hopkins and Whewell clashed directly over private teaching. Whewell was loth to cooperate with the commission at all as he believed Parliament had no right to interfere in the internal affairs of the ancient universities. He nevertheless took the opportunity to express strong opinions on several issues, one of which was the evils of private teaching. Reiterating Peacock's remarks a decade earlier, Whewell described coaching as a pernicious practice which increased the cost of undergraduate studies and undermined the authority of college and university teachers. Private teaching "enfeebles the mind and depraves the habits of study," he wrote, and he flatly denied that the way to undermine coaching was, as some suggested, to bring the methods of the best coaches into the public system. This "would not be *improving*, but *spoiling* College Lectures," Whewell argued, as coaches lacked the "more general or dignified tone" of public teachers, not least because such teaching was "given in a familiar and companionable tone, without any restraint from the presence of others."[148]

In his evidence to the commissioners, Hopkins roundly denied Whewell's accusations that coaching was overly concerned with the mere technicalities

---

147. On the "reform coach," see Bentley 1996, 78–90. On the Whig sympathies of Cambridge mathematical reformers, see Becher 1980a and 1995, and Smith and Wise 1989, 65, 151–55, 168–92.
148. *Commissioners' Report*, 75.

of mathematics and lacked the breadth and moral content of public teaching. As we shall see in chapter 4, Hopkins considered a tough competitive training and examination in higher mathematics a much better guide to "high moral feeling" than were the platitudes and potential "hypocrisy" of philosophy and theology.[149] With senior members of the university divided in their support for Whewell and Hopkins—though broadly more sympathetic to the latter—the commissioners concluded that it would be impractical to ban private teaching.[150] Such a move, they felt, might undermine the standard of the Mathematical Tripos, would deprive smaller colleges of access to the finest teachers, and would be very hard to enforce. Following, in effect, Hopkins's line that a tough competitive examination in higher mathematics was the best vehicle for preserving appropriate standards in the university, the commissioners concluded that students should be left to choose whichever teachers they felt would give them the best chance in the Senate House.[151] Peacock, who played a leading role in the commission, had clearly softened his position on the relative merits of public and private teaching. He and his fellow commissioners still considered it inappropriate that private tutors had such a monopoly on undergraduate studies, but conceded that any serious reform of the public system would have to take account of the methods used by the best coaches. They therefore recommended that a new staff of "Public Lecturers" be formed whose job would be to teach students from all colleges using the "catechetical" methods of the private tutors.[152]

Encouraged both by the commissioners' vindication of good coaching practice and their apparent enthusiasm for change, Hopkins quickly published a detailed plan for the wholesale reform of public teaching in the university. His aim was to introduce the best coaching methods into public teaching in a way that would enable it to displace private teaching by virtue of merit. Hopkins began by establishing the long-standing superiority of the private tutor; he reminded his reader that it had been private tutors, "unauthorized and unrecognised, but possessing that native energy and power which the Public System has never manifested," who had accomplished the

149. Ibid., 248–49.

150. All those who gave written evidence to the commission were required to state their views on the merits of private tuition.

151. *Commissioners' Report,* 79. The commissioners also urged that the cost of private teaching be reduced and that there be more contact between private tutors and the colleges.

152. Ibid., 81. The term "Public Lecturer" was explicitly chosen to distinguish the interactive style of teaching from the traditional monologue of the professors. Whewell was particularly incensed by the proposal to tax the colleges to pay for an expansion of university teaching. See Williams 1991, 129.

analytical revolution, massively expanded the advanced mathematical content of undergraduate teaching, and established and preserved the high standard of merit in the Tripos examination. He added that, by the early 1850s, public teaching had become so inadequate that even non-honors men routinely resorted to coaches.[153] Despite his obvious pride in the accomplishments of private tutors, however, Hopkins freely admitted that it was not a healthy state of affairs that the teaching offered by the university had almost no influence whatsoever in the education of its undergraduates. He heartily supported the commissioners' suggestion that a new staff of Public Lecturers be appointed, but offered a number of suggestions based on his own years of experience: the classes would have to be small—not more that "fifteen or sixteen"—and composed of men of roughly the same ability; the annual stipends would need to be increased from the hundred pounds or so earned by a college lecturer to the eight or nine hundred pounds earned by a good coach; the students would have to begin attending the public lectures not later than the beginning of their second year (before they had become reliant on a coach); and the lectures would have to run for part of the long summer vacation, a period that the coaches had found especially productive in the third year. Hopkins calculated that the scheme could be funded through a combination of student fees and college contributions, and suggested that the Board of Mathematical Studies would be the proper body to appoint the new staff. He also pointed out that provided the staff was sufficiently large, there could be a useful division of labor between the advanced classes, each lecturer concentrating on his own areas of special interest. Finally, he insisted that no attempt should be made to suppress private teaching, a move which he felt would in any case be "regarded as inconsistent with the spirit of the age." He was convinced that "free competition with this Private Teaching would afford the very best test by which to try the merits of a Public System."[154]

Hopkins's proposal for the appointment of a new force of well paid, non-clerical, university mathematics lecturers, whose success would be assured through small classes, a catechetical style, and free-market competition, clearly displays the new ideals, practices, and reforming potential of the coaching movement. His differences with Whewell and Peacock on these matters also help to clarify the mixture of personal circumstance and wider political sympathy which shaped their respective positions. Appointed by Robert Peel to the Mastership of Trinity College in 1841, Whewell worked

---

153. Hopkins 1854, 6–8.
154. Ibid., 10–41.

from the heart of the Cambridge establishment to defend the Anglicanism and institutional autonomy of the university, and to preserve the place of permanent mathematical studies at the foundation of an elite liberal education. Whewell saw no virtue in Hopkins's schemes for training small classes of evenly matched men to high levels of technical proficiency. On the contrary, in Whewell's opinion large classes were necessary because they required lecturers to concentrate on metaphysical aspects of the foundations of geometry and mechanics, and enabled the "slowest and most thoughtless" students to recognize and learn respect for "better intellects."[155] Hopkins, also a Tory, had spent his career as a professional mathematics teacher who relied on student fees for his income.[156] He agreed with Whewell that the inspirational value of mathematics was greatest when it revealed the divine order of the physical universe, but, unlike Whewell, Hopkins saw this as justification for teaching the most advanced rather than the more elementary parts of mathematical physics. Hopkins also disagreed with Whewell's fundamental belief that the moral content of undergraduate teaching relied heavily on religious indoctrination through public lectures. For Hopkins, hard work, self-discipline, competitive study, and mathematical rigor provided a more reliable hallmark of moral virtue.

From a political perspective, Whewell's and Hopkins's respective approaches can be seen to display the breadth of Peelite Toryism from the mid 1830s to the early 1850s. As a conservative Anglican, Whewell resisted any directly utilitarian or commercial view of undergraduate mathematical studies which would weaken the moral and theological dimension of public teaching and encourage reformers to make the university more accessible to religious dissenters.[157] The opposition he saw between the kinds of mathematics fit for "men of business" and that fit for the "gymnastics of education" highlights the boundary he sought to preserve in undergraduate teaching between that which was merely useful and that which was true and permanent. It is noteworthy in this context that, as we saw above, both Whewell and Peacock used the term "culture" to describe the broad collection of values and sensibilities which they believed should lie at the heart of a liberal education. This notion of culture implied the possibility of moral, aesthetic,

---

155. Whewell 1837, 105–8. Whewell felt that students should understand the foundations of mathematics rather than take them on trust.

156. For evidence of Hopkins's Tory sympathies, see Smith and Wise 1989, 66.

157. On this conservative faction of Peel's supporters, see E. J. Evans 1996, 262. Whewell's relentless defence of the role of mathematics in a liberal education is best understood as part of a dialogue with reformers both within and beyond the university. See Williams 1991, 119–21.

and reasoned judgments according to criteria that transcended contingent social relations, especially those inherent in the emergent industrial and commercial worlds of early Victorian Britain.[158] What Whewell saw as the indisputable truths of Euclid's geometry and Newton's mechanics served this pedagogical ideal extremely well, but progressive topics such as the mathematical theories of thermodynamics and electromagnetism were a quite different matter. They not only lacked secure, axiomatic foundations, but held what Whewell would have regarded as increasingly unwelcome associations with industry and manufacture. On this showing, these progressive topics were to be regarded as *acultural* because they were neither permanent nor independent of their social context.

Hopkins, at least implicitly, defended a very different notion of "culture" and of the moral and technical role of mathematics in a pedagogical scheme. He represented an emergent, progressive, middle-class Toryism, which embraced a much more utilitarian and technically meritocratic view of mathematical studies.[159] In his address as president of the British Association in 1853, Hopkins explicitly linked Britain's continued industrial and commercial success to the proper application of both experimental and mathematical science. The "great duty" owed to the public by the British Association was, he urged, to "encourage the application of abstract science to the practical purposes of life—to bring, as it were, the *study and the laboratory into juxtaposition with the workshop*" (1853, lvii, my italics). Hopkins had also done more than anyone else to see that these values, built on a tough, competitive, technical training, actually permeated undergraduate mathematical studies in Cambridge.[160] Far from simply cramming his students for the Senate House examination, as his critics claimed, Hopkins was "conspicuous for encouraging in his pupils a disinterested love of their studies."[161] Indeed, he probably did more than any other contemporary mathematics teacher in Cambridge to encourage the most able students to embark upon original research. During the second half of the 1830s, Hopkins had established himself as an outstanding researcher in the mathematical theory of physical geology, and the style of mathematical physics he developed in

---

158. On the use of the term "culture" as a collection of values above the social relations of industrializing Britain, see Williams 1958, xviii, and Golinski 1998, 163–64.

159. Robert Peel was described as an "ultra whig" in his dealings with the universities in 1845. See Smith and Wise 1989, 47.

160. Becher (1984) reveals that during the period 1830–60 more than half of all wranglers came from middle class and professional backgrounds.

161. See entry on Hopkins in the *DNB*.

these investigations provided an important model for those of his students who embarked upon original investigations of their own.[162] Ever willing to encourage his students in this direction, it was Hopkins who suggested to the young George Stokes in 1841 that he begin research in hydrodynamics, and who, in 1845, gave the young William Thomson a copy of Green's *Essay* on electricity and magnetism which, within days, had proved the inspiration for the method of electrical images.[163]

Hopkins's views on private teaching also make an interesting comparison with those of Peacock, the political Whig. Peacock too was in favor of including higher mathematics in undergraduate studies and in encouraging the more able students to engage in research, but agreed with Whewell that undergraduate teaching should preserve a broad moral and theological component and encourage students to develop at their own pace. In 1841, he was vehemently opposed both to private teaching and to the "increased severity of the examinations" as, in his opinion, they encouraged those students "whose industry is rather stimulated by their fears than by their ambition and love of knowledge" (1841, 162). Hopkins would have objected against this elitist and aristocratic vision of mathematical studies, that it failed to take account of the sheer volume and level of mathematics which, by the 1840s, students had to master in order to embark upon research directly after graduation.[164] He would have agreed with Peacock that students ought to attend the professors' lectures, but, he would have added, students had first to be thoroughly drilled in mathematical technique if they were to find such advanced monologues comprehensible.[165] The remarks on private teaching made by Peacock and his fellow commissioners in 1853 were a tacit admission that Hopkins's view had prevailed.

The attacks on private teaching mounted by Whewell, Peacock, and others during the 1840s and 1850s have generally been accepted with little or no criticism by recent historians of Cambridge mathematics.[166] These attacks appear, at least superficially, to accord with a widely held belief that rigorous

---

162. On Hopkins's style of mathematical physics and its legacy to his students, see Smith and Wise 1989, 195–98.

163. Stokes 1907, 8; Smith and Wise 1989, 215–16, 231.

164. It is a measure of the raised standard of undergraduate mathematics by the mid 1840s that William Thomson could criticize Airy for making mathematical errors that would now be "repulsive to a 2nd or 3rd year man." See Smith and Wise 1989, 769.

165. Even in 1841 Peacock (1841, 155) had been prepared to admit, as had Whewell, that private teaching during the first three terms was a very effective way of bringing students to a common level of mathematical knowledge.

166. See, for example, Searby 1997, 132.

preparation to solve cunningly contrived problems trivialized the study of mathematics and stifled that spirit of playful curiosity often seen as crucial to original research. The many outstanding mathematicians and mathematical physicists who emerged from Cambridge during the Victorian era are understood to have succeeded *despite,* rather than *because of,* the remarkable training they received. I shall argue in chapter 5 that this view of the relationship between training and research is deeply flawed, but for the moment I wish merely to draw attention to the fact that early Victorian criticism of coaching does not support such a view. Although Whewell and Peacock had serious reservations concerning the technical skills and educational values nurtured by private teaching, their objections were not those of a modern pedagogue. They believed that coaching undermined the moral and spiritual authority of college and public teachers, subverted the ideals of a liberal education, and left no leisure for able undergraduates to pursue their own intellectual interests. William Hopkins, by contrast, not only espoused a highly competitive, research orientated, and utilitarian vision of higher mathematics, but helped to develop the training techniques through which his vision could be realized. It is, moreover, variations on these training methods (discussed further in chapter 5) that have remained throughout the twentieth century at the heart of mathematical pedagogy in elite secondary and tertiary education in Britain.

2.8 Free Trade and the Evolution of Mathematicians

Hopkins's career had been devoted almost entirely to the training of the mathematical elite, and his 1854 plan for the reform of teaching in Cambridge, written as he approached retirement, was intended to bring the training methods he had pioneered within the formal structure of the university. In his plan he borrowed Whewell's terminology by differentiating between "elementary" lectures, intended for those who had little chance of advancing beyond the first three days of the examination, and "progressive" lectures which would cover the subjects of the last five days. In Hopkins's scheme it was the latter that would constitute the main work of the new lecturers. This extension of the meaning of the word "progressive," to cover not just the advanced mathematical topics to which original contributions might still be made but also the methods by which such topics were most effectively taught, captured those achievements of the full-time coach that reformers hoped the public teachers of the university would emulate. The confluence of opinion between Hopkins and Peacock on the best method of preparing students to undertake research in the 1850s can be read in political terms to reflect the progressive and reforming sympathies of the Whig-Peelite coalition of this period.[167] The Parliamentary Commissioners had themselves used the language of free trade in recommending that coaching should not be suppressed and that students should be left to choose whichever teachers they thought would best prepare them for the Senate House. Hopkins's vision of the content, form, and purpose of undergraduate mathematics, together with his suggestion that market forces be used to stimulate competition not only between the students but between public and private teachers, would have appealed strongly to these sentiments. In practice, however, his scheme for the wholesale reform of undergraduate teaching was too ambitious, even with support from the commissioners, to be implemented in the 1850s.

Hopkins's free-market defence of private teaching continued, nonetheless, to find defenders in Cambridge well into the 1870s. The commercial and political ideals from which his arguments drew credibility were propagated through the Liberal administrations of the mid-Victorian era, bolstered by the economic stability and prosperity of the period. With the rise in popularity of Darwin's theory of evolution in the 1860s, moreover, the defence of coaching and competitive examination shifted, perhaps predictably, from the grounds of competition and free trade to those of natural selection and the survival of the fittest. Evolutionary theory in the form of social Darwinism appealed to the same "ultra-Whig" constituency that had originally found Hopkins's arguments on free-market teaching so persuasive.[168] An excellent illustration of a debate conducted in precisely these terms is provided by an exchange on the merits of private teaching between two senior wranglers, Isaac Todhunter (SW 1848) and P. G. Tait (SW 1852), both of whom had been pupils of Hopkins. By the early 1870s, the paper-based examination system pioneered in Cambridge was beginning to pervade not just the secondary and tertiary education systems in Britain, but many areas of civil, professional, and military selection as well.[169] Concerned that the proliferation of examinations and examination boards was rapidly marginalizing the teacher's traditional role as assessor not just of a student's technical proficiency but also of his character, potential for leadership, and power of originality, Tait ar-

---

167. On the Whig-Peelite coalition in the 1850s, see Parry 1993, chap. 8, and Bentley 1996, chap. 4.
168. On the political reception of Darwin's theory by the ultra-Whigs, see Desmond and Moore 1992, 267.
169. On social Darwinism in the late Victorian education system, see Mangan and Walvin 1987. On the spread of science examinations in Victorian England, see MacLeod 1982.

gued, in an article published in 1872, that examinations constituted a form of "Artificial Selection." Just as animal breeders produced "fancy tulips and pigeons" rather than robust species when they interfered with the process of natural selection, so, Tait argued, impersonal examinations were quite unsuited to selecting students with proper "fitness for positions of trust and difficulty." Indeed, in Tait's opinion the question of how fitness for responsible office was to be properly assessed had become "enormously more important to the future of the empire than any mere speculation in science" (Tait 1872, 416).

Tait also noted, pursuing the evolutionary theme, that the prevalence of written examinations had led to the "development of a new and strange creature, the 'Coach.'" This "mysterious creature," he continued, "studies the Examiner, feels his pulse as it were from time to time, and makes a prognosis (often very correct) of the probable contents of the papers to be set" (1872, 419). Tait's critique, in short, was that examinations were an artificial test of student ability and that private tutors exacerbated the problem by preparing students simply to pass examinations. Isaac Todhunter, who had twice acted as an examiner for the Mathematical Tripos and had for some years been an outstanding coach in Cambridge, broadly agreed with some of Tait's comments on the dangers of poorly monitored examinations and bad teachers, but he was outraged by Tait's imputation that Cambridge mathematics coaches were paid to divine the contents of an examination which was, in any case, ill suited to test the true qualities of the candidates. In a thoughtful and carefully measured reply to Tait he pointed out that the Mathematical Tripos (now but one of four Triposes in Cambridge) no longer functioned as a general education for the nation's elite, but was intended for the "specific production of mathematicians." The rapid expansion of secondary and tertiary education in Britain and her empire from the late 1860s was providing a steady supply of jobs for high wranglers, and if one granted that "the aim of [the Cambridge] system [was] to produce eminent mathematicians, and to evolve them annually in the order of merit" then, Todhunter suggested, one had surely to admit that "the process is attended with conspicuous success" (Todhunter 1872, 63).

On the specific question of the efficacy of coaching as a method of training mathematicians, Todhunter made further appeal to Tait's own apparent enthusiasm for Darwinian evolution. The singular "peculiarity" of the Cambridge coaching system that ought on these grounds to recommend itself to Tait was "that it illustrates in a striking manner what may be called the principle of *Natural* Selection." Echoing Hopkins's disparaging remarks on the inefficiency of public teaching in the university earlier in the century, Todhunter observed that while the appointment of professors was a complicated, inexact, and sometimes even corrupt process, "the private tutor gains pupils neither by accident nor by favour, but by the sole recommendation of his ability to teach." Todhunter neatly summed up the Darwinian interpretation of this process with the remark: "many enter on the occupation; comparatively few attain eminence: these, I presume, exemplify the law of the survival of the fittest" (1872, 67).[170]

The brief exchange between Tait and Todhunter—of which the latter certainly got the better—represented in some respects a mid-Victorian expression of the earlier debate between Whewell and Hopkins. Tait saw examinations as an "artificial" form of selection because they took no account of the broader qualities of the candidates and because they seemed to empower the examiner rather than the professor. Coaching as a form of teaching was similarly artificial in his opinion because its primary aim was simply to secure students a high place in the examination for which they were being prepared. Referring directly to the case of Cambridge, Todhunter responded to Tait by pointing out that the purpose of a mathematical education had gradually changed since the early 1850s along the lines practically and vociferously supported by Hopkins. Todhunter informed Tait that he ought not to be surprised that it had become rare for high wranglers to become bishops, judges, and leading statesmen since the Mathematical Tripos was now regarded by many as a training for mathematicians. He also made the important point that, in Cambridge at least, many of the leading private tutors "may fairly rank with the professors of any university in Britain" (1872, 66). Here Todhunter had in mind men like Hopkins and the great mid-Victorian coach, Edward Routh, both of whom had made outstanding original contributions to mathematical physics and encouraged their best pupils to do the same. Far from being crammed by narrow and second-rate teachers as Tait implied, the most able undergraduates in Cambridge were much more rigorously trained and closely supervized by a first-rate mathematician than were mathematics students at any other institution in Britain.

Todhunter was nevertheless prepared to admit that there remained serious problems in the Cambridge system, and those he considered the most pressing were very similar to the ones identified by Hopkins twenty years earlier (1872, 68–69). The first was to find a way of bringing the coach's suc-

---

170. Todhunter also claimed that "there is scarcely a distinguished mathematician in the university who has not at some period been engaged in private tuition" (1872, 61).

cessful training methods—perhaps even the coaches themselves—within the public-teaching system of the university. As we have seen, the combination of competitive, written examination and expert private teaching had generated new levels of technical proficiency among the best students, but college and public teachers had played very little part in the process. The industry and skill exhibited by the best coaches were understood by many to be a product of an educational free market or natural selection process, and generating a body of comparably industrious and competent Public Lecturers would not, as Hopkins had discovered, be an easy matter. The majority of officially sanctioned teaching in the university remained college-based, and the only truly public teachers, the professors, had no desire to expand their lecturing duties or to assume responsibility for tutoring students in small groups. The second, and closely related problem, was that of actively encouraging and enabling mathematical research in the university. Although private tutors like Hopkins, Routh, and others were able to exert enormous influence over their pupils, they only taught undergraduates, and their complete dominance of teaching meant that professors and college fellows had little opportunity to stimulate interest and expertise in their own special fields of research. Once the top wranglers graduated and took up college fellowships, they therefore received official guidance neither from their coaches nor from more senior members of the university. What was required was a system of public lectures (at undergraduate and graduate level) delivered in the companionable, interactive, and progressive style of the coaches, yet covering a wide range of specialist and potentially researchable topics. As we shall see in chapter 5, the process by which this was gradually accomplished in the late Victorian period involved a mixture of Parliamentary intervention and initiatives by reformers within the university.

~ 3 ~
A Mathematical World on Paper
The Material Culture and Practice-Ladenness of Mixed Mathematics

> We are perhaps apt to think that an examination conducted by written papers is so natural that the custom is of long continuance.
> W. W. ROUSE BALL, 1889 [1]

3.1. The "Familiar Apparatus" of Pen, Ink and Paper

When James Clerk Maxwell delivered his inaugural lecture as the first Professor of Experimental Physics in Cambridge in 1871, he contrasted the new instruments and machines that would soon occupy the proposed physical laboratory with what he referred to as the "familiar apparatus of pen, ink, and paper" (Maxwell 1890, 2:214). Having been employed to deliver courses on experimental physics, a new departure for the university, Maxwell sought to reassure his audience that public provision for experimentation was now a necessity both for teaching and for research. Henceforth, students preparing for the Mathematical Tripos would be required not merely to master higher mathematics and physical theory but to exercise their "senses of observation" and their "hands in manipulation," activities that smacked too much of the artisan and the workshop to endear them to many of his new Cambridge colleagues. Pressing home his defense of the soon-to-be-built Cavendish Laboratory, Maxwell insisted that students and teachers of experimental physics would "require more room than that afforded by a seat at a desk, and a wider area than that of the black board." Maxwell's analogy between the material apparatus proper to mathematical training and that

---

1. Rouse Ball 1889, 193.

required to teach experimentation was ingenious. Few would deny, once they were reminded of the fact, that pens, paper, desks, and blackboards were everyday tools of the mathematician's trade; it followed that if the university wished to continue to "diffuse a sound knowledge of Physics" (214) it would have to accept the physical laboratory despite its connotations of industry and manufacture.[2]

Maxwell's rare allusion to the material culture of mathematical training provides an appropriate point of departure for this chapter. Historians of science have recently begun to explore the remarkable changes in scientific practice wrought by the so-called laboratory revolution of the late Victorian era, yet very little scholarly attention has so far been paid to the changes in *theoretical* practice wrought by the introduction of written examinations in mathematics a century earlier.[3] This may sound a somewhat surprising statement given the several scholarly studies that exist on the origins of written examination in Cambridge, but these studies are mainly concerned with the wider sociopolitical significance of competitive assessment rather than with the effects of its style and material medium on the content and practice of mixed mathematics. Teaching and assessing mathematical disciplines on paper now seems such a self-evidently correct practice that it is difficult to imagine that it was once otherwise. Even by Maxwell's day paper-based learning in mathematics was so commonplace that he could make a point simply by acknowledging the fact; and, as Rouse Ball remarked in the epigraph above, these pedagogical practices appeared so "natural" to mid-Victorian mathematicians that they were assumed, when considered at all, to have been in use for centuries.

At the beginning of the nineteenth century, however, the whole apparatus of paper-based learning and examining was not only new in Cambridge but seemed as peculiar to undergraduates as would galvanometers and Wheatstone's Bridge to Maxwell's first students at the Cavendish. And like the laboratory revolution, the introduction of written examinations dramatically altered the skills and competencies prized in undergraduate study. As we saw in chapter 1, mathematics had always been part of the education offered by the ancient universities but had generally been learned only to an elementary level and in connection with the practical use and construction of mathematical instruments. By the early Victorian period the written

---

2. On resistance to a physical laboratory in Cambridge, see Warwick 1994.
3. For an exception to the lack of work on the material apparatus of mathematical practice, see Brock and Price 1980. On the laboratory revolution, see Gooday 1990.

examination and its attendant pedagogical apparatus had rendered the study of mathematics a purely paper-based activity, and, as we saw in chapter 2, were producing young graduates whose technical mastery of mathematical physics far outshone the capabilities of earlier generations of students.

In the first half of this chapter I explore the impact and development of paper-based learning and examination in Cambridge from the perspective of some of the students first exposed to such practices. Their testimony is especially valuable as it reveals both the profound shift in student sensibilities that accompanied the introduction of written examinations as well as the change in the content of undergraduate studies prompted by the new culture of working on paper. Our understanding of the nature of college and public teaching circa 1800 is surprisingly thin, and the sources from which the history can be written are sparse. We must assume that university professors and college lecturers taught mainly by emulating their own teachers and by introducing whatever innovations seemed appropriate. What can be learned regarding the actual content and conduct of mathematical lectures and examinations has to be reconstructed from occasional remarks in the memoirs, correspondence, and diaries of students and teachers, and from occasional descriptions of important Senate House ceremonies published by the university.

My analysis of these sources will focus on two specific and closely related aspects of the shift from oral to written examination during the early decades of the nineteenth century. The first concerns the very different personal and intellectual qualities tested by oral and written examinations respectively, and the ways in which student sensibilities were gradually reshaped to enable them to deal effectively with the latter form of assessment. The second concerns the introduction of what Maxwell referred to as the "apparatus of pen, ink, and paper" to mathematical education. As I have already noted, these now seem the obvious tools for teaching a mathematics-based discipline, but it was only with the introduction of written assessment that paper-based techniques of calculation began routinely to be employed in undergraduate study. Prior to this, mathematical education, like other branches of study institutionalized in the medieval university, was based mainly on private reading, oral debate, and catechetical lectures. The introduction of paper-based mathematical study gradually displaced these long-established pedagogical traditions, instituting instead tough regimes of competitive technical training. An important aspect of these regimes was the requirement that students be able both to *write out* their mathematical knowledge from memory and to solve difficult mathematical problems on paper. This aspect of the shift from oral to written study represented much more than a mere change in the medium by which mathematics was learned. Paper-based problem solving had hitherto been a technique employed by those engaged in original mathematical investigations, and many of the problems in analytical geometry and dynamics that gradually became standard student exercises were ones that had originally challenged the greatest mathematicians of the eighteenth century. The methods of mathematical reasoning proper to the solution of such problems could not easily be mastered by oral debate, but required long and assiduous practice on paper. The introduction of paper-based learning and examination therefore marked a major and profound period of cultural transformation in the history of the exact sciences, during which the working practices used by the founders of modern mathematical physics became central to undergraduate training in the universities.

In the second half of the chapter I explore the changing style and content of Tripos-examination questions within the new regime of private teaching and paper-based study. The first questions to which written answers were required were relatively simple exercises taken more or less directly from standard treatises. But, during the early decades of the nineteenth century, and especially during the analytical revolution, these questions gradually became both more technically demanding and more original. By the 1830s, the examiners were routinely drawing upon their own research to construct the most difficult problems, while students were increasingly assessed by their ability to solve such novel problems under severe constraints of time. These developments played a major role in altering the relationship between the books from which students learned and the examination questions by which they were assessed. Where examination questions had originated as examples from mathematical and natural philosophical treatises, a new generation of what might properly be called mathematical "textbooks" incorporated large numbers of these questions as exercises for the reader. This convergence between training and assessment reinforced the ideals of competitive written examination within an emergent textbook tradition and provided a virtually endless supply of exemplars for those who aspired to master mathematics and mathematical physics in the Cambridge style. These developments were to some extent an instantiation of pedagogical changes taking place at several European institutions during this period, and I conclude by considering the extent to which new forms of undergraduate training altered not only the academy but the wider study of mathematical physics.

3.2. Oral and Written Cultures of Examination

We saw in chapter 2 that the introduction of written examinations played a major role in changing the nature of undergraduate studies in Cambridge in the late eighteenth century. These examinations were instrumental in the rise of private tutors, provided what came to be seen as an impartial, efficient, and objective means of assessing student ability, and gradually altered the entire pedagogical culture of the university. But, despite their importance, we actually know surprisingly little about precisely when and why they were introduced. The first definite evidence that written answers to questions were being required in the Senate House examination dates from the early 1770s, but it is quite possible, even likely, that such answers had by this time been required for a decade or more.[4] The *reasons* for the introduction of written examinations are similarly elusive. It is generally agreed that the Senate House examination was conducted orally until at least the middle decades of the eighteenth century, but thereafter the extant evidence reveals only that written answers to questions gradually became an increasingly important mechanism for assessing the relative performances of the most able candidates.

One factor which almost certainly played a major role in this process was the high status enjoyed by Newton's works in Cambridge.[5] As I noted in chapter 2, it was originally in the form of natural theology that Newtonian science entered undergraduate studies, but from the middle decades of the century more technical aspects of his mathematics and mechanics began to be emphasized.[6] Part of the explanation for this important shift probably lies in the appearance in the 1740s and 1750s of a large number of introductory treatises on the fluxional calculus.[7] Although it was Newton who had originally invented and developed the fluxional calculus, he neither published an introductory text on the subject nor showed how it could be systematically applied to physical topics in natural philosophy. In the first half of the eighteenth century, Newton's mechanics and theory of gravitation was therefore taught in Cambridge in the geometrical form in which it was presented in the *Principia*, while optics, astronomy, pneumatics, and hydrostatics were

---

[4]. Rouse Ball 1889, 190. The first definite evidence comes from John Jebb's (2W 1757) account (reproduced by Rouse Ball) of the Senate House examination of 1772.

[5]. On the role of Newton's works in eighteenth and early nineteenth-century Cambridge, see Gascoigne 1988 and 1989, and Yeo 1988.

[6]. On the difficulties experienced even by the best Cambridge mathematicians in understanding mathematical aspects of Newton's works, see Stewart 1992, 101–8.

[7]. These books are discussed in Guicciardini 1989, chapter 4 and Appendix A.

generally taught as qualitative and experimental subjects.[8] The new books which appeared in the mid-eighteenth century offered a systematic introduction to the fundamental operations of the fluxional calculus and showed how it could be applied to the solution of a wide range of mathematical and physical problems.[9] These books appealed widely to the many so-called philomaths who frequented the coffee houses and mathematical societies which had sprung up in several English towns, and who regularly sent solutions to the tricky mathematical problems posed in such periodicals as the *Ladies' Diary*.[10] The strongly problem-orientated presentation in the treatises may well have been intended specifically to appeal to this audience, but this orientation also made it much easier for university students to master the fluxional calculus and its applications, helped to define a new field of mixed-mathematical studies, and provided Cambridge examiners with a ready supply of standard problems.[11] And unlike problems in natural theology, natural philosophy, and even elementary Euclidian geometry and Newtonian mechanics, the solutions to complicated problems involving the calculus were far more readily worked out on paper than stated or debated orally.

The earliest known examples of Senate House problems to which written answers were required date from the mid 1780s, but we can gain some insight into the nature and content of written examinations in Cambridge in the late 1760s from the first papers set for the Smith's Prize examination.[12] First held in 1769, this competition took place two weeks after the Senate House examination and was contested by a small handful of the top wranglers. It was intended, as I noted in chapter 2, specifically to test the candidates' knowledge of natural philosophy and mathematics.[13] From the first year the examination was held, moreover, the candidates were required to submit at least some of their answers in written form.[14] Examples of what appear to be

---

8. Guicciardini 1989, 65.

9. For a summary of the kinds of problems solved in these books, see the table of contents of William Emerson's *The Doctrine of Fluxions* (1743, 2d ed. 1757), reproduced in Guicciardini 1989, 143–46.

10. On the role of Newton's works in coffee-house culture, see Stewart 1992, 143–51.

11. Guicciardini 1989, 22–23, 64.

12. The problem papers for 1785 and 1786 are in the Challis papers (RGO-CUL) and are reproduced in Rouse Ball 1889, 195–96.

13. The written form of the Smith's Prize examination was based upon the Senate House examination and the fellowship examinations at Trinity College (discussed in section 3.3).

14. Commenting on the character of the examination as conducted by one of the first examiners, Edward Waring (SW 1757), an anonymous obituarist remarked that the students were occupied "in answering, *viva voce,* or writing down answers to the professor's questions." This com-

problem papers (with answers) set for the years 1769–72 provide a useful indication of the breadth and level of mathematical knowledge of the most able questionists (fig. 3.1).[15] Assuming that the professors who set the questions had an accurate sense of what the top three or four wranglers might reasonably be expected to tackle, these students must have had a very sound knowledge of arithmetic, geometry, algebra, pneumatics, optics, acoustics, astronomy, and Newton's *Principia,* as well as an advanced knowledge of the operations of the fluxional calculus and its application to mathematical and physical problems.[16] The Smith's Prize papers also indicate—as do most of the earliest-known Senate House question papers—that it was problems involving the fluxional calculus that were the most technically demanding and that most necessitated the use of paper and pen. It should be kept in mind that candidates for the Smith's Prize examinations were the very best of each year and would probably have gained their advanced mathematical knowledge through private study guided by one or more college fellows. The vast majority of students would have had a much narrower and more elementary grasp of mixed mathematics.[17]

The *content* of the questions to which written answers were first required was by no means the only factor driving their introduction; other, more

---

mentator also judged that perhaps no other institution in Europe "affords an instance of so severe a process" (Anon. 1800, 48).

15. RGO 4/273 (RGO-CUL). The first page of this small collection of sewn sheets contains a list of candidates for the Smith's Prize for the years 1769–75 and 1777 (the candidates for 1777 being wrongly listed as those for 1776). The following pages contain questions (with answers) marked for the years 1769–72. It seems extremely likely that these questions were used in the Smith's Prize examination for those years (though they are not explicitly marked as such). The collection is kept in the Neville Maskelyne papers, but the handwriting strongly suggests that they were written by Edward Waring (whose papers are also in the collection), who was one of the examiners for these years.

16. The answers sketched in fig. 3.1 apply to the following questions : (13) To find the cube roots of 8. (14) What are the Geometrical, Arithmetical and Harmonical means between a and b? (15) To form an equation whose roots are 1, 2, 3, 4. (16) What surd multiplied by $\sqrt{2} + \sqrt{3}$ will make the product rational? (17) To find the length of a parabolic curve. (18) To determine the modular ratio. (19) Suppose the point A moves in the right line AT, and draws the body B by the string AB or TP of a given length, what curve will B describe? (20) To find the nature of the harmonical curve which a musical string forms itself into. (21) What was the origin of Fluxions, how does it differ from Archimedes's method of Exhaustions, Cavalieri's method of indivisibles, Dr. Wallis's Arithmetic of Infinites; and what advantages has it over them? (22) What force is necessary to make a Body describe a Conic Section? (23) The velocity, center of force and direction being given, to find the Trajectory described.

17. The questions just referred to are certainly more difficult than those that appear more than a decade later on the earliest known Senate House papers (note 12). Questions 19 and 23 (fig. 3.1), for example, require a higher knowledge of fluxional calculus and Newton's *Principia* respectively than do similar Senate House questions from the mid 1780s.

13. Let the cube root of $8 = 2+x$ then $8 = 8+12x+6x^2+x^3$ And $x^3+6x^2+12x=0$, one value of x is 0, And $x^2+6x+12=0$, $x = -3 \pm \sqrt{9-12} = -3 \pm \sqrt{-3}$ Therefore the three roots are $2, -1+\sqrt{-3}$ And $-1-\sqrt{-3}$.

14. The arithmetical mean is $\frac{a+b}{2}$, the Geometrical $\sqrt{ab}$ and the harmonical $\frac{2}{\frac{1}{a}+\frac{1}{b}} = \frac{2ab}{a+b}$

15. $1+2+3+4 = 10$, $1\cdot2+1\cdot3+1\cdot4+2\cdot3+2\cdot4+3\cdot4 = 35$, $1\cdot2\cdot3+1\cdot2\cdot4+1\cdot3\cdot4+2\cdot3\cdot4 = 50$, $1\cdot2\cdot3\cdot4 = 24$ And the Equation is $x^4 - 10x^3 + 35x^2 - 50x + 24 = 0$.

16. $\sqrt{3} - \sqrt{2}$

17. Cotes Schol. gener.

18. Cotes Logom:

19. $AO = x$, $OP = y$, $PT = a$ then $PT^2 = y^2 \times \frac{\dot{x}^2+\dot{y}^2}{\dot{y}^2} = a^2$ And $y^2\dot{x}^2 + y^2\dot{y}^2 = a^2\dot{y}^2$
$\dot{x}^2 = \frac{a^2\dot{y}^2 - y^2\dot{y}^2}{y^2}$ | $\dot{x} = \frac{\dot{y}}{y}\sqrt{a^2-y^2}$ of the 3d Form.
$x = \sqrt{a^2-y^2} - a\left|\frac{R+T}{S}\right|$, where R, T, S are as $a, \sqrt{a^2-y^2}, y$.

20. Smiths Harmonics.

22. If the force tends to the focus it ought to be inversly as the Square of the distance.

23. Let the Velocity be such a could be acquired by falling down CJ, towards the center S, let fall CJ perpendicular to the direction PJ and make $TH = CJ$, then H & S are the two focus's and SC the major Axis of the conic Section.

FIGURE 3.1. Outline solutions to Smith's Prize problems, probably by Edward Waring (SW 1757). The relevant questions are reproduced in footnote 16. The precise purpose of these solutions is unclear, but there were probably used either as an aide mémoire while marking examination scripts or as model answers by other examiners or Waring's own students. These are possibly the earliest extant examples of mathematical problems for Cambridge students to which written answers were required. RGO 4/273 (RGO-CUL). (By permission of the Syndics of Cambridge University Library.)

practical, considerations were also in play. During the latter eighteenth century, increasing student numbers together with greater emphasis on finely grading and ranking the candidates gradually made oral examination an excessively time consuming business and one that was vulnerable to accusations of partiality by the examiners (who might also be college or private tutors). The practice of reading out problems to which written answers were required eased the load of the examiners and facilitated a more impartial system of evaluation. Instead of engaging each student in lengthy discussion, an examiner could dictate a question to a large group of students and leave them to write out the answers as quickly as they could. Another important step in the same direction was taken in the late 1780s with the introduction of sheets of written problems which were handed out to the most able candidates. These sheets were taken by the student to a convenient window-seat in the Senate House, where he wrote out as many solutions as he could in the time allotted. After 1791 these problems sheets were printed and, in 1792, specific numbers of marks began to be allocated to specific questions.[18]

The gradual shift from oral to written examination described above should not be understood simply as a change in the *method* by which a student's knowledge of certain subjects was assessed. The shift also represented a major change both in *what* was assessed and in the skills necessary to succeed in the examination. Consider first the major characteristics of the oral disputation. This was a public event in which a knowledge of Latin, rhetorical style, confidence in front of one's peers and seniors, mental agility, a good memory, and the ability to recover errors and turn the tables on a clever opponent were all necessary to success. The course of the examination was regulated throughout by its formal structure and by the flow of debate between opponent, respondent, and moderator. The respondent began by reading an essay of his own composition on an agreed topic and then defended its main propositions against three opponents in turn according to a prescribed schedule (fig. 3.2).[19] The public and oral nature of the event meant that the processes of examination and adjudication were coextensive. The moderator formed an opinion of the participants' abilities as the debate unfolded and the nature and fairness of his adjudication were witnessed by everyone present. The examination was also open-ended. The moderator could prolong the debate until he was satisfied that he had correctly assessed the abilities of opponents and respondents, and would generally quiz the respon-

---

18. Gascoigne 1984, 552.
19. For a account of a disputation in the latter eighteenth century, see Rouse Ball 1889, 166–69.

FIGURE 3.2. A rare depiction of a disputation as seen by the audience. The student being examined (the "respondent") is shown (right) reading out his essay. On the left are the first "opponent" (below), who will shortly oppose the propositions advanced by the respondent, and the moderator (above). The second and third opponents are shown sitting in the background, one listening to the respondent's essay, the other apparently reading his notes. Huber, *English Universities*, 1843. (By permission of the Syndics of Cambridge University Library.)

dent himself. Finally, once the disputation was completed, the only record of the examination was the recollections of those involved.

The economy of a written examination was quite different in several important respects. First, the oratorical skills mentioned above became irrelevant, as the examination focused solely on the reproduction of technical knowledge on paper and the ability to marshal that knowledge in the solution of problems. Second, the process of examination was separated from that of adjudication, a change that destroyed the spontaneity of the public Act. Without the formal procedure of a disputation and the rhythm of debate between opponent and respondent, each candidate was left to work at his own pace with little sense of how well or how badly he was doing. And, once the written examination was completed, the student had no opportunity either to wrangle over the correctness of an answer or to recover errors. The advent of written examinations also brought substantial change to the role of the moderators. Instead of overseeing and participating in a public debate, they became responsible for setting questions, marking scripts, and policing the disciplined silence of the examination room (fig. 3.3). This brings us to a third important difference between oral and written assessment. As I noted above, setting identical questions to a group of students made it much

FIGURE 3.3. A written examination in progress in the Great Hall of Trinity College, circa 1840. This rare, possibly unique image, strikes a powerful comparison with figure 3.2. Oral wrangling has given way to paper-based problem solving while the moderator's role has altered from adjudicating a public disputation to preserving the disciplined silence of the examination room. The status of the shadowy figures watching the examination from the gallery is unclear, but they may represent a transitional remnant of the audience at a public disputation. Huber, *English Universities*, 1843. (By permission of the Syndics of Cambridge University Library).

easier directly to compare and rank the students, especially if each question was marked according to an agreed scheme. Furthermore, the written examination scripts provided a permanent record of each student's performance which could be scrutinized by more than one examiner and reexamined if disagreements emerged. These differences between written and oral assessment altered the skills and competencies required of undergraduates and completely transformed their experience of the examination process. In the balance of this section I explore that changing experience over the first half of the nineteenth century.

It is a measure of the relative novelty yet rapidly increasing importance of written examinations circa 1800 that this was the moment when the university published the earliest firsthand accounts of the manner in which they were conducted. In 1802, the annual Cambridge calendar, which described the constitution and customs of the university, provided a detailed commentary on the progress of an examination in the Senate House. Unlike earlier accounts of this examination, known from student correspondence and private journals, this one focused not on the *oral* questioning of candidates but on the peculiarities of providing written answers to questions. Following some brief remarks on the ceremony that announced the start of the examination, the anonymous commentator discussed the seating of the various classes of students (as determined by the disputations) at appointed tables, emphasizing that upon each table "pens, ink and paper are provided in abundance." The commentary then offered the following account of the method by which questions were posed and answered:

> The young men hear the proposition or question delivered by the examiners; they instantly apply themselves; demonstrate, prove, work out and write down, fairly and legibly (otherwise their labour is of little avail) the answers required. All is silence; nothing heard save the voice of the examiners; or the gentle request of some one, who may wish a repetition of the enunciation. (Rouse Ball 1889, 198)

This commentary, like those on other university customs, was partly intended to inform younger undergraduates of the conduct proper to university ceremonies. The new written form of the Senate House examination must have posed special problems in this respect, as most colleges did little to prepare their students for such examinations, and, unlike oral disputations, undergraduates could not witness the procedure for themselves.[20]

I have already noted that one peculiarity of written examinations is that they lacked the prescribed order and interactive rhythm of the oral disputation. The unusually breathless narrative of the above account was probably intended to convey this point to undergraduates and to impress upon them the importance of working as quickly as possible without the prompting of an examiner. The separation of examination and assessment also raised the possibility that a student might not fully appreciate that the act of writing the examination script did not of itself constitute the completion of the examination. Thus our commentator reminded undergraduates that "although a person may compose his papers amidst hurry and embarrassment" he must remember that his answers "are all inspected by the united abilities of six examiners with coolness, impartiality and circumspection" (Rouse Ball 1889, 199). One final point of difference, hinted at by reference to the "embarrassment" a candidate might experience, was the great breadth and detail of technical knowledge expected of candidates providing written answers. Although in an oral disputation the respondent could not anticipate precisely what objections his opponents would raise, he could choose and

---

20. On the introduction of written examinations in Cambridge colleges, see Gascoigne 1984, 556–57.

prepare the propositions he wished to defend. In written examinations by contrast, a student could be questioned on any topic considered appropriate by the examiners and had to tackle problems based on those topics. In the above commentary the student was warned accordingly that in a written examination "no one can anticipate a question, for in the course of five minutes he may be dragged from Euclid to Newton, from the humble arithmetic of Bonnycastle to the abstruse analysis of Waring" (Rouse Ball 1889, 199). Even the most able students had therefore to prepare themselves for the fact that they might be partially or entirely unable to tackle some of the questions posed.

During the first three decades of the nineteenth century, undergraduates were gradually familiarized with the techniques of written assessment by the introduction of regular college examinations. The first examinations of this kind had been introduced in the latter eighteenth century by St. John's and Trinity in an effort to stem the rising prestige of the Senate House examination and preserve the independence of college teaching.[21] The power of the examiners to choose the topics examined in the Senate House meant that any college wishing its students to do well was obliged to teach those subjects that appeared regularly on the Tripos papers. Efforts to resist this centralization of power seem eventually to have backfired, as the longer-term effect of college examinations was simply to drive the level of technical performance in the Senate House ever higher. By 1831 all colleges had introduced regular undergraduate examinations of their own, and, as we shall see in chapter 4, these examinations appear to have been almost as competitive as the Senate House examination itself. College examinations established a student's reputation among his immediate peers and provided an important indication of how well he could be expected to do in the Senate House.

By the early 1840s, the use of written examinations had become so commonplace in Cambridge that it was only to visitors from overseas that the system seemed sufficiently peculiar to warrant special comment. One such American visitor was Charles Bristed, who, as we saw in chapter 2, studied at Trinity College in the early 1840s. Bristed identified the emphasis on written examinations as one of the most striking characteristics of a Cambridge education and suggested that the "pen-and-ink system of examination" had been adopted "almost entirely at Cambridge, in preference to *viva voce*, on the ground, among others, that it is fairer to timid and diffident men" (Bristed 1852, 1:97). This explanation of the demise of public disputations,

---

21. Gascoigne 1984, 556–57.

probably common by the early 1840s, is somewhat implausible since, as I have already noted, oral and written examinations were not simply alternative methods of testing the same abilities. Bristed's remark nevertheless highlights the extent to which the new examinations had altered student sensibilities by the early Victorian period. The ideal of a liberal education prevalent in Georgian Cambridge was, as we have seen, one in which students were supposed to learn to reason properly through the study of mathematics and to acquire appropriate moral and spiritual values through the subservient emulation of their tutors. This kind of education, which had long been seen as an appropriate preparation for public life, valued the qualities of good character, civility, and gentility above those of introspection, assertiveness, and technical expertise.[22] The oral disputation formed an integral part of this system of education as it tested a student's knowledge through a public display of civil and gentlemanly debate. Timidity and diffidence were indeed qualities that handicapped a student in this kind of examination, though not because they prevented him from answering technical problems in mathematics, but because they were seen as undesirable qualities in a future bishop, judge, or statesman. During the early decades of the nineteenth century, the increasing reliance on written examinations as a means of ranking undergraduate performance began to undermine the ideal of a liberal education from within. As we saw in chapter 2, ambitious students aiming for high placing in the order of merit abandoned college and professorial lectures during the 1820s and 1830s, turning instead to the leading mathematical coaches. By the early Victorian era, as Bristed's comments make clear, the now defunct oral disputations had come to be regarded as *unfair* examinations because they required students to display qualities beyond the purely technical.

It is nevertheless important to notice that although the written examination was not a *public* ritual, it was by no means an emotionally neutral one. Candidates for written examinations might have been spared the trials of stage fright and of making intellectual fools of themselves in front of their peers, but they were subject to a whole new set of pressures and anxieties.[23] These subtle changes in student sensibilities were made very visible by those who sought to preserve the broader ideals of a liberal education. Most vocal among these was William Whewell, who pointed explicitly to competitive examinations as the source of what he saw as some new and highly undesirable traits in undergraduates. According to Whewell, these examinations

22. Rothblatt 1982, 5.
23. On the anxieties generated by oral examination, see Rothblatt 1982, 7.

aroused in students a desire for "distinction" and "conquest," and evoked a "play of hopes and fears, sympathy and novelty"; all sensibilities which in Whewell's opinion encouraged them to become *active* instead of properly *passive* participants in the educational process. As he revealingly stated in 1837, the competitive written examination made any candidate who was tipped to do well "one of the principal actors in the piece, not a subordinate character, as he is in the lecture room" (1837, 59). Whewell saw the oral disputation as the form of examination that best preserved the social order of the lecture room; it was a public event during which the candidate addressed an audience and remained visibly under the authority of the moderator. In a written examination each candidate worked alone and introspectively to the sole end of out-pointing his competitors. As we shall see in chapter 4, to succeed in these circumstances a student had not only to work for several years at the very limit of his intellectual ability but to deal with the emotional pressures induced by the build up to a few days of gruelling examination upon which the success or failure of his entire undergraduate career depended.

Bristed's undergraduate experience confirms that sitting a written examination could be just as much an emotional trial as could participating in a public disputation. His initial discussion of the relative pedagogical merits of written and oral assessment suggests that he supposed the former to be a more natural and less stressful form of testing mathematical knowledge. But once he had actually experienced a written examination for himself, he acquired a strange virtual nostalgia for the oral disputation. I have just noted the frantic preparations made by ambitious students prior to the examination, but Bristed claimed that even entering and working in the examination hall could be stressful and disorientating experiences. He recalled that the early impression of Trinity Great Hall imbibed by most freshmen was that of a friendly dining room in which a common collegiate spirit was developed through convivial meal-time conversation. That impression was quickly dispelled when he entered the hall for his first college examination: "The tables were decked with green baize instead of white linen, and the goodly joints of beef and mutton and dishes of smoking potatoes were replaced by a profusion of stationery"(1852, 1:97). According to Bristed, the atmosphere in the hall became yet more disconcerting once the examination got under way. The enforced silence was broken only by the ominous footfall of the examiners "solemnly pacing up and down all the time," while the peculiar sound of "the scratching of some hundred pens all about you [made] one fearfully nervous" (fig. 3.3). Bristed also lamented both the lack of personal interaction between student and examiner characteristic of written examinations, and the permanency and precision of the written script. He observed that any "little slips" made *viva voce* might be recovered, allowed for, or even pass unnoticed, whereas "everything that you put down" in a written examination would be "criticised deliberately and in cold blood. Awful idea!" (1852, 1:98).

The comments by Whewell, Bristed, and others cited above provide some insight into the new cultural values associated with the written examination in early Victorian Cambridge. The qualities of civility, gentility, and public presence valued in the older ideal of a liberal education had been superseded by those of competitiveness, individual coaching, solitary study, and the reproduction of technical knowledge on paper. The pressures and anxieties experienced by the students during the actual examination had also altered substantially. A public disputation lasted at most a couple of hours and during this time the respondent was required to display a range of abilities in defending a prepared topic.[24] By the early 1840s, the written Senate House examination lasted six days (papers being sat morning and afternoon) and focused narrowly on the writing-out from memory of mathematical proofs, theorems, and definitions, and the solution of progressively difficult mathematical problems.[25] Moreover, the written examination was considered just as much a test of moral qualities as had been the oral disputation. The protracted intellectual and emotional stress induced by consecutive days of examination proved an unendurable trial to some students, and, as we shall see in chapter 4, the ability to remain calm and to reason quickly and accurately under these conditions gradually became a new hallmark of good character.

3.3. A Mathematical World on Paper

Another extremely important change in undergraduate mathematical studies prompted by the introduction of written examinations was the shift it engendered from a culture of reading and oral discourse to one of writing on paper. The use of paper and pen as a medium of mathematical learning is now so commonplace that, like written examinations themselves, it is accepted as the natural and obvious way of training students in the technical disciplines. At the beginning of the nineteenth century, however, when this practice was still relatively new, it made a sharp contrast both with earlier techniques of learning in the university and with contemporary ones in el-

---

24. Rouse Ball 1889, 182.
25. Glaisher 1886–87, 15.

ementary schools. We have seen that, until the mid-eighteenth century, undergraduates continued to be assessed by oral Acts, and that in preparation for these Acts they learned from a range of sources: by private reading guided by a tutor, by attending college and, occasionally, professorial lectures, by attending formal disputations and debates, and by discussions with their tutors and peers.[26] In connection with these activities students also copied out, annotated, and summarized important texts, but, in a pedagogical culture of oral teaching and examination, they were required neither to *write out* what they had committed to memory nor to solve technically difficult problems that could only be tackled with the aid of paper and pen. The kinds of natural-philosophical proposition that students were required to defend from the *Principia* reflected the oral mode of examination. Rather than reciting long mathematical proofs and demonstrations or solving problems, students were required to attack or to defend such qualitative propositions as "whether the cause of gravity may be explained by mechanical principles" or "whether Newton's three laws of nature were true."[27]

During the second half of the eighteenth century, following the introduction of the Senate House examination, the emphasis in undergraduate studies began to shift away from qualitative natural-philosophical and natural-theological aspects of Newton's work towards more technical aspects of mixed mathematics. It was in this context, as we have seen, that students taking the Senate House examination were first required to provide *written* answers to orally dictated questions. This was not in fact the very first use of written examinations in the university, as Richard Bentley (3W 1680) had required written answers to questions from candidates for fellowships at Trinity College since 1702. In this case each candidate was given a different set of questions to tackle, but it is nevertheless likely that Bentley's innovation provided the model for the subsequent introduction of written examinations in the Senate House.[28] Once such a system had been introduced, moreover, it was a straightforward matter to make the questions more numerous and more difficult year by year as the competition intensified and the students' level of technical competence rose. By the first decade of the nineteenth cen-

---

26. For a discussion of undergraduate pedagogical practice in sixteenth-century Cambridge, see Sherman 1995.

27. Gascoigne 1989, 177.

28. Rouse Ball 1889, 193. This was probably the first occasion in Europe on which written answers to mathematical questions were required. It is likely that the questions were relatively elementary since, as I noted in chapter 1, Bentley himself had difficulty understanding technical aspects of the *Principia;* see Guicciardini 1999, 125.

tury, when the written papers had become the most important factor in determining a student's place in the order of merit, the most difficult questions had become too technical and too complex to be solved orally.[29] From this point on, learning and examining mathematics were primarily paper-based activities, and the formal disputations and oral questioning in the Senate House gradually declined in importance until they were discontinued completely in the late 1830s.[30]

We know very little about the methods by which school children and undergraduates were instructed in mathematics in the early decades of the nineteenth century, but there is no doubt that those arriving to study in Cambridge at this time were surprised by the emphasis upon working on paper.[31] When J. M. F. Wright arrived at Trinity College in 1815, he was admonished by the tutor who examined him when he confessed his inability to "write down" his mathematical knowledge on paper. The tutor then warned him sternly that at Cambridge:

> all things—prizes, scholarships, and fellowships, are bestowed, not on the greatest readers, but on those who, without any assistance, can produce most knowledge on paper. You must hence-forth throw aside your slate . . . and take to scribbling upon paper. You must *"write out"* all you read, and read and write some six or eight hours a day. (Wright 1827, 1:96)

This experience was echoed by Wright's contemporary Solomon Atkinson (SW 1821), who, as we saw in the previous chapter, arrived at Queens College in 1818. This very ambitious and largely self-taught son of a Cumbrian farm laborer hoped that he would be able to impress the president of the college, Isaac Milner (SW 1774), sufficiently to gain entrance as a sizar.[32] Atkinson had mastered a good deal of arithmetic and geometry before arriving at the university and, although unable to recall every proposition in Euclid by number, greatly pleased Milner by being able to scrawl the proof of the "proposition about the square of the hypothenuse" [sic] on a "slip of paper" (Atkinson 1825, 496). Although well informed about undergraduate studies at Cam-

---

29. Compare, for example, the problem papers for 1785 and 1802 reproduced in Rouse Ball 1889, 195, 208–9.

30. Ibid., 183.

31. For a discussion of elementary mathematical education in the nineteenth century, see Howson 1982, chap. 6.

32. A "sizar" was student who paid only a fraction of the usual college fees in return for acting as a college servant. See Searby 1997, 70–71.

bridge, Atkinson was still astonished by the emphasis on writing-out mathematical knowledge, and he reflected that most freshman would be "shocked" by the "University practice of scribbling everything on paper." To an impoverished student from a laboring community, this unrestrained use of paper to acquire skill with a quill pen together with speed and accuracy in mathematical technique seemed wanton. Paper was an expensive, hand-made commodity in the early nineteenth century, and one which peaked in price around 1810.[33] Charting the development of what he termed "the progress of a poor man's extravagance," Atkinson noted how what seemed at first the very expensive process of mastering mathematics in the Cambridge fashion soon became commonplace. In the case of textbooks, for example, he noted that at first a poor student was unwilling to spend what seemed to him the extravagant sum of five shillings on a "Treatise on Conics or Algebraic Equations" whereas, by the end of the second year, he would not "scruple to bestow 10 or 15 guineas" in the form of a college prize on a book of little or no direct value to his studies. Atkinson also discussed the student's changing perception of the value of paper, noting that the freshman felt "more reluctance at wasting an inch square of white paper" than he would subsequently feel "in throwing half a dozen sheets into the fire" (1825, 507).

The Cambridge emphasis on mastering mathematics on paper is highlighted from a different perspective by one of Wright and Atkinson's most brilliant younger peers, George Airy. Although Airy displayed some ability for mathematics early on in his school career, he was initially identified as a potential scholar by his excellent memory for Greek and Latin poetry. It was only after it had been decided that he would be sent to Cambridge that, as we saw in chapter 2, Airy was taught mathematics privately by a recent graduate of the university, Mr. Rogers. One of the important things that Airy learned from Rogers was that he had not only to grasp the gist of required theorems and techniques, but had also to be able to "write out any one of the propositions which [he] had read in the most exact form." It was this ability to reproduce his mathematical knowledge on paper that Airy remembered as standing him in particularly good stead when he arrived at the university (Airy 1896, 19–20).

The wide gap between the pedagogical practice anticipated by most freshmen and that actually encountered in Cambridge at this time is further

---

33. Dykes Spicer 1907, 86–87. An average quality paper cost around one shilling and sixpence a pound in 1810, while a farm day-laborer earned at best around ten shillings a week. The cost of a couple of pounds of paper a week would therefore have seemed considerable to a boy like Atkinson.

underlined by Wright's account of his first mathematical lecture at Trinity College. I mentioned above that a great deal of mathematical teaching from the late sixteenth century focused on the use of practical mathematical instruments. It was widely considered that a basic knowledge of the techniques of accounting, surveying, navigation, military fortification, gunnery, and so on was important in the education of a young gentleman.[34] This tradition appears to have persisted beyond Cambridge in the nineteenth century to the extent that most of the roughly one hundred students present at Wright's first lecture supposed that they would be required accurately to construct geometrical figures. Most had accordingly come equipped with a case of mathematical instruments for the purpose. But the lecturer, Mr. Brown, immediately informed the class that these instruments were superfluous as "the theory we are about to expound" required nothing but freehand sketching and writing.[35] Brown demonstrated to the class, from his rostrum, how the only figure that needed sometimes to be drawn with tolerable accuracy, the circle, could be produced by using one hand as a pair of compasses while rotating the paper with the other. Wright's account of his first Trinity lecture also reveals another important difference between elementary methods of teaching mathematics and that emergent in Cambridge. In addition to the case of mathematical instruments, most students had brought with them—as had Wright to his first interview at Trinity—a "slate" on which to draw and erase geometrical figures and calculations. However, in the Trinity lecture room each student was provided with an ample supple of "pens, ink and foolscap [paper]."[36] Trinity students were required from the very beginning of their undergraduate studies to master the art of writing rapidly and accurately with an ink quill on paper. Brown accordingly announced that "those gentlemen . . . who have slates before them, will be pleased to bring them no more, paper being the only thing scribbled upon, in order to prepare you for the use of it in the Examinations."[37]

The arrival of paper and pens in the Trinity College lecture room was, as Brown's comment makes clear, a concession to the rapidly increasing importance of written examinations in Cambridge, but it should not be taken as an indication that the teaching methods employed by college lecturers in

---

34. Turner 1973; Feingold 1997, 372–79.

35. Wright 1827, 1:118.

36. Ibid., 116. The use of slates in classrooms remained commonplace well into the mid-Victorian period. See Howson 1982, 121.

37. Wright 1827, 1:118. For additional comments on taking notes in the Trinity lecture room, see Bristed 1852, 1:23.

general had altered a great deal from the previous century. According to Wright (1827, 1:120–50), even at Trinity the primary function of a college lecturer was still to inform students what they should *read* and to test their recall and understanding of that reading by catechetical inquisition. Although by the 1820s the formal disputations played only a minor role in deciding a student's place in the order of merit, they nevertheless remained an important hurdle to be cleared, and college lecturers continued to prepare their classes accordingly.[38] In order to find out whether students had learned the definitions, proofs, and theorems they were required to know, the lecturer would go round the class asking them in turn to enunciate propositions and even to solve simple problems orally.[39] The paper and pens provided in the lecture room were not therefore central to the teaching process, but enabled students to take notes as they saw fit during the oral exchanges between the lecturer and individual members of the class.[40] A rare firsthand account by Charles Prichard of mathematical lectures at St. John's College reveals that paper and pens had not entered that college's lecture rooms even in the late 1820s. Prichard confirms that lectures remained "*viva voce*" and "strictly catechetical" in his time, the students sitting "round the walls of the room" with nothing in their hands "excepting an unannotated copy of some classical author." The only visual aid employed by the mathematical lecturer was "a cardboard, on which diagrams were drawn relating to the mathematical subject before us" (Prichard 1897, 36). This cardboard was handed from student to student as the lecturer went round the class and seems to have functioned as a kind of primitive blackboard. Instead of going through a geometrical demonstration in purely oral form, the student could refer to a diagram that was visible to the teacher and to the other students.

These oral exchanges between individual students and the lecturer could be of considerable help to other members of the class, especially when it came to gaining a firm grounding in more elementary mathematics. As Prichard recalled, the value of each of the "logomachies between the tutor and the undergraduate" depended mainly on the ability of the student being quizzed. If he were dull the interaction was at best amusing but if he were bright and well informed it was "frequently very instructive to those who came seriously to

---

38. Rouse Ball 1889, 183. For an account of a tough disputation in 1826, see DeMorgan 1872, 305.

39. Wright (1827, 1:147) notes that students practiced the enunciation of such demonstrations aloud in their rooms.

40. Wright only implies that this was the use of paper and pens in the Trinity lecture room, but it is confirmed by Airy (1896, 23–25) and Bristed (1852, 1:23).

learn" (1897, 36).⁴¹ But in the main, college lectures were run at the pace of the least able members of the class, and the most ambitious students soon resorted to private tutors. Atkinson was very disappointed at the elementary level of mathematical lectures at Queens College, claiming that apart from informing reading men of the "proper subjects of study" they were "so much time wasted."⁴² Trinity College appears to have been something of an exception in encouraging students to take notes on paper in the latter 1810s, and it is likely that this practice would have given Trinity men a significant advantage over their peers from other colleges when it came to the Senate House examination.⁴³

The rapid rise of the private tutor made paper-based learning in mathematics commonplace from the 1830s, and explicit references to the practice become correspondingly sparse thereafter. Occasional remarks by students in private correspondence nevertheless suggest that the imperative to learn with pen and paper still surprised some freshmen well into the Victorian era. Thus in a letter to his father in October 1840, even the well-heeled Francis Galton remarked: "I waste paper fearfully, i.e. scribble over both sides of it innumerable x, y's and funny looking triangles." Galton added that his "bedder" told him it was always easier to keep the room of a "reading gentleman" because she was never short of paper with which to light the fire (Pearson 1914, 1:143). The young William Thomson (2W 1845), who went on to become one of Cambridge's most original mathematical physicists, was similarly struck by the material apparatus of Cambridge mathematics. A few weeks after he began studying with the outstanding private tutor, William Hopkins, he recorded in a letter to his father that he had taken to doing all his mathematical work with a quill pen because they were used in examinations and one "must get into the habit of being able to write with them" (Thompson 1910, 1:31). This habit of learning mathematics through writing out and problem solving on paper was also one which shaped the style of successful wranglers when they undertook original investigations. In his second undergraduate term at Cambridge, George Airy commenced what he called the "most valuable custom" of always having "upon [his] table a quire of large-

41. Prichard considered this form of catechetical instruction preferable to the professorial style of simply reading lectures.

42. Atkinson 1825, 501. It is very likely that there was a wide disparity in the quality and level of lectures delivered at different colleges.

43. Trinity prided itself on the quality of its college lectures, and there is some evidence that Trinity students were marginally less reliant on private tutors than were students from other colleges at the beginning of the Victorian period. See Searby 1997, 131.

sized scribbling paper sewn together: and upon this paper everything was entered: translations out of Latin and into Greek, mathematical problems, memoranda of every kind (Airy 1896, 25). Airy also used his scribbling paper to "put [his] lecture notes in order" and to write out summaries of the more difficult mathematical topics with which he was expected to be familiar. On one occasion, having wrestled in vain with a textbook explanation of the phenomenon of "Precession," he "made out an explanation for [himself] by the motion round three axes." This manuscript, with "some corrections and additions," was subsequently printed as one of Airy's famous *Mathematical Tracts* (1826). William Thomson was famous for always carrying a "copy-book" which he would produce at "Railway Stations and other conveniently quiet places" in order to work on his current mathematical problems. He had begun keeping a research notebook as an undergraduate, and, as we shall see in chapter 4, would sometimes set to work on occasions that seemed highly inappropriate to his companions.[44]

Some very distinguished Cambridge mathematicians of the mid-Victorian period also preserved a copy-book style of teaching long after blackboards had become commonplace in college lecture and private teaching rooms. J. J. Thomson (2W 1880) recalled that when the Sadeirian Professor of Pure Mathematics, Arthur Cayley (SW 1842) gave lectures in the late 1870s he "sat at the end of a long narrow table and wrote with a quill pen on large sheets of foolscap paper" (Thomson 1936, 47). This, as Thomson remarked, "made note-taking somewhat difficult," but he insisted that the lectures were nevertheless "most valuable" in teaching students "not to be afraid of a crowd of symbols." Having launched into a difficult problem using the first method that came to mind, Cayley would frequently develop "analytical expressions which seemed hopelessly complicated and uncouth." Confident of his own manipulative powers, Cayley "went steadily on" and, to the satisfaction of the students, "in a few lines had changed the shapeless mass of symbols into beautifully symmetrical expressions, and the problem was solved" (1936, 47).[45] Thomson also recalled that although the Lowndean Professor of Astronomy and Geometry, John Couch Adams (SW 1843), delivered his lectures orally, they were "read from beautifully written manuscripts which he brought into the lecture room in calico bags made by his wife." Adams's lectures in fact combined both teaching and research, as the

---

44. Smith and Wise 1989, 180; Thompson 1910, 2:616. Maxwell too "always" carried a "notebook" in which he worked out "solutions of problems" (Campbell and Garnett 1882, 368).

45. Cayley's teaching method was probably taken from William Hopkins, who had coached him to the senior wranglership in 1842.

manuscript from which he read contained his "own unpublished researches on Lunar Theory" (1936, 47–48).

By the 1880s, the link between paper-based mathematical training and subsequent research practice was so well established that most commentators were concerned not by the novelty, but by the quality, of such training. When George Darwin (2W 1868) delivered his inaugural lecture as Plumian Professor of Astronomy and Experimental Philosophy in Cambridge in 1883, he drew the attention of his audience to the importance of teaching students to solve mathematical problems neatly and systematically on paper. Drawing upon his experience as a Tripos examiner, he expressed the opinion that too little attention was paid both to "style" and to the "form on which the successful, or at least the easy, marshalling of a complex analytical development depends." This was, he argued, a serious oversight, as some otherwise outstanding Cambridge mathematicians "seem never to have recovered from the ill effects of their early training even when they devote the rest of their life to original work" (Darwin 1907–16, 5:5). In order to illustrate his point, Darwin offered a comparison between, on the one hand, the exemplary copy-book work of William Thomson and John Couch Adams, the latter's manuscript being "a model of neatness in mathematical writing," and, on the other, James Clerk Maxwell (2W 1854), who "worked in parts on the backs of envelopes and loose sheets of paper crumpled up in his pocket." Darwin deplored this slovenly practice which, he suggested, was responsible for the fact that the first edition of Maxwell's well-known *Treatise on Electricity and Magnetism* "was crowded with errata, which have now been weeded out one by one." In Darwin's opinion, Maxwell had only been saved from more serious error "by his almost miraculous physical insight, and by a knowledge of the time when work must be done neatly" (1907–16, 5:6).

George Darwin was by no means the first person to comment on Maxwell's shortcomings as an analyst, but it is significant that he attributed them to poor training.[46] The reputations of many successful Cambridge mathematical physicists of the Victorian period were based on their finely honed skills of mathematical manipulation and problem solving, skills developed

---

46. One of Maxwell's examiners at Edinburgh University, J. D. Forbes, complained of Maxwell's "exceeding uncouthness" as a mathematician. Forbes confided to William Whewell that the "Drill" of Cambridge was "the only chance of taming [Maxwell]," but added that Maxwell was "most tenacious of physical reasonings of a mathematical class, & perceives them far more clearly than he can express." Maxwell's private tutor at Cambridge, William Hopkins, later remarked in a similar vein that although it appeared "impossible for Maxwell to think incorrectly on physical subjects," in his analysis he was "far more deficient." Forbes to Whewell, 2 May 1852, Add. Mss A.204/103 (WHP-TCC); and Glazebrook 1896, 32.

through years of progressive training under the tutelage of a handful of brilliant mathematical coaches. This was especially true of men like Couch Adams and Darwin, whose outstanding work in planetary motion and geodesy (discussed further below) relied heavily upon their ability to solve extremely difficult and analytically complicated mathematical-physical problems. The culture of reading and oral disputation prevalent in undergraduate studies a century earlier was neither intended nor suited to prepare graduates for this kind of original investigation, but, following the transformation of that culture wrought by the switch to working on paper, and with the rise of a research ethos during the last third of the nineteenth century, this became virtually its only purpose. In the eighteenth century a graduate who wished to make an original contribution to mixed mathematics would have had either to train himself to manipulate symbols reliably or to work under the guidance of a more senior member of the university. By the time of Darwin's inaugural address, it was taken for granted that undergraduate training should develop the paper-based skills of mathematical manipulation necessary for research, and that the career even of a mathematician of Maxwell's undoubted brilliance could be blighted if that training was inadequate.

3.4. Bookwork, Problems, and Mathematical Textbooks

Having explored the change in material culture and student sensibilities that accompanied the gradual switch from oral to written examinations, I turn now to the structure and content of the examination questions themselves and to their relationship with the wider pedagogical apparatus of undergraduate mathematical teaching. Just as the introduction of paper-based examination in mathematics was a gradual and historically specific process, so too the form of questions to which written answers were required was peculiar to Cambridge and changed over time. The first printed question-sheets, which, as we have seen, appeared in the 1780s, contained only short and relatively simple mathematical questions, generally only one or two lines in length. The likely purpose of these questions was, as I noted above, to provide a convenient, efficient, and uncontroversial method of comparing the abilities of the most able students, but the introduction of these questions marked the beginning of two very important changes in the way mixed-mathematical studies were approached in Cambridge.

Before outlining these changes, however, we should note that, almost from the time of their introduction, questions requiring written answers were divided into two kinds.[47] The first kind, which from the early nineteenth century would be known as "bookwork" questions, required students to write out standard definitions, laws, proofs, and theorems from memory.[48] These questions were answered by all the students and, until the late 1820s, were dictated orally by the examiners.[49] The second, and more difficult kind of question, was the "problem." These questions were also dictated in the 1770s and 1780s but, from the mid 1790s, were given to the more able students in the form of printed sheets. It was these problem sheets that provided the most severe test of a candidate's technical ability and, from around 1800, effectively decided his place in the order of merit.[50] The precise origin of these two types of question is unclear, but it seems extremely likely that they reflected the style of presentation of such classical mathematical texts as Euclid's *Elements* and, especially, Newton's *Principia*. Newton began the *Principia* by stating the definitions and laws which would apply throughout the text, and then developed the detailed geometrical-mechanical content of his argument as an extended series of propositions. Each proposition consisted either of a theorem concerning the motion of bodies, or of a specific problem that was solved by the application of the foregoing theorems. Given the *Principia*'s extremely high status in eighteenth-century Cambridge, it seems reasonable to conclude that, at least in mechanics, the bookwork questions reflected the canonical status of Newton's propositional theorems while the problem questions reflected the ingenious application of the theorems as exemplified in the propositional problems.

Returning now to the changes wrought by the introduction of written examination papers, the first concerns the status of the various physical-mathematical definitions, laws, and theorems taught to undergraduates. We have seen that when students were examined by oral disputation on, say, Newton's *Principia*, they were generally required to *defend* qualitative propositions against the carefully contrived objections of several opponents.[51]

---

47. Little is known of the form and content of questions posed and answered orally in the mid-eighteenth century, but the range of topics covered was roughly that of early written questions. See Rouse Ball 1889, 190–92.

48. It is unclear when the term "bookwork" was introduced, but it was definitely in use by the latter 1810s. See Wright 1827, 2:62.

49. Rouse Ball 1889, 199, 212. The new regulations of 1827 were the first to require that all examination papers be printed and given to the students.

50. Ibid., 194–96. The earliest examples of the problems set orally to candidates in the Senate House examination date from the mid 1780s.

51. Ibid., 182. Although the opponents were the undergraduate peers of the respondent, they were supplied with difficult objections by their college tutors.

These verbal encounters implied that there *were* seemingly plausible objections to Newton's celestial mechanics and required the student to locate the fallacies in such objections from a Newtonian perspective.[52] In providing *written* answers to questions, by contrast, students were required either to reproduce as bookwork the laws and propositional theorems found in such books as the *Principia,* or to *assume* their truth in tackling questions on the problem papers.[53] The move from oral disputation to written examination was therefore accompanied by a far more dogmatic approach to the physical foundations of mixed mathematics.

The second way in which written examinations altered the kind of mathematical knowledge prized in Cambridge concerns the technical application of physical principles and mathematical methods. Although many problems requiring written answers were at first sufficiently short and simple to have been solved orally—in this sense being a straightforward substitute for the more time-consuming oral examination—they gradually developed to become a test of quite different skills. In the mid 1780s, for example, a typical question might require the student to find the "fluxion" or "fluent" of a simple algebraic expression in one variable, or to calculate the velocity with which a body must be thrown to "make it become a secondary planet to the earth."[54] The steps in the solution of these simple questions could have been rehearsed orally by a well-prepared student, but, once the practice of requiring written answers had been established, it was a straightforward matter to increase the quantity and complexity of such questions to a level that made the solutions virtually impossible to obtain without the aid of paper and pen. By the turn of the century, typical questions could require the student to tackle one of the following: to find the fluxions or fluents of three or more complicated expressions; to solve one or more cubic equations; to extract several cube roots; or to sum several complicated arithmetic and/or geometrical series. All of these exercises would have been impossible for the vast majority of students to tackle orally.[55] The rise of written examinations,

---

52. Ibid, 180–81. The *Principia* makes an interesting contrast with Euclid's *Elements* in this respect since, as Rouse Ball points out, students were seldom allowed to defend propositions from the latter on the grounds that they were too difficult to argue against.

53. Ibid., 174–78, 195. Compare the specimen disputation of 1784 and the written problem papers for 1785, both reproduced by Rouse Ball. The 1785 paper reveals that half the problems set drew either explicitly or implicitly upon Newton's mechanics and fluxional calculus.

54. Ibid., 195. Questions 3 and 8 from the paper of 1785. Finding the "fluxion" or "fluent" of an expression were equivalent respectively to differentiating or integrating the expression.

55. Ibid. See paper for 1802 (208–9). By 1802, the Senate House problem papers set to the most able students were at least as difficult, perhaps more so, as the Smith's Prize questions circa 1770.

especially problem papers, during the last two decades of the eighteenth century therefore brought about a very significant change in the skills possessed by successful graduates. Instead of preparing to defend general principles verbally against standard objections, to recite short proofs and theorems from memory, and to solve simple problems orally, they were required to reproduce lengthy proofs on paper, and to learn how to apply them to technically complicated examples. It was this important shift from oral to paper-based examination that opened the way for a subsequent dramatic rise in the level of mixed-mathematical knowledge and problem solving skill demanded of the more able undergraduates.

Despite the gradual increase in the number and complexity of problems set from the 1770s to the early 1800s, their form remained stable and their content relatively elementary. As Rouse Ball (2W 1874) noted in the late 1880s, many of the "so-called problems" set at the beginning of the nineteenth century counted as little more than bookwork when compared with the problems set in his day (1889, 209). Why, then, did the form of problem questions remain relatively stable until roughly the end of the second decade of the nineteenth century but thereafter begin to become much more demanding? The simple answer to this question is that it was the analytical revolution circa 1820 that marked the beginning of a quarter of a century of rapid inflation in the conceptual and technical difficulty of Tripos problems, but the onset of this inflation needs to be understood in the context of a more subtle change that had been under way since the turn of the century. This change concerns the relationship between Tripos problems and the textbooks commonly used in undergraduate teaching. We saw in chapter 2 that until the end of the second decade of the nineteenth century, the teaching of undergraduates in all but the first few terms of residence remained substantially in the hands of college lecturers. Since the majority of these teachers had little interest in making original contributions to mixed-mathematical studies, they were heavily reliant, especially in teaching their more able pupils, on textbook expositions of standard topics and the accompanying worked examples.

The term *textbook* needs to be treated with considerable caution in this context, as most of the books from which advanced mixed-mathematical topics were studied from the 1770s until at least the 1830s were somewhat different from the textbooks familiar to the modern reader. Rather like the classical mathematical texts expounded by college lecturers—Euclid's *Elements* and Newton's *Principia*, for example—they made few concessions to pedagogy and were generally aimed as much at the experienced mathematician as at the student.⁵⁶ Even in the mid 1830s, as Augustus DeMorgan (4W 1827) pointed out (1835b, 336), there was no book (or small collection of books) that covered anything like a syllabus of undergraduate mathematical studies at Cambridge, so that any student who tried to study without the guidance of a tutor would end up with a very idiosyncratic view of mathematics. The word "textbook" never in fact appeared in the title of teaching texts in this period and was used mainly (and then only occasionally) to refer informally to whichever mathematical treatises happened to contain sections that approximated closely to courses typically given by college lecturers.⁵⁷ These books sometimes reproduced worked examples of some of the principles and propositions with which they dealt, but seldom gave systematic accounts of the range of problems that could be solved using specific techniques and certainly gave no problems for the students themselves to tackle as exercises.⁵⁸ The only exception to all but the last of these points were the treatises on the principles and applications of fluxions, but since books on other branches of mixed mathematics could not assume that the reader was familiar with the fluxional calculus these latter books generally presented their subject matter in a fairly nonmathematical form.⁵⁹ Moreover, even problems that required the use of the fluxional calculus tended to be straightforward reproductions of the worked examples given in the treatises.⁶⁰ The relative simplicity and uniformity of the problems set on the annual Tripos papers around 1800 can therefore be understood as a direct reflection of the kind of examples discussed in the mathematical treatises most commonly recommended by college lecturers.⁶¹ These examples formed the basis of college teaching, informed students of the type of question likely to be encountered in examinations, and could be reproduced more or less from memory.

56. The expository material in the earlier sections of many "treatises" appears to have been intended more to lead the reader to the original contributions of the author than as a general introduction to the topic in question.

57. DeMorgan 1832, 277. Books such as Evans 1834 were "textbooks" in the traditional sense that they were based on manuscripts used by undergraduates and "printed with the view of saving to the student the time and trouble, which it has hither to been necessary to bestow in copying them" (Preface).

58. For an example of a book widely used in undergraduate studies, see Wood 1790–99.

59. These books were only partial exceptions, however, as they too tended to lead up to the original contributions or expositions of the author. See Guicciardini 1989, 125.

60. Ibid., 125–26. Late eighteenth-century textbooks on fluxional calculus intended specifically for Cambridge undergraduates were much more elementary than those published mid-century.

61. Rouse Ball 1889, 209; Guicciardini 1989, 125. Rouse Ball attributed the elementary level of questions circa 1800 to the "inferior and incomplete" nature of the current textbooks.

During the first two decades of the nineteenth century a number of factors conspired slowly to alter the relationship between Tripos problems and textbook examples. We saw in chapter 2 that the rules restricting the use of private teaching were relaxed in 1807 and 1815, and that an extra day of mathematical examination was added to the Tripos in 1808. These developments, together with rising student numbers and increasing competition for high placing in the order of merit, meant that in order to tax the abilities of the most able candidates the examiners began to devise more novel variations on the standard problems. Over the first two decades of the nineteenth century, the Tripos problems gradually became a little more difficult, or, at least, more diverse, than the standard examples given in the textbooks.[62] This process altered the role of the examiners and accelerated the changes discussed above. The ability to set novel but neatly soluble problems now became an important part of the examiner's art, while the need to solve such problems quickly and accurately drove students to spend long hours working through past examination papers and to seek the help of private tutors to guide their efforts. Furthermore, it was the accumulating archive of locally manufactured problems drawn from the annual Senate House examination, rather than the standard examples in treatises, that gradually became exemplary of the breadth and level of performance expected of would-be wranglers.

A public manifestation of this very important shift was the commencement by the university of regular publications of complete collections of Tripos problems. The prime purpose of the first of these, which contained the problem-papers for the years 1801 to 1810 inclusively, appears to have been to counter criticism of the quality of undergraduate studies in the university.[63] Those contributing to the "din of obloquy" against Cambridge were invited to try solving the problem papers for themselves; but the collection was also intended to assist undergraduates in "acquiring a facility of investigation" that would eventually enable them to "command success" in the Senate House.[64] A second volume of past-papers, covering problems from the previous twenty years, appeared in 1820, the latter papers containing the

---

62. Compare the difficulty of the problems set in the Triposes of 1802 and 1819 reproduced in Rouse Ball 1889, 200, and Wright 1827, 2:63, respectively.

63. Anon. 1810; Williams 1991, 119.

64. Anon. 1810, xi. Wright considered himself extremely lucky as an undergraduate to have been lent a collection of past college examination papers by a professor. These papers were "not purchasable" and enabled him to get "direct by the nearest route, to academical distinction" (Wright 1827, 1:152).

first problems set by examiners such as Peacock and Gwatkin to promote the incipient analytical revolution.[65] The gradual establishment of the Tripos problem as exemplary of the kind of question that a would-be high wrangler had to be able to solve also altered the kind of mixed-mathematical knowledge prized in Cambridge. The growing archive of questions both defined an effective syllabus of mathematical studies and provided exercises through which students could hone their problem-solving skills.[66] Unlike textbook examples, Tripos problems were not worked exercises, but required the students to work out for themselves how to apply the physical principles and mathematical methods they had learned. This change naturally encouraged the more ambitious students to seek the assistance of private tutors. The best of these tutors had prepared manuscripts that not only explained the highly compacted proofs and problem solutions given by Newton in Book 1 of the *Principia,* but showed how these and other examination problems could be tackled using the techniques of the fluxional calculus. None of these manuscripts has survived, but it is a measure of their utility that when several young wranglers published technical exegeses of relevant sections of the *Principia* in the 1820s and 1830s, they found these manuscripts the most useful resource for explaining Newton's work.[67] The availability of these pedagogically useful manuscripts and books increased the technical ability of the undergraduates, a development which, in turn, encouraged the examiners to contrive ever more ingenious and difficult variations on the propositional problems given by Newton. By the 1830s, as we shall shortly see, it was the ability to solve highly original problems and to recapitulate the production of new mathematical knowledge (in the form of problems) that had become the most severe test of the top students. Before discussing this development we must examine the impact of the analytical revolution on the paper world of Cambridge mathematical studies.

3.5. Problems, Textbooks, and the Analytical Revolution

The problem-solving tradition of the early nineteenth century presented both potential obstacles and major opportunities to supporters of the analytical revolution. In addition to providing new textbooks and getting ques-

---

65. Anon. 1820. See, for example, 355–56 for questions on differential equations and D'Alembert's principle.
66. On examination questions as the best guide to the "staple of mathematics at Cambridge," see DeMorgan 1835b, 336.
67. See the prefatory comments in Carr 1821, Wright 1828, and Evans 1834.

tions involving analysis onto the Tripos papers, these men realized that it was expedient to present the new mathematics in a manner that could easily be incorporated within this tradition. It is noteworthy, for example, that the first book intended to introduce analytical methods to Cambridge, Robert Woodhouse's *Principles of Analytical Calculation* of 1803, was aimed at accomplished mathematicians rather than undergraduates and contained no examples of how these methods were applied to practical problems. These shortcomings would have severely limited the book's value to students and made it hard for tutors and examiners to formulate difficult but neatly soluble problems in analytical form.[68] When Babbage, Herschel, and Peacock translated Lacroix's *Elementary Differential Calculus* in 1816, they addressed this issue directly by assuring the reader that a "collection of examples and results" would shortly follow "indicating such steps in the processes as cannot be expected to be discovered by an ordinary student" (Peacock 1820, iv). In order to appreciate the full significance of this remark it is important to keep in mind that the inclusion of Continental mathematical methods in the Tripos papers of the late 1810s represented rather more than a mere switch of notation in the calculus. The examination papers set by Peacock and Gwatkin, for example, assumed that the students were familiar with a range of new analytical techniques including D'Alembert's principle and the theory of differential equations.[69] The reformers appreciated that numerous special techniques were required to apply these analytical methods to the solution of difficult problems and that such techniques were only learned through practice with graded examples. As Peacock astutely remarked in the preface to his book of worked examples in differential and integral calculus in 1820, many families of then standard problems had originally taxed the abilities of the greatest mathematicians of the eighteenth century and appeared straightforwardly soluble to contemporary readers only because the solutions had "been borrowed with little scruple or acknowledgement by succeeding authors" (1820, iv).[70]

The problem-solving approach to the new mathematics was further promulgated through the latter 1820s by the publication of William Whewell's *Treatise on Dynamics* in 1823. Whewell's book is especially relevant to our present concerns because in addition to providing a concise guide to dynamics in analytical form, it also discussed the introduction of the new

---

68. Subsequent analytical works by Woodhouse that did include examples were used extensively as a source of examination questions. See Becher 1980b, 393–94.

69. Wright 1827, 2:63–92.

70. Peacock had Leonard Euler especially in mind with this comment.

mathematics from an expressly pedagogical perspective.[71] Although Whewell is frequently, and correctly, depicted as a supporter of the analytical revolution in the 1820s, it is important to remember that, as we saw in the previous chapter, he envisioned a rather different role for analysis in Cambridge to that proposed by more radical reformers such as Babbage and Peacock. The latter, at least initially, saw analysis as a pure-mathematical discipline in which analytical techniques were studied in isolation from their physical applications.[72] Whewell's vision was more conservative in two respects: first, he introduced analytical methods as useful tools in the formulation and study of Newtonian dynamics; and, second, he presented this mixed-mathematical approach to analysis in the form of progressive propositions each followed by concrete examples. As far as Whewell was concerned, the abstract formulations of analytical mechanics due to analysts such as Lagrange had made dynamics a form of pure analysis from which such physically intuitive notions as force had all but disappeared. These formulations were not well suited to the mixed and problem-solving style of Cambridge mathematics and, in any case, were virtually impenetrable to most undergraduates and, no doubt, to many tutors. Rather than exploring the generality of such formulations, descending only occasionally to specific physical examples, Whewell proposed to start with physically intuitive propositions and then gradually to build via examples and problems to more generalized equations. The advantage of this approach, he argued in his *Treatise* (1823, vi), was that it made an analytical account of the "Newtonian System" accessible at some level to undergraduates of all abilities while providing an appropriately problem-based foundation for those brilliant students who would eventually go on to master the more abstract formulations.

One of the most striking aspects of Whewell's *Treatise*, especially when compared with other contemporary books intended to introduce Continental mathematical methods to Cambridge, is the several ways in which it was written explicitly for pedagogical purposes. First, the book was aimed specifically at undergraduates. That this was unusual at the time is highlighted by the fact that Whewell felt it necessary to acknowledge that his aim was to expound such books as Laplace's *Mécanique céleste* at "the level of the ordinary readers" and to warn any "mathematician" who consulted the book that he should expect to find little "except what he will consider as elementary."

---

71. Whewell had already published a treatise on analytical mechanics (1819), but this book was more elementary and less sophisticated from a pedagogical perspective.

72. Becher 1995. Peacock's *Examples* (1820) focused on problems in analytical geometry rather than physics.

Second, Whewell had taken some trouble to divide the material presented into progressive propositions which could easily be followed by students. By "breaking up the reasoning into distinct and short propositions," he assured the reader, the subject had been "rendered more easily accessible" (1823, v–vi, ix–x). Not only was it easier to follow a book that developed a topic proposition by proposition, but it was also clear to a student precisely at what point he failed to follow the argument. Third, Whewell thought it was the duty of the textbook writer not merely to reproduce the presentation of a Continental author or to offer an idiosyncratic compendium of analytical techniques and formulae, but to act as a reliable, informed, and selective guide to the most interesting and important results obtained by *all* authors to date. According to Whewell, students currently wasted a great deal of time "searching and selecting through a variety of books," engaging throughout in "unprofitable and unsystematic reading." The *Treatise on Dynamics*, he hoped, would "make the extent of our [mathematical] *encyclopedia* less inconvenient" (1823, iii). Fourth, Whewell made sure that each of the propositions he discussed was "elucidated by a considerable collection of mechanical problems, selected from the works of the best mathematicians, and arranged with their solutions under the different divisions of the science" (vi). His purpose was both to display the kinds of problem that could be tackled using analytical mechanics and to teach students the many "artifices which have been employed" (ix) in solving such problems. This emphasis on the use of analysis to solve difficult problems was calculated both to facilitate the inclusion of analytical methods within the Cambridge mixed-mathematical tradition and to make it easy for examiners to set, and for students to tackle, questions requiring analytical methods on the Senate House examination papers. Lastly, and following on from the previous point, Whewell thought it appropriate to use the problem-solving approach to analytical mechanics as a vehicle for teaching "the application and utility of some of those particular cases and branches of analysis, which might otherwise be considered as merely subjects of mathematical curiosity" (x).[73] In Whewell's book, analytical techniques would be taught not because they were fundamental to the study of pure analysis, but because they were useful mathematical methods for solving specific physical problems.

Whewell's presentation of analytical dynamics can be understood both as a political statement about the proper role of mathematics in undergrad-

---

73. For example, problems in which force varied "inversely with the distance" required students to "obtain the *definite integral* of exp $(-x^2)$," while the "motion of a complex pendulum" required a knowledge of "*simultaneous integration* of n differential equations."

uate studies and as an expedient for integrating analysis into the Cambridge mixed-mathematical tradition as smoothly as possible. It should also be read as an insightful commentary on the contemporary difficulty of teaching relatively advanced mathematics to large groups of mixed-ability students. Whewell was perhaps the first Cambridge author to write a book at this level specifically for undergraduates and to devote such attention to pedagogical issues. An ideal textbook presentation, according to Whewell, developed the student's mathematical knowledge systematically and progressively, presented relevant results obtained by a wide range of authors within the traditional divisions of mixed mathematics, illustrated every division with numerous worked examples, and introduced advanced analytical methods as and when they were required to solve specific problems. The various readings of Whewell's book referred to above were thus not only consistent, but mutually supportive. Whewell realized that a carefully designed textbook could be a powerful resource in his campaign to incorporate the power of analysis within the Cambridge mixed-mathematical and problem-solving traditions. For our present purposes the most important aspect of Whewell's book is that offered able undergraduates a concise account of the key elements of analytical dynamics, and made it easier for them to tackle difficult dynamical problems on the examination papers. Ambitious students no longer needed to struggle unguided through the original works of Continental authors, wrestling simultaneously with a wide range of disparate physical principles and analytical techniques. Drawing upon the "synthetical" style of such classical mathematical texts as Euclid's *Elements,* Whewell presented a carefully chosen selection of the works of such diverse authors as D'Alembert, Euler, Lagrange, Laplace, Maclaurin, Newton, Poisson, and Simpson, as a single, consistent discipline, developed in easy-to-follow propositions (1823, viii). As books of this kind recast the traditional branches of mixed mathematics in analytical form over the next two decades, examiners could exploit the analytical power of the calculus to devise a virtually endless supply of increasingly ingenious and technically demanding problems.[74]

Despite the evident utility of many of the new textbooks, not all were as accessible as Whewell's *Dynamics* and none (Whewell's included) contained worked examples of actual Tripos problems.[75] Both of these factors almost

---

74. For examples of other important texts introducing physical subjects in analytical form see Hamilton 1826, Airy 1826, Hymers 1830, and W. H. Miller 1831.

75. On the idiosyncratic style of many textbook writers and the resulting confusion on the part of the reader concerning the details and relative importance of various mathematical subjects, see DeMorgan 1835b, 336.

certainly contributed to the complete takeover of undergraduate teaching by private tutors during the latter 1820s and 1830s. There was an unwritten rule in the university that examination questions should not be set on new topics until they had been covered in a treatise suitable for use by Cambridge students.[76] In practice, however, many of the books that technically fulfilled this condition continued to be written, as DeMorgan had pointed out, in a style more suited to the needs of experienced mathematicians than to those of undergraduates.[77] It became a major responsibility of the private tutor to master and to teach the new topics covered in recent textbooks and to show how the techniques discussed were used in the solution of ever more intricate problems.

One contemporary indication both of the rising status of Tripos problems in the 1820s and of the role of private tutors in teaching students how to solve them was the publication in 1825 by J. M. F. Wright of the first collection of *solutions* to Cambridge mathematical problems.[78] College lecturers, teaching large, mixed-ability classes, had neither the time nor the inclination to drill the brightest students in the art of problem solving, and Wright, who, as we saw in chapter 2, made it his business to make the aids of the private tutor accessible to all students, believed that a volume of solutions to the published Tripos papers would enable poor but keen students to master problem solving on their own. As he advised in a second volume of *Hints and Answers* to questions set in college examinations, published in 1831, on those occasions when the student's "own patient efforts [had] entirely failed," a reference to his solutions would be "equally advantageous with the help of a tutor" (1831, iii). Like a private tutor, moreover, Wright's volumes not only provided model solutions but referred, where appropriate, to the sections of various textbooks in which the physical principles and mathematical methods in question were most clearly discussed. A bright student who found all or part of a solution difficult to follow was thus directed immediately to the best source of further instruction. Wright did not publish any more volumes of problem solutions after 1831, nor were his innovative publications initially emulated by members of the university.[79] This was probably because, by the 1830s, those ambitious students who could have

---

76. Rouse Ball 1889, 128.

77. Airy's *Mathematical Tracts* (1826), for example, did not discuss the solution of specific problems except in the case of the calculus of variations. See Whewell's comments on Airy's *Tracts* in Becher 1980a, 26, and DeMorgan's (1835a) review of Peacock's *Treatise on Algebra*.

78. Wright 1825. These solutions were intended to complement the volume of examination papers published by the university.

79. Wright 1831 was the last of his volumes of problem solutions.

made most use of such volumes were already working with private tutors. The content of the most difficult Tripos problems was in any case changing so rapidly at this time due to the influx of new analytical topics that past papers from all but the last couple of years would have been of little use to the most able students.[80]

The analytical revolution of the 1820s and 1830s witnessed a dramatic increase both in the level of mathematical knowledge required of the most able students and in the difficulty of the problems they were expected to solve. Where the final paper for the Tripos of 1819 contained twenty-four questions, some of which required no more than the calculation of logarithms, a description of how to measure refractive index, or an account of the principle and construction of an achromatic telescope, the final paper of 1845 contained only eight questions, none of which could be tackled without an advanced knowledge of one of either algebra, mechanics, integral equations, the wave theory of light, potential theory, the calculus of variations, differential geometry, or the mathematical theory of thermo-optics.[81] This enormous increase in technical competence was generated by the new paper-based pedagogical economy which had developed in Cambridge over the first third of the nineteenth century. In the vanguard of this training revolution were private tutors such as Hopkins and Hymers, but we should also note the important role played by new textbooks and the increasingly positive input being made by examiners. The more elementary textbooks by Lacroix, Whewell, and others enabled students quickly to master a wide range of basic analytical techniques that were generally applicable to most branches of mathematical physics. These books provided a progressive survey of useful mathematical methods taken from numerous far less accessible memoirs and treatises, and were designed to prepare students for higher studies in general, rather than the specific contributions of the author.

Remarking on the importance of the many excellent elementary mathematical textbooks employed in France at this time, DeMorgan observed that such books had "caused the road to be smoothed and levelled till the examiners are able to ask and obtain such a degree of mathematical acquirements from candidates of sixteen years of age, as would have made a prodigy a century ago" (1835b, 333).[82] This comment neatly captures the constructive

---

80. DeMorgan 1835b, 336; Becher 1980a, 19–23.

81. The Tripos examination papers for 1817 and 1845 are reproduced in Wright 1827, 2:87–92, and Bristed 1852, 2:426–28, respectively.

82. DeMorgan highlighted the pedagogical shortcomings of mathematical treatises by remarking that it "needs many treatises to make a good book" (1835b, 333).

power of a well-written textbook to organize, summarize, and communicate a previously disparate collection of mathematical methods and applications. Its analogical reference to road transport also highlights the pedagogical similarity between the teaching methods employed by good private tutors and the attributes of a good textbook. Both accelerated the learning process by selecting and explaining the most generally applicable techniques due to many authors, teaching in a well-ordered and progressive style (only assuming a knowledge of what has actually been covered), and giving numerous examples to build confidence and technical expertise. The Tripos examiners also made an increasingly positive contribution to this process in the 1810s and 1820s. Whereas the questions set in the latter eighteenth century were closely based on examples in well-known mathematical and natural-philosophical treatises, those set in the early decades of the nineteenth century offered ever more novel and ingenious variations on standard exercises. The publication of collections of these questions enabled students to hone their problem solving skills to a very high degree, especially when set and marked by private tutors as illustrative of techniques discussed in the new textbooks.

3.6. Cambridge Mathematics and "Tacit" Knowledge

Before pursuing the development of Tripos questions into the early Victorian period, it will be useful briefly to consider some of the ways in which, even by the early 1830s, the mathematical skills and competencies of Cambridge-trained undergraduates differed substantially from those possessed by students trained at another site. In order to make this comparison I draw upon an essay written by DeMorgan in 1835, in which, although ostensibly reviewing a recently published treatise on algebra, he wrote a wide-ranging commentary on the Cambridge system of mathematical training in general and the relationship between Cambridge textbooks and Tripos examination problems in particular. DeMorgan was one of the few people ideally placed to notice and to reflect upon the peculiarities of the Cambridge system. As an undergraduate at Trinity College in the mid 1820s he had first-hand experience of Cambridge pedagogy, while his subsequent career teaching mathematics at the recently established University College London had made him acutely aware of the difficulties of using Cambridge textbooks to teach non-Cambridge students. What is particularly striking about DeMorgan's remarks is the subtle relationship he detected between the Cambridge system of examination and the structure and content of the textbooks written by Cambridge authors.

DeMorgan began his essay with the interesting observation that books written *by* Cambridge authors *for* Cambridge undergraduates were virtually incomprehensible to students at other institutions (1835a, 299).[83] He attributed this strange phenomenon to two principal features of the Cambridge system of written examinations. The first was the setting of numerous *bookwork* questions. Conscious that they might be asked to write out any proof, theorem, or principle with which they were supposed to be familiar, Cambridge undergraduates committed all such material to memory. As a result, the authors of Cambridge textbooks troubled neither to recapitulate material covered in more elementary books nor to reproduce steps in a derivation or example with which a Cambridge undergraduate should be familiar. Non-Cambridge readers, with a more tenuous grasp of elementary mathematics, found the arguments impossible to follow and had no means of knowing even what additional material they should refer to in order to made the arguments comprehensible.

The second characteristic of Cambridge examinations that made Cambridge textbooks opaque to non-Cambridge readers was their emphasis on problem solving. DeMorgan noted that Cambridge students were "supposed to acquire facility of application" such that they could solve "problems which are given by the examiners, and which are *not to be found in the books.*" This remark underlines two very important aspects of Cambridge pedagogy to which I have already referred: first, that by the early 1830s it was the problems on examination papers, rather than exercises in textbooks, that defined the standard to which ambitious students aspired; and, second, that solving such problems was a highly skilled activity only acquired through extended practice. Given that a Cambridge undergraduate was supposed to begin cultivating these skills as soon as possible, DeMorgan concluded, it was "even desirable" for him "that the elementary works should not lead him all the way, but should indicate points between which he may be expected to travel for himself" (1835a, 299). Another difference between Cambridge and non-Cambridge students, therefore, was that the former were not only expected to find their way through the merest sketch of an example, but were taught to regard such exercises as useful preparation for tackling difficult problems in examinations. Conversely, regular practice at

---

83. The article was published anonymously but DeMorgan 1882, 402, reveals him as the author.

solving past examination problems prepared Cambridge students to deal with sketchy examples in textbooks. DeMorgan did not point out, perhaps because he was unaware of the fact, that it was the job of the private tutor to provide the intermediate steps when necessary and that these tutors could also supply model answers to past examination questions if one of their pupils was stuck.[84] What he did note with some insight was that the many steps in Cambridge books left as exercises to the reader fostered in Cambridge students a singular attitude towards mathematical problems. When a Cambridge student tackled a textbook exercise or past examination problem he "[did] not permit himself to believe in impossible difficulties; but consider[ed], or should consider, that he *must* conquer whatever any one who knows the public examinations thinks it expedient to write" (1835a, 299–300).

These fascinating and unique comments on Cambridge mathematical training take us to the heart of much of what can be labeled "tacit" in Cambridge pedagogy by the early 1830s. First, the requirement that ambitious students be able to *write out* on demand virtually all the formal proofs, theorems, and definitions they were supposed to know gave them an encyclopedic mathematical knowledge unmatched by students at other British institutions. Second, the emphasis on problem solving required the students to develop a rare facility for applying physical principles and mathematical methods to the branches of physical science studied in Cambridge. But, since the many examination questions through which this facility was developed were not reproduced in the textbooks, they were unknown to non-Cambridge students. It is also worth noting in this context that the inability of non-Cambridge students to solve Tripos problems did not derive simply from their lack of access to examples. As we shall see in chapter 5, even when Tripos questions became widely available in popular textbooks, they remained insoluble to non-Cambridge readers because the techniques by which they were tackled were known only to Cambridge tutors.[85] Lastly, and closely connected with the previous point, Cambridge students approached any problem set by a Cambridge examiner in the certain belief that they *could* and *must* solve it using the mixed-mathematical methods in which

---

84. DeMorgan might have been unaware of rapid expansion in private tutoring that had taken place since his undergraduate days in the mid 1820s.

85. Tripos problems were, strictly speaking, available in published form from 1810, as were Wright's solutions for the years 1801–20. However, these volumes seem to have been little known or used beyond Cambridge.

they had been trained. Students accordingly devoted a great deal of time and effort to trying to solve difficult problems, and readily attributed any failure to do so to their own insufficiencies.

These attributes of Cambridge mathematical education are appropriately labeled "tacit," not because they were inescapably invisible or intangible, but because they had come to be taken for granted within the university, and made the exercises and problems by which undergraduates were trained and tested extremely difficult for students at University College to understand. Furthermore, I have made these attributes visible by pointing to the pedagogical incommensurability that existed between these two geographically separated places of mathematical training. It was DeMorgan's comparative perspective that enabled him to identify important and site-specific characteristics of the Cambridge system, characteristics of which Cambridge tutors were probably not themselves consciously aware (this being the reason that these attributes are appropriately labeled "tacit"). Comparisons of this kind are extremely useful because they highlight the crucially important connection between local regimes of training and the specific kinds of mathematical skill and competence that such regimes produce. If students at University College had been able, without the aid of the pedagogical resources specific to Cambridge, to solve difficult Tripos problems with the same speed and ease as their Cambridge peers, then it would be virtually impossible to establish a relationship between training and technical competence. But, as DeMorgan's remarks make clear, the style and content of Cambridge textbooks embodied the pedagogical ideals and practices of the university to such an extent that they were of little use at other sites.

Even more interesting, however, is DeMorgan's conclusion that the purpose of the emphasis on problem solving in Cambridge was to foster mathematical *originality* in the best students. There was, he remarked, "certainly no place where original effort [was] so much the character of education" as at Cambridge, no undergraduate achieving marked success "until he became capable of original mental exercise" (1835a, 300). The hallmark of this originality was that a student could rapidly solve a large number of difficult problems, each of which was not exactly like (but not completely unlike) any problem he had previously encountered. An obvious implication of this aspect of Cambridge pedagogy is that tacit skills acquired in undergraduate training played a major role in shaping the kinds of original investigation in which graduates of the Mathematical Tripos subsequently engaged. Discussion of this important issue is taken up in chapter 5, in which I also explore the conditions under which the locally specific skills generated by a Cambridge training could be propagated to new and distant sites. In the balance of this chapter I return to the development of Tripos problems in the mid-Victorian period, paying special attention to the forging of a new relationship between Tripos problems, textbooks, and the research undertaken by Tripos examiners.

3.7. Research, Problems, and Senate House Solutions

> An inextensible heavy chain
> Lies on a smooth horizontal plane
> An impulsive force is applied at A,
> Required the initial motion of K.
> JAMES CLERK MAXWELL, FEBRUARY 1854[86]

We have seen that during the first two decades of the nineteenth century, the growing archive of Tripos questions became both definitive of the range of topics undergraduates were required to master and constitutive of the practical exercises through which technical proficiency and problem-solving skills were acquired. Up to and including the early 1820s, these problems remained relatively elementary, generally being little more than variations on the standard examples given in the commonly used mathematical treatises. But, during the late 1820s and 1830s, the upward spiral of technical proficiency, driven by a combination of private teaching, intense competition, and the regular introduction of new and advanced analytical methods, gradually altered the job of the examiners. In order to stretch the abilities of the most able and well-prepared candidates, some examiners began to draw upon their own research as a source of advanced Tripos questions. Remarkable though this development may sound, it was in many respects a predictable outcome of the new pedagogical economy discussed above. We have seen that it was widely believed in Cambridge that the best way of teaching mathematics, including the new analytical methods, was through practical examples and problems, and, by the mid 1830s, some of the first generation of young college fellows to have been taught higher analysis this way were beginning both to undertake their own research and to be appointed as Tripos examiners.[87] It is not surprising therefore that these men sometimes

---

86. These opening lines of a "A Problem in Dynamics," penned by Maxwell shortly after completing the Tripos of 1854, enunciate a problem that is solved in rhyme in subsequent verses. The poem offers a rare and amusing insight into the way Maxwell tackled such problems. The poem is reproduced in Campbell and Garnett 1882, 625–28.

87. In the mid 1830s, the examiners were on average only six years from graduation.

approached research as a problem-solving exercise and could easily turn such exercises into Tripos problems.

The best-known case of a new theorem being announced in this way is that due to George Stokes, who set the derivation of what would become known as "Stokes's theorem" as a question in the Smith's Prize examination of 1854.[88] This example is sometimes seen as an amusing oddity but it is actually quite typical of the way Cambridge examiners tested the mettle of the most able students throughout the Victorian era. An early example of a Cambridge mathematician famous for his "unrivalled skill in the construction and solution of problems," especially those requiring the "application of complicated analysis," is Thomas Gaskin (2W 1831).[89] A private tutor and textbook writer through the 1830s and 1840s, Gaskin possessed such "extraordinary power" in constructing problems that he acted six times as a Tripos examiner.[90] As his obituarist pointed out, Gaskin not only based his questions on his own research, but made it "his custom to put any new theorem that he discovered in the form of a problem, rather than in that of a paper in a mathematical journal."[91] A Cambridge mathematician of the next generation who routinely announced new results in the form of Tripos problems was Joseph Wolstenholme (3W 1850). Despite the fact that he published a number of original mathematical papers in standard journals, Wolstenholme's fame as a mathematician rested chiefly upon the "wonderful series of mathematical problems" he composed as an examiner, many of which announced "important results, which in other places or at other times would not infrequently have been embodied in original papers."[92] Wolstenholme acted seven times as a Tripos examiner from the 1850s to the 1870s, and was so proficient at the "manufacture of problems" that he was able to pass on questions superfluous to his own needs to less able colleagues.[93]

As these remarks suggest, the most advanced Tripos problems were understood, at least in Cambridge, to embody new results every bit as important

---

88. Cross 1985, 144. Arthur Cayley (SW 1842) regularly published solutions to the (sometimes original) questions he set for the Smith's Prize. See *Quarterly Journal of Pure and Applied Mathematics* 8 (1867): 7–10; *Oxford, Cambridge and Dublin Messenger of Mathematics* 4 (1868): 201–26; 5 (1871): 40–64, 182–203; and *Messenger of Mathematics* 1 (1872): 37–47, 71–77, 89–95; 3 (1874): 165–83; 4 (1875): 6–8; 6 (1877): 175–82.

89. Routh 1889, ii.

90. Gaskin was a Tripos examiner in 1835, 1839, 1840, 1842, 1848, 1851.

91. Routh 1889, iii.

92. A. R. Forsyth (SW 1881) quoted in the *DNB* entry on Joseph Wolstenholme.

93. Wolstenholme 1878, v. Wolstenholme acted as a Tripos examiner in 1854, 1856, 1862, 1863, 1869, 1870, and 1874.

as those appearing in the standard research journals. Thus Edward Routh insisted that some of the most original problems set in mechanics were frequently "so good as to rank among the theorems of science rather than as among the examples," while Isaac Todhunter, defending the examination as a unique test of intellectual originality, claimed that the Tripos papers "abound in new results which are quite commensurate in importance and interest with the theorems previously established and studied."[94] By the late Victorian era, William Shaw (16W 1876) was prepared to go as far as to claim that, in the case of mathematical physics, the "original contributions to the subject were not papers in the Royal Society Proceedings or the mathematical journals but the questions set in the problem papers of the Tripos."[95] Shaw's claim is somewhat exaggerated, as many Cambridge mathematicians did publish important work in the standard mathematical journals, but there is one sense in which his comment might be taken literally. The announcement of a new result in the form of an examination question was an invitation to those who read the paper to try to prove the result (along the lines sketched by the examiner) for themselves. This was true not just for the students sitting the examination but for students from subsequent years and for senior members of the university who wished to master such problems either for teaching purposes or as a matter of interest or pride. One great advantage of this method of publication, therefore, was that it virtually obliged both the rising generation of wranglers as well as many established Cambridge mathematicians to explore the technical details of the examiners' original investigations.

Such explorations could also result in original questions becoming the inspiration for a research publication. If an examiner did not publish a solution to an interesting problem, solutions were sometimes proposed in such publications as the *Cambridge Mathematical Journal (CMJ)*. Established in 1837 by the founding editor, Duncan Gregory (5W 1837), and strongly supported in its early volumes by Archibald Smith (SW 1836) and Leslie Ellis (SW 1840), the *CMJ* regularly contained short articles proposing solutions to questions from past Tripos papers.[96] An interesting example of this kind is discussed by Routh (1898a, 80), who revealed that a special technique (the method of "infinitesimal impulses"), devised and taught by William

---

94. Routh 1891, v; Todhunter 1873, 7. Routh's textbooks reveal numerous new results first announced in the form of Tripos problems. See, for example, Routh 1892, 80, 141, and 1898a, 265, 361.

95. Shaw, "Twice Twenty and Four," Box 3, WNS-CUL.

96. On the establishment and early development of the *CMJ*, see Smith and Wise 1989, chaps. 3 and 6.

Hopkins for solving certain difficult questions in dynamics, had been made public by one of Hopkins's pupils in the form of a Tripos problem in 1853. Following the examination, three recently graduated wranglers, Arthur Cayley (SW 1842), P. G. Tait (SW 1852), and William Steele (2W 1852), each published an elegant solution to the problem. Routh himself on at least two occasions took Tripos problems, possibly of his own manufacture, as the starting point for research papers.[97] In these cases it appears that a line of investigation begun with an examination problem in mind turned out to lead to more general results or techniques that warranted publication in a journal.

On some occasions, examination questions could aid or inspire research on much more wide-ranging topics. William Thomson found inspiration of this kind while actually sitting the Tripos of 1845. In the late summer of 1844, just prior to the beginning of his final term's coaching, Thomson encountered a serious difficulty in attempting to establish the equilibrium conditions on which a mathematical theory of the distribution of electrostatic charge on a conductor could be built. In the Senate House, Thomson tackled a problem set by Harvey Goodwin (2W 1840) which "suggested some consideration about the equilibrium of particles acted on by forces varying inversely as the square of the distance." Developing these considerations in the days immediately following the examination, Thomson was able to establish the equilibrium condition he sought.[98] A question set in 1868 by Maxwell further illustrates the role of the Tripos papers as a medium for the local circulation of new results. Maxwell's problem concerned the dynamical interaction between waves traveling through the ether and the atoms of a material medium. One of the expressions to be derived by the candidates predicted that, under certain conditions, an "irregularity of refraction" would occur, a prediction that Maxwell retrospectively claimed as an explanation of the phenomenon of the anomalous dispersion of light.[99] In 1899, when Lord Rayleigh (SW 1865) edited a paper he had written in 1872 on the reflection and refraction of light for inclusion in his collected works, he stated that Maxwell's question "may have been in [his] mind when the text of this paper was written."[100] Rayleigh's vague and belated acknowledgement of his possible debt to Maxwell's problem suggests that he, like many other Cambridge mathematical physicists of the period, found it extremely difficult to keep track of

97. Marner 1994, 19–20.
98. For a technical discussion of these events, see Smith and Wise 1989, 213–16.
99. Harman 1990–95b, 11–12.
100. Rayleigh 1899–1920, 1:156. Maxwell drew attention to his question when refereeing a paper by Rayleigh in 1873.

the many results and techniques picked up from examination questions that subsequently found their way into his research. As we shall see in chapter 6, one of the problems Rayleigh himself set for the Tripos of 1876 began J. H. Poynting (3W 1876) on a line of investigation that eventually led to some very important new results in electromagnetic theory.[101]

Yet another important route by which recent research could become the basis of a Tripos problem was that an examiner could borrow a new technique or result from another Cambridge mathematician. An excellent example of this kind from the late 1850s, and one that beautifully illustrates just how intimate the relationship between undergraduate teaching, new mathematical results, and Tripos examining could become, was brought to light in the autobiographical reminiscences of James Wilson (SW 1859). A few days before the commencement of the Tripos examination of 1859, Wilson took an evening stroll to a Cambridge bookseller to read the papers. When he arrived, Wilson noticed the latest edition of the *Quarterly Journal of Pure and Applied Mathematics (Quarterly Journal)* lying on the counter. Flicking through the pages, he came across an article by the previous year's senior wrangler, George Slesser (SW 1858), on "the solution of certain problems in Rigid Dynamics by a new artifice, referring the data to movable axes." Wilson mastered the method of moving axes out of curiosity, but thought no more about it until a few days later when he was sitting an examination paper. He then had the "singular good fortune" to come upon an extremely difficult problem in rigid dynamics that was very similar to an example given in Slesser's paper and which could easily be solved using Slesser's method. Wilson applied the method to the problem and quickly "got the result stated in the paper."[102]

At first sight this seems, as it did to Wilson, a remarkable piece of happenstance, but a little research quickly reveals the predictable course of events by which Slesser's work ended up as a Tripos problem. Slesser was the first student coached to the senior wranglership by Edward Routh (SW 1854), and, after graduation, Slesser began research on the theory of moving axes, a subject he had been taught by Routh. We know that the method had originally been shown to Slesser by Routh (whose career is discussed in chapter 5), as the latter seems to have been annoyed that one of his pupils had developed and published a technique apparently shown to him confidentially

---

101. Rayleigh considered his Tripos problems so original that he published them; Rayleigh 1899–1920, 1:280–86.

102. Wilson and Wilson 1932, 46; Slesser 1858. George Slesser became Professor of Mathematics at Queen's College, Belfast, in 1860 but died in 1862 aged just 28.

as a useful device for solving a certain class of Tripos problems. Routh had himself published in the *Quarterly Journal* on the subject of moving axes and when he explained the technique in a textbook some years later he noted, rather pointedly, that the general equations had first been published by Slesser to whom "two special cases . . . had been previously shown by the author, together with their application to the motion of spheres."[103] Slesser, it seems, had abused his privileged access to Routh's special techniques—perhaps because he had not discussed his intended research with his old coach—but the episode nevertheless highlights the way coaches could, albeit sometimes unwittingly, shape the early research interests of their pupils. The path from Slesser's training to his first publication was thus extremely smooth, and the further step by which his research found its way onto an examination paper was equally direct. One of the editors of the *Quarterly Journal,* Norman Ferrers (SW 1851), was also an examiner for the Tripos of 1859. Having received and accepted Slesser's paper for publication, Ferrers turned its central technique and one of its special cases into a difficult Tripos problem.[104]

I shall return in chapter 5 to the role of examination questions as models of original mathematical research, but for the moment it is sufficient to note that it was the increasing diversity, novelty, and difficulty of the problems set through the 1830s that prompted the first widespread publication of solutions to Tripos problems. We have seen that Wright's attempts to use problem solutions for pedagogical purposes in 1825 did not catch on during the latter 1820s and 1830s, but, by the early 1840s, two new factors greatly increased the demand for the publication of solutions in some form. First, by this time all but the most elementary examination problems had ceased to be obvious variations on the examples routinely reproduced in textbooks. Any undergraduate hoping to obtain a respectable honors degree had to be able to tackle a wide range of problems case by case, a skill that could only be learned through practice with numerous examples. Those students who were fortunate enough to study with one of the leading private tutors were almost certainly supplied with answers and model solutions to selected Tripos problems, but even these tutors could not guarantee that the solutions they provided followed the mathematical principles or precise line of argument envisaged by the examiners. There was thus a growing market in Cam-

---

103. Routh 1892, 4, 162. Routh observed of Slesser's original contribution that he "uses moving axes, and his analysis is almost exactly the same as that which the author independently adopted."

104. Compare Slesser 1858 and *Cambridge Examination Papers,* Mathematical Tripos, Friday (morning) 21 January 1859, question 5. Ferrers was a Tripos examiner eleven times from 1855 to 1878.

bridge through the latter 1830s for books containing answers and model solutions to past Tripos problems, especially if the solutions were supplied by the examiners themselves. The second difficulty posed by the absence of any formal mechanism for the publication of Tripos solutions concerned the examiners' recently acquired habit of introducing new analytical methods, and even completely novel mathematical results, in the form of Tripos problems. This practice made the more advanced Tripos papers comparable in some respects to research articles or sections of a mathematical treatise. As Edward Routh subsequently pointed out, the more difficult Tripos problems set in mechanics each year provided an excellent guide to "recent directions in dynamical thought" in the university (1898a, vi). Yet apart from occasional articles in research journals, those who studied the Tripos papers prior to the mid 1840s had little or no indication of the techniques of proof or solution envisaged by the examiners.

The first steps toward the resolution of these difficulties were taken in the latter 1830s, when new textbooks began to contain increasingly large numbers of worked examples on each topic. This was clearly an attempt to help students master the many applications of the mathematical methods discussed, but it is important to note that even these books did not reproduce actual Tripos problems, and rarely provided additional, unsolved problems (with answers) for students to tackle for themselves.[105] It was only in the early 1840s that the enormous archive of past examination questions began to be exploited to provide numerous exercises for aspiring wranglers. The first example of this practice occurred in 1841 when John Coombe (4W 1840) published a complete collection of solutions to the Tripos problems of 1840 and 1841. This collection, written with the assistance of the examiners, was explicitly intended to provide exercises for high-flying students during their final term of Tripos preparation. With this aim in mind, Coombe offered solutions that illustrated "general principles applicable to a class of problems" so that students were prepared to tackle new variations on the problems they had mastered. Coombe also noted that he had deliberately provided solutions to problems from the most recent Tripos examinations because they offered the best guide to "the present character of the Senate House Examination" (Coombe 1841, iii–iv).

During the next decade the use of past Tripos problems for pedagogical purposes rapidly became commonplace. When William Walton (8W 1836)

---

105. Circa 1840 textbooks sometimes left the proof of a few propositions or results as exercises for the reader. See, for example, Gregory 1841.

published a new collection of student exercises in mechanics in 1842, it was considered a special promotional feature of his book that many of the problems had been "selected from the Cambridge Senate-House papers."[106] Two years later, in 1844, Matthew O'Brian (3W 1838) published one of the first Cambridge mathematical treatises (in this case on coordinate geometry) to contain "various Problems and Examples, many of which are taken from the Senate-House Problems... added at the end of each chapter" (O'Brian 1844, iv). That same year, 1844, O'Brian was a moderator in the Mathematical Tripos, and he and his co-moderator, Leslie Ellis (SW 1840), established another new tradition in the university by publishing a complete set of solutions to the problems they had set (Ellis and O'Brian 1844). In the introduction to their volume, O'Brian and Ellis drew attention to the fact that although the chief value of a mathematical problem was to illustrate a general mathematical principle or analytical process, the solutions to Tripos problems that had occasionally been published hitherto "did not indicate the point of view contemplated by the proposer." It was in order to provide a set of exemplars for future reference that the moderators had on this occasion been "induced to publish a collection of their own solutions" (1844, Preface).

Seven years later a complete collection of Tripos-problem solutions for the years 1848 through 1851 was published by Norman Ferrers (SW 1851) and J. Stuart Jackson (5W 1851) (Ferrers and Jackson 1851). It is unclear whether the examiners from these years played a significant role in the preparation of the volume, but this project has to be understood in the context of the reforms of 1848. As we saw in chapter 2, this was the year in which the examination was extended from six to eight days and divided into two parts. It was also the year in which the Board of Mathematical Studies was established, its first job being to prescribe a stable syllabus for the Mathematical Tripos. The volume by Ferrers and Jackson provided a definitive guide both to the kinds of problems examiners were henceforth likely to set and to the most appropriate forms of solution to such problems. As the authors of the solutions to the Senate House problems of 1860, Edward Routh and Henry Watson (2W 1850), noted, it was very important that problem solutions were provided by an authoritative source, preferably the "framer," so the that the student learned "the manner in which he is expected to proceed in the Senate House."[107]

---

106. Walton 1842, Preface. Walton's book also contained a small number of problems (with answers) for students to solve for themselves.

107. Routh and Watson 1860, Preface. Routh and Watson were the moderators for the examination of 1860.

Further evidence of the demand for definitive solutions to Tripos questions in the wake of the reforms of 1848 is provided by the publication of a book of solutions to "Senate-House 'Riders'" for the years 1848 to 1851 by Francis Jameson (6W 1850). In view of the often uncomprehending way in which less able students reproduced from memory the proofs and theorems required in bookwork questions, the Board of Mathematical Studies recommended in 1849 that every such question be accompanied in future by a "rider" which required the student to apply the bookwork to solve a short problem.[108] The rider carried around half the marks of the question and was probably intended also to reduce the effect of cheating among the less able candidates.[109] These questions were known from around 1850 as "bookwork-and-rider" questions, the problems that constituted the riders frequently being as ingenious and as difficult as many of the questions set on the easier problem papers. As Jameson sternly advised his undergraduate readers, no student should consider a proposition or piece of bookwork thoroughly mastered until he had "diligently practised examples connected with it, so as to be able, when called upon in an Examination, to apply it readily to any required purpose" (1851, viii).

It was, as we have seen, a mark of originality in Cambridge students that they could apply their mathematical knowledge to the rapid solution of novel examination problems. Indeed, by the early Victorian period it was this ability that had become the definitive product of the new paper-based pedagogy peculiar to the university. The closest we can now come to glimpsing in real time the skills of speed and application that an able student had to display is to study exemplary problem-solutions actually written by candidates in the Senate House. My comments on Poynting's answer to Rayleigh's question of 1976 (discussed in chapter 1) showed the sort of information that can be gleaned from examination scripts; here I analyze one more example in order to look more closely at the technical skills demanded by typical Tripos problems. The following solution by William Garnett (5W 1873) was written in answer to a difficult bookwork-and-rider question set by Maxwell for the Tripos of 1873.[110] Based on rigid dynamics (a topic at the heart of mixed

---

108. Jameson 1851, v.

109. It was fairly common until at least the mid 1850s for some of the least-able candidates to steal paper from the Senate House a few days before the examinations and to write out some propositions they thought likely to appear. These were then smuggled into the Senate House and "shown up" at the end of the examination. On this and other common methods of cheating, see Harrison 1994, 75–77.

110. This example is also chosen as it might equally have been set as an easy problem. Both Maxwell and Lord Rayleigh used old examination scripts as rough paper when drafting their sci-

mathematical studies), Maxwell's question concerned a wheel on an axle suspended by three strings, one wound round the wheel and two wound the opposite way round the axle on each side of the wheel (fig. 3.4).[111] The student was required to determine the relationship between the physical dimensions of wheel and axle such that when the string around the wheel was drawn up or let down, the tension in the other two strings remained the same. This part of the question, whose detailed form was unlikely to have been anticipated even by the best coach, tested the student's ability (i) to visualize the arrangement described, (ii) to analyze the kinematics and dynamics of the arrangement, (iii) to decide what kinematic and dynamical relationships satisfied the special conditions given, and (iv) to show that the conditions led to the required expression without making an error in algebraic manipulation.[112]

Garnett's attempt at this question is typical of those submitted by the top six wranglers. Maxwell allotted a total of 51/66 marks for the bookwork and rider respectively. Garnett scored full marks for the bookwork, crossing out only a couple of minor errors in two pages of neat, textbook-like prose. His attempt at the rider was slightly less successful. He made a three-dimensional sketch of the arrangement described by Maxwell but did not take sufficient care in defining the length of the axle. It is not clear in Garnett's sketch whether the length "a" of the axle includes the part running through the wheel (as Maxwell's result requires) or whether it is a measure only of the parts protruding on either side of the wheel (as Garnett's calculation of the total mass "M" assumes). Garnett correctly identified the forces keeping the arrangement in equilibrium, the accelerations generated by raising the string round the wheel, and the dynamical conditions that should lead to the required result. In setting up the equations from which the required expression could be derived, however, he made a second, more serious error (penultimate line on fig. 3.4) by associating the additional tension (τ) produced by pulling on the string around the wheel with the acceleration (f) of the string itself (rather than the acceleration of the center of

---

entific papers and books. Numerous answers to Tripos problems are to be found on the backs of their manuscripts.

111. *Cambridge University Examination Papers*, Mathematical Tripos, Thursday (morning), 2 January 1873, question v.

112. Maxwell's rhymed solution to the chain problem (see epigraph to this section) follows a similar sequence of operations. Having sketched the problem and labeled the key variables, he commences his analysis with the lines: "In working the problem the first thing of course is / To equate the impressed and effectual forces." He then writes down the dynamical and kinematic relationships, solves the resulting equations, and works out several special cases (Cambell and Garnett 1882, 625–28).

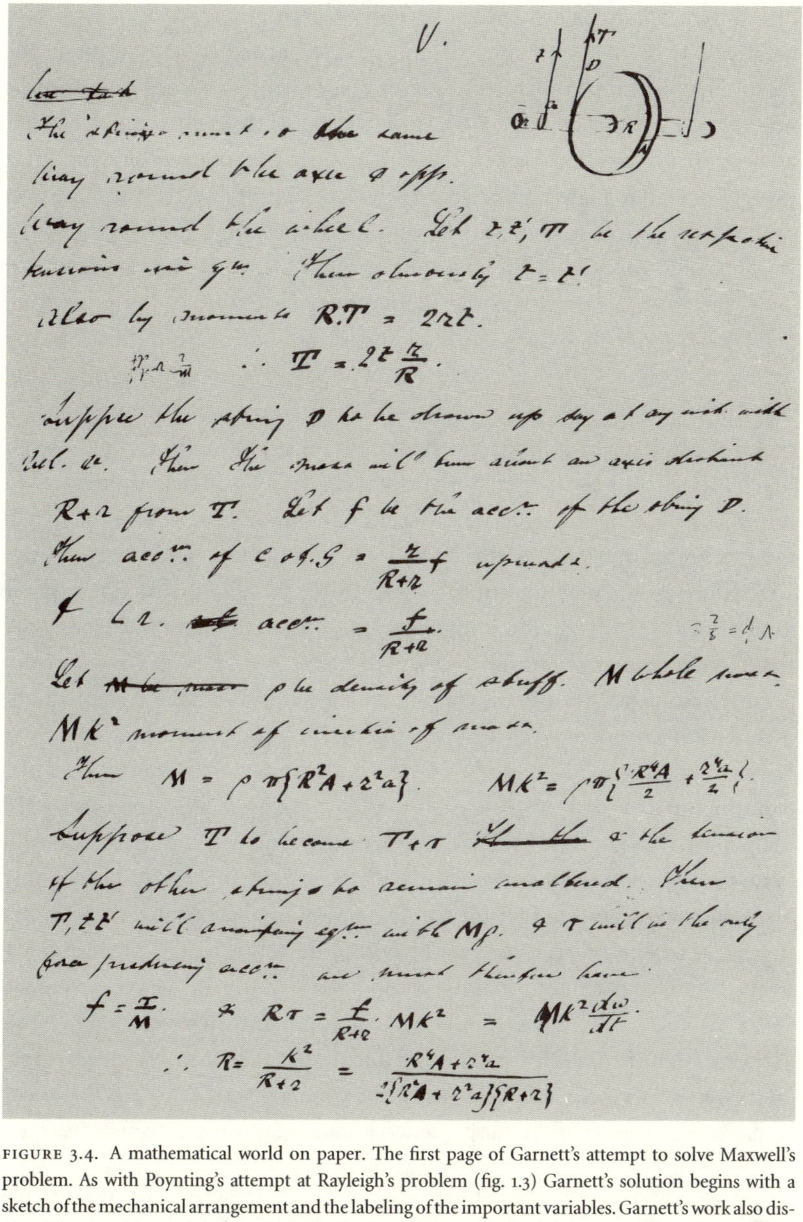

FIGURE 3.4. A mathematical world on paper. The first page of Garnett's attempt to solve Maxwell's problem. As with Poynting's attempt at Rayleigh's problem (fig. 1.3) Garnett's solution begins with a sketch of the mechanical arrangement and the labeling of the important variables. Garnett's work also displays his corrections and errors as he struggles in real time to obtain the required result. The two equations jotted upside down are rough calculations made by Maxwell at a later date. Box 5 Add 7655/vk/8, verso (JCM-CUL). (By permission of the Syndics of Cambridge University Library.)

mass of wheel and axle together).[113] This error, together with the incorrect definition of the length of the axle, meant that Garnett could not obtain the required expression. When this became clear after a few lines of algebra he noted that he had probably made "an error in estimating the angular acceleration of the mass" (which he had not) before moving on the next question. Maxwell deducted ten marks overall for Garnett's errors, leaving him with a total of 51/56 for bookwork and rider respectively.[114]

The questions set on the most difficult problem papers were far more complicated and mathematically demanding than the example just discussed, but scripts like Garnett's nevertheless provide a rare glimpse of a top wrangler deploying the skills that the new pedagogical apparatus was designed to produce.[115] The fact that Garnett attempted all nine questions on the paper and was unwilling to waste time trying to locate his errors suggests that he spent little more than ten minutes on the rider. This would have left him very little time to ponder or experiment with solutions; having launched upon a line of analysis, he was bound to see it through and then to move straight on to the next question. This example highlights that fact that the kind of originality required of mathematics undergraduates was largely defined by the form of the problems they were expected to solve. They were not required to invent and deploy new physical principles or mathematical methods, nor even to analyze novel or unfamiliar physical phenomena. They were required, rather, to show that they could understand the enunciation of a well-formulated problem, analyze the physical system described using the principles and techniques they had been taught, and use that analysis to generate specific mathematical expressions and relationships. It was, as we have already seen, this aspect of undergraduate training that enabled some recent graduates to publish solutions to the more interesting Tripos problems. Indeed, the ability to solve problems rapidly and elegantly was so well developed in some students that they could even outperform their examiners while working in the Senate House. The standard of excellence by which examination questions were marked was set by the proofs given in the best textbooks and the solutions to problems provided by the examiners. It sometimes hap-

---

113. Garnett has correctly calculated the acceleration $[rf/R + r)]$ of the center of mass in the middle of the page.

114. vk/8/8 (xii), JCM-CUL. The bulk of the marks were probably deducted for the error in dynamics as the definition of the axle is somewhat ambiguous.

115. The top two wranglers scored full marks for the question, but below the seventh wrangler the marks tail off dramatically. Only seven of the 41 students who attempted the question scored more than half the marks, no one below the fourteenth wrangler scoring any marks at all for the rider (vk/8/8 [xii], JCM-CUL).

pened, however, that an outstanding student offered one or more solutions that were judged better than those provided by the examiners. In these cases the student could score more than full marks for the question (and even for the whole paper) and have his original answers published.[116] If the examiners produced a set of solutions to their Senate House problems, they would generally use those student answers they considered superior to their own.[117]

Collections of solutions to Senate House problems and riders were published at regular internals during the mid-Victorian period, generally appearing in the wake of alterations to the Tripos syllabus in order to show students and tutors how problems and difficult riders on new topics were likely to be framed and were properly solved.[118] From the mid 1840s, it also became increasingly commonplace for Cambridge textbook-writers to reproduce large numbers of past Tripos problems in illustration of virtually every branch of mixed mathematics. Thus Isaac Todhunter (SW 1848), one of the most influential Cambridge textbook writers of the mid-Victorian period, selected his examples "almost exclusively from the College and University Examination papers," sometimes reproducing more than seventy problems at the end of a chapter.[119] Some tutors and examiners produced books of selected Tripos and college examination problems (with or without solutions), collected under the various divisions of the Tripos syllabus, for students to work through either by themselves or under the guidance of a private tutor.[120] And, as we shall see in chapter 5, these textbooks and books of problems helped to carry the techniques and ideals of Cambridge mathematics into schools and institutions of higher education throughout Britain and her Empire. The use of the examination papers as vehicles for the announcement of new mathematical methods or results also accelerated the ongoing change in relative status between Tripos problems and textbook examples. Where examination questions had once been mere variations on the standard problems discussed in mathematical treatises, they now became definitive of many of the techniques and examples discussed in textbooks. Look-

---

116. If the student scored more than full marks, he was said to have "beaten the paper" (Wright 1827, 1:290). Bristed reports that Robert Ellis beat a paper in the Tripos of 1840 (Bristed 1852, 1:238–39).

117. See, for example, Walton and Mackenzie 1854, Preface; and Greenhill 1876, vii.

118. In addition to those mentioned above, volumes of compete solutions to problems and riders were also published by Walton and Mackenzie (1854), Campion and Walton (1857), Walton et al. (1864), Greenhill (1876), and Glaisher (1879).

119. Todhunter 1852, v. Todhunter acted as a Tripos examiner in 1865 and 1866.

120. For examples of books of college and Tripos examination problems, see respectively Morgan 1858 and Wolstenholme 1867.

ing back in the late 1880s on the development of mathematics in early and mid-Victorian Cambridge, Edward Routh pointed out that the problems set in the Senate House had "generally [been] absorbed into the ordinary textbooks, and become the standard examples by which successive generations of students acquire[d] their analytical skill" (1889, iii). Referring to the papers of just one prolific examiner, Thomas Gaskin, Routh noted that it now seemed remarkable "how many of [Gaskin's] problems [had] been taken, and may be recognized as old friends" (1889, iii).

The inclusion of illustrative exercises and problems at the end of chapters in textbooks of mathematical physics is now so commonplace as to seem unexceptional, but it is important to appreciate that this pedagogical device is of relatively recent origin and was introduced in a specific historical context. The books from which students learned higher mathematics in Cambridge made no use of these devices prior to the 1840s, and, when they were introduced, they were taken from an extant and local tradition of competitive written examination. This tradition already enshrined the belief that mathematics was best learned through problems and examples, and ranked student performance by the ability to solve such problems under severe constraints of time. The introduction of Tripos problems to mathematical treatises was an important step in the invention of the modern technical textbook and one which assumed and embodied the ideal of the written examination as the natural means of testing student ability. The immediate outcome of this innovation was to raise the standard of technical performance in Cambridge still higher by making a wide selection of worked exemplars and student exercises far more readily available. In the longer term, the appropriation of examination questions as textbook exercises and end-of-chapter problem sets played an important role in shaping undergraduate notions of the nature of mixed mathematics and, as we shall see in chapter 5, of how it was properly advanced through research.

3.8. The Material Culture and Practice-Ladenness of Theory

Commenting more than twenty years ago on the rapid mathematization of physical theory which took place in Europe from the mid-eighteenth century, historian Enrico Bellone noted that "the 'paper world' of theory experienced a very lively growth between 1750 and 1900."[121] Few would disagree

---

121. Bellone 1980, 5. Bellone emphasizes that mathematical theories of the physical world generated an increasingly autonomous intellectual space over the latter eighteenth and nineteenth centuries.

with this statement, but Bellone, like other historians of mathematical physics following his lead, uses the expression "paper world" to refer to an abstract intellectual space rather than to a process of practical reasoning involving the material apparatus of pen, ink, and paper. In this chapter I have explored the historical development of the material culture of mathematical work by highlighting the importance of paper-based calculation not merely as a record of what goes on inside a theoretician's head, but as a skilled activity that developed coextensively and inextricably with technical innovations in mixed mathematics. It is by charting the gradual institutionalization of these skills in undergraduate studies that what might be termed the "practice-ladenness" of mathematical theorizing can most effectively be brought to light.[122]

Historians of seventeenth-century university education have rightly pointed out that one reason the new physico-mathematical sciences of that period did not initially flourish at postgraduate level in Oxford and Cambridge is that these were primarily undergraduate institutions whose collegiate structure and liberal educational ideals were conducive neither to effective communication between professors and undergraduates nor to the specialized study of technical subjects.[123] The vast majority of undergraduates were hopelessly ill equipped—as were their tutors—to deal with Newton's celestial mechanics or the fluxional calculus and had little incentive to master the technicalities of this new knowledge. They were therefore extremely unlikely to try to contribute to the development of these subjects after graduation. What is clear from this chapter is that even if there had been a widespread desire to foster the new subjects in undergraduate studies, such a change could not have been accomplished simply by altering the syllabus. As long as the bulk of formal teaching was based on reading, tutorial discussion, catechetical lectures, and public disputations, training students to a level of technical competence from which they might follow a professor's lectures on anything but elementary mathematics remained virtually impossible.[124] Those few undergraduates who did reach a level from which they could embark upon original mathematical investigations shortly after graduation acquired their learning mainly through private study aided by those senior members of the univer-

---

122. For a similar approach to the practice ladenness of mathematical computation, see Warwick 1995b.

123. See, for example, Feingold 1997, 447–48.

124. For examples of the increasing long and complicated, paper-based analytical calculations being undertaken by Continental analysts in the mid-eighteenth century, see Greenberg 1995.

sity who recognized and encouraged their exceptional enthusiasm and ability. The propagation of the advanced mathematical sciences therefore remained capricious in the early modern universities in this period in the sense that it relied almost entirely on the personal interests, abilities, and reading of individual students as well as the goodwill and expertise of a handful of tutors.

What made the skills acquired by these students more visible and accessible in the latter eighteenth century was the change in pedagogical practice that accompanied the introduction of written assessment. As success in the Senate House examination came to depend upon the reproduction of mathematical knowledge on paper, the majority of students began to adopt learning techniques that had once been the preserve of a tiny minority of enthusiasts and academicians. As we saw in chapter 2, the rise of competitive examination prompted large numbers of students to turn to private tutors who were prepared, for a fee, to pass on their own technical expertise through personal instruction, varied explanations, progressive teaching, expository manuscripts, and graded exercises. The gradual institutionalization of once informal and rare processes of technical education also opened the way to further pedagogical developments that would enable paper-based mathematical training to transcend the boundaries of its eighteenth century origins. As we have seen, the relative speed of the analytical revolution in Cambridge from the late 1810s to the early 1830s was driven by the combined effects of competitive examination, private teaching, new textbooks, and innovative problem setting by examiners; the breadth and level of technical competence achieved by the top wranglers toward the end of this period would have been unimaginable just two decades earlier.

The outcome of this pedagogical revolution is best summarized as a change in the *scale* of mathematical training at one specific site. The doubling of the number of students achieving honors in the Mathematical Tripos during the analytical revolution was accompanied by a sharp increase in the breadth and level of mathematical knowledge that had to be mastered in just ten terms. The size, competence, and eminence of the mathematical community generated by this system also attracted students with mathematical ability from all over Britain, a development that further heightened both the atmosphere of competitive study and the standard of technical performance. By the latter 1820s, Cambridge was fast becoming not just the most important center of mathematical study in Britain, but an almost obligatory passage point for British students who aspired to make original contributions to mixed mathematics.[125] When a new journal was started at the beginning of the 1860s to encourage original publications from young mathematicians at the three centers of mathematical learning in Britain—Oxford, Cambridge and Dublin—it turned out to be unsatisfactory because, despite the best efforts of the editors, more than 90 percent of the contributions came from Cambridge.[126] The journal was refounded in the early 1870s with several new Cambridge-based editors and thereafter became an important vehicle for the early publications of many young wranglers.[127]

The fact that the Cambridge training system was problem oriented and paper based also shaped the skills possessed by Cambridge graduates. Long hours spent solving novel examination problems—which, from the 1840s, were embodied in a new tradition of textbooks—gave them a degree of originality in the application of mixed mathematics that was not possessed by students from other institutions. The inclusion of original results and new mathematical techniques in the problems set by some examiners also enabled the most able undergraduates to become familiar with the topics and methods being employed by at least some, especially Cambridge-trained, research mathematicians.[128] Most important to our current concerns, however, is that the development of the new pedagogical economy from a play of written questions and answers led to paper-based calculation becoming the common currency of private teaching, private study, and competitive assessment. The importance of the material culture of mathematical work lies not so much in the simple materiality of pen, ink, and paper, as in the many activities that are bound together by their use. Mastering mixed mathematics was not just a matter of learning the appropriate notation and writing out proofs and theorems until they could be reproduced from memory—though these activities

---

125. The few outstanding analysts whose careers appear to confute this claim actually support it on closer inspection. The great Irish mathematicians William Rowan Hamilton and James MacCullagh were bringing French analytical mathematics to Dublin in the 1820s, and the fact that they did not establish a major school was partly due to the lack of a training system comparable to that in Cambridge. The early studies of the Nottingham miller and analyst, George Green, were almost certainly supported by two Cambridge-trained mathematicians, John Toplis and Edward Bromhead. See Cannell 1993.

126. *The Oxford, Cambridge and Dublin Messenger of Mathematics* 1 (1862): 1.

127. *Messenger of Mathematics* 1 (1872): Preface; A. R. Forsyth 1929–30, x.

128. The emphasis on "physico-mathematics" remained very strong in Britain in the mid-nineteenth century. The subjects recommended by the editors of the *Oxford, Cambridge and Dublin Messenger of Mathematics* as especially worthy of investigation were the moon's motion, the propagation of sound, the wave theory of light, and the application of quaternions to physical problems.

were important—but of learning through the very act of repeated and closely supervised rehearsal on paper to manipulate mathematical symbols according to the operations of, say, algebra or the calculus, to apply general laws and principles to numerous physical systems, and to visualize, sketch, and analyze a wide range of geometrical and physical problems.

It is in the relationship between this complex pedagogical economy and the specific range of skills, competencies, and attitudes that it produced that what I referred to above as the practice-ladenness of mathematical theorizing is most clearly seen. William Garnett's solution to Maxwell's wheel-and-axle problem (fig. 3.4) provides a real-time example of a small fraction of these abilities in action. Readers with a training in mathematical physics might find these abilities too elementary to be of any real significance, but it is precisely their taken-for-granted status that I wish to emphasize. Histories of mathematical physics are generally concerned with novel and innovative contributions to the discipline, but without the shared skills and competencies generated by years of intense training, it would not be possible to sustain a large community of expert practitioners within which highly technical, innovative work, could be generated, assessed, disseminated, and advanced.

The above remarks also raise a number of important issues concerning collaborative activity within and between research communities, especially with respect to some of Thomas Kuhn's influential remarks on the role of "paradigms" and "normal science" in the development of mathematical physics. One of the most important insights of Kuhn's philosophy of science, at least for this study, is the recognition that a physicist's knowledge cannot adequately be described through the formal statement of the laws and principles he takes for granted and the mathematical methods he employs. Kuhn argued that a physicist's knowledge is intimately linked to innumerable other skills and competencies acquired not through miraculous insight into the essential meaning of, say, Newton's laws of motion, but through years of practice at solving canonical exemplars in each of the areas where Newton's laws had been successfully applied (Kuhn 1970, 46–47, 188–89). This admirably practice-oriented approach to mastering science nevertheless sits somewhat uneasily with Kuhn's more idealistic notion of a ubiquitous paradigm that can guide the development of a technical discipline as a normal-scientific activity. Kuhn was fond of referring to the exercises found at the end of chapters in mathematical textbooks to support his claim that physics students learned by practice rather than by precept, but he also acknowledged that such books did not become "popular" until the early nineteenth century (1970, 10). Prior to this, he suggested, it was Newton's

Principia that acted as a canonical text for the development of celestial and terrestrial mechanics. Those who read the *Principia* were supposed to have acquired, from the laws, proofs, theorems, and solved problems in the book, the specific tools and techniques necessary for the normal-scientific development of "Newtonian mechanics" during the eighteenth century.

The development of celestial and terrestrial mechanics over the eighteenth century has long stood as the classic example of normal scientific activity, a process that allegedly witnessed the "mopping up" of problems implicit in Newton's *Principia*. Yet several historians of the mathematical sciences have insisted that this is a mistaken view which derives from a simplistic understanding of the work in question and an artificial separation of the histories of mathematics and mechanics. Men like Euler and D'Alembert were not just formalizing and extending Newton's work in a more appropriate mathematical language, but were actively and simultaneously creating and applying new mathematical methods and mechanical principles. It is a mark of the novelty of their work that they often failed to follow each other's technical arguments and sometimes disagreed strongly about the legitimacy of each other's problem solutions.[129] In short, the range of mathematical methods and mechanical principles they invented and employed is far too disparate and diverse to be meaningfully understood to have originated in the examples given in the *Principia*.[130] While we might therefore acknowledge that the *Principia* was a paradigmatic text to the extent that it offered, in Kuhn's words, "a criterion for choosing problems that ... can be assumed to have solutions" (1970, 37), it did not provide the very technical apparatus through which those solutions were eventually accomplished. What is missing from Kuhn's account of normal science is a historical explanation of the way a community or subcommunity of practitioners comes to agree on what can be taken for granted as the foundational principles and techniques to be deployed in the solution of problems. Kuhn's only comments on this crucial issue were a few vague references to the importance of textbooks, but the content and use of such books cannot be properly analyzed in isolation from the pedagogical regimes in which they were used.

129. Trusdell 1984, 98–101; Guicciardini 1989, 141; Greenberg 1995, 621. Greenberg suggests that eighteenth-century mechanics has served as the "paradigm of a 'paradigm.'"

130. See Greenberg 1995, 620–24, and Garber 1999, 35–53. The different variational principles used by Continental mathematicians to tackle mechanical problems were quite alien to the Cambridge understanding of Newton's mechanics and could barely be expressed in the notation of the fluxional calculus. See Rouse Ball 1889, 97, and Guicciardini 1989, 89–91, 131, 141. Guicciardini discusses the difficulties experienced by Edward Waring (SW 1757) in understanding Euler's works.

What began to bring a measure of collective order to the disparate works of many Continental analysts at the end of the eighteenth century was the new regime of teaching and examination instituted at the Ecole Polytechnique in Paris.[131] It is not my purpose here to discuss pedagogy at this major center, but there are many important similarities between the training regimes instituted circa 1800 at the Ecole Polytechnique and Cambridge respectively that can usefully be emphasized. Although the former was organized according to military rather than monastic discipline and brought the professors into regular and effective contact with the students, it shared with Cambridge strong emphases on the teaching of small classes of students of roughly equal ability and on the thorough mastery of mathematical technique by supervised rehearsal on paper.[132] Likewise when Franz Neumann was struggling in the mid 1830s to establish the first physics seminar in Prussia, he felt that his efforts were being undermined by, among other things, the lack of a paper-based system of training and examining for physics students. Eventually Neumann himself developed a common system of graded exercises that introduced students to a hierarchy of essential mathematical skills and techniques, and (in the absence of problems in textbooks) began to construct his own problem sets through which his students could learn their craft.[133]

The common thread linking these otherwise very different centers of education in mathematical physics is a new concern with training students in the minutiae of analytical technique and its application to the solution of exemplary physical problems. Such skills were a prerequisite not just to higher studies in mathematics and mathematical physics but also, when held collectively, to the kind of collaborative activity that Kuhn identified as normal science. On the model presented here, then, normal-scientific activity is the product not of a ubiquitous paradigm originating in a canonical text, but of specific and localized pedagogical regimes. It was the emergence of these

---

131. Grattan-Guinness 1990, 1: chap. 2; Hodgkin 1981.
132. For a detailed contemporary account of the teaching methods used at the Ecole Polytechnique, see Barnard 1862, 59–87. Barnard actually remarked upon the similarity between the Polytechnique and a large Cambridge college.
133. Olesko 1991, 111, 113–14. Neumann remarked that without such training the study of physics was reliant on the vagaries of "inclination and talent." The German academic, V. A. Huber, who compared undergraduate studies in Cambridge with those at German universities in the mid 1830s, noted that while "English examinations" in classics were "somewhat superior to our own" those "in the Mathematics [were] altogether beyond us." He attributed much of he superiority of Cambridge training to the use of "private study" (as opposed to formal lectures) and to the "method of paper examination" (Huber 1843, 2:358–63).

regimes in the early nineteenth century that generated communities within which the rapid growth of the paper world of mathematical-physical theory could take place. It is also implicit in this model that different training regimes will produce students with different technical skills, interests, and priorities, but this does not imply that their respective mathematical knowledge will be radically incommensurate. There was certainly a good deal of overlap in the range of competencies inculcated at each of the institutions mentioned above, due in large measure to the common ancestry of many of the mathematical methods and problem solutions taught. To what extent the production and interpretation of new knowledge in mixed mathematics in late nineteenth- and early twentieth-century Cambridge was specifically tied to the pedagogical regime by which wranglers were trained is a question that is explored in the second half of this study.

~ 4 ~
Exercising the Student Body
Mathematics, Manliness, and Athleticism

> A German student is not a pleasant companion he is unclean, eats "Blut-und-Leber Wurst" and is usually not "begeistert." On the other hand the English Undergrad, is as a rule clean, lives healthily and learns in a healthy fashion. The aesthetic ideal mind and body combined, is on our side.
>
> KARL PEARSON, 1880 [1]

4.1. The Body-built Mathematician

Shortly after being placed third wrangler in the Mathematical Tripos of 1879, the twenty-two-year-old Karl Pearson (future statistician and eugenicist) set off to the recently united Germany to further his education.[2] Convinced of the superiority of German professors, German universities, and the sheer scale of German learning, he studied physics and metaphysics in Heidelberg, and law and evolutionary theory in Berlin. But, as he wrote to his friend and mentor Oscar Browning back in Cambridge, his intellectual quest seemed to have ended in disappointment. Pearson found German intellectual life uncultured and overly concerned with "facts." In Berlin he found the art, the architecture, and the people "barbaric"; the streets of Berlin flowed not with "Geist," he lamented, but with "sewage."[3] Pearson's comments should not, of course, be taken at face value. After eight months of itinerant study in a new land and a new language, he was tired, homesick, and increasingly anxious about his professional future. His youthful hopes of finding intellectual enlightenment beyond what he had seen as the competitive mathematical myopia and outdated religiosity of Anglican Cambridge had

---

1. Karl Pearson to Oscar Browning, 16 January 1880 (OB-KCC).
2. For an overview of Pearson's career, see E. S. Pearson 1938.
3. Pearson to Browning, 16 January 1880 (OB-KCC).

faded.[4] The collapse of his belief in the intellectual superiority of the new German Reich had reduced him to racist insults.

In another sense, however, Pearson had been enlightened. He confided to Browning that his studies in German universities had taught him the merits of the very Cambridge system he had so recently abandoned. Pearson now reckoned Oxbridge professors every bit as cultured and learned as their German counterparts. He had, moreover, acquired a new respect for Cambridge mathematical discipline. Pearson's remarkable technical proficiency in mathematics had given him a satisfying edge over his German peers when it came to tackling problems in mathematical physics.[5] In Berlin, he often found himself idly "scribbling" such problems when he ought to have been studying Roman Law, longing once more to be "working with symbols and not words."[6] He even conceded to Browning that mathematical physics seemed somehow "true" in a sense that eluded other disciplines, as if in physics one were "struggling with nature herself."[7] Pearson's German adventure had given him a new perspective on his undergraduate experience in Cambridge. As a student he had reluctantly accepted the tough regime of coaching, knowing from the start that he would be assessed solely by the number of marks he could accrue in nine days of gruelling examination.[8] What had been less visible to him as a student was the Cambridge ideal of combining both bodily and intellectual endeavor. Pearson had always balanced hard mathematical study against such physical activities as walking, skating, and hockey, but only in Germany did he recognize this as an especially beneficial practice and one that was peculiar to English universities.[9] As he wrote to his mother in May 1879:

> I used to think that Athletics and sport generally was overestimated at Cambridge, but now I think it cannot be too highly valued. The Germans here are surprised at my knowing an Englishman always by sight and want to know what it is, but I can hardly tell them that it is a compactness, a look of strength which they never possess. If I set out with one over the hills, I almost kill him

---

4. Pearson to his mother, 22 July 1880 (CP-UCL).

5. For example, Pearson (1936) strongly supports the training provided by the Mathematical Tripos.

6. Pearson to Browning, 16 January 1880 (OB-KCC).

7. Ibid.

8. Pearson's undergraduate correspondence reveals his discomfort under the competitive training system; see Pearson to his mother, 15 December 1878 (CP-UCL).

9. On Pearson's undergraduate sporting activities, see Pearson to his brother Arthur, 22 December 1878 (CP-UCL).

before half-an-hour is over and this is the case where he has served his time as a soldier and one would think ought to be enured to fatigue.[10]

It now seemed to Pearson that regular exercise and organized sport were important undergraduate activities that endowed English students with a strength and stamina unknown to their Continental peers.

Pearson's insistence on the importance of an "aesthetic ideal" of "mind and body combined" in a university that prized mathematical studies above all others might seem peculiar to the modern reader. Mathematics and mathematical physics are, after all, generally regarded as archetypical of disembodied disciplines whose efficacy ought not to depend on the physical fitness or state of health of the practitioner. Furthermore, the culture of athleticism which became so characteristic of British educational establishments, including Cambridge, during the mid-Victorian era has generally been *contrasted* with scholarly culture, as if these two cultures straightforwardly reflected a long-standing division between serious scholars committed to academic excellence and the sons of the aristocracy who sought a good time and a pass degree.[11] Yet, as the epigraph above reveals, Pearson had come to recognize the *combination* of fully developed body and fully developed mind as constitutive of the robust Cambridge intellect. Nor was Pearson in any way original in making this connection, especially with respect to men of science. Twenty years earlier a British medical man had cautioned against the creeping effects of "namby-pambyism" and "effemination" caused by a lack of physical training among students. Those who felt sheepish about participating in games of tennis or cricket, or who sought direct evidence of the benefits of physical fitness, were recommended to: "Go to the British Association for the Advancement of Science—that parliament of the intellect! and observe there who are the science-Hercules. You will find them amongst the men who are most genial, best body-built, least conventional."[12]

In this chapter I explore the relationship between training in the coaching room and on the playing field in order to reveal a new economy of the student mind and body that became definitive of mathematical study in early Victorian Cambridge. In the sections that follow I begin by discussing the relationship between mathematics and the ideals of a liberal education, and explain how these ideals enjoined students both to study extremely hard and

---

10. Pearson to his mother, 11 May 1879, (HHP-UCL), Box 8.
11. On sport in the Victorian education system, see Haley 1978, Mangan 1981, and Curthoys and Jones 1995.
12. *British Medical Journal*, 23 January 1858, 79.

to compete fiercely with their peers. This change in undergraduate life, described by contemporaries as an effective *industrialization* of the learning process, helped to drive the increase in technical performance that took place during the 1830s and 1840s, but it was not accomplished without cost. Using contemporary student diaries and correspondence, we shall see that undergraduates privately railed, though never publicly rebelled, against a system that incited them to study to the point of emotional and intellectual breakdown. It was, I shall argue, in the context of this demanding system, especially after 1815, that undergraduates began to employ regular physical exercise both to regulate the working day and in the belief that it preserved a robust constitution.

We shall also see that, during the 1830s, physical and intellectual endeavor became even more closely linked as new team sports added a complementary, competitive dimension, to physical exercise. By the 1840s, the combined pursuits of mathematics and sport had become constitutive of the liberal education through which good undergraduate character was formed. In the coaching room, on the playing field, and through spectacular public ceremonies celebrating the success of the top wranglers, undergraduates were constantly reminded of the virtues of mathematical study, physical exercise, and competition. Emphasis on both bodily and mental development in a university that prized manly individuals of rational intellect also enabled the respective ideals of physical and intellectual endeavor to pervade one another: success in competitive sport became a hallmark of the rational body, while the hard study of mathematics became a manly pursuit. The research style of Cambridge mathematicians was also shaped, in part, by these same ideals of competition, fair play, and manliness. In conclusion I shall argue that, just like written examinations and many of the teaching methods developed by mathematical coaches, the links between competition, physical fitness, and intellectual endeavor forged in Victorian Cambridge remain embodied in elite technical training in the Western academy.

4.2. Mathematics, Competition, and a Liberal Education

We have seen in the previous two chapters that the introduction of competitive examinations and private teaching in mathematics in late Georgian Cambridge completely transformed undergraduate studies in the university. For our present purposes the most important aspect of this transformation is the change it wrought in the aspirations, attributes, and lifestyle of the ideal undergraduate. The respecification of university education as a reliable mechanism of self-improvement for bright students from the middle classes proved a powerful incentive to hard study. Those who came to Cambridge in search of meritocratic advancement were obliged to submit themselves to its tough regime of disciplined learning. Indeed, in a system based upon competitive study and open-ended examination papers, the academic standard was defined, and continually inflated, by the combined efforts of the examiners, the most able students, and their private tutors. By the 1830s, as we have seen, mathematical studies had displaced virtually all other disciplines in the university, even those students who intended to sit the Classical Tripos being required first to obtain honors in mathematics.

This strong emphasis on mathematics in what was regarded in early Victorian Cambridge as a liberal education requires further comment. We saw in chapter 2 that one of the chief architects of the Mathematical Tripos during the 1830s and 1840s, William Whewell, defended the place of mathematics at the heart of a liberal education by arguing that this discipline alone offered students reliable "fundamental principles" from which they could learn to argue to equally reliable conclusions. One might imagine logic to be the obvious pedagogical means to this end, but, Whewell pointed out, although logic enabled one to draw proper conclusions from given premises, it could not guarantee the truth of the premises themselves. A premise could only become a reliable fundamental principle if it contained an intuitive dimension that rendered it self-evident to the student. Whewell firmly believed that mathematics—especially Euclidian geometry and Newtonian mechanics—provided both the soundest fundamental principles and the most certain means of proceeding through "strict reasoning" to certain conclusions. Consider the case of mechanics. Here the mathematical definition of "force" enabled one to use the calculus to solve difficult mechanical problems while the familiar sensation of muscular effort enabled the student literally to "feel" the concept of force and, thereby, to shift its status from a mere algebraic variable to a practical conviction.[13] In passing correctly through the steps of the solution to a difficult problem in mechanics, a student was therefore able to display the ability to make his speculative inferences coincide with his practical convictions—an ability that supporters of liberal educational values thought to be invaluable in any walk of life.

The place of mathematics at the heart of a Cambridge liberal education was justified, then, not as a vehicle for training professional mathematicians, but as the best discipline through which to teach students, via practical

13. Whewell 1835, 5, 24–25.

examples, to reason properly and reliably. It was, as we have already seen, for this reason that Whewell and some of his contemporaries were hostile to the wholesale introduction of Continental analysis into the undergraduate curriculum. They believed that some analytical methods—such as the calculus of variations—obscured the proper logical steps to be followed in solving a physical problem, and so preferred to retain geometrical and mechanical methods in which the student never lost sight of the physically intuitive principles behind a problem. In this sense, mixed mathematics was a language through which rational students could be identified and ranked relative to their peers. Furthermore, those who excelled in the Mathematical Tripos revealed themselves as potential leaders who would occupy positions of responsibility as teachers, clergymen, lawyers, statesmen, imperial administrators, and men of business. This role of mathematics, to prepare the nation's intellectual elite, lent an extremely important moral dimension to the study of the discipline. It was the duty of every student—both to himself and to his country—to display his ability through mathematical attainment and to strive for high placing in the order of merit. But, as we have also seen, during the analytical revolution an ever widening gap opened between the ideal of a liberal education promoted by men like Whewell and actual undergraduate experience. Whewell believed that undergraduate teaching should preserve a strong spiritual and moral component based on Anglican theology, and that bright undergraduates ought not to concentrate excessively on higher mathematics merely in order to perform well in a purely technical examination. William Hopkins, by contrast, believed that the study of higher mathematics was a virtuous activity in its own right because it revealed the mathematical principles of God's creation and was of practical use in its commercial and industrial applications. In Hopkins's scheme a student displayed his moral worthiness through mathematical ability and by his willingness to work to the very limits of his intellectual capabilities without regard to his own suffering.

We saw in chapter 2 that the analytical revolution and the rise of the private tutor enabled the de facto triumph of Hopkins's scheme, and, in the ruthlessly competitive environment that ensued, ambitious students explored every avenue for improving their performance. An obvious strategy was to work extremely long hours, but it was soon discovered that prolonged periods of intense mental activity could produce unpleasant and even dangerous side effects. Promising undergraduates, and their tutors, gradually learned that hard study was most efficiently and safely accomplished when interspersed with periods of more leisurely activity and recreation. Some students relaxed by socializing or playing musical instruments, but, for reasons not entirely clear, the most ambitious undergraduates gradually transformed the traditional afternoon ramble or promenade into a daily regimen of measured physical exercise.[14] This exercise became the recognized complement of hard study, and students experimented with different regimes of working, exercising, and sleeping until they found what they believed to be the most productive combination. As the Mathematical Tripos became yet more demanding and competitive through the 1820s and 1830s, these regimes of exercise were transformed into a parallel culture of competitive sport. In the following two sections I discuss the route by which physical and mental exercise became complementary undergraduate activities, and argue that both were pursued most vigorously by the so-called reading men.

4.3. Mathematical Rigor and the "Wrangler Making Process"

> [A] man must be healthy as well as strong—"in condition" altogether to stand the work. For in the eight hours a-day which form the ordinary amount of a reading man's study, he gets through as much work as a German does in twelve; and nothing that [American] students go through can compare with the fatigue of a Cambridge examination. If a man's health is seriously affected he must give up honours at once.
>
> CHARLES BRISTED, 1852 [15]

I noted in chapter 2 that most historical studies of Cambridge mathematics in first third of the nineteenth century are concerned either with what is seen as the belated arrival of Continental methods in the university in the 1810s and 1820s, or else with the subsequent struggles between radicals and conservatives to control the content of the undergraduate syllabus. The eventual assimilation of analytical methods within undergraduate studies is invariably portrayed in both cases as a wholly positive event that enabled Cambridge to emerge as an important center of mathematical research in the Victorian era. But if we are to understand how the student's body became explicitly implicated in mathematics pedagogy during the this period, it is important to appreciate that the new regimes of teaching and examining through which Continental methods were effectively introduced had *destructive* as well as *productive* aspects. From the productive perspective, the Mathematical Tripos offered the most able students a thorough training in

---

14. On the leisurely activity of regular walking in eighteenth-century Cambridge, see Rothblatt 1974, 247. For an example of a senior wrangler from around 1815 who relaxed by playing the flute, see Atkinson 1825, 510.

15. Bristed 1852, 1:331.

a range of mathematical techniques and provided them with the motivation to work hard. Some students thrived in this competitive environment, at least for a while, finding pleasure in the rapid mastery of new skills and the timed solution of difficult problems. From the destructive perspective, students lacking the requisite qualities were alienated, dispirited, and sometimes even damaged by the system. Even those who were successful in the early years often found it hard to live up to the promise they had shown and, in their final year, almost always found themselves being driven at an exhausting pace.[16]

An informative contemporary account of both aspects of competitive study occurs in the undergraduate correspondence of Francis Galton. We saw in chapter 2 that Galton was initially exhilarated by the intellectual adventure and sheer pace of Hopkins's coaching sessions and that he wrote home to his father claiming excitedly that he had "never enjoyed anything so much before." Galton claimed to "love and revere" Hopkins like no other teacher (Galton 1908, 65), but he soon discovered that enthusiasm for his discipline and submission to his tutelage were by no means enough to guarantee success. In the middle of his second year, Galton's letters assumed a very different tone as he informed his father that two of Hopkins's pupils were leaving because they could not take the pace, and that his own health was beginning to suffer. A few weeks into his third year, Galton's health began to fail completely and he informed his father "my head very uncertain so that I can scarcely read at all." This letter also reported that the three best mathematicians in the college in the year above him were all graduating as poll men because their health had broken down under the pressure of hard study. Galton concluded that the unremitting emphasis on competition in Cambridge undergraduate studies was in desperate need of reform because the "satisfaction enjoyed by the gainers is very far from counter-balancing the pain it produces among the others" (K. Pearson 1914, 166). He subsequently suffered a complete mental breakdown and had to leave Cambridge for a term. At home Galton experienced an "intermittent pulse and a variety of brain symptoms of an alarming kind"; a "mill seemed to be working inside my head," he complained, and he could not "banish obsessive ideas" (Galton 1908, 79).

The American Charles Bristed was also surprised by the extraordinary toll the work took on the health of many students. As the epigraph at the be-

---

16. On the rise of the fear of failure that accompanied the development of meritocratic examination, see Rothblatt 1982.

ginning of this section reveals, Bristed soon realized that in order to have a reasonable chance of withstanding the pressure of intense coaching, undergraduates needed to possess and retain strong and healthy constitutions. His remarks are the more poignant as he too suffered a mild mental breakdown shortly before the Tripos examination of 1845:

> About ten days before the examination, just as I was . . . expecting not merely to pass, but to pass *high* among the Junior Optimes . . . there came upon me a feeling of utter disgust and weariness, muddleheadedness and want of mental elasticity. I fell to playing billiards and whist in very desperation, and gave myself up to what might happen. At the same time, one of our scholars who stood a much better chance than myself, gave up from mere "funk," and resolved to go out in the poll. (Bristed 1852, 1:229)

These rare personal comments by Bristed and, especially, Galton, begin to reveal the extent to which the training process in Cambridge made very tough demands on the students. The coaches and examiners had clearly been successful in establishing what was widely regarded as an objective scale of intellectual merit. Neither Galton nor any of his similarly disaffected contemporaries challenged the fairness of the Tripos examination, but they learned, by painful experience, that this was a system that celebrated a handful of winners—especially the "upper ten" wranglers—while remaining largely indifferent to the fate of the rest.

The suffering produced by hard study was by no means confined to the mediocre. Archibald Smith (SW 1836) informed his sister at the beginning of his final term's coaching with Hopkins that he was: "getting heartily sick of mathematics—and the pleasure I anticipate from being again at home is much increased by the thought that I shall by that time have done for ever with the drudgery of mathematics and be able to apply myself to more pleasant and more profitable studies." Three months later, having been through the gruelling examination, he further confided to his sister that he was "quite tired of, I might almost say disgusted with, mathematics."[17] The outstandingly brilliant Robert Ellis (SW 1840) wrote even more damning comments on his undergraduate experience in a private journal kept daily during the academic year 1839–40, his final undergraduate year. Although exceptionally well prepared in mathematics before entering Cambridge, Ellis had been warned by J. D. Forbes that his delicate health would make the Mathemati-

---

17. Both quotations from Smith and Wise 1989, 56.

cal Tripos a risky venture.[18] And, despite his head start, Ellis did indeed suffer appallingly as an undergraduate, especially during his final year of hard coaching with Hopkins. He recorded in his journal how early success in college examinations had marked him out as a potential senior wrangler, and how his subsequent attempts to live up to these expectations had gradually replaced his "freshness and purity of mind" with "vulgar and trivial ambition."[19] Ellis privately expressed his sense of apprehension and despondency upon returning to Cambridge in February of his final undergraduate year: "And so here I am again, with a little of that sickening feeling, which comes over me from time to time, and which I can but ill describe, and with some degree of[,] harness bitter dislike of Cambridge and of my own repugnance to the wrangler making process." Like Smith before him, Ellis longed for the Tripos to be over. He confided despairingly to his journal "this must and will pass away—if not before—when I leave this [place] and shake the dust off my shoes for a testament against the system."[20] Ellis's private comments are especially interesting because they reveal that where Galton wondered whether the wrangler-making process was worth the suffering it produced in the losers, Ellis clearly doubted that it was worth the suffering it produced in the winners.

The two most outstanding mathematical physicists produced by Cambridge in the mid-nineteenth century, William Thomson (later Lord Kelvin) and James Clerk Maxwell, were similarly disaffected by their undergraduate experience. More than nine months before he sat the Tripos examination of 1845, the extraordinarily able and energetic Thomson informed his father that "three years of Cambridge drilling is quite enough for anybody."[21] During the equivalent year of Maxwell's undergraduate career, 1853, he was taken ill while working "at high pressure" for the Trinity College summer-term examinations.[22] Struck down by what was described as a "brain fever," Maxwell was initially left unable to "sit up without fainting" and remained disabled for more than a month. During his first two years at Cambridge, Maxwell had resisted pressure to concentrate solely on preparations for the Mathematical Tripos and had continued to read and discuss literature and philosophy. Upon his return to Cambridge, the still weakened Maxwell abandoned

---

18. Forbes wrote: "I hope you will not go to Cambridge unless your are equal to the fatigue of such a career." Forbes to Ellis, 14 February 1836 (JDF-UAL).
19. Ellis Journal, 1 June 1839 (RLE-CUL).
20. Ibid., 8 February 1839.
21. Smith and Wise 1989, 56.
22. Campbell and Garnett 1882, 167, 170.

all but his mathematical studies, doing only what "Hopkins prescribe[d] to be done, and avoiding anything more."[23]

It was not only the years of hard study that tested the mettle of ambitious students. The examinations themselves were intended partly as tests of endurance, taking place on consecutive mornings and afternoons for four and five days together. Bristed experienced the communal sense of mounting apprehension as examinations approached, and noted that students were especially prone to "nervous attacks from over work just before, or excitement at an examination" (1852, 1:331). In the weeks immediately preceding the examination the most able students had to cope with frenzied speculation among their peers, tutors, and servants regarding their likely place in the order of merit. This pressure is nicely illustrated in the correspondence of Karl Pearson (3W 1879), who, as the examination approached, felt that his peers had far too high an estimation of his mathematical talents.[24] Like Ellis and Smith, Pearson longed for the "confounded" examination to be over, revealing in a letter to his mother that he was regularly accosted in the street by strangers with such comments as "Oh! you are going to take such and such a place are you not? Am I safe in backing you for —— place?" On one occasion when he was returning home from a day's ice skating in a village near Cambridge, Pearson overheard a group of strangers "busily employed in giving the odds as to whether [he] or a Pembroke man whose existence [he] never even knew of before would be higher." Pearson felt it to be the final straw when he was approached by the servant of a college friend who, touching his cap, enquired "I beg pardon sir[,] I suppose I am all right in backing you for the first twelve [wranglers]"; he admitted shamefacedly to his mother that he had lost his temper at this remark and snapped back, "D[amn] you, what has that to do with you?"[25]

The earliest references to the frantic nature of the examinations themselves occur in the mid 1810s, just at the time the analytical revolution was getting under way. Describing the beginning of an examination at Trinity College in 1816, for example, J. M. F. Wright reported that "the utmost anxiety [was] depicted upon the countenances of the Reading-men," and that some entered the examination hall clutching "a handful of the *very best pens*, although there is an ample supply upon every table, so fearful are they lest a moment should be lost in mending the same" (Wright 1827, 1:218). Wright

---

23. Ibid., 190, 193.
24. Pearson to his brother, 22 December 1878 (KP-UCL).
25. Pearson to his mother, 15 December 1878 (KP-UCL).

also noted that the examinations heaped further stress upon these men who were already exhausted by months of tough preparatory study. After several days of consecutive examinations from dawn to dusk, he described how the "martyrs of learning," already "pale and death-like ... from excessive reading before the Examination commenced," grew "paler and paler" as the examination proceeded (1827, 1:249). For some students the examinations proved the final straw, and they physically collapsed in the examination room. Wright also gives a rare firsthand account of such a happening—known as a "funking fit"—in 1817. Passing by the open door of Trinity College Hall, he chanced to see a student collapse to the floor "as lifeless as a corpse." The student was "carried off to his rooms by his fellow-candidates," who immediately rushed back to the examination hall leaving their "inanimate competitor" to be revived with ether (1827, 2:17).

Yet another affliction feared by would-be high wranglers during examinations was insomnia. Henry Fawcett, for example, another of Hopkins's subsequently famous pupils, was tipped to be senior wrangler in the Tripos of 1856, but his chance slipped away due to "over excitement." As his contemporary and biographer, Leslie Stephen, records, "In the Tripos, for, as I imagine, the first and last time in his life, Fawcett's nerve failed him. He could not sleep, though he got out of bed and ran round the college quadrangle to exhaust himself." Fawcett slipped to sixth place, a shame Stephen still felt it necessary to mitigate thirty years later by analyzing his examination marks day by day to show how lack of sleep had gradually undermined his technical skill.[26] Fawcett was by no means the only wrangler to experience difficulty sleeping during the Tripos examinations. John Hopkinson, senior wrangler in 1871, attributed an attack of insomnia in the examinations to the fact that "having had one's brain in vigorous activity it is hard to get the blood out of it." Following a night in which he lay awake until five in the morning, he acquired some "chloral" which, the following night, "secure[d] immediate sleep" (Hopkinson 1901, 1:xxiii). J. J. Thomson, second wrangler in 1880, barely slept at all during the last five days of the examination, and pointed out how the insomniac's misery is exacerbated in Cambridge by the several clocks that audibly chime each quarter of an hour. Thomson's exhaustion was such that he actually fell asleep during one paper of the subsequent Smith's Prize examination.[27] Lord Rayleigh, senior wrangler in 1865,

---

26. Stephen 1885, 32. On the common use of exercise to induce sleep at examination time, see Bristed 1852, 1:314.

27. J. J. Thomson 1936, 63. The Smith's Prize examinations do not appear to have been as strictly timed as the Tripos examination.

tells the other side of the story. He partly attributed his success to his ability to remain relaxed throughout the examination and, especially, to the possibly "unique feat" of taking a short nap before the afternoon examination papers (Strutt 1924, 34). The contrast between Fawcett's failure of nerve and Rayleigh's cool head under pressure is an important one to which we shall return in the section on athleticism.

Perhaps the single most remarkable response to the protracted rigors of hard study and examination was that experienced by James Wilson (SW 1859). Shortly after completing his final examination paper, Wilson experienced a severe mental and physical breakdown such that he "could not walk 50 yards." Having spent three months recuperating on the Isle of Wight, he made the surprising discovery that the illness had "entirely swept away my higher mathematics, indeed, all that I had learnt at Cambridge was completely gone. I could not differentiate or integrate; I had forgotten utterly all lunar theory and Dynamics; nearly the whole of trigonometry and conic sections was a blank" (Wilson and Wilson 1932, 48–49). Wilson's case nicely illustrates the point that the study of mathematics in mid-Victorian Cambridge was seen by many as a means of developing and testing a student's powers of ratiocination under competitive duress. Although Wilson found it inconvenient that he was unable to return to Cambridge to coach undergraduates and to seek a job as a mathematical lecturer, his subsequent career as a schoolmaster appears otherwise to have been little blighted by the fact that he had become a senior wrangler who knew no higher mathematics.[28]

One final point of interest concerns the heightened emotional state experienced by some students in the Senate House and the air of mystique and sanctity lent to the whole examination process by reports of such experiences. As the most heroic performances became surrounded by tales of altered states of consciousness or physical and mental collapse from overexertion, a student folklore gradually emerged concerning how best to prepare to withstand these trials. This aspect of the examination was never commented upon formally by the university, but it can occasionally be glimpsed in the reminiscences and biographies of the students themselves. Thus James Clerk Maxwell's biographers claimed that when Maxwell entered the Senate House for his first examination "he felt his mind almost blank," but that once the examination began "his mental vision became almost preternaturally clear." As he left the Senate House, Maxwell was also said to be "dizzy and

---

28. Wilson's career as a reforming mathematics teacher is discussed in Howson 1982, chap. 7.

staggering, and was some time in coming to himself."[29] Likewise the following recommendations on the way an ambitious student should conduct himself during the examination provide a possibly unique insight into undergraduate folklore. A few days before Donald MacAlister (SW 1877) sat his first paper in the Senate House, he received a letter from R. F. Scott (4W 1875) offering what he claimed to be the "customary" advice to a prospective champion about to enter the "battle." Scott sought to comfort MacAlister with the assurance that the relatively elementary nature of the questions on the first few papers would produce "a quite balmy soothing effect on the nerves." Regarding the remainder of the examination, Scott offered MacAlister the more practical injunctions: "Go to bed early, don't read during the exam, keep your feet dry, & wear flannel next to your skin. These maxims, never before printed, are A oners in their way" (MacAlister 1935, 32–33).

The training process developed in Cambridge during the early decades of the nineteenth century clearly worked ambitious undergraduates to the limits of their emotional and intellectual tolerance. To what extent the system was consciously designed or developed this way is difficult to assess, but visitors to the university saw skilled coaching and competitive examinations as an effective industrialization of the learning process which reflected the wider manufacturing ideals of British culture. Bristed viewed the strong emphasis on the mastery of advanced mathematical technique as characteristic of a country "where the division of mental labour, like that of mechanical labour, is carried out to a degree which must be witnessed and experienced to be conceived." He also used explicitly industrial terms to describe the activities of the students: those making special preparation for an examination were "getting up steam," while the perfectly trained high wrangler could solve mathematical problems with the "regularity and velocity of a machine" (1852, 1:13, 88, 319). Another American visitor to both Oxford and Cambridge during the late 1840s, Ralph Waldo Emerson, was also struck by the distinctly industrial flavor of undergraduate teaching. The English "train a scholar as they train an engineer," he exclaimed, and he emphasized the importance of selecting good raw materials for the manufacturing process; when born with "bottom, endurance, wind" and "good constitution," he concluded, Cambridge students made "those eupeptic studying-mills, the cast iron men, the *dura ilia* [men of guts], whose powers of performance compare with [an American's] as the steam-hammer with the music-box"

---

29. Campbell and Garnett 1882, 176. William Garnett, was the fifth wrangler of 1873.

(Emerson 1856, 131, 135). Cambridge students often choose similarly industrial terms to describe their experiences. When William Thomson arrived in Cambridge from industrial Glasgow, he described how Hopkins had invited the freshmen to an informal party at the beginning of term at which he could assess their potential as "raw materials for manufacture" (Thompson 1910, 32). Likewise Galton, reflecting upon his breakdown in 1842, felt that it was as if he had "tried to make a steam-engine perform more work than it was constructed for, by tampering with its safety valve and thereby straining its mechanism" (Galton 1908, 79).

Galton's evocative notion of the Cambridge training process tampering dangerously with the human safety valve is in some respects very appropriate. The system of financial inducements in the form of scholarships and fellowships, the pervasive atmosphere of competition regulated by examination, and the disciplined training mechanism provided by the best coaches incited and enabled students to participate willingly in the manufacturing process. Furthermore, even the most able young mathematicians knew that high placing in the order of merit depended crucially upon a man's ability to judge how much sustained hard work he could tolerate without suffering a debilitating mental breakdown. Bristed certainly believed the "excessive devotion" demanded of undergraduate mathematicians to be responsible for the "disgust and satiety" which they frequently felt for their "unattractive studies," a comment he illustrated with the example of a "high wrangler" who, just before his Tripos examination, was so sick of mathematics that he wished he had never opened a mathematical book and "never wanted to see the inside of one again" (1852, 1:307). Submission to the training process and stoicism under its rigors were, as we have seen, qualities of character wranglers were expected to display. For some, the drawn features of an exhausted scholar even became an aesthetic hallmark of scholarly piety and intellectual strength. As Robert Ellis walked the length of a packed Senate House to tumultuous applause to receive his degree in 1840, William Walton was awed by the way his "pale and ill" countenance enhanced his "intellectual beauty." Even more strikingly, another onlooker remarked to Walton "pithily" that had he seen Ellis before the examination he would have known him to be unbeatable.[30]

The code of stoicism, discussed further below, also prohibited most students from speaking publicly of their undergraduate suffering. After win-

---

30. Walton 1863, xvii. Conversely, George Darwin (2W 1868) was said to have been "unfairly handicapped in being in such robust health and such excellent spirits" just prior to his examination (Darwin 1907–16, 5:xii).

ning the coveted senior wranglership, Ellis became a coach, examiner, and staunch advocate of the very "wrangler making process" he had privately despised. Likewise Galton, who eventually left Cambridge as one of the hoi polloi, later cited the order of merit as objective proof of the uneven distribution of natural intellectual ability among men, without mentioning his own miserable experience as an undergraduate (Galton 1869, 16–22). Only rare contemporary comments—generally made under duress and in private journals or correspondence—and the ethnographical observations of strangers such as Bristed and Emerson (and several others quoted below) give personal insight into actual undergraduate experience. That experience produced extraordinary technical proficiency in those who survived the wrangler-making process while, necessarily, placing enormous strain on all students and making a strong constitution a virtual prerequisite to successful study. But just as hard reading was prone to push students beyond their natural capacities, so they found ways to increase that capacity and to place an effective physical governor on the manufacturing machine.

4.4. Exercising the Student Body

> In short, it is a safe rule to lay down, that, to keep a student in good working order for a length of time, *the harder he applies himself to his studies while studying, the more diversion he requires when taking exercise.*
> CHARLES BRISTED, 1852[31]

The most popular method employed by hard-reading students to manage their working day was the taking of regular physical exercise, a practice which emerged in the late 1810s. At the turn of the nineteenth century, sporting activity in Cambridge was largely confined to such aristocratic outdoor pastimes as hunting, shooting, and angling.[32] In only a handful of English public schools were competitive games encouraged, mainly for amusement and to develop physical fitness and a manly animality. These latter qualities were not highly regarded in the universities, where the cultivation of gentlemanly manners remained an important component of the education provided.[33] "What use is the body of an athlete," wrote Sydney Smith in the *Edinburgh Review* in 1810, "when a pistol, a postchaise, or a porter, can be hired for a

---

31. Bristed 1852, 2:30.
32. Haley 1978, 123.
33. Rothblatt 1974, 259. Rothblatt notes that in the eighteenth century undergraduate sports and games were associated neither with "moral or character formation" nor with "physical exercise."

few shillings? A gentleman does nothing but ride or walk" (Smith 1810, 328–29). Smith also considered it most inappropriate that young gentlemen, even at school, should spend time mastering the technical skills of what were seen as subservient, laboring activities. A gentleman had no need to "row a boat with the skill and precision of a waterman" when, as Christopher Wordsworth confirms, rowing was viewed solely as a means of transport in early nineteenth-century Cambridge (Wordsworth 1874, 175).

But as the analytical revolution intensified progressive study during the late 1810s, regular physical exercise began to be regarded as a necessary and appropriate companion to hard study. The first and most enduring form of physical exercise used for this purpose was daily long-distance walking (fig. 4.1). When the eighteen-year-old George Airy arrived in Cambridge in October 1819, one of the chief advocates of the analytical revolution, George Peacock, not only offered to oversee his mathematical studies but "warned" Airy to "arrange for taking regular exercise, and prescribed a walk of two hours every day before dinner."[34] Airy thereafter followed Peacock's advice religiously, later ascribing his life-long good health to regular exercise. Peacock's warning to the ambitious young mathematician was also novel in 1819, as Christopher Wordsworth records that the use of regular walking as a means of exercise and relaxation was a nineteenth-century "refinement."[35] Walking remained the main form of daily exercise in the 1820s and 30s, the roads around Cambridge being thick with students and college fellows between two and four o'clock in the afternoon as they took their daily "constitutionals." Several walks became so well established that they were given names such as the "Granchester grind" and "wrangler's walk."[36] Another example of an outstanding mathematician who recorded his walking habits is George Stokes. Arriving in Cambridge in 1837, Stokes made sure only to read for eight hours each day and, like Airy, always took his constitutional walk in the afternoon. As his son later wrote, "This habit [Stokes] maintained in youth, and until long past middle life long walks were the custom, both

34. Airy 1896, 22–23. J. M. F. Wright also took exercise expressly to ward off the effects of hard study in 1818. See Wright 1827, 2:48–50.
35. Wordsworth 1874, 170. Atkinson (1825, 508) believed himself to be unusual in using afternoon rambles to regulate his studies circa 1815.
36. The "Granchester grind" was a walk which included the village of Granchester about three miles south of Cambridge. For an interesting account of Thomson and Hopkins pacing "wranglers' walk" together, see Thompson 1930, 113. "Senior wrangler's walk" retained the air of a promenade along which final-year students displayed their ambition for the senior wranglership. See Wright 1827, 1:57.

FIGURE 4.1. Roget's humorous sketch puns on the ambiguity already inherent in the word "exercise" in Cambridge in the 1840s. Roget, *Cambridge Customs and Costumes*, 1851. (By permission of the Syndics of Cambridge University Library.)

summer and winter, at a pace of nearly four miles per hour. At eighty-three years of age he still went the Granchester 'Grind,' of three or four miles, and other equally long walks as his afternoon exercise" (Stokes 1907, 1:7).

By the 1840s, regular physical exercise had become such an established aspect of wrangler life that Bristed believed it to be the "great secret" of scholarly success (1852, 2:27). For eight or nine months of the year the Cambridge undergraduate was now in a "regular state of training," his exercise being "as much a daily necessity to him as his food" (1852, 2:327–28). Bristed also emphasized that it was the hardest reading men who took the hardest exercise (1852, 2:57), a comment supported by contemporary student satire which ridiculed the "studious" rather than the idle freshman as one who "taketh furious constitutional walks."[37] Emerson too noticed that it was not the lazy or unintelligent students but the "reading men" who were kept "at the top of their condition" by "hard walking, hard riding and measured eating and drinking."[38] The range of physical activities engaged in by reading men had also expanded to encompass a wide range of competitive sports. The serious scholar now took:

> "Constitutionals" of eight miles in less than two hours, varied with jumping hedges, ditches, and gates; "pulling" on the river, cricket, football, riding twelve miles without dra\wing bridle; all combinations of muscular exertion and fresh air which shake a man well up and bring big drops from all his pores; that's what he understands by his two hours exercise. (Bristed 1852, 1:328)

Students were also discouraged from exercising alone. The solitary walker or rider might find his mind drifting back to his Tripos problems, and to ward off this possibility men would exercise in twos and threes or participate in team games. According to Bristed, talking "shop" on long walks was restricted to discussions about the relative academic abilities of one's peers (1852, 1:331–32).

It was during the 1830s that rowing began to rival walking as the method of taking regular exercise. As we have seen, in the early nineteenth century the sports pursued in a few public schools had not penetrated the universities, but the increasing enthusiasm for healthy exercise altered the situation. During the mid 1820s, occasional recreational rowing on the Cam was transformed when several ex-public school boys formed a boat club at Trinity College for the purpose of competitive rowing.[39] St. John's responded within a few months by setting up the Lady Margaret boat club and importing an "eight" (with which to race against Trinity) directly from Eton College.[40] Through the 1830s many other colleges set up boat clubs, and intercollegiate and intervarsity races became formalized and commonplace (fig. 4.2). Unlike walking, rowing combined extreme physical exertion with keen inter-

---

37. "Characters of Freshman," *Cambridge University Magazine* 1:176–179, at 176.
38. Emerson 1856, 131. Curthoys and Jones (1995) emphasize that hard reading and hard exercise were similarly linked in Victorian Oxford.
39. For a concise critical discussion of the origins of competitive rowing in Oxbridge, see Curthoys and Jones 1995, 306, and Rothblatt 1974, 258–61.
40. Rouse Ball 1899, 157–58. On the origins of competitive rowing in England, see Halladay 1990.

Boat Races on the Cam, 1837.

FIGURE 4.2. This scene, about a mile down the Cam from Cambridge (seen in the background), depicts intercollegiate rowing between "eights" as a popular spectator sport in the late 1830s. Rouse Ball, *A History of the First Trinity Boat Club*, 1908. (By permission of the Syndics of Cambridge University Library.)

collegiate competition. An example of an outstanding Cambridge mathematician who balanced hard reading with hard rowing is William Thomson. Already an accomplished mathematician when he arrived in Cambridge in 1841, Thomson was recognized from the start as a potentially outstanding scholar. His father, a mathematics professor in Glasgow, was soon troubled to learn that his son had bought a boat to row on the river Cam and was contemplating joining the college boat club. William quelled his father's fears by pointing out that his coach, Hopkins, not only approved of rowing but actively recommended it as a means of exercise and diversion from study. By the end of his second term, Thomson claimed that his general health had been greatly improved by rowing, and that he could "read with much greater vigour than [he] could when he had no exercise but walking."[41] In his second year, Thomson went into a period of intense physical training, ending up number one oarsman in the first Trinity boat (fig. 4.3). By the 1860s, rowing was described as a "mania" in Cambridge which, together with cricket, shared the "honour" of being the "finest physical exercise that a *hard reading* undergraduate can regularly take."[42]

41. Quoted in Smith and Wise 1989, 73.
42. "The Boating Mania," *The Eagle* 2 (1861): 41–44, at 42; my italics.

FIGURE 4.3. This sketch, from a plate entitled "The Boats," puns on the word "training" and captures the physical exertion (compare with fig. 4.1) associated with competitive rowing. Roget, *Cambridge Customs and Costumes*, 1851. (By permission of the Syndics of Cambridge University Library.)

The most brilliant mathematical physicist produced by Cambridge in the 1850s, James Clerk Maxwell, also used walking, swimming, and rowing to regulate his studies. Indeed, Maxwell actively experimented on his own mind and body by trying different regimens of work, exercise, and rest. One daily routine involved working late at night and then taking half an hour's vigorous physical exercise. A fellow student who shared his lodgings in King's Parade recorded the downfall of this system as follows: "From 2 to 2:30 A.M. he took exercise by running along the upper corridor, *down* the stairs, along the lower corridor, then *up* the stairs, and so on, until the inhabitants of the rooms along his track got up and lay *purdus* behind their sporting-doors to have shots at him with boots, hair-brushes, etc. as he passed." Maxwell's father was partially successful in dissuading his son from engaging such unsociable practices, but Maxwell continued his experiments after moving into rooms in Trinity College. In the summer he exercised in the river Cam: Maxwell would run up to the river and "take a running header from the bank, turning a complete somersault before touching the water." Alternatively, as P. G. Tait records, Maxwell would "go up on the Pollard on the bathing-shed, throw himself *flat on his face* in the water, dive and cross, then ascend the Pollard on the other side, [and] project himself *flat on his back* in the water."[43] Maxwell claimed, humorously, that this method of exercise improved his circulation.

Yet another example of an outstanding wrangler who made use of physical exercise is W. K. Clifford, second wrangler in 1867. Clifford took a "boyish pride in his gymnastic prowess," drawing upon his "great nervous energy" to perform "remarkable feats": he could "pull up on a bar" with one hand and, on one occasion, "hung by his toes from the cross-bar of a weathercock on a church tower."[44] Praise of Clifford's athletic excellence apparently gratified him "even more than official recognition of his intellectual achievements"; he once described himself as in "a very heaven of joy" because an athletic exercise he had invented—the "corkscrew"—was publicly applauded. These antics are more than mere curiosities because, as his biographer pointed out, Clifford undertook athletic exercises as "experiments on himself" intended to find the best way of "training his body to versatility and disregard of circumstances."[45] Clifford shared the psychophysiological belief popular among mid-Victorian intellectuals that the health and strength of mind and body were intimately related and reliant one upon the other. This belief enabled Clifford to rationalize the Cambridge athletic tradition as a method of "making investments" in nervous energy which, in his case, sustained such habits as studying through the night without reduction in daytime activities. His exploits turned out to have been sadly ill advised, as he suffered with "pulmonary disease" which, exacerbated by extreme physical exertion and long hours of work, led to his early death at the age of thirty-four (Clifford 1879, 25).

These examples could easily be multiplied, as virtually every high wrangler (for whom records exist) participated in some form of regular physical exercise to preserve his strength and stamina. As a final example, however, it is perhaps more instructive to look in greater depth at the sporting activities of the honors graduates of the Mathematical Tripos for a typical year. The Tripos of 1873 was of special academic significance as a new set of regulations came into force that year expanding the range of topics studied.[46] These changes heightened both local and national interest in the order of merit, prompting one Cambridge newspaper to provide in-depth coverage

---

43. Campbell and Garnett 1882, 153, 164, 165. A "pollard" was a pollarded tree.
44. See Leslie Stephen's entry on Clifford in the *DNB*.
45. Clifford 1879, 1:7, 25. The "corkscrew" consisted in "running at a fixed upright pole which you seize with both hands and spin round and round descending in a corkscrew fashion."
46. Wilson 1982, 335.

of the successful candidates. Almost two thirds of the article was devoted to coverage of the sporting activities of the honors candidates, thereby providing unique insight into the variety and popularity of undergraduate sport. Both the senior and second wranglers, Thomas Harding and Edward Nanson, were keen oarsmen and swimmers (Nanson being a university champion), while the third wrangler, Theodore Gurney, was president of the Athletic Club, an outstanding athlete, and a first class racquets and tennis player.[47] The article also provides a useful indication of the popularity of rowing. Of the undergraduate mathematicians interviewed, there were more than "three dozen" notable oarsmen, all of whom had rowed since first coming into residence in Cambridge. Indeed, of all those who had passed the Mathematical Tripos with honors in 1873, more than forty percent were identified as "notable" practitioners of at least one sport. Bearing in mind that the reporter did not interview all the students, identified only those who were especially notable sportsmen, and had not included those with interests in football, cricket, and canoeing (or walking), it is quite reasonable to infer that a very high percentage of undergraduate mathematicians, certainly a substantial majority, participated in some form of regular physical activity.

This conclusion is supported by the comments of visitors to Cambridge in the late 1870s and 1880s. Samuel Satthianadhan, an Indian mathematics student at Corpus Christi College from 1878 to 1882, observed with astonishment that his fellow students payed "as much attention to their bodily as to their mental development." Satthianadhan reckoned that nothing would more surprise a visitor to Cambridge than the "fine, stalwart, muscular figure of English students" who so loved sport that any of their number who did not take two-hours exercise each afternoon would be looked upon as "an abnormal character and snubbed by other members of the College" (Satthianadhan 1890, 38, 90) (fig. 4.4). Likewise Michael Pupin, an American immigrant from Europe who came to Cambridge in 1883, echoed Bristed's comments in noting that every student "took his daily exercise just as regularly as he took his daily bath and food." Upon commencing his studies with Edward Routh, Pupin was initially "bewildered" to find that the same students who studied hard in the mornings "like sombre monks" came out cheerfully in the afternoons to take their "athletic recreation," an apparent ambiguity which Pupin resolved by referring to his peers as "mathematical athletes."

---

47. The details in this paragraph are taken from *The Cambridge Chronicle*, 25 January 1873, 8. Harding subsequently became a mathematics master at Marlborough School, while Nanson and Gurney both became professors of mathematics in Australia.

FIGURE 4.4. St. John's College Rugby team, 1881. Examples of what Samuel Satthianadhan described as the "fine, stalwart, muscular figure of English students." Of the twelve men depicted, seven graduated with honors degrees including four in mathematics, two of whom were wranglers: R. W. Hogg (6W 1883) and A. C. Gifford (14W 1883). Hogg (seated on a chair on the right) went on to become a mathematics master at Christ's Hospital and revised Isaac Todhunter's textbook on plane geometry. St. John's College Library, Photographic Collection, Album 21. (By permission of the Master and Fellows of St. John's College, Cambridge.)

On the advice of his King's College tutor, Oscar Browning, Pupin too adopted this "universal custom" and soon won a place in the college boat. Mastering the skills of the mathematician in the morning and those of the oarsman in the afternoon, Pupin described the daily events of his academic life in Cambridge as shaped by "Routh and rowing."[48]

The practice of combining hard study with regular, often competitive physical exercise was a tradition deeply entrenched in undergraduate life in early and mid-Victorian Cambridge. Physical exercise was believed to develop all-round stamina, provide a forced and wholesome break from study, and leave students sufficiently tired to get a good night's sleep. Students were

---

48. Pupin 1923, 170, 173, 177. Like Pearson, Pupin was tutored by Browning and coached by Routh.

convinced that, with the aid of this daily regimen, they would be able to withstand ten terms of highly competitive learning, without succumbing to the mental anguish or "funk" described by Smith, Ellis, Galton, and Bristed. Furthermore, student participation in daily physical activity was, as we have just seen, effectively policed through peer pressure; any student who did not participate was treated as antisocial and abnormal. Equally significant in this respect was the example set by some of the teachers the students most revered—the coaches. The three leading coaches at St. John's, William Besant, Robert Webb, and Percival Frost, were all keen sportsmen: Besant was a mountaineer, Webb a walker, and Frost was extremely proficient at cricket, tennis, running, and swimming. On one occasion when Frost was walking with a pupil along King's Parade, he astonished the student by demonstrating the pace required to complete a five-minute mile "though hampered by cap and gown and weighted by the books he had been using at lectures."[49] Edward Routh, the greatest of the Cambridge coaches, was well known for taking two-hour's constitutional exercise each working day at exactly two o'clock. As J. J. Thomson later recalled, "the regularity of Routh's life was almost incredible . . . every fine afternoon he started out for a walk along the Trumpington Road; went the same distance out, turned and came back."[50] As we shall see in chapter 5, these coaches provided important role models whose actions were carefully emulated by ambitious undergraduates.

Students and their tutors also became familiar with the symptoms of overwork so that serious illness could be averted or else attributed to incorrect habits of study. For example, Maxwell's illness of 1853 was made public by his biographers in 1882, but attributed to his having studied topics not prescribed by his coach—a practice warned against by "grave and hard-reading students [who] shook their heads at [his] discursive talk and reading."[51] The importance of student participation in an increasing range of competitive sporting activities in mid-Victorian Cambridge was not confined to balancing the effects of hard study. During the 1850s and 1860s, athleticism in Cambridge began to assume a more complex and a more public series of meanings which cannot properly be understood in isolation from broader developments taking place in the English education system.

49. Morgan 1898, 405; *The Eagle* (St. John's College, Cambridge) 39 (1918): 245.
50. Thomson 1936, 40–41; Turner 1907–8, 240–41.
51. Campbell and Garnett 1882, 173.

4.5. Mathematics and Public Spectacle

> The custom stands of immemorial years—
> When Alma Mater decks her favourite son
> With her fair crown of Honour, fairly won,
> In the assembled presence of his peers,
> The deed they ratify with triple cheers,
> That in the spirit's ear all burst in one,
> Spontaneous irrepressible "Well done!"
>
> WILLIAM NIND, SONNETS OF CAMBRIDGE LIFE, 1852 [52]

Before placing the mid-Victorian development of wrangler values in a wider educational context, it will be helpful to discuss two Cambridge ceremonies that reinforced the ideals of competitive, mathematical study by staging undergraduate performance for a much broader public. We saw in chapter 2 that the advent of written examination gradually displaced the oral disputation as a means of assessing student ability, thereby depriving the university of a local public display of the knowledge it prized and nurtured. The publication (within Cambridge) of the annual order of merit partially compensated for this loss, but during the early nineteenth century the university began to place increasing emphasis on two ceremonies intended respectively to celebrate the announcement of the order of merit and the awarding of degrees to the honors students. Although these ceremonies were initially intended mainly for the consumption of members of the university and their families, in the early Victorian period the announcement of the order of merit became increasingly elaborate and was presented to a national audience through the medium of the popular press. The emergence of this national celebration of academic success forged an important and reciprocal relationship between the public's perception of Cambridge mathematical studies and the wrangler's own sense of duty and self-importance.

The ceremonies took place in quick succession a few days after the completion of the Senate House examination. As soon as the examiners had finished marking the scripts, the marks were passed to the University Registry, who drew up the order of merit. In the early decades of the nineteenth century, the names of the top wranglers were then communicated informally to senior members of the university while the complete list was posted outside the Senate House at eight o'clock the following morning.[53] Mem-

---

52. Nind 1854, 24. William Nind (17W 1832).
53. Wright 1827, 1:96–98. The "Registry" kept the university records and was responsible for the practical business of awarding degrees (Searby 1997, 55–58).

bers of the wider British public during this period had relatively little idea of the nature and content of Cambridge examinations—or, indeed, of university business generally—but in 1825 the *Times* newspaper took the novel step of publishing the order of merit.[54] On this occasion the list was followed by a mock examination paper offering an informed mathematical satire of national life, but the humorous preamble to the paper made it clear that the Senate House examination was actually a very tough test of mathematical knowledge.[55] The *Times* thereafter published the order of merit annually until the latter was abolished in 1909 (fig. 4.5). As the ceremony became more elaborate and national interest in the outcome more intense, the accounts in the *Times* and other newspapers became longer and more detailed. By the mid 1840s, the announcement of the order of merit—especially the naming of the Senior Wrangler—was becoming a far more dramatic affair. The list was now kept strictly secret until it was ceremonially posted on a pillar in the Senate House at exactly nine o'clock in the morning on the Friday following the completion of the examination.[56] In the early 1860s, the ritual was further elaborated: as the bells of Great St. Mary's church (opposite the Senate House) struck nine o'clock, the senior moderator would begin reading out the order of merit from the east balcony in the Senate House (fig. 4.6).[57] By this time the national press was publishing not just the list of names, but short biographies of many of the wranglers, while the new wave of illustrated journals which appeared around the same time often reproduced likenesses of the top three or four men (see chapter 8).

The announcement of the order of merit had now become a severe emotional trial for every candidate, and it was especially nerve wracking for those tipped for high honors. Of the numerous accounts published in newspapers and the biographical memoirs of successful wranglers, the one that best conveys the ceremony's powerful mixture of spectacle, excitement, and theater

---

54. Atkinson 1825, 491. It is unclear why the *Times* began publishing the order or merit at this time, but it was probably related to the rapid increase in student numbers and the university's attempts to present an image as a center of more secular learning and meritocractic examination.

55. The *Times*, 25 January 1825. The questions were of the kind: "(1) Find the actual value of zero, and from hence explain the general expression of a man sending a circular letter to his creditors. ... (10) Prove all the roots of radical reform to be either irrational or impossible." This paper appeared in Gooch 1836, 1–3, where it is dated 1816.

56. Bristed 1852, 1:236.

57. The results were still being posted in 1861 but were being read by 1865. See "Cambridge Wrangler," *Once a Week* 4 (1860–61): 153, and Strutt 1924, 34.

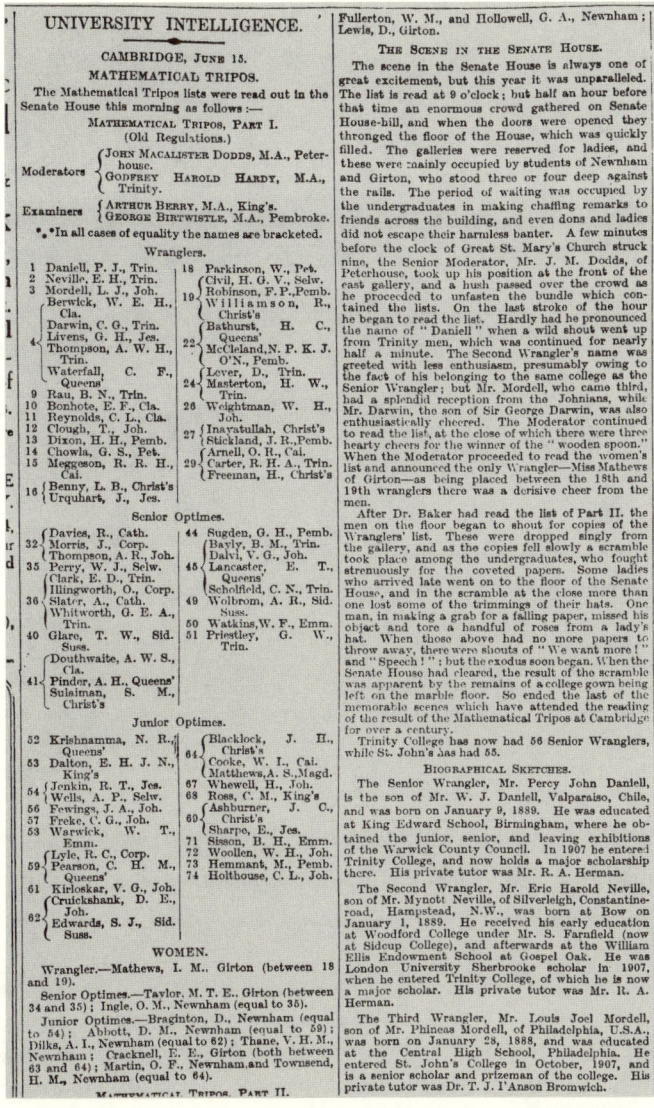

FIGURE 4.5. The announcement of the final mathematical order of merit in the *Times* newspaper. The bracketed fourth wrangler, C. G. Darwin, was the son of George Darwin (see chapter 5) and grandson of Charles Darwin. The last of the junior optimes is Cuthbert Holthouse, who is depicted in fig. 4.9. Note how the women candidates are unofficially ranked below the junior optime according to their relative standing with respect to the men. The column on the right provides a firsthand account of the scene in the Senate House during the final reading of the order of merit, and continues with thumbnail biographies of the leading wranglers. *Times*, 16 June 1909. (By permission of the British Library.)

FIGURE 4.6. The reading of the order of merit from the east balcony of the Senate House in 1905. The senior moderator (center) on this occasion was R. R. Webb, the great coach at St. John's during the 1890s (chapter 5). In the illustration, Webb has just concluded the reading of the order of merit and is in the process of throwing lists of the names to the crowd beneath. He is flanked by two of his fellow examiners, but note that the galleries were otherwise reserved on these occasions for women students. (By permission of the Cambridgeshire Collection, Cambridgeshire Libraries.)

was written by a first-year undergraduate in January 1874.[58] Writing to his father shortly after the ceremony, Donald MacAlister (SW 1877) reported:

> The wildest rumours have been flying about regarding the Senior Wrangler, and every one has got some favourite man of his own settled for the mighty dignity. Two men, Barnard of St. John's, and Ball of Trinity are however generally thought, one or other, certain of the place. The Johnians of course favor the first. A perfect stillness falls as the clock approaches 9. The body of the hall is full of students all staring up at one man who holds a paper folded in his hand: beside him sit ladies and dons, and Mr Routh, the great coach, is immediately to the right. It is a serious moment for him. For if Barnard is first, his unparalleled run of success is broken through (remember Routh

---

58. MacAlister 1935, 20–22. For other especially detailed and interesting accounts of this ceremony in 1890 and 1901, see Grattan-Guinness 1972a, 118–19; *The Cambridge Graphic,* 15 June 1901, 7; and *The Cambridge Chronicle,* 14 June 1901, 7.

has coached 19 Senior Wranglers running, and Barnard was with Mr Besant).[59] The great bell of St. Mary's strikes. [The moderator] lifts his finger, and almost in a whisper, yet heard to the farthest corner of the great hall,—says "Senior Wrangler—Calliphronas of Caius"—. A sudden gasp seemed to rise from the whole multitude, for nobody had heard of this man before, so to speak. Then a ringing cheer that lasts many minutes—a cheer taken up by the crowd outside and spread far down Trinity Street—while the bells of St. Mary's struck up a merry peal in honour of the lucky man (he has to pay £5 to the bell-ringers for it though). "2nd Wrangler, Ball of Trinity"—another big cheer, soon over though,—then "3rd, 4th,—9th, 10th," and Barnard hasn't come yet.—"11th, 12th, Barnard"—poor fellow, that is as much as ruining his hopes for life: almost the sorest disappointment he could have.

MacAlister went on to describe how the moderator eventually flung copies of the list to the undergraduates below, who fought frantically for them, tearing off each other's academic hats and gowns in the process. The ceremony turned out to be a triumph for Routh (who must have been informed of the outcome) as he had coached twenty of the first twenty-three wranglers, including the top six. Macalister also described the bemused state of the senior wrangler, Calliphronas, who, after being carried shoulder high for some while—scarcely sure whether or not he was awake—bowed repeatedly to his peers, "nearly fainting," before being carried off to his rooms.

Within twenty-four hours the names of the top wranglers, their coaches, schools, and athletic accomplishments would appear in national and provincial newspapers throughout the land. The name of the senior wrangler was held in especially high esteem, his home town often holding special celebrations in his honor and his old school sometimes granting its pupils a day's holiday.[60] A nationally reported ceremony of this kind played an important role in shaping both public perceptions of Cambridge University and the sensibilities of undergraduates towards their studies. The British public came to regard the study of mathematics as the pinnacle of intellectual achievement, and the term "senior wrangler" became synonymous with academic supremacy. For the undergraduates, the annual ceremony played several more subtle roles. It provided a powerful and recurrent reminder of the high status of mathematical studies in the university as well as a stark vision of the impending trial each student would himself have eventually to face. The

---

59. Besant and Routh's careers are discussed in chapter 5.
60. The celebrations surrounding Joseph Larmor's winning of the senior wranglership are discussed in chapter 7.

national reporting of the ceremony also added a strong sense that it was not only members of the university who witnessed the event but also the whole British nation. The anxiety generated in the mind of a student who anticipated being caught between national humiliation and local notoriety was used to great effect by W. M. Thackeray (who had been an undergraduate at Cambridge in the late 1820s) in *Pendennis*. Having failed his final examinations in mathematics at "Oxbridge," the eponymous hero waited with a deep sense of foreboding for the examination lists to appear in the national press:

> [P]oor Arthur Pendennis felt perfectly convinced that all England would remark the absence of his name from the examination-lists . . . his wounded tutor, his many duns, the skip and bedmaker who waited upon him, the undergraduates of his own time and the years below him, whom he had patronised or scorned—how could he bare to look any of them in the face now?[61]

Some virtual experience of these emotions must have haunted every candidate for the Mathematical Tripos once he had attended the ceremony and seen it reported in the *Times* the following day.

The reading of the order of merit was, then, a major event in the national and university calendars, filled with emotion and social meaning. It was partly through this ceremony that the university defined its relationship with its own members, the townsfolk of Cambridge (many of whom worked for the university), and the wider British nation and empire beyond.[62] For the finalists themselves, the potent combination of grand architecture, ceremonial dress, and seemingly immemorial ritual made the reading of the order of merit a severe trial—judgment upon years of work would be dispensed, publicly, in seconds. It was also in this charged atmosphere that the sensibilities of younger students were shaped towards their studies. Albion and alma mater expected them to give their best, and their success, or failure, would be duly acknowledged in future enactments of the ceremony.[63] This was an extremely important focal point in university life, and one that spoke to different audiences in different voices. While cheering dons and under-

---

61. Thackeray 1994, 240–41, 243. A "dun" was a hired debt collector. The "skip" and the "bedmaker" were college servants.

62. On the relationship between the university and the local townsfolk during the nineteenth century, see Mansfield 1993.

63. Nind's sonnet (the epigraph to this section) goes on to describe how "Albion" hears the intellectual freedom in the students' cheers. In another sonnet, "Plucked," he speaks darkly of the shame of failed students who slink away "hope-wrecked and sad from disappointed vows" (Nind 1854, 27).

FIGURE 4.7. The senior wrangler of 1842, Arthur Cayley, is presented by his College (Trinity) Father to the Vice Chancellor of the University. The drawing conveys the great ceremonial importance of the moment when the senior wrangler is acknowledged by senior members of the university and colleges as well as by his peers and younger undergraduates. The women present are probably relatives of the top wranglers. The character standing with his right hand on the table (on the left) is probably William Hopkins, Cayley's coach. Hopkins was the "senior" of the two Esquire Bedells, his mace of office being visible on the table by his hand. The junior Bedell (too young to be Hopkins) is overseeing the presentation (holding his mace). Huber, *The English Universities*, 1843. (By permission of the Syndics of Cambridge University Library.)

graduates reassured the British public that Cambridge fostered a healthy intellectual meritocracy, the aspirations of future wranglers were fashioned beneath a nation's gaze.

On the Saturday following the reading of the order of merit came the second great ceremony in the university calendar—the conferring of degrees by the Vice Chancellor in the Senate House. Although less public than the reading of the order of merit, this ceremony still formed an extremely important focal point in undergraduate life. Each candidate was presented (by a designated member of his college) to the Vice Chancellor, before whom he knelt while the degree was conferred (fig. 4.7). The Senate House was packed on these occasions with students, parents, and senior members of the university, and despite the official nature of the proceedings it was generally an extremely boisterous, and sometimes even unruly, occasion.[64]

A rare firsthand account of the ceremony from the perspective of the senior wrangler occurs in the private journal of Robert Ellis (SW 1840). Hav-

---

64. Senior members of the university were often worried about preserving order in the Senate House. See Bury and Pickles 1994, 41.

ing attended a special celebratory breakfast at Trinity College, Ellis made his way, in hood, gown, and bands, to the Senate House. There he was congratulated by people waiting for the ceremony to begin, until William Whewell came up and "hinted at the impropriety" of the senior wrangler mingling with the crowd. Once everyone was seated in their proper places and the galleries filled with students, Ellis was led the full length of the Senate House by Thomas Burcham to thunderous applause. We saw in an earlier section how some members of the crowd saw beauty in his pale and drawn features, but Ellis's account describes the event from the other side:

> When all was ready [William Hopkins] and the other esquire bedell made a line with their maces. Burcham led me up [the Senate House]. Instantly my good friends of Trinity and elsewhere, two or three hundred men, began cheering most vehemently, and I reached the Vice [Chancellor's] chair surrounded by waving handkerchiefs and most head rending shouts. Burcham nervous. I felt his hand tremble as he pronounced the customary words "*vobis praesento hunc juvenem.*" [65]

Having knelt before the Vice Chancellor to have his degree conferred, Ellis walked slowly to the back of the Senate House—to further loud cheers—where, overcome with emotion, he was sat down and revived with a bottle of smelling salts provided by a young woman from the crowd.[66]

The effect of the degree ceremony on most undergraduates was, no doubt, to reinforce the impressions gained a day or two earlier at the reading of the order of merit. The university now lent its official sanction to the former proceedings and further celebrated the unique achievement of the senior wrangler. There was however one special ritual that took place during the awarding of degrees that was not represented at the former ceremony. After the senior wrangler had been presented to the Vice Chancellor, the other degree candidates likewise received their degrees one at a time according to the order of merit. As the last of the junior optimes approached the Vice Chancellor, it was customary for the students to lower a large wooden spoon—on a rope strung between the galleries—until it dangled above the candidate's head (fig. 4.8). Having been admitted by the Vice Chancellor, the

---

65. Ellis Journal, 15 January 1840 (RLE-CUL). The job of the two Esquire Bedells—each of whom carried a silver mace—was to attend the Vice Chancellor in the Senate House and to see that formal ceremonies were properly conducted.

66. For accounts of similarly rowdy scenes during the awarding of degrees in 1819 and 1874, see respectively Wright 1827, 2:100–1, and MacAlister 1935, 22–23.

FIGURE 4.8. Presentation and parading of the wooden spoon. *(Above)* The lowering of the wooden spoon from the Senate House Gallery as the last Junior Optime, Henry Coggin of Trinity College, has his degree conferred by the Vice Chancellor. *(Below)* Coggin carries the wooden spoon outside the Senate House before parading it back to Trinity College. This sketch also depicts the facade of the Senate House. (Courtesy of the Cambridgeshire Collection, Cambridgeshire Libraries.)

FIGURE 4.9. The last Wooden Spoon, 1909. This is a rare, possibly unique photographic image of the celebration of the wooden spoon. The photograph depicts the last Wooden Spoon, Cuthbert Holthouse (St. John's College), being carried by his peers along Trinity Street to his college. (By permission of the Master and Fellows of St. John's College, Cambridge.)

student—known as "wooden spoon"—jumped and caught the spoon, paraded it round the Senate House, through the town center and back to his college (fig. 4.9). In the eighteenth century the term "spoon" referred to someone who was silly or lovesick, and something of this meaning seems to have been attached humorously to the last of the junior optimes in the late eighteenth century and reified in wood in 1804.[67] It is not known in which year the spoon was first presented in the manner described above, but, like the reading of the order of merit, it was probably during the mid-Victorian period.[68]

This informal ceremony, organized and executed by the undergraduates themselves, represented a mildly defiant and consolatory act on behalf of the majority of students who were not among the celebrated upper-ten wranglers. Senior members of the university evidently disapproved strongly of these proceedings, as they undermined the Vice Chancellor's dignity and mocked the presentation of the senior wrangler with a boisterous celebra-

---

67. As I noted in chapter 2, the fact that the "spoon" was sometimes a malting shovel also carried connotations of beer drinking.

68. Wright 1827, 2:28. Wright implies that the first physical wooden spoon was made for Clare College in 1804, as this was the third year in a row that a student from the college had come last in the order of merit. The earliest reference to the spoon ceremony I have found is in Joseph Ward's Diary, 1 February 1875 (JTW-SJCC). Ward gives no indication that this was the first year the ceremony was held.

tion of mediocrity.[69] A poignant illustration of the strength of feeling raised on both sides by this ceremony occurred in June 1882, when a senior member of the university successfully intercepted the spoon as it was being smuggled into the gallery and locked it away in the Vice Chancellor's robing room.[70] Not to be beaten by their loss, the undergraduates quickly obtained a second spoon which was successfully smuggled into the gallery. As the second spoon was being "got into place," however, a "rush was made at it" by several senior members and, in the ensuing struggle, the spoon fell striking the head of a woman in the crowd beneath.

After the second spoon had also been confiscated, a resourceful group of undergraduates "rushed" to a nearby ironmongers and purchased a "large shovel" which they managed successfully to smuggle into the gallery. At this point a fight began between the students and the Proctor's men for possession of the shovel.[71] The disturbance had now become so severe that the Vice Chancellor ordered the proceedings suspended while the upper galleries were cleared. The makeshift spoon, meanwhile, had been handed out of the window and, as soon as the ceremony was restarted, the students burst in through the doors of the Senate House. A "free fight of the most discreditable character" then ensued between the senior members of the university, aided by the Proctors' men, and undergraduates determined that the spoon ceremony should go ahead. The students having finally been ejected, the awarding of degrees was completed. Still the students were resolved that some form of ceremony should take place. When the last of the junior optimes emerged from the Senate House, he was "seized and hoisted on the shoulders of some of the men and borne off, preceded by the shovel." At the same time, the senior members of the university believed by the students to have led the opposition to the spoon ceremony were jostled and verbally assaulted. Although the students deemed responsible for these unseemly and embarrassing events were severely punished by the university, there appears to have been no attempt in future years to prevent the ceremony from taking place.[72]

69. Ward's diary reveals that the spoon was "bagged" by the Proctor's men in 1875 and that a second one was "extemporised" from the wainscoting in the Senate House gallery; 1 February 1875 (JTW-SJCC). For similar comments on the ceremony of 1887, see *The Graphic,* 8 October 1887, 402.
70. My account of these events is taken from the *Cambridge Independent Press,* 24 June 1882, and the *Times,* 21 June 1882. The *Times* report played down the provocative actions of university authorities. We shall see in chapter 5 that the regulations for the Tripos examination were altered substantially in 1882, and it is likely that the university hoped to stamp out the spoon ceremony as part of the reform process.
71. The Proctor's men, or "Bulldogs," were roughnecks hired to enforce university law.
72. Both the *Cambridge Independent Press* and the *Times* (see note 71) reported that those students judged to be the ringleaders would be disciplined severely by the university.

The extraordinary scenes described above reveal the strength and complexity of sentiment and social meaning unleashed at ceremonies in the Senate House. At one level, the reading of the order of merit and the awarding of degrees can be understood as formal ceremonies which respectively announced the examination results and marked the passage of degree candidates to Bachelors of Arts. Certainly these were memorable and meaningful occasions for the graduands and their families, but they had even deeper significance for the university and for the other undergraduates.[73] The celebration of the order of merit simultaneously made visible and reinforced both the high status of mathematical studies in Cambridge as well as normally unspoken codes of tough, competitive study. The powerful sentiments raised and focused on these occasions were sublimated by aspiring wranglers in greater resolution to succeed. For the majority of students, those who formed the bulk of the crowd, it was partly fear of having no place in the proceedings in years to come—of being conspicuous by their absence—that spurred them to make some effort. The violent events surrounding the attempt to stop the spoon ceremony of 1882 also highlight the potential volatility of a ceremony at which the success of a few was validated by the plaudits of the less-successful majority. The credibility of the order of merit depended on the fact that many candidates had to work hard even to become senior or junior optimes, and this effort required some recognition. In failing to tolerate the ritual awarding of the wooden spoon, senior members of the university seriously misjudged the subtle relationship between the competitive rigors of undergraduate study and the outwardly good-natured public spectacle of a Senate House ceremony.

4.6. Mathematics, Athleticism, and Manliness

> [S]uch manly exercises as cricket, boating, &c... are among the most important means of bringing into existence and fostering those grand moral qualities patience, perseverance, pluck and self-denial, without which a nation can never excel in anything.
> SAMUEL SATTHIANADHAN, 1890 [74]

I have already noted that at the beginning of the nineteenth century the encouragement of competitive games in education was largely confined to a few public schools. As Mangan has shown, it was during the 1850s and 1860s

---

73. On recent attempts by anthropologists to understand the meaning of public rituals, see Marcus and Fischer 1986, 61–62.

74. Satthianadhan 1890, 95.

that a number of reforming headmasters used such games to transform student life in the public schools and to foster a more liberal notion of Christian manliness.[75] Competitive sport was employed to improve discipline, health, and appetite while keeping students away from illicit activities when not in the classroom. Sport was also used to build bonds of respect and admiration between masters and pupils and, crucially, as a medium through which to reconcile qualities of manliness which, to an earlier generation, would have seemed incompatible: personal strength with compassion for others, self-interest with team loyalty, and a determination to win with respect for rules and established authority. These were the virtues of liberal manliness captured pithily by Thomas Sandars in 1857 in the phrase "muscular Christianity."[76] These public school ideals, which made the river and playing field as important as the classroom and chapel as sites for moral education, were soon felt in Cambridge. The public schools provided students for the ancient universities which, in turn, provided masters for the public schools, a circulation which legitimated and mutually reenforced such values in both institutions. Many Cambridge dons recognized that students coming to the university from around 1860 were "morally superior" to their predecessors.[77] Traditional pastimes of gambling, drinking, horse racing, and womanizing were being replaced by a love of exercising and training bodies to physical endurance. The new students also brought new sports to Cambridge, and the mid-Victorian period witnessed the rapid formation of new sports clubs through which intercollegiate and intervarsity competitions could be organized: first cricket and rowing, then racquets, tennis, athletics, rugby, soccer, cycling, skating, boxing, swimming, and fencing; by the end of the century, twenty-three recognized sports were contested annually between the ancient universities.[78] The sporting boom in Oxbridge also played an important role in popularizing athleticism throughout Victorian society. As Bruce Haley (1978, 127) has shown, during the 1850s the Oxford and Cambridge boat race became the first mass spectator sport, while its coverage in the national press marked the beginning of modern sporting journalism.

75. Mangan 1981, chap. 1. On changing notions of manliness in mid-Victorian England, see Hilton 1989.
76. Haley (1978, 108) explains how the phrase "muscular Christianity" was inspired by the writings of Charles Kingsley.
77. Mangan 1981, 122
78. Ibid., 125.

The clean living, discipline, and competitive ideals of muscular Christianity found obvious resonances in wrangler life. The fashioning of adolescent sexuality away from heterosexual desire towards an ideal of manly love between master and student provided a means of managing an obvious distraction from hard work while enhancing student loyalty to coaches and tutors. Bristed's account of Cambridge undergraduate life reveals that, as early as the 1840s, sexual abstinence was becoming a factor in Tripos preparation. In a town containing a disproportionately large number of young men, many officially sworn to celibacy, several brothels had appeared, most notoriously in the nearby village of Barnwell.[79] Bristed was appalled to find that many students thought nothing of visiting prostitutes after hall in the evening, and he recounted with disgust a conversation in which a don dismissed a group of poor young Cambridge girls as nothing more than "prostitutes for the next generation." Students similarly talked about "gross vice" in a "careless and undisguised" way, considering abstinence from "fornication" an oddity practiced only by those of "extraordinary frigidity of temperament or high religious scruple." But, for the would-be wrangler, abstinence from fornication, indeed from all sexual activity, was already being practiced as a means of "training with reference to the physical consequences alone" (1852, 2:40, 48–49). It was a striking example to Bristed of how "physical considerations" were apt to overcome all others in Cambridge that students would frequently remain chaste because common medical opinion held that sexual activity affected their working condition. Late night hours spent in brothels deprived a man of valuable sleep and exposed him to the risk of disease or arrest by the university Proctors. Moreover, masturbation and other irregular sexual activity (associated with extramarital sex) were widely believed to induce precisely the kind of nervous debility most feared by ambitious students.[80] Bristed recounts the case of George Hemming (SW 1844), fellow oarsman of William Thomson, who "preserved his bodily purity *solely and avowedly* because he wanted to put himself head of the Tripos and keep his boat head of the river."[81] In this climate, team games provided a medium through which to develop the ideal of manly love untainted by eroticism: on the playing field, physical contact was channelled into the

---

79. Venereal diseases were referred to collectively in Georgian Cambridge slang as the "Barnwell ague" (Anon. 1824, 14). Very little has been written about the Cambridge underworld, but see Desmond and Moore 1991, 54, and Mansfield 1993, 187–91. Wright (1827, 2:144) mentions the university's generally tolerant attitude towards prostitutes.

80. Mason 1995, 205–27.

81. Bristed 1852, 2:41; my italics. Bristed does not name Hemming in this passage, but his identity is easily inferred from other references; Bristed 1852, 2:24, and Thompson 1910, 36.

maul, emotion into self-sacrifice and comradeship, while, in the classroom, the coach or tutor took the place of the heroic schoolmaster.[82]

As Parliamentary pressure for Oxbridge reform increased during the 1850s and 1860s, some young college fellows began to employ both the leadership qualities pioneered by reforming headmasters and the teaching methods long employed by successful coaches. These men eschewed the aloof donnishness of traditional college fellows to become the manly and avuncular student leaders who promoted both competitive sports and intellectual endeavor. An exemplary life in this respect was that of the wrangler and reformer Leslie Stephen. As a child, Stephen was believed to have such a "precociously active brain" that it affected his physical health and hindered his academic progress. Only when he entered Cambridge and began to combine the study of mathematics with a regimen of hard physical exercise did his condition improve. Along with a reputation as a scholar in mathematics and literature, Stephen distinguished himself as a long-distance runner, a walker of unusual endurance, and a fanatical oarsman. Stephen was twentieth wrangler in 1854 and, following his success, became a tutor at Trinity Hall. Like many of his contemporaries, he reckoned that hard physical exercise had restored the inner balance between his mind and body while increasing the stamina of both. As a college tutor, Stephen was in the vanguard of introducing team sports and encouraging intercollegiate and intervarsity competition; he joined in enthusiastically with student games, eventually earning himself the reputation of first of the muscular Christians in Cambridge.[83]

This combination of intellectual and athletic activity, together with the increasing emphasis on sport as a means of moral education, gradually made both competitive sport and academic examination complementary tests of character. From an intellectual perspective, the Mathematical Tripos was understood to be the most severe and accurate test of the rational mind ever devised. Successful candidates had to learn rapidly in a highly competitive environment and accurately reproduce what they had learned under stressful examination conditions. Likewise on the river or playing field a man had to display a rational control of the body that transcended adversity, exhaustion, or pain. Just as Bristed described the ideal wrangler as solving problems hour after hour with the "regularity and velocity of a machine," so the ideal rowing crew achieved and maintained an "entire uniformity and machine-like regularity of performance."[84] But in order to ensure that the successful

82. Mangan 1981, 186
83. Leslie Stephen, *DNB*; Mangan 1981, 124.
84. Halladay 1990, 46.

candidates were really made of the right stuff, competition in the examination room, as on the playing field, had to push some competitors to the breaking point. Contemporary observers noted that it was not the primary purpose of the Tripos to produce mathematicians, nor even to test for a sound knowledge of mathematics; it was, rather, a test of the rational mind under duress. As Robert Seeley wrote revealingly in 1868, the Tripos was as much intended to detect weakness as it was to detect ability (1868a, 9). Seeley also made explicit the link between examination stress and good character. For him, the "supreme quality of great men is the power of resting. Anxiety, restlessness, fretting are marks of weakness" (1868b, 207). These comments explicate the moral dimension to Rayleigh's claim to have slept between examination papers in 1865 and, by contrast, the shame felt by Fawcett who, despite becoming a Cambridge professor and successful member of Parliament, could never forget that his nerve had failed him in the Tripos of 1856. The ideal product of the Mathematical Tripos displayed a calmness of spirit together with a robust and unshakable rationality of mind and body. Characterizing Cambridge in the 1860s, Henry Jackson recalled succinctly that it was "the man who read hard, the man who rowed hard, and let me add the man who did both, whom I and my contemporaries respected and admired."[85]

An excellent contemporary illustration of this point is John Hopkinson (SW 1871), an outstanding oarsman and runner who was an inspiration to other undergraduates. Hopkinson was most admired by his peers for the "tremendous energy with which he would row a hard race on the river and then, on the same day, go in and win a scholarship" (Hopkinson 1901, 1:xiv). In middle age he became an outstanding mountaineer, a sport in which he continued to set an example of the ideal wrangler type. As his son recalled:

> [H]e was a truly magnificent sight upon a mountain. I have seen him ploughing through soft snow up to his waist on the slopes of the Aletschhorn and tiring out men ten years his junior who had but to follow in his tracks. On other occasions he would lead, carrying the day's provisions, over rocks where his companions would scarcely follow him unloaded. He seemed to need no sleep, he would do two long ascents in consecutive days, and enjoy both thoroughly. (Hopkinson 1901, 1:xix)[86]

---

85. Quoted in Rothblatt 1968, 228.

86. The mid-Victorian enthusiasm for competitive mountaineering led to the founding of the Alpine Club in 1857. See Haley 1978, 131–32. Many wranglers—including Arthur Cayley and Leslie Stephen—were members of the club, a founder member of which was H. W. Watson (2W 1850).

Hopkinson's son also made an explicit link between his father's bodily and intellectual activities. He pointed out that his father's energetic but modest "sentiment" in climbing mountains "was much the same as in [his] scientific discovery." In both cases he "cared quite as much for his own feeling that he stood where no man had been before, as for public knowledge of that fact" (xix).[87] As these comments imply, intellectual toughness had become as much a "manly" virtue as was physical strength. Bristed had observed in the 1840s that coaching developed "manly habits of thinking and reading," such that students acquired a "fondness for hard mental work" (1852, 2:13). Subsequent commentators, such as Isaac Todhunter, similarly defended the substantial rewards offered to those who excelled in mathematics in Cambridge on the grounds that they provided "incentives to manly exertion" (1873, 242). The hard reading and competitive examinations in mathematics through which intellectual success was measured thus made mastering the discipline an overtly manly accomplishment.

The educational and moral economies described above remained intact in Cambridge throughout the mid-Victorian period. Todhunter was in no doubt that the "self-denial, the systematic application, and the habit of struggling with difficulties, which may be learned by preparation for examinations," could not fail to "be of service when the energy that was once devoted to schoolboy contests shall be employed in the serious applications of life" (1872, 69). And as the epigraph at the beginning of this section shows, the visitor to Cambridge in the late 1870s and early 1880s soon recognized both the combination of mental and physical endeavor which defined the ideal undergraduate and the national and imperial ends it was supposed to serve. When he returned to Madras, Satthianadhan sought to inspire the same qualities in his students so that they too would be ideally suited to serve their country. Like Bristed forty years earlier, Satthianadahan's four years at the intellectual hub of the British Empire had taught him that physical exercise was "the secret of success in mental advancement." He was now convinced that if there was one thing Indian students could learn from their Cambridge peers it was "to pay as great an attention to their bodily as to their mental development" (Satthianadhan 1890, 39).

The qualities of manliness associated with hard mathematical study were also made starkly visible when women sought to study the discipline. From

---

87. In 1898 Hopkinson and three of his children were tragically killed while climbing "without guides" in the Val d'Herens *(DNB)*.

the early Victorian period, even progressive mathematicians such as Augustus DeMorgan—who were broadly in favor of women's education—cautioned that women's minds and bodies were not strong enough to deal with the rigors of higher mathematics.[88] In the last quarter of the century these assumptions were made even more explicit when women sought to study mathematics at Cambridge. Apart from the fact that women were believed to lack the emotional stability and intellectual power to compete with men, it was considered highly inappropriate for women to participate in the competitive physical games that necessarily accompanied hard study. The new women's colleges nevertheless made provision for competitive sport and adopted the coaching methods employed by the men, a strategy which paid off spectacularly in 1890 when Philipa Fawcett (daughter of Henry Fawcett) caused a national sensation by being placed "above the senior wrangler."[89]

4.7. Mathematics and Athleticism after 1875

During the last two decades of the nineteenth century, the Mathematical Tripos began to lose its elite status. New Triposes, especially the Natural Sciences Tripos, rivaled the Mathematical Tripos in popularity, and high placing in the order of merit ceased to be a unique passport to college or professional success. Wranglers also began to view their training more vocationally. In the 1880s and 1890s they sought work as mathematicians and mathematical physicists in the expanding secondary and tertiary education systems throughout Britain and the Empire. These changes gradually undermined the notion of the wrangler as the ideal intellectual who studied mathematics merely as the most appropriate discipline to hone the rational faculties of his mind. As mathematics was studied for its own sake and for vocational purposes, healthy exercise became more explicitly associated with the regulation of the working day and the maintenance of a healthy body. Ambitious students sought fitness, amusement, and mental stability rather than a physical manliness and outstanding athletic achievement.

The problem of striking an appropriate balance between physical and intellectual endeavor had long been a problem for wranglers. From the late

---

88. Winter 1998, 216–20.

89. On the success of Philipa Fawcett, see the *Times*, 9 June 1890, 7. For an account of the remarkable scene in the Senate House as Fawcett was announced to be top, see Grattan-Guinness 1972a, 118–19. On women in physics in late Victorian Cambridge, see Gould 1993.

1830s, when rowing became the sport that best promoted extreme physical fitness and the competitive spirit, ambitious scholars were aware that physical exercise, although crucial, could easily absorb too much of their precious time and energy. Keen oarsman George Hemming, for example, raced in a private boat because he preferred to work to his own training schedule and did not want to be associated with the dissolute ways of some boat club members. His fellow oarsman William Thomson, by contrast, did join the boat club but kept clear of the "dissipated men" and avoided competition during his final year.[90] By the 1880s ambitious wranglers were taking advantage of the wide range of undergraduate sporting activities now available to preserve physical fitness with less concern for competitive sporting success. J. J. Thomson contrasted the closing years of the century with his own experience in the late 1870s, when wranglers often found it difficult to find an appropriate from of exercise. He recalled that those who did not row hard enough to win a place in a college boat were probably not getting enough exercise, while those who did win a place were required to train so hard that they would not be able to manage a second stint of reading in the early evening. It was for this reason that some scholars, himself included, had walked rather than rowed. But the rise in popularity of new games such as tennis, golf, and squash, together with the proliferation of mediocre teams for football, hockey, and cricket, meant that scholars could find opponents "nearly as bad" as themselves, thereby obtaining competitive exercise without having to train too hard (Thomson 1936, 67).

This sporting ethos remained popular among mathematics undergraduates in Cambridge throughout the 1890s and well into the Edwardian period. G. H. Hardy (4W 1898), described by his Trinity colleague C. P. Snow as "the purest of the pure" mathematicians, acquired a very Cambridge attitude towards mathematics at Winchester. By the 1890s, many public schools had prominent mathematics departments run by Cambridge graduates and, even as a schoolboy, Hardy learned to think of mathematics "in terms of examinations and scholarships" and studied hard because he "wanted to beat other boys, and this seemed to be the way in which [he] could do it most decisively." As an undergraduate, Hardy played real tennis and cricket, games which, according to Snow, he continued to play competitively into his fifties.[91] Hardy's famous collaborator, J. E. Littlewood (SW 1905), likewise acquired the Cambridge style at St. Paul's, a school which produced a crop of

---

90. Thompson 1910, 1:36, 61; Smith and Wise 1989, 71–78.
91. Hardy 1967, 9, 18, 144.

FIGURE 4.10. Bertrand Russell as cox of the Trinity College first boat for the Lent Term races of 1891 (Russell is in the center, seated on the ground). Coxing is an example of the sort of less physically exhausting, competitive sport pursued from the 1890s by students with high academic ambitions. One of the oarsmen depicted, T. E. Hodgkin, nevertheless obtained first class honors in the Natural Sciences Tripos of 1893. (By permission of the Master and Fellows of Trinity College, Cambridge.)

high wranglers around 1904. Mentally exhausted by his first term's hard coaching in Cambridge, Littlewood became a keen oarsman in his second term; he later became an accomplished rock climber, practicing difficult maneuvers around the masonry of Trinity College.[92] Other famous wranglers from this period, Bertrand Russell (7W 1893) and J. M. Keynes (12W 1905), for example, were also sports enthusiasts: Russell played tennis, walked, and swam as well as coxing the Trinity College first boat (fig. 4.10); Keynes rowed and played tennis, racquets, and football.[93]

During the same period, Cambridge mathematics itself became increasingly fragmented. Coaches such as Edward Routh and analysts such as Joseph Larmor (who was coached by Routh) strongly defended the tradition of "mixed mathematics" in which mechanics provided the foundation for a comprehensive account of the material universe. These men also maintained

92. Littlewood 1986, 80–84; Wiener 1956, 152.
93. Moggridge 1992, chap. 3; Moorehead 1992, 37.

Whewell's notion that fundamental principles should be intuitive, Larmor arguing as late as 1900 that "muscular effort" remained the "chief concept" of mechanics (Larmor 1900, 271). The common association between muscularity and the conceptual foundations of mathematical physics formed a subtle link between the sensibilities of the coaching room and those of the playing field, and one that was propagated in such important texts as Thomson and Taits's *Treatise on Natural Philosophy* (1867, 1:220). But among the younger generation of mathematicians, men like Russell, Hardy, and Littlewood sought to bring the new Continental tradition of "pure mathematics" to Cambridge. They envisaged a self-contained discipline that did not look to mechanics for foundational security. This was a more aesthetic, even narcissistic, vision of mathematics that distanced itself from the physical and athletic sensibilities of muscular exertion. Pure mathematics was characterized by an obsession with proof, rigor, beauty, and elegance, and sought its foundations in the disembodied worlds of logic or intuition. Far from being coextensive with physics, pure mathematics could be "applied" only after it had been made foundationally secure by the purists, a shift exemplified in Edmund Whittaker's (3W 1895) famous textbook on modern analysis of 1902 in which analytical methods were founded on number theory and only briefly "applied" to the "equations of mathematical physics" (Whittaker 1902, chap. 13). In the words of H. R. Hassé (7W 1905), the reformers replaced the "problem age," in which students were drilled by coaches to solve problems by the proper application of advanced mathematical methods, with the "examples age," in which students attended lectures and were required to display an understanding of the foundations and limitations of the methods themselves (Hassé 1951, 156). But despite this shift in intellectual orientation, the reformers retained other deeply embodied sensibilities of wrangler culture. The belief that a healthy mind was the complement of a healthy body remained commonplace, and students were still expected to work hard, display aggressive individualism, submit themselves to regular competitive examination, and practice sexual restraint.

J. J. Thomson's comments, that the new and more genteel sports of the 1880s were very appealing to reading men, also suggest a gradual separation between college hearties who pursued sport for its own sake—the "athletocracy" as Mangan has dubbed them—and those who saw themselves as the intellectual elite. In the latter nineteenth century the mid-Victorian ideal of manliness, with its emphasis on earnestness, selflessness, and integrity, gave way to a new cult of neo-Spartan virility with explicit overtones of patriotism, militarism, and anti-intellectualism.[94] This version of manliness, also cultivated in the public schools, undermined the holistic ideal by celebrating the playing field as the primary site for character formation while marginalizing the "highbrow" scholar who showed no special talent for athletics.[95] But although competitive contact sport at the highest level began to acquire unwelcome associations for some aspiring wranglers, regular participation in tennis, cricket, golf, and walking were still regarded by the majority as proper and necessary diversions from hard study.

The robust and competitive qualities nurtured in wranglers also played a role in shaping the research style of those who did go on to become mathematicians. We have seen that William Thomson always carried a mathematical notebook about with him and would begin calculating as soon as he had an idea. When the great German physicist Hermann von Helmholtz was being entertained on Thomson's private yacht in 1871, the two men studied the theory of waves together. Helmholtz was surprised that his host treated the work as a "kind of race" to such an extent that on one occasion Helmholtz was forbidden to work on the problem while Thomson went ashore (Thompson 1910, 2:614). The deeply embedded sensibilities of competitiveness and fair play also appear to have survived the transition from the problem age to the examples age and to have persisted well into the twentieth century. The American mathematician Norbert Wiener, who first visited Cambridge just before the Great War, found a style of mathematics very different the one he had learned in Cambridge, Massachusetts. Surprised by Hardy's skill on the tennis court and Littlewood's prowess as a rock-climber, Wiener noted that the young Cambridge mathematician "carried into his valuation of mathematical work a great deal of the adolescent 'play-the-game' attitude which he had learned on the cricket field."[96] Wiener was even more explicit about this Cambridge style in discussing his subsequent collaboration with Raymond Paley, a student of Littlewood's, who worked with him at MIT during the early 1930s. Wiener observed that Paley approached a piece of research mathematics as if it were a "beautiful and difficult chess problem, completely contained within itself." He also offered a telling field-sport analogy to capture the difference between his and Paley's style of mathematics; Paley did not regard Wiener's pragmatic approach to problem solving as "fully sportsmanlike," and was "shocked" by Wiener's "willingness to shoot a mathe-

---

94. Mangan 1981, chap. 5, 127; Mangan and Walvin 1987, 1.

95. Mangan 1981, 107. Curthoys and Jones (1995, 315) detect a similar weakening of the relationship between high academic achievement and sport in Oxford around 1900.

96. Wiener 1956, 152; see also idem 1953, 189.

matical fox if [he] could not follow it with the hunt" (Wiener 1956, 168). Paley's notion that a properly formulated problem had a proper elegant solution—as if set by a cunning examiner—and that resorting to short-cut solutions was unsportsmanlike was fully consistent with both the Cambridge approach to applied mathematics and his broader athletic view of life. He was a keen skier for whom "any concession to danger and to self-preservation was a confession of weakness which he dared not make in view of his desire for the integrity of a sportsman." Paley's collaboration with Wiener was cut short, tragically but appropriately, when the former was killed in an avalanche while deliberately skiing across "forbidden slopes" in the Canadian Rockies (Wiener 1956, 170).

4.8. The Body of Mathematical Knowledge

> Where religions once demanded the sacrifice of bodies, knowledge now calls for experimentation on ourselves.
> MICHEL FOUCAULT [97]

Historians of the body have shown that concerns about both the presentation of the scholarly self and what lifestyle best promoted intellectual endeavor were commonplace in Cambridge well before the nineteenth century.[98] There are, nevertheless, several senses in which the athletic tradition discussed in this chapter represents a new, modern, economy of the scholarly mind and body. First, there is little evidence that students—as opposed to college fellows—in early modern Cambridge were especially concerned with the care of the scholarly self, nor that they engaged collectively in daily physical exercise to promote hard study. In the early nineteenth century, by contrast, athleticism became a widespread student activity—continued by many scholars into later life—and one which was preserved and passed on within student culture. Second, the novel regimens of daily exercise followed by students appear to have originated not as a new twist in fashionable self-presentation but, at least in part, in response to the development in the 1810s and 1820s of what I described in chapters 2 and 3 as recognizably modern practices of training and examining students in technical disciplines. Only in the early Victorian period did the competitive study of mathematics become a hallmark of cerebral manliness. Third, such activities as daily walking and rowing, together with clean living and sexual restraint,

---

97. Quoted in Rabinow 1984, 96.
98. On the care of the scholarly self in seventeenth century Cambridge, see Iliffe 1998.

gradually implicated the scholar's body in the learning process itself: athleticism preserved a robust constitution, regulated the working day, precluded what were considered illicit activities, and fueled the competitive ethos. But despite the complementary nature of physical and intellectual endeavor in Cambridge, it is important to remember that both enterprises retained a large measure of autonomy. Wiener saw the competitive spirit and sense of mathematical fair play exhibited by the young Paley as derivative of an English playing-field mentality, yet it was, as we have seen, an accomplishment of the early Victorian era to make the playing field a key site of moral education and the competitive study of mathematics a manly pursuit. What Wiener did not appreciate was that Paley's competitive spirit and well-defined sense of how a mathematical problem should be formulated and solved was as much derivative of the Cambridge style of teaching mathematics as it was of the ideals of competitive sport. In this sense, Paley's attitudes to mathematics and skiing respectively should be seen as mutually supportive, rather than as causal in either direction.

The intimate relationship between hard training both in the coaching room and on the playing field also indicates that we should treat the traditional distinction between "reading" and "rowing" men—between aesthetes and athletes—with caution.[99] While it is certainly the case that many rowing men were not scholars, most reading men *were* sportsman. Indeed, from the 1840s until at least the 1870s, it was the hardest reading men who took the hardest exercise. Only in the climate of anti-intellectual athleticism towards the end of the century did the distinction between the elite mathematician and the elite sportsman begin to become a distinction between the serious scholar and the college hearty. The athletic accomplishments of men like Hardy, Littlewood, and Paley remind us, nevertheless, that many successful mathematicians did continue to participate in competitive college sport to the highest level. The combination of physical and intellectual endeavor which typified wrangler studies requires us also to exercise caution in assessing the practical skills possessed by the most able graduates. Conservative ideologues such as Whewell defended a "liberal education" as a means of preserving an elite Cambridge intellectualism against the radical, materialist, and commercial forces which they felt were displacing traditional aristocratic and Anglican values in England. Yet in practice the regimens of hard

99. The terms "reading" and "rowing" man were current in mid-Victorian Cambridge. The ambiguity in the latter term (where "row" is pronounced to rhyme with "cow") captured the stereotypical image of the boisterous, quarrelsome, non-reading man; see Bristed 1852, 1:23, and Smith and Wise 1989, 71.

training through which men prepared for the competitive ordeal of the Mathematical Tripos produced graduates with precisely the narrow technical skills against which Whewell and Sidney Smith had cautioned. The claim by contemporary observers in the 1840s that Cambridge had effectively industrialized the learning process confirms that ranking students through highly competitive written examinations in mathematics had enabled the very forces of secular, technical meritocracy, which Whewell had resisted, to undermine the ideal of a liberal education by another route. What this system of tough technical training did produce in the latter decades of the century was a remarkable school of outstanding mathematical physicists whose mastery of mathematical technique and problem solving was unrivaled elsewhere in Britain and the envy of other major European centers of mathematical physics.

Finally, the disciplining of the student body in ways generally conducive to a new social order in the academy, and to progressive study and competitive examination in particular, became a distinctive feature of technical education in many elite institutions in the late nineteenth and early twentieth centuries. Curthoys and Jones (1995) have shown that undergraduate athleticism developed in Victorian Oxford in broadly similar ways to those discussed above; though apparently with less direct link between intellectual and physical endeavor.[100] Likewise, Larry Owens (1985) has shown that athleticism began to be deployed in American colleges in the 1860s in ways that paralleled earlier developments in Oxbridge. Harvard and Yale both encouraged competitive sport in order to improve student discipline, preserve physical fitness, balance intellectual endeavor, and nurture competitiveness, perseverance, individualism, and respect for established authority. Owens does not discuss the origins of these developments in depth, but, bearing in mind that Emerson and Bristed wrote about British universities with educational reform in mind, it seems likely that Cambridge played some role as a model for comparable American institutions. Recent ethnographical studies of contemporary student life also suggest that the values of competitive technical training consolidated in early Victorian Cambridge remain embodied in student learning practices in the elite scientific institutions of the late twentieth century.[101] Traweek has shown that would-be high energy physicists submit themselves to years of progressive training, are extremely competitive, hard working, and contemptuous of mediocrity, and discountenance

---

100. Oxford developed written examinations around classical rather than mathematical studies and placed much less emphasis on the ranking of students by marks; see Rothblatt 1974, 261, 280.

101. Traweek 1988, 74–105; White 1991.

extramarital liaisons as "an unworthy distraction of vital energies" (1988, 84). There are, likewise, few modern higher-educational establishments that do not provide extensive facilities for competitive sport and the physical development of the individual. These embodied values of modern academic life are now so commonplace as to be virtually invisible, yet their origin has both historical and epistemological significance.

It is tempting to see the emergence of progressive teaching and competitive, written examination in the early nineteenth century as an inevitable process in which the mathematical sciences finally shed all cultural baggage to become purely technical disciplines. But, as we have seen, these historical events cannot be understood in isolation from the changing role of the universities and of the nature and content of undergraduate life in Georgian England. From an epistemological perspective, the cultural values associated with the establishment of a large community of technically expert mathematicians and mathematical physicists can also be explored at the level of everyday undergraduate experience. The private anguish expressed by Ellis, Smith, and Galton makes visible the painful process by which student sensibilities were slowly fashioned to the needs of industrialized learning. In the process, athleticism became an adjunct to competitive study and mathematics became a manly pursuit that selected and shaped the minds of the intellectual elite. Conversely, the mathematics prized in Cambridge came to reflect the ideals of competition, examination, and fair play prevalent in the coaching room and on the playing field. In exploring the minutiae of everyday student activity we are, therefore, dealing simultaneously with historical questions of local and national significance, a fact acknowledged by Pearson who, viewing the training of Cambridge mathematicians in the context of Anglo-German competition, saw the "aesthetic ideal, mind and body combined" as a weapon for the British national cause.

～5～

Routh's Men
Coaching, Research, and the Reform of Public Teaching

> For reasoning—a practical art—must, I think, be taught by practice better than by precept, in the same manner as fencing or riding, or any other practical art would be.
> WILLIAM WHEWELL, 1835[1]

> Mathematics cannot be learned by lectures alone, any more than piano playing can be learned by listening to a player.
> CARL RUNGE, 1912[2]

5.1. Training and Research

The previous three chapters have provided a broad account of the dramatic changes in undergraduate studies in Georgian and early Victorian Cambridge that enabled the emergence of a large community of very technically competent mathematicians. I have emphasized the importance of the pedagogical resources that drove these changes, paying special attention to new methods of teaching and examining as well as the regimes of physical and intellectual exercise engaged in by the students themselves. In this chapter I turn to two important aspects of Cambridge pedagogy which, although touched upon several times already, must now be explored in greater depth. The first of these concerns the relative significance of individual components of the training process. We saw in chapters 2 and 3 that the high level of paper-based mathematical skill acquired by many Cambridge undergraduates was a product of a complex pedagogical economy that included small-class teaching, textbooks containing past examination questions, and long hours of solitary study. In this chapter I shall argue that the success of an economy of this kind was heavily dependent not only on the organizational skills of the coach but on regular face-to-face interaction between the coach and his pupils. Textbooks and private study were certainly vital components of the training process, but, as we shall see, their role was in many

1. Whewell 1835, 6.
2. Hobson and Love 1913, 2:601.

respects ancillary to that of the teacher. This focus on the importance of personal interaction will also emphasize that the oral tradition of learning that prevailed until the late eighteenth century was not so much displaced by the rise of paper-based learning as modified and re-sited. From the early 1850s, students learned mathematical technique not just through practice but by hearing and watching a brilliant coach talk and chalk his way through problems and examples on a blackboard.

These issues are addressed in the following sections in three stages. I begin by describing the teaching methods used by the greatest of all the Cambridge coaches, Edward Routh. As is shown in Appendix A1, Routh trained many more distinguished wranglers than did any of his contemporaries, and a comparison of his techniques with those of his competitors goes some way to explaining his success. I next embark upon an ethnographic study of coaching-room practice. By drawing upon firsthand accounts of Routh's teaching as well as the coaching manuscripts of his successor, R. R. Webb (SW 1872), I reconstruct something of the manner in which an outstanding coach addressed his pupils while teaching at the blackboard. It was through watching and listening to these performances that students picked up part of the coach's unique approach to mathematics, an approach that could not be learned simply from the unsupervised use of textbooks and private study. Finally, I attempt to make some of the skills learned in face-to-face encounters visible, at least indirectly, by exploring the circumstances in which they were reproducible at sites beyond Cambridge. This exploration adds another historical dimension to my study. The sites at which these skills were most successfully reproduced were the English public schools. I shall show that many of the most prominent public schools not only taught mathematics in the Cambridge style, but sent their most able students to Cambridge to become wranglers. The effect of this extension of the wrangler-making process was to exclude students who had not received pre-university training and further to inflate the standard of technical performance expected in the Senate House.

The second important aspect of Cambridge pedagogy explored further in this chapter concerns the relationship between training and research. We have already seen that examiners often drew upon original investigations when setting difficult questions and that the most able students were trained to solve these questions with remarkable speed and accuracy. But to what extent did this training actually shape wranglers' subsequent attempts to make original contributions to mathematical physics? Discussion of this intriguing question will be central to the whole second half of this book and is initiated in this chapter along two separate but related lines of enquiry. The first concerns the ways in which students trained to solve mathematical problems took problem solving as a model for research after graduation. This is a somewhat complicated issue because the problem-solving model had both positive and negative components. Many young wranglers initially failed to appreciate that the well-formulated, neatly soluble problems they had encountered on examination papers were the contrivances of skillful examiners rather than natural products of the physical world. To become a successful researcher, a wrangler had to learn how to construct rather than simply to solve interesting problems; but, as we shall see, these abilities were intimately and inextricably related.

The second line of enquiry concerns the rise of the research ethos in Cambridge during the mid-Victorian period and attempts to reform public teaching in mathematics so as to foster contact between students, professors, and the most able college lecturers. This reform was initiated during the 1870s by the informal establishment of courses of *intercollegiate* lectures which were open, for a fee, to students from all colleges. These courses made it possible for young college fellows to teach advanced mathematical topics to graduates and undergraduates alike, and intercollegiate classes soon became a focus for the discussion of original investigations and possible research projects. This was also the period during which Parliamentary reform led to the establishment of five new university lectureships in mathematics. Moreover, the coordination during the 1880s of the courses given by these lecturers with those given by the intercollegiate lecturers and the university professors represented the reestablishment of an effective system of *public* mathematical teaching within the university. These developments gradually reduced the coach's status from master of undergraduate mathematical studies to supervisor and remedial tutor in basic mathematical technique. The success of the emergent system of public teaching was nevertheless built firmly on the training techniques the coaches had pioneered.

5.2. Edward Routh and the Art of Good Coaching

> Is it not somewhat excessive to derive such power from the petty machinations of discipline? How could *they* achieve effects of such scope?
> MICHEL FOUCAULT[3]

I mentioned towards the end of chapter 3 that a serious shortcoming of Thomas Kuhn's account of scientific training is its excessive emphasis on the

---

3. Foucault 1977, 194.

importance of textbooks. According to Kuhn, physics students learn to ascribe meaning to the formal laws and principles stated at the beginning of textbook chapters by working through the exercises and problems—or "exemplars"—found at the end of each chapter. The problem with this claim is that having argued persuasively for the fundamental importance of the skills and competencies developed during the process of problem solving, Kuhn offers no account of how students reliably establish the correct relationship between the formal statement of laws and principles and the solution of exemplars. A student might decide that he had established the relationship correctly once he began routinely to obtain the right algebraic or numerical answers to exemplars. But this criterion provides no guarantee that the solutions are being obtained in an appropriate way (assuming they are obtained at all), let alone that the method employed is efficient or elegant. In any institutionalized training system in which the student is expected rapidly to learn a wide range of mathematical methods and their applications, the final arbiter of proper practice is a competent teacher.

We can easily make these points concrete for the case of Victorian Cambridge by drawing upon the testimony of Isaac Todhunter, who was not only a successful coach but, as we saw in chapter 3, one of the greatest mathematical textbook writers of the period. Todhunter would certainly have agreed with Kuhn that problem solving is a skilled activity and one that can only be mastered through long hours of practice. Writing in the early 1870s, he warned prospective undergraduates that "the power of solving mathematical problems is drawn out, *I might almost say is created*, by practice," and, he added, the "practice must be long and assiduous before any decided success can be gained" (Todhunter 1873, 75; my italics). Todhunter was also painfully aware that learning by heart the formal statement of laws and principles given in textbooks did not enable students to master the exercises and problems that followed. In terms remarkably redolent of Kuhn's comments a century later, Todhunter wrote: "I hear repeatedly from solitary students that although they believe they understand what they read in an ordinary Cambridge book of good reputation, yet they find themselves completely baffled by the exercises and the problems, to their very great discouragement" (1873, 74; cf. Kuhn 1970, 189). *Pace* Kuhn, Todhunter did not conclude that "solitary students" should struggle harder with the examples given, but that, on the contrary, they were better advised to ignore such exercises entirely until they could be guided by a properly qualified teacher (1873, 74–76). These comments point directly to private tutors as the arbiters of good mathematical practice in Cambridge, textbooks being but one device in the complex pedagogical economy by which they passed on their craft. In the balance of this section I illustrate such an economy in action by describing the teaching methods of Edward Routh.

During the mid 1850s, even as he was lobbying for the complete reform of undergraduate teaching, William Hopkins, now in his sixties, was beginning to lose ground as the leading mathematics coach in Cambridge. A new generation of private tutors, including Percival Frost (SW 1839), Stephen Parkinson (SW 1845), Isaac Todhunter (SW 1848), William Besant (SW 1850), William Steel (SW 1852), and Edward Routh (SW 1854), began regularly to place their pupils at, or near the top, of the order of merit. This second generation of outstanding young private tutors display two broad lines of coaching descent: Todhunter, Steele, and Routh had all been coached by Hopkins; Besant had been coached by Parkinson who, in turn, had been coached by John Hymers; it is not known who coached Frost, but it was almost certainly either Hopkins or Hymers, probably the latter.[4] What is certain is that, from the mid 1850s, the coaching lineage of the most outstanding private tutors could be traced, either directly or indirectly, to Hymers or, most frequently, to Hopkins (Appendix A2). For a couple of years in the late 1850s, Parkinson and Todhunter appeared the likely successors to Hopkins, but, by the early 1860s, it was Routh who had emerged as the new "senior wrangler maker" (fig. 5.1).[5] Routh was by far the most successful of all the Cambridge coaches and was probably the most influential mathematics teacher of all time.[6]

Born in Canada in 1831, the son of a distinguished British Army officer, Routh came to England at the age of eleven and was sent to University College School and University College London. At University College he was taught by Augustus DeMorgan (4W 1826), under whose tutelage he emerged as a brilliant mathematics student. Winning an exhibition to Peterhouse College in 1850, Routh was coached first by Todhunter and then by Hopkins for the Tripos of 1854. That year Routh beat James Clerk Maxwell into second place, the two young mathematicians subsequently sharing the Smith's Prizes.[7] At

---

4. Hymers, Frost, Parkinson and Besant were all students at St. John's, and it is possible (assuming Frost was coached by Hymers) that the succession represented a local, college-based, coaching tradition.

5. In 1861, for example, Routh was recommended to Lord Rayleigh as the best coach as "Todhunter was no longer taking pupils" (Strutt 1924, 24). For a discussion of the "coaching wars" of this period see Marner 1994, Appendix A.

6. Thomson 1936, 39.

7. Biographical details of Routh's life are from the *DNB*, Larmor 1910–11, and Rouse Ball 1907. According to the *DNB*, Maxwell migrated from Peterhouse to Trinity "in anticipation of future rivalry" with Routh.

FIGURE 5.1. Edward Routh at the height of his powers as a mathematician and private tutor (circa 1870). Routh was by far the most influential mathematics teacher in mid and late Victorian Britain. Photograph, AA.3.62, Wren Library, Trinity College. (By permission of the Master and Fellows of Trinity College, Cambridge.)

the time Routh completed his degree there were no suitable fellowships or lectureships available in Cambridge, and he was obliged, unusually for a senior wrangler, to consider career possibilities beyond the university. This short period of uncertainty was cut short when the leading coach and lecturer at Peterhouse, Steele, fell ill, and Routh agreed to take over his private pupils and college lectures as a temporary measure. It quickly became evident that Routh was an outstanding exponent of the coaching methods pioneered by Hopkins, and, following the complete failure of Steele's health and his untimely death in March 1855, Routh was elected to a fellowship and a mathematical lectureship at Peterhouse. The lectureship remained his official job in Cambridge until 1904.[8] Routh's first pupil, Charles Clarke, was placed third wrangler in 1856 and, two years later, Routh's pupils were placed senior and second wrangler and first and second Smith's Prizemen respectively. In 1862 Routh's reputation was established beyond doubt when seven of his students were among the top ten wranglers. Over the next quarter of a century Routh enjoyed unparalleled success as a private mathematics tutor: of the 990 wranglers who graduated between 1862 and 1888, 48 percent were coached by Routh (including 26 senior wranglers) while between 1865 and 1888, 80 percent of the top three wranglers were his pupils.[9]

Routh's training methods were modeled broadly on those of his own teacher, Hopkins, but due to the larger number of students that Routh taught and the ever-rising levels of technical competence necessary for Tripos success, Routh developed a more intensive, systematic, and protracted teaching regime than Hopkins had employed. Where Hopkins would not accept freshmen as pupils, Routh would teach any student who showed enthusiasm, a willingness to work hard, and who could afford to pay his fees. He was even prepared to coach men before they entered the university. When Karl Pearson (3W 1879) first called on Routh in his coaching rooms in the summer of 1874, finding him "covered with chalk . . . his coat off and his shirt sleeves rolled up, with the sponge and duster in his hand," Routh agreed to coach the young Pearson starting at 7 A.M. the following morning even though he would not be entering King's College for another year.[10] The consummate mathematics coach, Routh made time to teach any promising student and was keen to begin developing the skills of a prospective high wrangler as early as possible. Until 1882, when the Tripos examination was shifted from January of the fourth year to June of the third year, Routh's undergraduate students were divided into four groups by academic year: three undergraduate years, and the questionists in their tenth term prior to the examination in January. Each year was further subdivided into three or four classes of between five and ten pupils of roughly equal ability. Classes were taught three

---

8. The celibacy requirements in the university forced Routh to vacate his fellowship and move out of Peterhouse College in 1864 when he married Hilda Airy, the daughter of George Airy. When the college statutes were reformed and the celibacy requirement abolished, Routh was the first person elected to an honorary fellowship by Peterhouse.

9. The figures not taken from Appendix A1 are from Rouse Ball (1889 and 1907) who, we must assume, got them from Routh.

10. Pearson 1936, 28. Routh taught Pearson "subjects not of first-class importance for the Tripos." Pearson spent the academic year 1874–75 in Cambridge being tutored by Routh (Magnello 1998, 3309).

times a week throughout the eight weeks of the three formal terms and for a further eight weeks during the long summer vacation, all lectures taking place in Routh's lecture room in Peterhouse College. In the busiest term (the first) of Routh's busiest years he therefore taught some 48 hours from Monday to Friday, beginning at seven or eight o'clock in the morning and, apart from his two hours constitutional exercise every afternoon, sometimes not finishing until ten o'clock at night.[11]

The primary method of teaching, around which Routh's whole system was built, was the one-hour lecture to a class of not more than ten pupils using blackboard and chalk.[12] The blackboard was a fairly recent pedagogical innovation in Cambridge at this time, private tutors having previously worked on paper with their pupils sitting next to them.[13] The new generation of coaches used the blackboard as a means of displaying the art of mathematical work on paper in a form that was readily visible to a relatively large class of students. Routh would begin with a "swift examination of exercise work" set for the class at the end of the previous meeting. These exercises generally required students to reproduce proofs and theorems, and to solve related problems, as they would have to in the examination, and Routh would quickly discuss any errors common to several members of the class and those of an individual from which he felt the class might learn. Despite the public nature of these corrections there was "no jesting, no frivolous word over a blunder," and Routh would neither give any "tips" on which exercises he thought likely to appear in forthcoming examinations nor express an opinion on the relative abilities of members of the class. Having corrected the work of the previous lecture he would at once launch into a "continuous exposition" of the material to be covered that day, each member of the class taking notes as fast as he was able.[14] Routh generally led students through what he considered the best textbook accounts of each subject—leaving

---

11. Where Pearson was seen at 7 A.M., J. W. Strutt was given extra tuition at 9 P.M. (Strutt 1924, 26).

12. The four extant volumes of Routh's lecture notes (EJR-PCC) cover the following subjects: (I) collected problems on various topics; (II) inversion, Newton's *Principia,* solid geometry, rigid dynamics, differential equations, and optics; (III) hydrostatics, hydrodynamics, figure of the earth, thermodynamics, theory of gases, planetary theory, precession and nutation; (IV) sound waves, elasticity, light, electricity and magnetism. These notes are dated in Marner 1994.

13. William Besant (discussed below) was reputed to have introduced the blackboard into Cambridge mathematical teaching in the early 1850s (*The Eagle* 39 [1918]: 243). If this information is reliable, fig. 2.5 must depict a very recent innovation in college teaching.

14. Forsyth 1935, 173. Forsyth records that it took a while for the top classes to learn to take notes fast enough but that this skill paid off in the examination.

especially clearly written sections for students to read for themselves—clarifying obscure points, shortening theorems, removing what he deemed superfluous material, and supplementing omissions. As one of his most successful students, J. J. Thomson recalled, to teach an "adequate knowledge of such a large number of subjects in so short a time required a time-table that had been most carefully thought out and thoroughly tested" (1936, 46). Using Hopkins's technique of teaching to small classes of men of roughly equal ability, Routh could pitch his lectures at a level and pace that seldom required him to pause to explain specific points or answer queries. On occasion he would deliberately adopt a catechetical teaching style for a few minutes to ensure that he was preserving "an average" in the class (Larmor 1910–11, xii). In most lectures he would "talk without hesitation for the allotted hour, rarely forgetting to begin just where he had left off, and seldom making even a slip in working on the blackboard" (Turner 1907–8, 240). In the final long vacation and the tenth term, the lectures became revision classes and the students were continually set examination questions to build speed and accuracy.

Revision sessions apart, Routh devoted little time to solving difficult examination problems in his lectures. At the end of each hour, rather, he would hand out about six problems "cognate to the subject" of the lecture which were to be solved and brought to the next class. The most able students were also referred to manuscripts "in his own handwriting" kept in a second teaching room (the "general pupil-room") which covered those advanced aspects of the current topic that were not dealt with adequately in the textbooks but on which Routh thought it possible an examination question might be set. Routh's pupil-room was frequently crowded with students trying to copy out or read these manuscripts, the material from which was to be "dovetailed" into the lecture notes at the appropriate place. During the first two years, students were also given additional questions designed specifically to build speed and accuracy in writing out standard proofs and theorems, and to develop skill and confidence in tackling a wide range of simple problems.[15]

Once a week Routh gave a common problem sheet to all his students, regardless of year or group. In one week the students were allowed as much time as they required to puzzle out the solutions, but, every other week, the problems had to be solved in a timed three hours under examination conditions. Each student was required to leave the problems in the pupil-room on Friday or Saturday in order that Routh could mark them over the weekend. The following Monday the marked scripts would reappear in the pupil-

---

15. Forsyth 1935, 173–74; Larmor 1910–11, xiii; Rouse Ball 1907, 481.

room together with Routh's model solutions (to save him having to waste precious minutes in the lecture) and a publicly displayed mark sheet ranking all students according to the marks they had scored. These biweekly "fights" gradually accustomed students to working at the pace required in the Senate House, incited and preserved an atmosphere of fierce competition, and provided that objective measure of relative merit upon which Routh himself took care never to comment.[16] Routh was extraordinarily scrupulous in marking student scripts to the extent that "it was one of his peculiarities that he was never wholly satisfied with any work shown up to him" (Moulton 1923, 24). On one memorable occasion when a brilliant student, Fletcher Moulton (SW 1868), mischievously prepared a problem paper "on which no criticism could be offered," Routh still found room for improvement by urging Moulton to "Fold neatly"—as J. J. Thomson recalled, Routh's interaction with individual pupils, although generally "conversational," was "never eloquent, never humorous" (Moulton 1923, 24).

The remarkable energy, enthusiasm, and attention to detail that Routh brought to his teaching goes some considerable way to explaining his success as coach. There is no doubt that once Routh acquired a reputation as the new senior-wrangler-maker he automatically attracted the most able and ambitious undergraduates, but even a cursory look at the coaching practice of his contemporaries reveals the relative thoroughness of Routh's methods. His closest rival, in so far as he had one, was the leading coach at St. John's College, William Besant, a student of Parkinson (and hence of the Hymers lineage). Like Routh, Besant made use of two teaching rooms and saw his students three times a week, but here the similarities end. Besant had only half as many pupils as Routh and preserved the coaching style of the 1830s: he taught students individually or in twos and threes, and preferred, whenever possible, to use the notation and methods of Newton's fluxional calculus.[17] Nor was Besant's teaching founded on formal lectures covering the whole of the Tripos syllabus. In Besant's scheme, the syllabus was approached via a large collection of past examination questions, painstakingly written out in manuscript form and left in the outer room.[18] The student or

---

16. Larmor, Thomson, and Forsyth reported that competitive problem solving was an important vehicle for inciting hard work and ambition. See Larmor 1910–11, xiii, Thomson 1936, 38, and Forsyth 1935, 174.

17. Forsyth 1935, 168.

18. Besant's scheme is now best represented in the *Schaum Outline Series* in which physical and mathematical methods are taught through brief statements of definitions, axioms, and principles, followed by large numbers of graded examples and problems.

students would arrive in this outer room an hour or more before they were due to see Besant and begin working, with ancient quills on the back of old examination papers, through the appropriate section of the manuscript.[19] At the allotted hour they were ushered into the "inner parlour," where the work submitted at the previous visit would be returned "corrected and annotated." Besant would then go over any difficulties the students had experienced on a blackboard before dictating problems for the students to try to solve. If there was time, Besant would invite suggestions from the class on how to tackle a specific problem and then write his own elegant solution on the board. Once a week all Besant's pupils were set problems to solve "at leisure," the list of marks being made "available for inspection" so that his students, like Routh's, could estimate their relative standing. Besant's scheme of teaching was generally more leisurely and more tailored to the individual than was Routh's; where Routh took classes of approximately eight pupils through the prescribed syllabus as quickly as possible and made ancillary use of manuscripts, textbooks, and examination questions, Besant left each student to approach the syllabus at his own pace via graded examples from past examinations.

The unique qualities of Routh's methods are further highlighted when compared with those of another famous coach, Percival Frost, whose long career straddled both the Hopkins and Routh periods.[20] Having forfeited a St. John's fellowship through marriage in 1841, Frost spent his career as a mathematical lecturer at Jesus College (1847–59) and King's College (1859–89). As with other lecturers barred by marriage from further college or university preferment, Frost also took private pupils, his most famous being W. K. Clifford (2W 1867).[21] Frost was an engaging, popular, and able teacher, but he never employed the organized training regimens used by Routh. Karl Pearson was able to compare the two coaches directly since, although formally a pupil of Routh, he was also taught individually by Frost in lieu of compulsory college lectures at King's. Pearson recalled that while Frost was a "dear old boy" and a charming conversationalist, he would often become absorbed in amusing illustrations of dynamical principles or be sidetracked into discussion of theology or mathematical conundrums (Pearson 1936, 30–31). Another pupil of Frost, Henry Dickens (29W 1872), sixth son of Charles Dickens, was convinced that he would have done a good deal better

---

19. The details of Besant's coaching methods are taken from *The Eagle* 39 (1918): 243–45.
20. Dickens 1934, 116; Morgan 1898, 405; Pearson 1936, 30–31.
21. Frost's pupils are discussed in *The Eagle* 20 (1899): 446.

in the Senate House if he had been coached by Routh. Dickens concluded that it was partly "keen competition" among Routh's students that kept them "up to concert pitch," whereas with Frost "there was nothing of this kind" (Dickens 1934, 116). Frost's ex-pupil, friend, and obituarist, H. A. Morgan (26W 1853), pointed to a further peculiarity of Frost's teaching that separated him from both Hopkins and Routh: Morgan insisted that Frost had a rare ability to make mathematics "interesting and attractive" to students, but was of the opinion that many more of "the ablest students would have sought the advantage of his tuition had he remained in Cambridge during the reading period of the Long Vacation." Frost preferred to keep the old custom, abandoned by Hopkins, of taking reading parties of students to "various places of interest" during the summer, a plan that "did not commend itself to many on the ground that it was not so conducive to hard study as when the time was spent in Cambridge" (Morgan 1898, 405).

The relative success of the training methods employed by Routh, Besant, and Frost from 1865 is clearly reflected in the Tripos lists given in Appendix A1. Besant's methods were adequate to enable the brilliant G. B. Mathews to beat Routh's best man in 1883—one of only two occasions on which Routh's men were beaten—and to produce at least four second wranglers. Frost too could occasionally produce outstanding mathematicians such as Joseph Wolstenholme (3W 1850) and Clifford, but, overall, neither of these men, nor any other contemporary mathematics coach, came close to equaling Routh's success.[22] It is also relevant in this context to recall the experience of James Wilson, who, as we saw in chapter 3, was able to solve a very difficult Tripos problem using the "method of moving axes" only because he had stumbled upon the method by chance (published by one of Routh's students) a few days before the Senate House examination of 1859. Wilson was not a pupil of Routh, but of Stephen Parkinson (SW 1845), one of the leading coaches at St. John's. It is a yet another indication of Routh's devotion and thoroughness as a teacher that he routinely taught his elite pupils powerful problem-solving techniques that were unknown to other mathematicians and which constituted original contributions to mathematics when published.[23]

---

22. Wolstenholme became Professor of Mathematics at the Royal Indian Engineering College and is discussed further below.

23. We saw in chapter 3 that Routh's coach, William Hopkins, had also taught unpublished techniques, such as the method of "infinitesimal impulses," for solving certain difficult problems.

```
STAGE I SUBJECTS
Euclid; arithmetic and the elementary parts of algebra, trigonometry and
conic sections
The elementary parts of statics; namely, the composition and resolution of
forces action in one plane at a point, the mechanical powers, and the
properties of the centre of gravity
The elementary parts of dynamics; namely, the doctrine of uniform and
uniformly accelerated motion, of falling bodies, projectiles, collision,
and cycloidal oscillations
The first, second and third sections of Newton's Principia; the
propositions to be proved in Newton's manner)
The elementary parts of hydrostatics; namely, the pressure of non-elastic
fluids, specific gravities, floating bodies, the pressure of the air, and
the construction and use of the more simple instruments and machines
The elementary parts of optics; namely, the laws of reflection and
refraction of rays at plane and spherical surfaces, not including
aberrations; the eye; telescopes
The elementary parts of astronomy; so far as they are necessary for the
explanation of the more simple phenomena, without calculation

STAGE II SUBJECTS
                         FIRST DIVISION
Algebra                              Differential equations
Trigonometry; plane and spherical    Statics, Hydrostatics
Theory of equations                  Dynamics of a particle
Analytical geometry; plane and solid Dynamics of rigid bodies
Finite Differences                   Optics
Differential and integral calculus   Spherical Astronomy

                        SECOND DIVISION
Higher parts of algebra and of the   Higher parts of differential
theory of equations                  equations
Higher parts of finite differences   Calculus of variations
Elliptic integrals                   Theory of chances, including
Higher parts of analytical geometry  combination of observations

                         THIRD DIVISION
Newton's Principia, book I,          Laplace's coefficients
sections IX and XI.                  Attractions
Lunar and planetary theories         Figure of the earth
Higher parts of dynamics             Precession and nutation

                        FOURTH DIVISION
Hydrodynamics                        Vibrations of strings and bars
Theory of sound                      Theory of elastic solids treated
Physical optics                      as continuous
Waves and tides

                         FIFTH DIVISION
Expression of arbitrary functions    Heat
by series or integrals involving     Electricity
sines and cosines                    Magnetism
```

FIGURE 5.2. Syllabus of the Mathematical Tripos for the period 1873–1882, including the newly introduced physical subjects of Heat, Electricity, and Magnetism (Fifth Division). The Stage I subjects were examined during the first three days of the nine-day examination. The fourth day was devoted to easy problems based on the Stage II subjects. All aspects of the Stage II subjects were examined during the final five days. From *The Student's Guide to the University of Cambridge*, 1874. (By permission of the Syndics of Cambridge University Library.)

Routh's career from the early 1860s to the late 1880s marked the high point of coaching in Cambridge. In no other major center of mathematical education, before or since, has one man exercised such comprehensive control over undergraduate training. Moreover, the methods that Routh perfected are exemplary of the new pedagogy which generated and sustained a large community of highly technically competent mathematical physicists in Cambridge throughout the Victorian period. In Routh's mature scheme, the content of ten terms of undergraduate mathematical study was broken down into progressive stages and exercises, each taught so that the student had command of just the mathematical methods and physical concepts necessary to move on to the next level. Although, as we shall see, the content of undergraduate mathematical studies was altered several times during the second half of the century, the syllabus published in 1873 (fig. 5.2) provides a useful guide to the major topics covered over the course of Routh's teaching career.[24] With the exception of the "Fifth Division," which contained new physical subjects excluded from the Tripos since 1848, the syllabus illustrates both the preeminence of such established mixed-mathematical topics as optics, hydrostatics, hydrodynamics, elastic solid theory, wave theory, astronomy, celestial mechanics, and the shape of the earth, and the continuing practice of teaching mathematical methods in close association with their physical applications.[25] Bearing in mind that the study of optics, the theory of chances, and hydrodynamics prepared the ground for discussion of astronomical instruments and observations and of the shape of the earth, it is clear that the pinnacle of wrangler studies in the mid-Victorian period remained the mastery of the ordered motion of the earth and heavens. The emphasis on teaching mathematics via physical applications is also exemplified in Routh's extant lecture notes from the early 1870s.[26] Each new method Routh introduced was accompanied by numerous special applications and past examination questions through which students were taught to formulate physical problems in terms of the mathematical tools to hand, and to solve such problems quickly and reliably. The pace, breadth, and level of learning were also carefully managed by teaching students in small groups according to year and ability. Using these methods, Routh was able to preserve an optimal balance between, on the one hand, catering for the needs of the individual, which militated for taking pupils individually or in twos and threes, and, on the other, lecturing to larger classes in which students could compete with, and learn from, each other.

24. The various syllabi for the years 1873 to 1907 are reproduced and discussed in Wilson 1982, Appendix 1.
25. See G. H. Darwin's lecture notes for the year 1864 (GHD-CUL and EJR-PCC). Routh's lecture notes are discussed in Marner 1994.
26. See EJR-PCC.

5.3. Talking, Chalking, and Trying It at Home

> [Routh's] quaint little touches of humour [were] often the quickest route to an explanation.... After explaining the method used for planetary aberration by antedating the observation—"Why cannot this method be used for a star?"—he would ask, and then reply, "because light may take a thousand years to reach us from the star, and during that time the path of the earth is sensibly curved." Those who looked up from their notes at this would catch the little twinkle which impressed the point permanently on their memory.
>
> H. H. TURNER (2W 1882)[27]

The foregoing account of the teaching methods employed by the most successful private tutors of the mid-Victorian period describes what might be called the *organizational economy* of successful coaching practice. Within this economy, both the syllabus and the body of students were organized and taught in ways that enabled a single tutor, most notably Edward Routh, to get the maximum number of students through the entire Tripos syllabus in ten terms of study. However, in addition to these mainly structural aspects of teaching and learning, there was a more personal dimension to successful coaching that is not so easily described. We saw in chapter 2 that one of the most notable characteristics of private tutoring was the intimate "intercourse" it occasioned between master and pupil. As Charles Bristed succinctly noted, it was during such face-to-face encounters that a good coach could "throw a part of himself into his pupil and work through him" (1852, 1:149), and this was an aspect of coaching that continued to draw comment throughout the Victorian era. According to W. W. Rouse Ball (2W 1874), a pupil of Routh, coaching was not only indubitably the best method of mathematical training but one that relied for its effectiveness on "intimate constant personal intercourse," the importance of which "cannot be overrated" (Rouse Ball 1912, 321). In insisting on the great relevance of personal interaction in mathematical training, men like Bristed and Rouse Ball were not referring simply to the practical advantages of small class teaching as outlined above, but to a more subtle manipulation of student sensibilities that seemed only to occur in the intimate atmosphere of the coaching room. In this section I draw on the testimony of Routh's students, and the coaching manuscripts of Routh's successor as the leading mathematics coach, Robert Webb (SW 1872), to explore the importance of personal interaction between master and pupil in the training process.

---

27. Turner 1907–8, 240.

In addition to prescribing a graded course of study and ensuring that students worked hard, one of the most important duties of a good coach was to teach the art of solving mathematical problems. For those lower down the coaching hierarchy this meant little more than cramming dull or recalcitrant pupils with required theorems and variant forms of standard problems, but for the top coaches a much more complex and interactive process was involved. A private tutor such as Routh had little need to convince his able and ambitious pupils of the need for hard work and devotion to study; the pace at which he ran his classes together with the local order of merit would quickly have informed students of the effort necessary to obtain high ranking in the Senate House. Routh's reputation assured him the admiration and attention of his pupils so that he could employ the intimate atmosphere of his coaching room to offer himself as an ideal-type of the Cambridge wrangler and to incite his students to emulate his performance.

This *performative* dimension to Routh's teaching was extremely important in imparting a particular approach to mathematical physics to his pupils. According to James Dixon (5W 1874), Routh's presence in the classroom "inspired his pupils with implicit confidence," and, Dixon added, one of the most important factors in Routh's success as a teacher was his "capacity for showing his pupils how to learn and how to use their knowledge."[28] This capacity is best illustrated with a couple of examples. When the young Lord Rayleigh (SW 1865) attended his first coaching sessions in 1861, he was surprised by the pace at which Routh drove his class and wrote home with some alarm that Routh was "very *rough*" and that he was clearly going to "grind [him] pretty hard." Over the years Rayleigh's memory of his struggles as an undergraduate faded, but he could still "vividly recall the amazement with which, as a freshman, [he] observed the extent and precision of [Routh's] knowledge, and the rapidity with which he could deal with any problem presented to him" (Strutt 1924, 26, 28). Like so many of Routh's most successful students, Rayleigh's abiding memory of his coaching sessions was "the extraordinary perfection with which Dr. Routh was able to bring his knowledge to bear on every subject with which he was dealing." This perfection could be admired and copied-out from the manuscript solutions to Tripos problems left in Routh's pupil-room, but it could be witnessed in action at the beginning of lectures and during revision sessions,

---

28. *DNB* entry on Edward John Routh. For similar comments on imitating a private tutor's style in the late 1820s, see Prichard 1897, 39.

when Routh developed solutions to difficult Tripos questions, sometimes at sight, quickly and elegantly on the blackboard.

Another student to comment explicitly on this aspect of Routh's teaching style was the American, Michael Pupin, who arrived in Cambridge in the early 1880s to study Maxwell's electromagnetic theory. A talented mathematician from Princeton, Pupin was sent to Routh to master the advanced mathematical methods he would need to tackle Maxwell's *Treatise on Electricity and Magnetism*. He quickly came to regard his coach as a mathematical "wonder," recalling of his early coaching sessions:

> Problems over which I had puzzled in vain for many hours he would resolve in several seconds. He was a virtuoso in the mathematical technique. . . . I never felt so small and so humble as I did during the early period of my training with Routh. Vanity and false pride had no place in my heart when I watched Routh demolish one intricate dynamical problem after another with marvellous ease. I felt as a commonplace artist feels when he listens to a Paderewski or to a Fritz Kreisler. (Pupin 1923, 177)

It was not only inspiration that Routh provided as he talked his students through a difficult problem on the blackboard. Each step was accompanied by a verbal explanation informing the class of why this was the correct way to proceed, acceptable alternative procedures, and errors that must be avoided. Most of the time Routh would of course have worked out the solutions in advance, but it was through these legendary performances that he enthralled his class and incited his pupils to try to find a similar performance in themselves.

The process of learning to emulate Routh's approach to problem solving was one that also contained a profoundly performative dimension. It could not be accomplished simply by attending his lectures and reading his model problem solutions but emerged, gradually, from long hours of repetitive practice combined with minute and exacting correction from Routh himself. I discussed in chapter 4 the carefully planned regimens of exercise and self-study employed by students to manage the competitive environment of Victorian Cambridge, but it must be emphasized here that these regimes often included an hourly timetable of daily work through which the student could keep up with Routh's relentless pace. Once again an example will be helpful. Some nine months before sitting the Tripos of 1876, a pupil of Routh, Joseph Ward (SW 1876), worked out a collective regime of weekly study with two of his fellow undergraduates at St. John's. Their plan was to support and to police each other's working hours via a system of rewards and penalties according to the following rules recorded in Ward's diary:[29]

1. To be out of bed by 7.35 (or on Sundays 8.45)
2. To do 5 hours work before hall
3. At least one hour's exercise after hall
4. Three hours' work after hall
5. Finish work by 11 and be in bed by 11.30 (except on Saturday when it is 12)
6. A fine of 3^d to be paid for the first rule broken on any day and 1^d every other rule broken on the same day
7. A halfpenny to be allowed out of the fund to every member waking another between 6.35 and 7.35 on weekdays and between 7.45 and 8.45 on Sundays
8. Work before 8AM may count either for morning or evening work of the day
9. Time spent at Church Society meeting count for half the same time's work and also allows the member attending to work till 11.30 and stay up till 12
10. These rules binding till further notice and any alteration of them requires unanimity

Whether the three friends kept strictly to this demanding schedule is impossible to say, but the schedule certainly represents the aspirations of ambitious students and, given that the hours of work dictated are not excessive by contemporary standards, probably provides a reliable guide to their actual working habits during term-time.[30]

It was through these long hours of solitary study, tackling example after example and problem after problem, that Routh's pupils strove to emulate his mathematical style, and in this context Pupin's analogy between virtuoso performance in music and mathematics is revealing in two respects (fig. 5.3). First, mastery of a musical instrument is a highly performative accomplishment, relying as much on repetitive exercises as on direct instruction by an expert teacher. Routh's scheme of teaching was similarly dependent on the *active* participation of the student and, judging by the rules above, his pupils spent roughly ten times as many hours practicing at home as they spent in his classroom. Second, the performance of a virtuoso musician displays a distinctive style of playing that is not easily captured in formal musical nota-

29. JTW-SJC, 14 April 1875.
30. For very similar regimes of daily work recorded in diaries and reminiscences, see Wilson and Wilson 1932, 37–39, MacAlister 1935, 23–26, and Searby 1997, 625–26.

FIGURE 5.3. A unique flashlight photograph of a senior wrangler posed working at his desk. This image of Alexander Brown (SW 1901) was taken in his rooms (Room 3, O Stair Case, Caius Court) in Gonville and Caius College shortly after the reading of the order of merit in June 1901. Note the large, wicker wastepaper basket (bottom right), into which sheets of bookwork and rough calculations would have been tossed, and the golf or tennis shoes under the desk by the shuttered window. Brown claimed in *The Cambridge Chronicle* (14 June 1901) to "dabble" in both sports. Photograph from *The Cambridge Graphic*, 15 June 1901. (By permission of the Syndics of Cambridge University Library.)

tion and which is best learned by direct emulation. As a master of analytical dynamics, Routh too had a distinctive way of approaching and solving problems in his field, and this style was gradually acquired by his most able students during his lectures and revision sessions and through his model solutions to numerous problems. The essence of such an approach is, by definition, impossible to convey succinctly in words, but it can be glimpsed indirectly through the comments of his students. The pupil of Routh who most evocatively captured the lasting power of association between Routh's presence, real or virtual, and the mastery of mathematical problem solving was Lord Rayleigh. Whenever during his prolific research career Rayleigh was stuck trying to solve a difficult problem in mathematical physics, he would take a notebook and pen and call to mind his old coach; he would then "write it out for Routh" and, during the process, the difficulties "generally melted away."[31]

As Rayleigh's remarks also imply, Routh's teaching played an important role in shaping the way many of his best pupils formulated and tackled research problems. The problem-solving approach employed by Routh embodied the kind of mathematics prized in Victorian Cambridge, and this approach was also evident in Routh's own research work and in the way he presented original results to his pupils. Despite the long hours he spent teaching from the early 1860s to the late 1880s, Routh found time during this period to write two major treatises on rigid dynamics, a brilliant Adams Prize-winning essay on the stability of motion (which introduced his famous "modified Lagrangian function"), and to publish original contributions on almost every major topic on the syllabus of the Mathematical Tripos (with the exception of the new physical subjects).[32] These contributions were often published in the standard mathematical journals of the period, but they were sometimes made public in the form of advanced Tripos problems or as original sections in his treatises. The treatises themselves were also replete with worked examples and past examination problems which, for Routh, were not merely exercises but the appropriate means of "illustrating and completing the theories given in the text" (Routh 1892, v). He believed that Tripos problems in particular were so representative of contemporary currents in research in Cambridge that it was important to include numerous examples in his treatises.

In attending Routh's lectures, studying his treatises, and solving Tripos problems, Routh's pupils were therefore learning not simply the methods and techniques that might be useful in a research career; they were at the same time learning what, in Cambridge, counted as researchable problems and how such problems were to be solved. The most outstanding of Routh's pupils were also given special manuscripts to study, which contained both examples of the most difficult Tripos problems of recent years and questions manufactured by Routh himself based on "memoirs in the home and mathematical journals" (Larmor 1910–11, xiii). Whenever Routh came upon a new mathematical technique or an important contribution to a branch of mathematical physics required by the Tripos syllabus, he would quickly prepare a succinct account in the form of a short exposition and/or a problem. These accounts sometimes appeared on a daily basis, his most able students being

---

31. Strutt 1924, 27. For similar comments on coaching in the late 1820s, see Prichard 1897, 39.
32. Routh's published papers covered lunar theory, wave theory, geometrical optics, fluid mechanics, rigid dynamics, differential equations, and algebraic geometry. For an analysis of Routh's publications, see Marner 1994, 16–30.

informed at the end of a lecture that there was "a little manuscript, which you had better copy out in the other room" (Turner 1907–8, 240). Routh's elite students were thus introduced to important research being undertaken in Britain and overseas, but this material too was generally recast by Routh in the form of Tripos problems.

A second route by which we can gain insight into the processes by which a specific, Cambridge approach to mathematical physics was shaped through coaching is by looking at the work of the coach who followed Routh as the senior-wrangler-maker of the 1890s, Robert Webb (SW 1872).[33] Appointed to a fellowship at St. John's College, and to lectureships at St. John's (1877–1911) and Emmanuel College (1878–93), Webb (fig. 5.4) supplemented his stipends by taking private pupils.[34] During the 1880s, Besant and Webb constituted Routh's only competition, and, following the retirement of Routh in 1888 and Besant in 1889, Webb emerged to dominate coaching as only Routh had before him (Appendix A1). Unlike Routh, Webb did not combine a successful career as a coach with important research or even textbook writing. Apart from a dozen or so minor papers written during the 1870s, Webb's professional life was devoted entirely to lecturing and coaching, which, during his heyday in the 1890s, occupied him for some sixty hours a week.[35] Although Webb had been a pupil of Routh, his coaching methods seem to have had more in common with those of Besant (his colleague at St. John's) than those of his old coach. Unlike Routh, Webb did not hold lectures on the whole Tripos syllabus, but got his students to work through manuscripts in his own hand covering all the required material. Webb also preferred to teach students in groups of two or three, a choice that, as we shall see, may well have reflected the changing role of the coach in the 1880s and 1890s. Sitting "by the side of his blackboard, a wet sponge in hand," Webb would work through numerous Tripos problems, using them to illustrate mathematical methods and physical applications, and to help his students develop skill in solving examination questions.[36]

Webb's extant coaching notes are especially interesting as they are one of only two surviving examples of material written by a coach to be copied out by his pupils.[37] They have in common with Routh's lecture notes that they

33. Obituary notices for Webb appeared in *The Eagle* 49 (1935–36): 272–73.
34. Forsyth 1935, 175. Forsyth also rated Webb's lectures on rigid dynamics better than those of Routh himself.
35. *The Eagle* 49 (1935–36): 272.
36. Cunningham 1969, 13.
37. The only other example of which I am aware is a collection of Stephen Parkinson's coaching notes (SP-SJCC) from the mid 1850s.

FIGURE 5.4. Robert Webb around the time (1877) he was appointed to a lectureship in mathematics at St. John's College. According to one of his pupils, P. J. G. Rose (20W 1901): "In his zeal for his pupils' success [Webb] drove them very hard, indeed almost to the point of terror. A Johnian who was Second Wrangler [almost certainly Frank Slator (2W 1902)] and rowed at 13 stone in the College boat once told me that he wakened with a shiver ever day that he was due to coach with Webb. Nevertheless Webb inspired lasting affection as well as admiration." *The Eagle* 49 (1935–36): 273. (By permission of the Master and Fellows of St. John's College, Cambridge.)

reveal precisely what wranglers were taught in the late Victorian period, but Webb's notes also provide unique insight into the form by which students were addressed indirectly through manuscripts and, very likely, directly by the coach himself. Unlike the formal textbooks of the period, the notes are filled with asides, recommendations, reminders, warnings, exhortations, and indications of the relative importance of specific techniques, results, and examples. Where the public lecturer or textbook writer would consider the clarification and repetition of a clearly stated theorem, technique, or minor point of notation redundant, Webb continually reminds and reassures his pupils at points where experience had taught him they might have difficulties.[38] Unfamiliar or forbidding expressions are transformed into "old friends," while important results which must be remembered are flagged with such epithets as "old but useful" and "time honoured." Webb often uses the word "mind" (as in "mind the proof of...") to indicate where the student should make a special effort to commit an expression or proof to memory in exactly the form given in the notes. In some cases Webb would require students to verify equalities, formulae, and corollaries for themselves, offering encouragement with such comments as: "You should do this carefully... and you will then quickly absorb the real point once for all." When students were required to complete long or awkward derivations for themselves, they were continually reminded to "be neat" so as not to miss important results through careless manipulation.[39] Webb also flagged the importance of specific results or techniques through a hierarchy of indicators, their relative significance being marked with single, double, triple, and even quadruple underlining. A similarly hierarchical terminology was used to distinguish techniques that were "quite useful" or "very useful to know" from those that were "really very important" or "most important."[40] The notes also reveal that Webb continued to teach symbolical and operational methods in the calculus that would have been frowned upon in the 1890s by the handful of mathematicians in Cambridge who were beginning to foster a rigorous pure mathematics in the university.[41] If used with care, however, such techniques were very useful in the rapid solution of whole families of problems, and Webb explicitly cautioned his pupils not to "be afraid" of the sometimes forbidding notation they involved.[42]

Some of the above points are illustrated in the extract from Webb's manuscript reproduced in figure 5.5. This page contains three examples (XIX–XXI) from a long series of subsections illustrating ways in which a change of independent variables can sometimes expedite the solution of a family of difficult problems. In this case the student is being taught to transform what is possibly the single most important equation in late Victorian mathematical physics, Laplace's equation, from Cartesian into spherical polar coordi-

---

38. RRW-SJC, 27. Webb begins one section with "Do remember that X, Y, Z are *each* functions of x, y, z" even though this is conventional in his notes.

39. RRW-SJC, 26, 4, 8.

40. RRW-SJC, 36, 5, 13.

41. RRW-SJC, 16. For an excellent example of one of Webb's students using a symbolical technique learned from these notes, compare the cited page with Love 1892, 131–32.

42. RRW-SJC, 15–16.

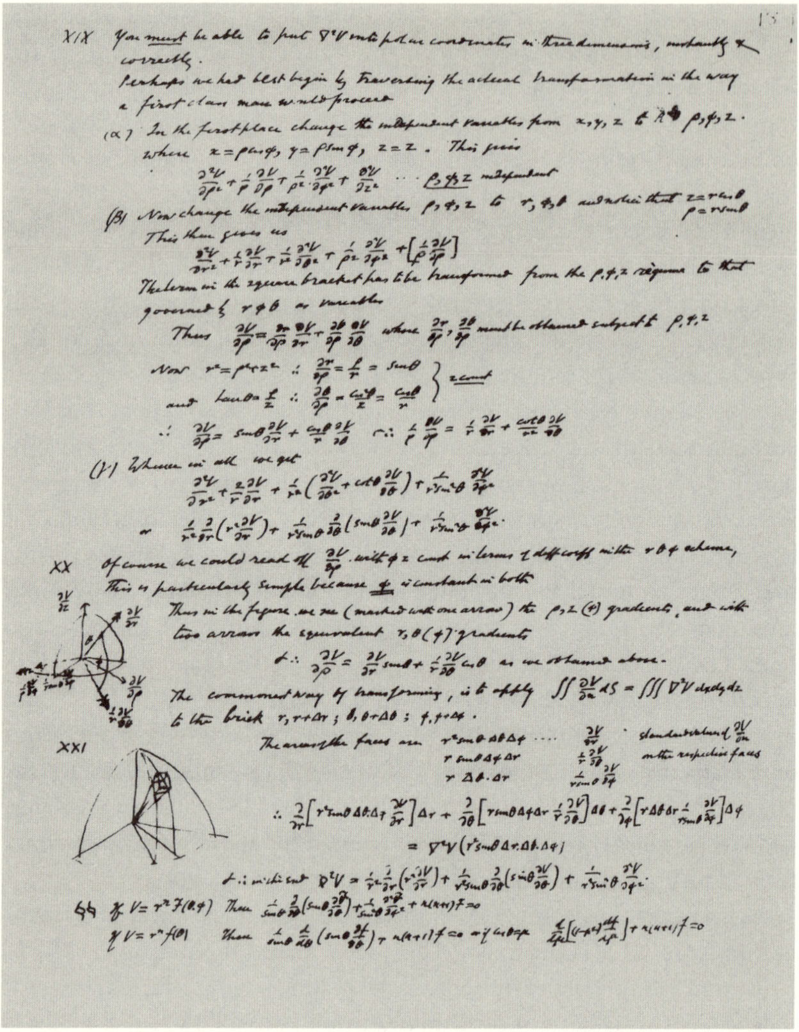

XIX You *must* be able to put $\nabla^2 V$ into polar coordinates in three dimensions, mechanically & correctly.

Perhaps we had best begin by traversing the actual transformation in the way a first class man would proceed.

(α) In the first place change the independent variables from x, y, z to ρ, ϕ, z. where $x = \rho\cos\phi$, $y = \rho\sin\phi$, $z = z$. This gives

$$\frac{\partial^2 V}{\partial\rho^2} + \frac{1}{\rho}\frac{\partial V}{\partial\rho} + \frac{1}{\rho^2}\frac{\partial^2 V}{\partial\phi^2} + \frac{\partial^2 V}{\partial z^2} \quad \ldots \rho, \phi, z \text{ independent}$$

(β) Now change the independent variables ρ, ϕ, z to r, ϕ, θ and notice that $z = r\cos\theta$, $\rho = r\sin\theta$

This then gives us

$$\frac{\partial^2 V}{\partial r^2} + \frac{1}{r}\frac{\partial V}{\partial r} + \frac{1}{r^2}\frac{\partial^2 V}{\partial\theta^2} + \frac{1}{\rho^2}\frac{\partial^2 V}{\partial\phi^2} + \left[\frac{1}{\rho}\frac{\partial V}{\partial\rho}\right]$$

The term in the square bracket has to be transformed from the ρ, ϕ, z régime to that governed by $r\,\phi\,\theta$ as variables

Thus $\frac{\partial V}{\partial\rho} = \frac{\partial r}{\partial\rho}\frac{\partial V}{\partial r} + \frac{\partial\theta}{\partial\rho}\frac{\partial V}{\partial\theta}$ where $\frac{\partial r}{\partial\rho}, \frac{\partial\theta}{\partial\rho}$ must be obtained subject to ρ, ϕ, z

Now $r^2 = \rho^2 + z^2$ ∴ $\frac{\partial r}{\partial\rho} = \frac{\rho}{r} = \sin\theta$
and $\tan\theta = \frac{\rho}{z}$ ∴ $\frac{\partial\theta}{\partial\rho} = \frac{\cos^2\theta}{z} = \frac{\cos\theta}{r}$ } z const

∴ $\frac{\partial V}{\partial\rho} = \sin\theta\frac{\partial V}{\partial r} + \frac{\cos\theta}{r}\frac{\partial V}{\partial\theta}$ ∴ $\frac{1}{\rho}\frac{\partial V}{\partial\rho} = \frac{1}{r}\frac{\partial V}{\partial r} + \frac{\cot\theta}{r^2}\frac{\partial V}{\partial\theta}$

(γ) Whence in all we get

$$\frac{\partial^2 V}{\partial r^2} + \frac{2}{r}\frac{\partial V}{\partial r} + \frac{1}{r^2}\left(\frac{\partial^2 V}{\partial\theta^2} + \cot\theta\frac{\partial V}{\partial\theta}\right) + \frac{1}{r^2\sin^2\theta}\frac{\partial^2 V}{\partial\phi^2}$$

or $$\frac{1}{r^2}\frac{\partial}{\partial r}\left(r^2\frac{\partial V}{\partial r}\right) + \frac{1}{r^2\sin\theta}\frac{\partial}{\partial\theta}\left(\sin\theta\frac{\partial V}{\partial\theta}\right) + \frac{1}{r^2\sin^2\theta}\frac{\partial^2 V}{\partial\phi^2}.$$

XX Of course we could read off $\frac{\partial V}{\partial\rho}$ with $\phi\, z$ const in terms of diff coeffs in the $r\,\theta\,\phi$ scheme, This is particularly simple because ϕ is constant in both

Thus in the figure we see (marked with one arrow) the ρ, z (ϕ) gradients, and with two arrows the equivalent r, θ (ϕ) gradients

∴ $\frac{\partial V}{\partial\rho} = \frac{\partial V}{\partial r}\sin\theta + \frac{1}{r}\frac{\partial V}{\partial\theta}\cos\theta$ as we obtained above.

The commonest way of transforming, is to apply $\iint \frac{\partial V}{\partial n}dS = \iiint \nabla^2 V\,dx\,dy\,dz$ to the brick $r, r+\Delta r$; $\theta, \theta+\Delta\theta$; $\phi, \phi+\Delta\phi$.

The areas of the faces are $r^2\sin\theta\,\Delta\theta\Delta\phi$ …, $r\sin\theta\,\Delta\phi\,\Delta r$, $r\,\Delta\theta\,\Delta r$; $\frac{\partial V}{\partial r}$, $\frac{1}{r}\frac{\partial V}{\partial\theta}$, $\frac{1}{r\sin\theta}\frac{\partial V}{\partial\phi}$ standard normal $\frac{\partial V}{\partial n}$ on the respective faces

XXI

$$\therefore \frac{\partial}{\partial r}\left[r^2\sin\theta\,\Delta\theta\Delta\phi\frac{\partial V}{\partial r}\right]\Delta r + \frac{\partial}{\partial\theta}\left[r\sin\theta\,\Delta\phi\Delta r\frac{1}{r}\frac{\partial V}{\partial\theta}\right]\Delta\theta + \frac{\partial}{\partial\phi}\left[r\Delta\theta\Delta r\frac{1}{r\sin\theta}\frac{\partial V}{\partial\phi}\right]\Delta\phi = \nabla^2 V(r^2\sin\theta\,\Delta r\,\Delta\theta\,\Delta\phi)$$

∴ in the limit $$\nabla^2 V = \frac{1}{r^2}\frac{\partial}{\partial r}\left(r^2\frac{\partial V}{\partial r}\right) + \frac{1}{r^2\sin\theta}\frac{\partial}{\partial\theta}\left(\sin\theta\frac{\partial V}{\partial\theta}\right) + \frac{1}{r^2\sin^2\theta}\frac{\partial^2 V}{\partial\phi^2}.$$

§§ If $V = r^n F(\theta,\phi)$ Then $\frac{1}{\sin\theta}\frac{\partial}{\partial\theta}\left(\sin\theta\frac{\partial F}{\partial\theta}\right) + \frac{1}{\sin^2\theta}\frac{\partial^2 F}{\partial\phi^2} + n(n+1)F = 0$

If $V = r^n f(\theta)$ Then $\frac{1}{\sin\theta}\frac{d}{d\theta}\left(\sin\theta\frac{df}{d\theta}\right) + n(n+1)f = 0$ or if $\cos\theta = \mu$ $\frac{d}{d\mu}\left[(1-\mu^2)\frac{df}{d\mu}\right] + n(n+1)f = 0$

FIGURE 5.5. A page (one third full size) from Robert Webb's coaching notes. Webb's pupils were required to copy out and to master these notes before being shown by him how they were used to tackle difficult problems. Note how Webb's narrative reproduces a virtual form of his own apparently severe teaching style ("You *must* be able to . . ."). Note also the range of techniques he offers for remembering the required transformation. (RRW-SJCC.) (By permission of the Master and Fellows of St. John's College, Cambridge.)

nates.[43] Webb announces the importance of what follows by assuring the student that he "*must* be able" to put Laplace's equation into polar coordinates in three dimensions "instantly and correctly," and then begins "by traversing the actual transformation in the way a first class man would proceed." The touchstone of proper procedure here is the technique typically adopted by a senior or high wrangler, but, having gone through the transformation step by step, Webb offers two additional, quicker and more visual methods of carrying out the transformation, which make use of easily sketched diagrams and, in one case, Gauss's theorem in integral form. These geometrical methods would have been less acceptable in an examination than would the analytical demonstration of the transformation, but they would have enabled a good student quickly to check that he had written Laplace's equation correctly in the new coordinates. At the very bottom of the page, Webb notes that, when expressed in the new variables, the equation can be separated into parts that are applicable to a wide range of physical problems and relatively straightforward to solve. Later in the manuscript, he begins a new section with the advice: "You cannot ever know too much about the solutions of [Laplace's equation]," and, starting with the equation in the now familiar spherical-polar form, develops the solutions that are most useful in physical applications.[44]

The two most striking aspects of Webb's coaching manuscript overall are the language of its presentation and its mathematical structure. Webb's notes are written in a familiar, avuncular style, that displays the craft of an experienced teacher anticipating points where a student is likely to get stuck, bored, make an error, lose confidence, miss an important point, or fail to grasp relevant applications. The manuscript contains relatively few problems showing precisely how the methods discussed were to be applied, but its structure reveals that is was nevertheless intended first and foremost to prepare students for the Tripos examination. The mathematical methods are not grouped according to general mathematical principles but as a collection of techniques useful in the solution of problems. The transformation of Laplace's equation, for example, is presented neither as part of a general discussion of non-Cartesian coordinate systems nor even as a direct prerequisite to the solution of Laplace's equation. It is presented, rather, as one of many techniques involving a change or transformation of independent vari-

43. Laplace's equation is given in section 8.5. It is often written in the form $\nabla^2 V = 0$ (pronounced "del-squared V equals zero") where ∇^2 is the differential operator $\partial^2/\partial X^2 + \partial^2/\partial Y^2 + \partial^2/\partial Z^2$.

44. RRW-SJC, 30.

ables, that must be learned by heart so that they can be reliably pressed into service at the right moment in a problem. Precisely when and how these techniques were brought to bear on problems was what Webb discussed and demonstrated in his coaching room as he worked his way through numerous examples on the blackboard.

5.4. Translating Mathematical Culture

> [W]herever his pupils have gone, notes of his lectures have gone with them, and classes that never heard Herman's name have heard the echo of his voice, though not the strange idiomatic "That part" with which he used to break off a piece of work and indicate that the difficulties were resolved and that his hearers could be trusted to reach the conclusion without further help.
>
> E. H. NEVILLE (2W 1909)[45]

An important difficulty raised implicitly in the previous section is that the subtle competencies painstakingly acquired by students during years of supervised private study and face-to-face encounters in the coaching room cannot, by definition, be succinctly described in words. If such a description were possible it would be a straightforward matter to circumvent the very pedagogical apparatus outlined above and to inscribe the essence of Routh's coaching technique in a textbook. But while we cannot enumerate the individual skills and sensibilities that generated the problem-solving ability of a high wrangler, we can obtain an indirect measure of their dependence on a specific system of training. The command of mathematical physics displayed by high wranglers was admired by many non-British mathematicians, some of whom were keen to nurture similar abilities in their own students. Yet in no cases of which I am aware did any mathematical center beyond Cambridge generate a community of students capable of emulating a wrangler's performance without hiring a wrangler to organize and to manage the training process. For example, the great German mathematician and leader of the Göttingen school, Felix Klein, was extremely impressed by what he saw as the close and productive relationship nurtured in Cambridge between the development of mathematical methods and their application to problems in mathematical physics, especially mechanics.[46] Klein was keen to preserve a similar relationship in Germany as a means of resisting what he saw as the

---

45. Neville 1928, 238.
46. Rowe 1989, 206. Klein commissioned three major works on mechanics by wranglers—Horace Lamb (2W 1872), A. E. H. Love (2W 1885), and Edmund Whittaker (3W 1895)—for his international *Encyklopaedie der mathematischen Wissenschaften*.

potentially unhealthy separation of pure and applied mathematics that was taking place during the 1890s.[47]

A regular visitor to Cambridge, Klein made a special study of mathematics teaching in the university in an explicit attempt to understand "why English mathematical physicists had so much practical command over the application of their knowledge."[48] He specifically identified Routh's training methods as the key to this special facility and decided that the most practical way of making Routh's approach accessible to a Continental audience would be to commission a German translation of Routh's *Dynamics of a System of Rigid Bodies* (1868, hereafter *Rigid Dynamics*). In his preface to the German edition, Klein explained that Routh's book was of special importance because it was exemplary of the dominant teaching method in Cambridge, of which Routh had been the "well-known master for many years." This method, Klein continued, which placed "great emphasis on the working through of individual examples, without doubt develop[ed] the abilities of the students in a specific direction to an extraordinary degree" (Routh 1898b, 1:iii). Klein was naturally obliged to acknowledge the virtues of the more abstract German approach to dynamics, but he insisted nonetheless that the problem-solving approach embodied in Routh's book should be seen as an "addition of extraordinary meaning" (iv). Klein's comments are interesting not least because they confirm that the qualities identified above as peculiar to the Cambridge approach to mathematics were recognized and valued at the time by mathematicians from other institutions. It is also interesting, however, that, despite the eminence of Klein and his school, the German translation of Routh's book appears to have had little if any impact on the German mathematical community. One reason for this lack of impact would surely have been that it was not possible to produce students in the Cambridge mold at a distant site simply by making an exemplary textbook available in the German language. To have accomplished such a geographical translation would have necessitated the recreation of something very like the system of coaching and examining painstakingly constructed by Cambridge coaches and examiners over the previous half century.

47. Ibid., 208. It is a measure of the importance attached in Germany to Cambridge work in mathematical physics that in addition to Thomson and Tait's *Treatise on Natural Philosophy* (1867), Maxwell's *Treatise on Electricity and Magnetism* (1873), and Rayleigh's *Theory of Sound* (1877–78), such books as Love's *Treatise on the Mathematical Theory of Elasticity* (1895) and Lamb's *Hydrodynamics* (1892) were also translated into German.
48. Larmor 1910–11, xvi. On Klein's regular visits to dine at Trinity College, see Whittaker 1942–44, 219. For a firsthand account of the way Klein taught in Göttingen in the 1890s, see Grattan-Guinness 1972a, 122–24, and 1972b, 375–84.

The culture in which Routh's students were fashioned certainly made use of textbooks, but they were only part of a carefully constructed training program which included the important element of "personal intercourse" between master and pupil. And, as we saw in chapter 3, even British students who had not studied in Cambridge found textbooks by Cambridge authors difficult to understand because they assumed a high level of mathematical knowledge as well as an aptitude for, and training in, problem solving on the part of the reader. These points are further underlined by the attempts made in Germany and elsewhere to use another popular Cambridge textbook, A. R. Forsyth's (SW 1881) *Treatise on Differential Equations* (1885). An appealing aspect of Forsyth's book was that it offered a good general overview of the subject, including many recently published techniques hitherto available only in mathematical journals. Following the Cambridge tradition, Forsyth, a pupil of Routh, presented much of this material in the form of more than eight hundred problems and examples, most "taken from University and College Examination papers set in Cambridge at various times" (1885, v). When the Berlin mathematician, Hermann Maser, translated the book into German to use as a teaching text, he experienced such difficulty solving the problems that he had very frequently to write to Forsyth asking for help. Some years later, following a lecture series held by Forsyth at the University of Calcutta, an Indian mathematician proposed to write a key to the problems in the most recent edition of the book, but Forsyth, anticipating a repeat of his extended correspondence with Maser, decided to publish a book of solutions himself (Forsyth 1918). In the preface to the book, Forsyth emphasized that his solutions were intended in the first instance to assist teachers rather than students. He anticipated that even professional mathematicians would routinely experience difficulty solving the problems, while their students would not necessarily be enlightened even by the skeleton solutions.

An alternative method of estimating the importance of site-specific pedagogical resources in the production of wrangler culture is to investigate cases in which the culture *was* successfully reproduced. The most striking example of this kind concerns the rise of mathematical teaching in English public and grammar schools during the latter nineteenth century. The period of Routh's dominance as a private tutor in Cambridge witnessed a further substantial increase in the level of mathematical expertise necessary for Tripos success. This was due partly, as it had been in the 1820s and 1830s, to more intensive coaching regimes and to the examiners' habit of setting more difficult and searching questions on the standard topics year by year. Of equal or possibly greater importance, however, was the rapid increase in the level of mathematical knowledge acquired by students before they entered the university.[49] We have already seen that some of the most successful wranglers of the early Victorian era had graduated in mathematics from London or Scottish universities before arriving in Cambridge.[50] This tradition was expanded during the second half of the century as new provincial universities—most notably Birmingham, Manchester, and Liverpool—also began routinely to send their most able mathematics graduates to Cambridge. Moreover, from the 1860s, the mathematics professors at virtually all of these universities were themselves high wranglers, many of whom had been coached by Routh.

An even more dramatic transformation in the level and status of mathematics in pre-Cambridge education took place at around the same time in many English schools. During the first half of the nineteenth century, mathematics was not normally a major subject of study in the school curriculum, and, as we saw in chapter 3, the mathematical knowledge of the majority of students arriving at the university during this period was confined to elementary arithmetic, algebra, and geometry.[51] Beginning in the 1850s, the gradual reform of the public, private, and grammar schools, and the founding of several examination boards to test students' knowledge of mathematics by written examination, raised standards considerably and established mathematics as an important and compulsory part of the school curriculum.[52] The details of these reforms are beyond the scope of the present study, but one consequence of them relevant to our current concerns is the impetus they provided to the establishment of a wrangler culture of mathematical training in several leading schools.[53] These schools traditionally recruited masters from the ancient universities, and as wranglers began to be hired explicitly to teach mathematics they naturally employed the methods by which they had been trained in Cambridge.[54] When Rawden Levett (11W 1865) was

49. Morgan 1871, 4.
50. The introduction of Cambridge-style written examinations in Glasgow and Edinburgh in the 1830s is briefly discussed in Wilson 1985, 20, 30.
51. The two most eminent of the English public schools, Eton and Harrow, had only appointed competent mathematics masters in 1834 and 1838 respectively (Howson 1982, 124).
52. For a brief overview of the development of examinations and examination boards in schools, see Howson 1982, 160–61.
53. The older public schools were increasingly concerned that they were rapidly losing ground to newer schools (patronized by the middle classes) that were preparing students intensely for mathematical scholarships and high mathematical honors at Cambridge (Howson 1982, 128–30).
54. James Wilson (SW 1859) was taught mathematics by John Harrison (3W 1828) at Sedbergh Grammar School before going to Cambridge. He later played a major role in reforming mathe-

hired to invigorate the study of mathematics at the King Edward VI Grammar School in Birmingham, he set about stimulating "interest and keenness" in the subject by establishing a local order of mathematical merit. Levett persuaded the headmaster to institute a regular arithmetic examination for all four hundred boys, the results of which were "read out to the whole school assembled in the Great School Room" (Mayo 1923, 326). The establishment of Minor Scholarships and Exhibitions by several Cambridge colleges around 1860 also encouraged wrangler-schoolmasters to coach their best pupils individually or in small groups.[55] These valuable and prestigious awards, open to schoolboys and first year undergraduates alike, provided an important vehicle by which bright students from lower middle-class backgrounds could fund a Cambridge education. By the mid 1860s, preparation for the Mathematical Tripos had effectively been extended to include either a preliminary degree in mathematics at a provincial university or the final two or three years of secondary schooling. As Henry Morgan (26W 1853) wrote in 1871, any schoolboy who showed an aptitude and liking for mathematics was likely to be "noticed and responded to by his mathematical master—probably a wrangler—who is only too glad to find a boy with whom he can read something beyond the drudgery of Arithmetic and Euclid's propositions, and who gives promise of bringing credit to the school" (1871, 4–5). Morgan estimated that in order to be reasonably certain of obtaining a mathematical scholarship to Cambridge, a talented student would need to devote at least two years to the almost exclusive study of mathematics. If a boy did win an award, he would return to school to find the "hopes of masters and boys . . . centred on him" and be "pressed on in more advanced mathematics" in preparation for a distinguished undergraduate career (5–6).

It is difficult to assess just how much mathematics an average scholarship candidate would have known by the 1860s, but we can be sure it was considerably more than all but the most exceptional students would have known a couple of decades earlier. A useful guide to the knowledge of the best-prepared students in this period is provided by the mathematical training of Lord Rayleigh (SW 1865) and the future economist Alfred Marshall (2W 1865). Rayleigh's undergraduate success was reckoned the more surprising as

---

matical education in English schools (Howson 1982, 123–40). Charles Prichard (4W 1830) was headmaster at Clapham Grammar School from 1834 to 1862, during which time sent several mathematical scholars to Cambridge including George Hemming (SW 1844) and George Darwin (2W 1868) (Prichard 1886, 89–96).

55. *The Student's Guide to the University of Cambridge*, 1st ed. (Cambridge: Deighton Bell, 1862), 43.

he entered Cambridge in 1861 "decidedly less advanced in mathematical reading than the best of his contemporaries" (Strutt 1924, 24). This comment might be taken to imply that Rayleigh was rather poorly prepared in mathematics when he arrived in Cambridge, but his biography reveals that he had actually received a good grounding in arithmetic and geometry by the age of ten (probably as much as most freshmen half century earlier) and studied mathematics continuously from the age of fifteen until he entered Cambridge a month before his nineteenth birthday. At school in Torquay he was taught by William Leeming (26W 1858), and, during the summer before he entered Trinity College, Rayleigh was intensively coached at his country house by the recently graduated Frederick Thompson (11W 1861) (Strutt 1924, 13–24). That this preparation still left him below the standard of the most knowledgeable freshmen in Routh's first-year class is an indication of the enormously increased level of technical competence expected of undergraduates by the early 1860s.

Marshall is an example of just such a freshman who was also a fellow pupil of Rayleigh's in Routh's coaching room. Trained at Merchant Taylors' School in London, Marshall too was prepared for Cambridge entrance by a wrangler, J. A. L. Airey (2W 1846). During the 1860s, Merchant Taylors' offered what was generally regarded as the best mathematics education provided by any school in the country.[56] In addition to the usual mathematical subjects of arithmetic, algebra, and geometry, the most able students were required to study the differential calculus, statics, dynamics, and hydrodynamics (from Cambridge textbooks) and to tackle problems from books by Cambridge authors as well as the published volumes of Senate House problems.[57] As a second wrangler from the mid 1840s, Airey was almost certainly an ex-pupil of William Hopkins and well aware therefore of how to prepare his pupils to succeed in Routh's elite class.

The intensity of pre-Cambridge mathematical coaching in schools increased relentlessly throughout the last quarter of the century. During the early 1880s, Tonbridge School became especially noteworthy for the number of future high wranglers it produced, while, in the latter 1880s and 1890s, Manchester Grammar School and the King Edward VI Grammar School assumed similar roles.[58] In all thee cases, the mathematical masters in question, Henry Hilary (11W 1870), Henry Joseland (15W 1886), and Rawden Levett (11W 1865) respectively, were successful wranglers who had very likely

56. Groenewegen 1995, 55–56.
57. Ibid., 68.
58. *The Graphic* 28 (1883): 62.

been coached by Routh.[59] There was also a steady increase in the number of "crammers" specializing in the preparation of students for formal written examinations—now common throughout the British educational system—including Cambridge entrance.[60] The young Bertrand Russell (7W 1893) was prepared in the late 1880s for a scholarship examination at Trinity College by the University and Army Tutors, and having won the scholarship was sent for the next eight months to Wren's Coaching Establishment, where he was intensively coached by H. C. Robson (6W 1882). A private diary kept by Russell during this period reveals him using Routh's *Rigid Dynamics* as a textbook while struggling under Robson's guidance to solve related Tripos problems.[61] When G. H. Hardy (4W 1896) studied at Winchester in the early 1890s, the Senior Mathematics Master, George Richardson (3W 1857), had successfully generated a regime of mathematical training very like the one he would have experienced at St. John's in the mid 1850s. Hardy was taught to think of mathematics "in terms of examinations and scholarships," and it was, no doubt, this emphasis on preparation for competitive examinations and university entrance that led to Winchester being regarded by some around 1890 as the "best mathematics school in England" (Hardy 1967, 17, 144).

A training regime bearing remarkable similarity to Routh's scheme was installed at St. Paul's School in London by another wrangler, F. S. Macaulay (8W 1882), who was especially adept at coaching students for Minor Scholarships to Cambridge (fig. 5.6). During the period 1885–1911 his students won thirty-four Minor Scholarships and eleven Exhibitions to the university, four going on to become senior wranglers, one a second and one a fourth wrangler.[62] Macaulay nurtured a "University atmosphere" in his classes, and, like Routh (who was very likely his coach), he gave his students weekly problems and revision papers (including past Tripos questions) to solve, and supplemented lectures and textbooks with his own manuscripts. If there was a problem that none of the senior students could solve, it was considered Macaulay's "duty to *perform* at sight at the blackboard," a tradition highly reminiscent of Routh's famous problem-solving performances at Peterhouse. Macaulay also found time to undertake significant research in mathematics and was always ready to suggest a "mild" topic of investigation

---

59. It is difficult to find out who coached these men as they were not in the celebrated "upper ten."
60. On science examinations in Victorian Britain, see MacLeod 1982.
61. Griffen and Lewis 1990, 53–54; Moorehead 1992, 29.
62. Baker and Littlewood 1936–38, 359.

FIGURE 5.6. Francis Macaulay (8W 1882) who recreated the culture of the Cambridge coaching system at St. Paul's School in London. Macaulay also undertook "pioneering work in the theory of algebraic polynomials" and inspired an interest in original mathematics in his pupils. Photograph and quotation from Baker and Littlewood 1936–38. (By permission of the Syndics of Cambridge University Library.)

to any of his pupils who wanted to try their hand at original work.[63] An informative account of life under Macaulay's tutelage is given by one of his most distinguished pupils, J. E. Littlewood (SW 1905), who entered St. Paul's in 1900 aged 14. Having just arrived in England from the Cape University in South Africa, Littlewood estimated his knowledge of mathematics—the first six books of Euclid, a little algebra, and trigonometry up to solution of triangles—"slight by modern standards." During the next three years Littlewood was intensively coached by Macaulay using standard Cambridge

63. Ibid., 359–60. Littlewood recalled of his time with Macaulay that "while a boy with any mathematical gift can do with extremely little direct help, yet *some* contact is necessary, and that of the right kind."

textbooks (including two of Routh's) to a level of mathematical knowledge comparable with that of a high wrangler three decades earlier.[64] His coverage of statics and gravitation, for example, included the use of spherical harmonics and an "exhaustive treatment of the attractions of ellipsoids," topics considered too specialized even for the Mathematical Tripos until the mid 1860s.[65]

The success of the wrangler schoolmaster in reproducing the coaching system in schools and colleges throughout Britain is evident in the escalating levels of technical competence expected of Cambridge freshmen. By the early 1890s, even the new edition of the *Student's Guide* felt it necessary to warn prospective mathematics undergraduates that it had become "rare" for a student to do well in the Mathematical Tripos unless he had received "considerable training at school, or elsewhere, before coming into residence." Most students now arrived with "some knowledge of Co-ordinate Geometry, Differential and Integral Calculus and Mechanics."[66] It is a further mark of the success of wrangler masters such as Macaulay that some of their most precocious students had virtually completed an undergraduate course of coaching prior to entering the university. These young men were so far ahead of their peers that, from the late 1890s, some of them, Littlewood included, elected to sit the Mathematical Tripos at the end of just six terms of undergraduate study.[67] G. H. Hardy and James Jeans (2W 1898) were probably the first to attempt this feat, both having been trained at school by leading wrangler masters; Hardy, as we have just seen, was a student of Richardson at Winchester, while Jeans worked under Samuel Roberts (7W 1882) at Merchant Taylors' School, which, as we have also seen, had a long-established reputation for offering a Cambridge-style mathematics education.[68]

These examples cast considerable light on the mechanisms by which the training system nurtured in Victorian Cambridge was successfully reproduced in numerous classrooms throughout Britain and her Empire. The suc-

64. Littlewood (1986, 80–83) gives a detailed account of the topics and books studied at St. Paul's School.

65. Glaisher 1886–87, 20–21.

66. *The Student's Guide to the University of Cambridge*, 5th ed. (Cambridge: Deighton Bell, 1893), 15–19.

67. Students were permitted to sit the Mathematical Tripos after six terms according to new regulations of 1893; *Reporter*, 20 May 1890, 734. These regulations were intended to enable students to study mathematics for two years prior to studying physics in the Natural Sciences Tripos. See Wilson 1982, 340.

68. *Cambridge Chronicle*, 17 June 1898, 7; Milne 1945–48, 574. Jeans and Hardy were recommended to take the Tripos after two years by their Trinity Tutor, G. T. Walker (SW 1889) (Milne 1952, 4–5).

cess of the wrangler schoolmasters derived not simply from the use of well-known Cambridge textbooks but from the re-creation of the whole apparatus of coaching and examining. Indeed it was largely Routh's teaching methods, Routh's books, and, above all, Routh's pupils, that were most responsible for this cultural translation. Reflecting upon Routh's extraordinary influence on mathematics during the period 1860–1910, J. J. Thomson wrote:

> for about half a century the vast majority of professors of mathematics in English, Scotch, Welsh and Colonial universities, and also the teachers of mathematics in the larger schools, had been pupils of [Routh's], and to a very large extent adopted his methods. In the textbooks of the time old pupils of Routh's would be continually meeting with passages which they recognised as echoes of what they had heard in his classroom or seen in his manuscripts. (Thomson 1936, 39)

In other words, schoolboys and students in the leading British schools and provincial universities received a form of virtual coaching by Routh. Routh's textbooks, as well as those written by his pupils, reproduced the content and structure of his Peterhouse lectures, and successive editions gradually incorporated the most important of his manuscript summaries of recent research. The examples and problems at the end of each chapter also provided a convenient guide to the kinds of examination question relevant to every topic. These books made it easy for wrangler schoolmasters to follow Routh's syllabus and order of presentation, but it was firsthand experience in his classroom, or that of another leading coach, that enabled them to re-create the atmosphere of competitive endeavor peculiar to undergraduate mathematical studies in Cambridge.

Wrangler masters also knew from firsthand experience how hard, for how long, and in what manner students needed to study in order to tackle difficult problems with the speed and confidence that would win them a Minor Scholarship to the university. The Tripos problems they set were generally taken from standard collections of the period—often the very ones with which they had themselves struggled as undergraduates—and these problems, like the ones reproduced in Routh's books, provided a practical introduction to the mathematical methods and topics of research interest current in Cambridge. The author of the best known and most widely used of these collections, Joseph Wolstenholme (3W 1850), deliberately ensured that his *Mathematical Problems* (fig. 5.7) was "unusually copious in problems in the earlier subjects" so that it would be useful to mathematical

A BOOK

OF

IV. 30 48

MATHEMATICAL PROBLEMS

ON SUBJECTS INCLUDED IN

THE CAMBRIDGE COURSE

DEVISED AND ARRANGED BY

JOSEPH WOLSTENHOLME,

FELLOW OF CHRIST'S COLLEGE; SOMETIME FELLOW OF ST JOHN'S COLLEGE;
AND LATELY LECTURER IN MATHEMATICS AT CHRIST'S COLLEGE.

"Deduct but what is Vanity or Dress,
"Or Learning's Luxury, or Idleness;
"Or tricks to shew the stretch of human brain,
"Mere curious pleasure, or ingenious pain;
. . . .
"Then see how little the remaining sum."

London and Cambridge:
MACMILLAN AND CO.
1867.

[*All rights reserved.*]

FIGURE 5.7. The title page of the first edition of Joseph Wolstenholme's *Mathematical Problems*, 1867. Books of problems of this kind were one pedagogical device used by Cambridge-trained mathematics teachers to take wrangler culture beyond the university. Woolstenholme deliberately included large numbers of elementary problems in his book so that it could be used to prepare senior school students for Cambridge entrance. (Reproduced by permission of the Science Museum/Science and Society Picture Library.)

students "not only in the Universities, but in the higher classes of public schools" (1867, v). Despite the fact that he published a number of mathematical papers, Wolstenholme's fame as a mathematician rested chiefly upon his mathematical problems, many of which, as we saw in chapter 3, contained new results that other mathematicians would have published in a research journal. These problems were acknowledged to have "exercised a very real influence upon successive generations of undergraduates" and, we might add, the pupils of wrangler schoolmasters.[69] It was from "Wolstenholme's Problems" that Macaulay set his weekly exercises at St. Paul's, performing solutions at the blackboard when necessary just as his coach had done for him.[70]

The above comparison between the relatively unsuccessful attempts by men like Klein to foster the wrangler's problem-solving skills in Göttingen and the highly successful establishment of those same skills by wrangler masters in the English public schools provides some insight into the complexities of learning particular approaches to mathematical physics. When the textbooks written by leading Cambridge mathematicians (and used by Cambridge students) traveled overseas to Europe, India, and elsewhere, they carried with them neither an explanation of the skills necessary to tackle the many Tripos problems contained within their pages nor any indication of the rich pedagogical economy from which they had emerged. What students left to struggle unguided with these books would sorely have lacked was the competitive ethos and daily routine of solitary study and physical exercise typical of Cambridge undergraduate life, as well as the minutely organized and brilliantly taught courses offered by men like Routh. When mathematics teaching in the English public schools was reformed in the last third of the nineteenth century, these difficulties were unwittingly circumvented by the recruitment of wrangler masters to teach new courses and by the provision of an environment very like the one in Cambridge. The fact that many of the most successful of these masters also instituted teaching regimes based closely on their own undergraduate experience also had important implications for the study of mathematics at Cambridge itself. As students arriving at the university had increasingly mastered the material that would once have occupied them for the first year or two of undergraduate study, coaches like Routh could push their best students to ever higher levels of technical competence. This continued inflation in the technical ability of some students would, as we shall shortly see, eventually play a role in undermining the

---

69. A. R. Forsyth (SW 1881) quoted in the *DNB* entry on Joseph Wolstenholme.
70. Littlewood 1986, 81, and 1953, 81.

coach's place in undergraduate teaching. But through the last third of the nineteenth century the mathematical form of life that had once been peculiar to Cambridge was effectively redistributed to numerous schools and colleges throughout Britain and her Empire.

5.5. The Reform of Public Teaching and the Research Ethos

> The fact is that we have, I apprehend, gradually, half unconsciously, altered our aim from the training of men for after-life to the specific production of mathematicians.
> ISAAC TODHUNTER, 1873[71]

During the 1870s, the coach's domination of undergraduate mathematical training began to be challenged by the gradual reform of public teaching. Even as Routh's methods were being adopted at centers of mathematical learning throughout Britain, the place of the coach in Cambridge itself was being redefined by the emergence of new courses of *intercollegiate* and *university* lectures in advanced mathematics. It is important to understand the origins and purpose of this challenge, and, in order to do so, we must go back to calls for changes in the syllabus of the Mathematical Tripos a decade earlier. In the mid 1860s, several prominent Cambridge-trained mathematicians voiced the opinion that the undergraduate syllabus, stable since 1848, still devoted too much time to the study of pure analysis and not enough to mathematical physics. Led by the Astronomer Royal, George Airy, these men urged that more attention be paid to several established physical topics and, most important of all, that the major new topics of electricity, magnetism, and thermodynamics be included in the Tripos syllabus. These calls for reform represented a continuation of William Hopkins's campaign of the early 1850s to make undergraduate mathematical studies more relevant to the commercial and industrial interests of the British nation.[72] During the subsequent decade, the work of men like William Thomson and James Clerk Maxwell, who combined brilliant theoretical investigations with both experimentation and participation in such major industrial projects as submarine telegraphy, realized Hopkins's vision. By introducing more, and more advanced, mathematical physics into undergraduate studies, Cambridge could ascend the intellectual high ground of an ever more industrially based physics and justify its oldest Tripos as contributing to the nation's wealth and power. As Tod-

---

71. Todhunter 1873, 63.

72. In 1864 William Thomson railed against Arthur Cayley's pure mathematical work, claiming that it would "possibly interest four people in the world" and that it was "too bad" such people did not "take their part in the advancement of the world" (Smith and Wise 1989, 189).

hunter's remarks in the epigraph above suggest, the Whewellian defence of mathematical studies as the best means of shaping men's minds for an "after life" in the church, law, or civil service was rapidly giving way to a more utilitarian and vocational ideal of preparation for useful research.

The changes called for by the reformers of the 1860s were accommodated in 1868 by new Tripos regulations that would come into effect in 1873. According to these regulations, an expanded syllabus of advanced topics was divided into five separate divisions and an extra day added to the examination (making nine in total). In recognition of this increase in content, it was agreed that advanced students would no longer be required to study every topic to the highest level, but could specialize in those they found the most interesting. These important changes, soon to be followed by further reforms, generated a serious crisis in undergraduate teaching. The coaches' grip on advanced training relied on the fact that they offered a complete course of study which prepared students for every topic likely to be set in the Senate House. But in the new scheme no single teacher could claim to have mastered every subject to advanced level. The most important of the new physical subjects (electricity and magnetism) were especially troublesome in this respect. They required a knowledge of new physical concepts, new instruments and experiments, and were made the more alien by their association with industrial applications. It was also widely believed that these topics could not be taught properly without practical demonstrations of the major instruments and experiments involved. This belief played an important role in the appointment of a new Professor of Experimental Physics (James Clerk Maxwell) and the building of the Cavendish Laboratory in the early 1870s.[73] As we shall see in chapter 6, neither the new professor nor the new laboratory actually had much immediate impact on undergraduate mathematics teaching, but their appearance added to the sense that the coach's half-century of domination of undergraduate teaching was under siege.

The new regulations presented similar difficulties for most college lecturers.[74] They were no better placed than the coaches to teach the new subjects and, in many cases, were actually less experienced at teaching advanced topics. As we have seen, the most able undergraduates had long since abandoned college lectures precisely because they were run at the pace of mediocre students and did not offer the carefully managed scheme of advanced

---

73. On the founding of the Cavendish Laboratory, see Sviedrys 1970 and 1976, and Warwick 1994.

74. The arrival of the new subjects caused some colleges to become even more reliant on private tutors; see *Cambridge University Reporter*, 21 November 1878, 49.

study provided by men like Routh. An obvious solution to this problem for college lecturers, and one that would have been much harder for coaches to adopt, was to organize themselves as a team of specialists, each concentrating on a small group of advanced topics. But since, by tradition, each college restricted its lectures to its own members, this solution was practicable only for a handful of the largest colleges that could afford a sizable staff of mathematical lecturers.

The sense of impending crisis that gathered in the late 1860s in anticipation of the introduction of the new regulations played into the hands of those who sought to reform public teaching from below. In the vanguard of the reformers was a young Trinity fellow, James Stuart (3W 1866), who, in 1868, having just been made an assistant tutor, felt it was an opportune moment to offer a course of *intercollegiate* lectures in physical astronomy and the mathematical theories of heat, electricity, and magnetism.[75] Stuart's plan, which must have been approved by senior members of Trinity College, was to organize coordinated courses of advanced lectures, first at Trinity and then at other colleges, which were free to members of the host college but which could be attended by any student for a fee. We must assume that these lectures were reasonably well attended as, two years later, Stuart and three of his colleagues organized a meeting of all the principle mathematics lecturers in the university to discuss a much wider collaboration in teaching advanced mathematics.[76] The outcome of the meeting was a joint scheme of intercollegiate courses given at Trinity, St. John's, and Peterhouse, which covered advanced aspects of most higher mathematical subjects, placing special emphasis on the new physical topics. The scheme was published in the *Cambridge University Reporter* together with a defense of the reformers' unofficial actions and an open invitation for "anyone who has taken up as his speciality any of these higher subjects" to participate in the collaboration. The main arguments advanced by the reformers hinged on what they saw as the crisis generated by simultaneously expanding the syllabus of advanced topics while allowing the most able students to study just a few topics in great depth. They argued that any lecturer teaching an advanced topic in these circumstances ought to have made it his "special study for some time."[77] Even the

75. Stuart's first course was held in 1868 and attracted fifteen students. Non-Trinity students were supposed to pay one guinea to attend but only one actually paid up (Stuart 1911, 152–53).

76. The meeting was organized by Stuart, W. K. Clifford, Coutts Trotter, and Percival Frost, and held on 25 November in Norman Ferrers' (SW 1851) rooms in Gonville and Caius College (Stuart 1912, 153). A complete list of those present is given in the *Cambridge University Reporter,* 7 December 1870, 144.

77. *Cambridge University Reporter,* 7 December 1870, 144–45.

larger colleges could not claim to have such specialist lecturers in every advanced topic, and, since such courses would attract not more than a dozen or so students, the only practical solution was intercollegiate collaboration.

The scheme proposed in 1870 incorporated just three colleges and covered only the last four terms prior to the Tripos examination, but during the next decade it expanded rapidly on every front. The list of courses advertised annually in the *Reporter* grew longer each year until, by 1880, there were thirty-seven courses offered by ten colleges (fig. 5.8). The relevant lectures offered by the Professor of Experimental Physics and other mathematics professors were also incorporated, and, following the practice of the leading coaches, courses were offered during the Long Vacation. This was a major achievement by Stuart and his fellow reformers. Prompted by the new regulations of 1868, they had taken just twelve years informally to institute a system of university-wide teaching in advanced mathematics. It was not, strictly speaking, a truly public system, as students still had to pay an additional fee to attend intercollegiate lectures, but it had opened the financial resources of the wealthiest colleges, and the intellectual resources of every college, to all mathematics students. The scheme also provided a coordinated framework of advanced courses within which the long-marginal professorial lectures could begin to return to center stage. The success of the intercollegiate courses soon provoked further and far-reaching reforms of the Tripos regulations and finally opened the way for the introduction of a genuinely public system of mathematics teaching.

By the mid 1870s it had become apparent that the gap between the mathematical attainments of the average and most able undergraduates was becoming too great for them to be ranked according to a single examination. If the questions were made hard enough to stretch those who were attending advanced intercollegiate lectures, they were incomprehensible to the majority of candidates. After protracted consideration of the problem, the Board of Mathematical Studies decided that, from 1882, the Senate House examination would be split into two parts. The order of merit would henceforth be drawn up according to a two-stage examination (four days plus five days) taken after nine terms of study (known as Parts I and II), the wranglers alone going on to take a new examination (Part III) at the end of the tenth term. It was also decided that the advanced subjects would be divided into four groups: Group A, pure mathematics; Group B, the astronomical subjects; Group C, hydrodynamics, sound, optics, and vibration theory; and Group D, thermodynamics, electricity, and magnetism plus related mathematical methods. Candidates for this part of the examination were allowed

Professorial and Inter-Collegiate Lectures in Mathematics, 1880–1881

MICHAELMAS TERM, 1880

Place	Lecturer	Subject
New Museums	Prof. Challis	Practical astronomy
"	Prof. Cayley	Theory of equations
"	Prof. Stuart	Mechanism and applied mechanics
Cavendish Laboratory	Lord Rayleigh	Electricity and electromagnetism
St Peter's College	Mr Dickson	Dynamics of a particle
"	Mr Allen	Magnetism
Gonville and Caius College	Mr Ferrers	Theory of the potential
Trinity Hall	Mr Bell	Higher algebra
Corpus Christi College	Mr Wallis	Definite integrals, including calculus of variations
King's	"	Mr Stearn	Hydrodynamics
Christ's	"	Mr Hobson	Rigid dynamics
Trinity	"	Mr Niven	Electrostatics
"	"	Mr Glaisher	Theory of elliptic functions
"	"	Mr Ball	Higher solid geometry

LENT TERM, 1881

Place	Lecturer	Subject
New Museums	Prof. Adams	Theory of Jupiter's satellites
"	Prof. Stuart	Theory of structures
St Peter's College	Mr Dickson	Thermodynamics
"	Mr Allen	Theory of sound
Pembroke "	Mr Burnside	Hydrodynamics
Trinity Hall	Mr Bell	Differential equations
Christ's College	Mr Hobson	Solid geometry
St John's "	Mr Webb	Higher dynamics
" "	Mr Pendlebury	Higher optics
Trinity "	Mr Niven	Electromagnetism
" "	Mr Appleton	Physical optics (with experiments)

EASTER TERM, 1881

Place	Lecturer	Subject
New Museums	Prof. Stokes	Hydrodynamics and physical optics
Cavendish Laboratory	Lord Rayleigh	Electricity and Magnetism
Pembroke College	Mr Prior	Attractions
Trinity Hall	Mr Bell	Lunar and planetary theories
Corpus Christi College	Mr Wallis	Functions of Laplace, Lamé and Bessel
King's	Mr Stearn	Conduction of electricity and heat
Queens' "	Dr Campion	Higher differential equations
" "	Mr Temperley	Higher solid geometry
St John's "	Mr Besant	Sound and vibrations
Trinity "	Mr Dale	Heat
" "	Mr Taylor	Higher plane curves
Emmanuel "	Mr Webb	The potential and Green's theorem

In reply to the invitation of the Mathematical Board (*Reporter*, p. 488, April 27, 1880) the following lectures have been offered for the Long Vacation, 1881.

Place	Lecturer	Subject
St Peter's College	Mr Allen	Electromagnetism
Clare "	Mr Mollison	Heat
Trinity "	Mr Niven	Elasticity
" "	Mr Ball	Determinants
" "	Mr Lewis	Vortex motion and viscosity

FIGURE 5.8. This advertisement for the forthcoming year's lectures in mathematics marks the collapse of the long established tradition of teaching solely by college affiliation in favor of a public system that integrated intercollegiate and professorial lectures as well as those given at the Cavendish Laboratory. The range of specialization in advanced subjects made possible by the system gradually diminished the coaches' control over the most able pupils in their final years of training. Note (bottom) the introduction of a short new term (the "Long Vacation"), which represented a direct parallel to the summer training offered by the best coaches. *Cambridge University Reporter,* 22 June 1880, 669. (By permission of the Syndics of Cambridge University Library.)

to specialize in just one of the four groups. The new form of examination soon proved unsatisfactory, as it generated a level of specialization that could not be properly tested by the extant Part III examination. The new arrangement also placed an intolerable burden on the examiners, who found themselves examining over a six month period. Finally, in 1886, Parts I and II (now jointly renamed Part I) were formally separated from Part III (now renamed Part II), the latter being made into an entirely separate examination taken at the end of the fourth year. The higher subjects for this examination were divided into eight optional divisions, the last of which contained electromagnetism (including the electromagnetic theory of light), thermodynamics, and the kinetic theory of gases. This division enabled candidates to specialize in pure or applied subjects in their fourth year by choosing in advance those divisions in which they wished to be examined.[78]

At around the same time that the long-established structure of the Mathematical Tripos was being called into question, the Parliamentary Commissioners, who had reported regularly on the universities since the early 1850s, were investigating the state of public teaching in Cambridge. Consulted in the mid 1870s on how mathematics teaching could best be improved, the Board of Mathematical Studies suggested that the system of intercollegiate lectures be further developed and organized. But, under pressure from other boards in the university, the commissioners decided that a new category of university employee, the "university lecturer," should be created. These new lecturers, who would offer courses in advanced subjects to students from all colleges, constituted the beginnings of a properly public system of teaching very much along the lines suggested by William Hopkins more than twenty years earlier. In the *Universities of Oxford and Cambridge Act* of 1877, the Commissioners required each college to contribute, according to its means, to a Common University Fund which would pay for the new public system.[79] The Board of Mathematical Studies applied for, and received, five university lectureships: four to teach one each of the divisions of the then Part III, and one to teach applied mechanics.[80] These lectureships provided the first permanent university posts for young mathematicians in Cambridge and, as we shall see in chapter 6, helped to support the men who would develop Maxwell's electromagnetic theory during the 1880s and 1890s.

78. The detailed changes in the content of the Tripos syllabus from 1873 to 1907 are listed in Wilson 1982, 367–71.
79. *Parliamentary Accounts and Papers* (Education; Science and Art) 59 (1876): 42.
80. Minute Book of the Board of Mathematical Studies (MB-CUL), 17 February 1883. The fifth lecturer was possibly appointed in anticipation of Routh's imminent retirement. A list of all the new lecturers is given in the *Cambridge University Reporter*, 3 June 1884, 791.

The reforms of the 1870s and 80s were not of course intended merely to update and expand the content of the Tripos syllabus. Their primary purpose was to establish a research ethos among young college fellows and to facilitate the formation of research schools around the most capable mathematicians.[81] It was a common criticism of the Cambridge system in the early 1870s that, although it produced a small number of researchers of "the very highest eminence," it did not enable these men to establish a school of lesser researchers capable of building on their work. In a typically unfavorable comparison with Britain's main international competitor, Germany, the Professor of Pure Mathematics at University College London claimed in 1873 that "while the number of mathematical students at Cambridge exceeds that at a large number of German universities put together, the proportion of these students who are pursuing their studies for any higher purpose than taking a good degree ... is very small."[82] Making a similar comparison with German universities, the Cambridge reformer J. W. L. Glaisher (2W 1871) noted, using an appropriately military metaphor, that "in mathematics we have in England generals without armies: the great men who are independent of circumstances have risen among us, but where are the rank and file?"[83] The major obstacles to the formation of research schools in Cambridge—the lack of advanced courses in higher mathematical subjects and the absence of contact between students and the leading mathematicians—were largely removed by the introduction of intercollegiate and university lectures and by the division of the Tripos in 1882. Several other reforms of the period further encouraged high wranglers to pursue research. In the mid 1880s the regulations for the Smith's Prizes were altered in order that they could be awarded on the strength of an original dissertation rather than by written examination. Between 1872 and 1886, Trinity, St. John's, and King's colleges ceased to appoint fellows according to the order of merit, relying instead on the submission of original pieces of mathematical research.[84] During the 1870s and early 1880s the Parliamentary Commissioners swept away the religious tests that required the holders of many college and university

81. Glaisher 1871 and 1886–87; Henrici 1873.

82. Henrici 1873, 492. Henrici had studied in Germany with Clebsch at Karlsruhe, Hesse at Heidelberg, and Weierstrass and Kronecker at Berlin (Henrici 1911, 71–72).

83. Glaisher 1886–87, 33. Glaisher aimed to encourage the study of pure mathematics, the study of mixed mathematics being encouraged by coaches like Routh.

84. Ibid., 37–38. See also Barrow-Green 1999. Glaisher identifies several of these investigations that were subsequently published in leading journals. Glaisher also encouraged the most able undergraduates to publish short articles in the *Messenger of Mathematics* and the *Quarterly Journal of Pure and Applied Mathematics,* both of which he edited (Forsyth 1929–30, x).

posts to be in holy orders. With these tests abolished, young college fellows could concentrate on academic research and aspire to be appointed to permanent college and university lectureships as well as to professorships in the expanding tertiary education system in London and the provinces.

During the 1870s and 1880s, then, public teaching was successfully reestablished in Cambridge as an important and independent adjunct to coaching. As we shall see in chapter 6, advanced intercollegiate courses provided a forum at which recent research from Britain and overseas could be interpreted and discussed by senior mathematicians, and taught to small groups of the most able students. This success was achieved moreover by college and university lecturers adopting several of the training methods pioneered by the coaches. Advanced lecture courses were delivered to small classes of students of approximately equal ability in which a catechetical style could be adopted in order to ensure that the majority understood what was being taught. The importance of getting students to work through examples and problems had also been acknowledged. When the first comprehensive system of intercollegiate lectures was announced in 1870, it was emphasized that, for an additional fee of one guinea, each lecturer would "set weekly papers on the subject of his lecture to be looked over privately."[85] These papers enabled students to see how examination problems on the subject were constructed and solved, and would no doubt have been discussed in subsequent lectures.

The new generation of lecturers were also well aware of the importance of close personal interaction with their pupils. Some colleges saw the rise of public teaching and the splitting of the Tripos in 1882 as an opportunity to weaken their students' reliance on private teaching, but they realized that in order to do so, college lecturers would have to offer the kind of individual tuition provided by the coaches. When, for example, J. J. Thomson was appointed to an assistant lectureship in mathematics at Trinity College in 1882, it was a condition of his appointment that, in addition to holding formal lectures, he would "take men individually, help them with their difficulties and advise them as to reading" (Thomson 1936, 81). For the rest of his long career, Thomson continued to urge that "the most important part of teaching," especially when it came to nurturing independence of thought, was that part "when the teacher comes into contact with his pupils not as a class, but as individuals." It was under these conditions, Thomson believed, that a teacher could "suggest points of view" and "sometimes even point out that

85. *Cambridge University Reporter*, 7 December 1870, 145.

things are not quite so clear as they seem to appear to the student" (1912, 399). The want of this "personal communication and interchange of ideas between teacher and pupil" had long been identified as the most serious obstacle to the development of research groups around the most able mathematicians in Cambridge (Henrici 1873, 492). With the advent of intercollegiate and university courses, the "personal intercourse" that had been the stock in trade of private tutors for more than half a century began to become an important part of advanced mathematical training as a preparation for research.

5.6. Problem Solving as a Model for Research

I have touched several times in this and previous chapters on the question of the extent to which the original publications produced by the most successful wranglers can be understood as a product of their undergraduate training. We saw in chapter 3 that Cambridge students were expected to display a degree of originality in answering difficult Tripos problems and that it was commonplace for examiners to draw upon their own research in formulating the most advanced questions. It will be recalled in this context that some students became so skilled at the rapid solution of Tripos problems that they sometimes wrote answers under examination conditions that were more elegant than those offered by their examiners. We have also seen that ingenious answers to the most original questions could become research publications in their own right and that new results or techniques announced in the form of questions could aid ongoing research projects. There can therefore be no doubt that the Tripos examination papers acted as a conduit of communication between Cambridge mathematicians, and to some extent provided undergraduates with examples of what counted locally as legitimate research problems. But to what extent, at a more general level, did the problem-solving experience inform the wrangler's choice of research topic, shape the way he posed a research problem and generated what seemed to him an appropriate solution? These are questions that could be explored historically at any point from the late Georgian period, but, for reasons of space, and in anticipation of the research topics discussed in subsequent chapters, I shall focus here on the mid and late Victorian periods. It was in any case only in the 1860s that undergraduate mathematical training in Cambridge began to be regarded explicitly as preparation for postgraduate research and so was discussed in those terms by contemporary commentators.

The first two points to emphasize regarding the relationship between training and research are that the problem-solving experience certainly did shape many wranglers' early attempts at original investigations and that this experience provided a somewhat misleading model of how significant research should be pursued. An inaugural editorial in the first issue of the *Oxford, Cambridge and Dublin Messenger of Mathematics*, which, as we saw in chapter 3, was intended to encourage young mathematicians to publish original papers, warned prospective contributors that the mere solving of problems, "unless they are of a high order of originality and conception," could "scarcely be looked upon as original investigations at all."[86] The editors were obviously keen to dissuade able undergraduates and young college fellows from submitting numerous solutions to variations on standard problems, which, as subsequent commentators confirm, many Cambridge-trained students would indeed have believed to constitute research. According to J. J. Thomson, training students by requiring them to derive numerous neat new relationships between the variables of a physical system (as in the wheel-and-axle example of chapter 3) led them to regard the "normal process of investigation" in mathematical subjects as little more than "the manipulation of a large number of symbols in the hope that every now and then some valuable result may happen to drop out" (1893, vi).

George Darwin observed in a similar vein that drilling students to solve questions with great rapidity left them with the impression that "if a problem cannot be solved in a few hours, it cannot be solved at all." Remarks of this kind confirm that many young wranglers, left more or less to their own devices after graduation, failed to appreciate the following fundamental difference between problem solving and significant research: when solving examination questions, students were always, in Darwin's words, "guided by a pair of rails carefully prepared by the examiner," whereas in undertaking a truly original investigation no such guiding rails existed ([1883] 1907–16, 5:4–5). Drawing upon his own experience to develop exactly this point, Darwin noted that when a mathematical physicist pursued an original line of enquiry, he did not encounter a collection of self-evident or ready formulated questions, but rather a "chaos of possible problems"; in order to make substantial progress he had to be prepared to keep the "problem before him for weeks, months, years and gnaw away from time to time when any new light may strike him." It was only through protracted work of this kind that some aspect of his enquiry was gradually and painstakingly "pared down *until its characteristics [were] those of a Tripos question*" (5, my italics). Darwin's remarks indicate that it would be a serious oversimplification to depict

---

86. *The Oxford, Cambridge and Dublin Messenger of Mathematics* 1 (1862): 1.

the methods of research employed by senior Cambridge mathematicians as little more than the continued application of the problem-solving skills they had learned as undergraduates. Indeed, on Darwin's showing the process of research had as much in common with the art of *setting* Tripos problems as with that of solving them, a point to which I shall shortly return.

There are nevertheless several specific and closely related ways in which the unique mastery of problem solving generated by a Cambridge training did shape many wranglers' subsequent research efforts. First, the ability quickly to prove new theorems, apply new techniques to the solution of problems, and derive special cases from general theories gave wranglers ready access to the original investigations of other, especially Cambridge, mathematical physicists. Articles published in treatises and research journals neither reproduced every step in the proof or demonstration of new techniques or results, nor did they explain in detail how such novelties were applied to the solution of specific problems. Filling in these steps was second nature to students drilled as undergraduates to tackle exercises of exactly this kind. Cambridge-trained graduates were thus far better equipped to assimilate and to build upon the research of other Cambridge mathematicians than were graduates trained at other British institutions of higher education.

The second way undergraduate training can be understood to have shaped wrangler research concerns the relationship between the formulation and the solution of problems. We have seen that experienced researchers such as Thomson and Darwin criticized their younger colleagues for expecting to encounter in the physical world the kind of well formulated and quickly soluble problem they had met on examination papers. But, despite this admonition, Darwin himself depicted the physical world as a "chaos" of possible *problems* and characterized research as a process by which physical phenomena were gradually reduced to the form of examination questions. What Darwin criticized in his younger colleagues was not their belief that research was essentially a problem-solving exercise, but their failure to appreciate that the researcher's skill lay not so much in solving ready-made problems as in formulating new or recalcitrant ones in such a way that they became soluble using the physical principles and mathematical methods commonly taught in Cambridge. It is interesting in this context that, despite their warning to students not to confuse mere problem solving with research, the editors of the *Oxford, Cambridge and Dublin Messenger of Mathematics* still advised that the "highest mathematical operation is the reduction of the elements of a problem to mathematical language, and the expression of its conditions in equations."[87] The skills and sensibilities developed during a problem-based mathematical training prepared wranglers to pursue this project by enabling them to conceptualize the physical world as a series of *possible* problems and to recognize when a particular problem had been expressed in a soluble form; that is, in a form suitable as the basis of an advanced examination question.

The third and closely related way in which undergraduate training shaped the research of Cambridge graduates concerns the specific mathematical skills they were taught. Despite the difficulties mentioned by Thomson and Darwin, the emphasis on problem solving equipped the most able students with remarkable fluency both in the application of analytical mathematical methods and in selecting or inventing the most appropriate physical models, boundary conditions, and approximations required to solve specific problems. These skills not only enabled some students to give better solutions to well-formulated Tripos questions than those provided by their examiners, but, when developed and applied over a period of years to much more complicated problems, they could also produce new results of major importance. Numerous examples of work of this kind are to be found in the original investigations of Cambridge-trained mathematicians, ranging from John Couch Adams's studies of planetary motion in the mid 1840s to A. S. Eddington's work on the internal constitution of stars during World War I.[88] In fact, George Darwin's own work on tidal friction and the evolution of the earth-moon system is itself exemplary of the Cambridge style of applying advanced analysis to produce precise quantitative solutions to extremely complicated physical problems. Darwin's initial interest in these issues was even prompted by a specific mathematical problem on the physical causes of glacial epochs announced in the Presidential Address to the Geological Society of London in 1876.[89] It was the ability successfully to tackle problems of this scale that so impressed mathematicians like Klein, who, as we have seen, sought to nurture the same skills in his own students at Göttingen.

87. *The Oxford, Cambridge and Dublin Messenger of Mathematics* 1 (1862): 1.
88. Couch Adams's explanation of the irregularities in the motion of the planet Uranus on the hypothesis of the disturbing effects of a more distant planet is discussed in Harrison 1994. The problem-orientated nature of Eddington's work on the inner constitution of stars is discussed in Crowther 1952, 144–60.
89. Kushner 1993, 201, 210. Couch Adams's calculation of the position of a new planet beyond Uranus was also made in response to a problem posed by George Airy in his 1832 British Association Report on the state of astronomy (Harrison 1994, 271).

The value of a training system so well suited to the preparation of mathematical physicists was strongly defended by several outstanding wranglers towards the end of the nineteenth century, frequently in response to criticism from reformers keen to abandon the system in favor of the study of pure mathematics. Responding to a retrospective critique by the greatest of these reformers, A. R. Forsyth, a near contemporary, Karl Pearson (3W 1879), insisted that the Mathematical Tripos had been an "excellent examination" precisely because it gave a "general review" of the principles of many branches of mathematics and, above all, "because the weight given by it to 'problems'" forced the teaching coaches to present mathematical methods as techniques for problem solving.[90] According to Pearson, the training he had received in the late 1870s had been invaluable to him as a researcher because:

> Every bit of mathematical research is really a "problem," or can be thrown into the form of one, and in post-Cambridge days in Heidelberg and Berlin I found this power of problem-solving gave one advantages in research over German students, who had been taught mathematics in theory, but not by "problems." The problem-experience in Cambridge has been of the greatest service to me in life, and I am grateful indeed for it. (Pearson 1936, 27)

Pearson's Cambridge training also gave him advantages over colleagues trained in other British institutions. His famous collaboration with W. F. R. Weldon in the 1890s—during which Weldon and Pearson laid the foundations of biometry—was characterized by Pearson solving numerous problems raised by Weldon concerning the application of statistics to heredity and evolutionary theory. Looking back on this collaboration at the time of Weldon's early death in 1906, Pearson recalled that Weldon, who had been trained in mathematics at University College London, had "never reached a high wrangler's readiness in applying analysis to the solution of new questions, possibly this requires years of training in problem papers" (Pearson 1906–7, 17).

Another illuminating example of a wrangler who strongly defended the applied and problem-solving approach to mathematics he had learned in Cambridge is Lord Rayleigh. Virtually the whole of Rayleigh's prolific career as a mathematical physicist was devoted to solving difficult physical prob-

---

90. It is interesting that although wranglers like Forsyth, and later even G. H. Hardy, saw themselves as proponents of "pure mathematics," they continued to teach by numerous examples in a distinctly Cambridge style. See Forsyth 1885 and Berry 1910, 305.

lems, his outstanding and lasting contribution to the discipline being his brilliant two-volume treatise, *The Theory of Sound*, published in 1877. In the first of these volumes Rayleigh developed a general dynamical theory of vibrating systems based on energy considerations and the principle of virtual work. It is a mark of his expertise in analytical dynamics that, as a reviewer of his book in the mid-twentieth century pointed out, "there is no vibrating system likely to be encountered in practice which cannot be tackled successfully by the methods set forth in the first ten chapters of Rayleigh's treatise" (Lindsay 1945, xxviii). In subsequent chapters, Rayleigh provided a further outstanding display of the wrangler's art by tackling numerous problems too complicated to be solved analytically, finding in each case the "best possible approximation to lead to a physically useful result" (xxviii–xxix).

The second edition of the treatise was published in 1894 just as debates over the respective merits of pure and applied approaches to mathematics were gathering pace in Cambridge (see section 5.7). In this atmosphere, Rayleigh felt obliged to comment on the fact that the advanced analytical methods introduced in the book were generally justified by appeal to physical intuition rather than by the rigorous methods of the pure mathematician.[91] But far from being apologetic on this matter, Rayleigh insisted that while it was appropriate to "maintain a uniformly high standard in pure mathematics," the physicist did well to "rest content with arguments which are fairly satisfactory and conclusive from his point of view." The mind of the physicist, Rayleigh reminded his reader, was "exercised in a different order of ideas," so that to him "the more severe procedure of the pure mathematician may appear *not more but less demonstrative*"; and, he concluded, if physicists were obliged to adopt the methods of the pure mathematician, "it would mean the exclusion of [mathematical physics] altogether in view of the space that would be required" (Rayleigh 1894, xxxv; my italics). Rayleigh's comments nicely illustrate the fact that the skills and sensibilities of the mathematical physicist were not only different from those of the pure mathematician but were in some respects incompatible. For the mathematical physicist, the meaning of mathematical variables, functions, and operations sprang from an intuitive understanding of the physical entities and processes they represented, an intuition forged through years of experience in setting up and solving problems as well as assessing the physical plausibility of the results obtained.

---

91. See, for example, Rayleigh's proof of Fourier's theorem "from dynamical considerations" in art. 135 of Rayleigh 1877–78, vol 1.

One final aspect of the problem-solving approach to mathematics that was of special significance to wrangler research is the technical unity lent to mathematical physics by the analogical application of common mathematical methods. Several historians have remarked upon the deep commitment to dynamical explanation displayed by Cambridge mathematical physicists in the last third of the nineteenth century, but it is important to appreciate that this was as much a commitment to a range of techniques for tackling problems as it was to an ideology of mechanistic explanation.[92] For example, I noted above that coaches such as Routh and Webb spent a good deal of time drilling their pupils in the solution of Laplace's equation. The reason for this was that the concept of "potential" was one that could be applied not just to gravitational theory and the various branches of mechanics but to thermodynamics, electrostatics, magnetostatics, and electromagnetism. Having learned a range of techniques for solving problems in, say, gravitational potentials, a coach could introduce a new topic such as electrostatics by developing the analogy between gravitational and electrostatic potential theory. Likewise Lagrangian dynamics and the principle of least action were powerful methods precisely because, although dynamical in origin, they could be applied to numerous physical phenomena while making only very general assumptions regarding the mechanical nature of the micro-processes involved. Undergraduate training thus encouraged students to approach physical phenomena as illustrative exercises in the application of common physical principles (especially the conservation of energy and the principle of least action) and mathematical methods. This attitude was of considerable importance for research in Cambridge since, as we shall see in chapter 6, when wranglers came across new methods for tackling problems in fields such as acoustics and hydrodynamics, they naturally assumed that the same methods would find useful application in, say, thermodynamics or electromagnetic theory.

The above remarks suggest that a definite relationship between training-style and research-style did exist in Victorian Cambridge, but that the relationship was complicated and potentially open ended. Thomson's and Darwin's comments reveal that although wranglers could not produce substantial research simply by continuing to solve problems in the way they had as undergraduates, their problem-solving skills nevertheless played a significant role in shaping the research they undertook. As Darwin succinctly

---

92. Compare, for example, Topper 1971 on J. J. Thomson's early research with my account of the same work in chapter 6.

pointed out, the challenge facing young researchers was not to solve Tripos-like problems presented by the physical world, but to reduce the physical world to Tripos-like problems. To accomplish this goal they had to establish similarity relations between tractable and currently intractable phenomena using the laws, principles, and mathematical methods they had been taught to take for granted. A problem-based training was especially useful in this process because it provided a ready supply of analogical examples and, crucially, because it helped the researcher to recognize when a problem had been expressed in soluble form. And, once it was so expressed, it was generally accessible to other Cambridge mathematicians and to the most able undergraduates.

All research enterprises associated with a Cambridge training nevertheless remained open ended in ways that could make their finished products inaccessible even to Cambridge-trained mathematicians. As Darwin noted, an ambitious project could take months or even years to come to fruition, during which time the author might incorporate conceptual components that were alien to normal wrangler sensibilities. An excellent example of this kind is Maxwell's dynamical theory of the electromagnetic field, which, as we shall see in the next chapter, was very difficult for Cambridge men to understand until it had been painstakingly recuperated by the combined efforts of several outstanding Cambridge mathematicians (including Maxwell himself). This important point reminds us that the problems associated with the translation of mathematical skill discussed in earlier sections of this chapter apply equally to the products of research. As we have just seen, it was a useful rule of thumb that a piece of research was considered complete when it could be—and sometimes actually was—expressed in the form of a Tripos problem. An obvious corollary of this homologous relationship is that research publications by Cambridge mathematicians were likely to be as difficult to understand to mathematicians outside of the Cambridge tradition as were solutions to difficult Tripos problems. Here again Maxwell's electromagnetic theory is an interesting case in point. As Buchwald and Hunt have shown, Continental mathematical physicists who studied Maxwell's work in the 1880s and 1890s adapted much of his mathematical formalism to their own purposes, but never understood Maxwell's physical conception of the electromagnetic field nor his methods for tackling specific problems.[93] As we shall see in the next chapter, it was these methods that were gradually mastered and modified in Cambridge before being propagated locally

---

93. Buchwald 1985, 177–201, and 1994, 193–99, 325–29; Hunt 1991a, 176–82.

through the coaching room, the intercollegiate classroom, and numerous illustrative Tripos problems.

5.7. The Demise of Coaching in Edwardian Cambridge

The revival of public teaching in Cambridge, discussed in section 5.5, severely weakened the coaches' influence over the most able mathematics undergraduates. The decision by Routh and Besant to retire in the mid 1880s was almost certainly prompted, at least in part, by the rapidly diminishing role of the private tutor as primary mentor of the intellectual elite. Students capable of research now looked to intercollegiate, university, and professorial lectures for specialist training and, during their fourth year, had no formal contact with coaches whatsoever. It would nevertheless be wrong to see this change as rendering the coaches irrelevant to undergraduate mathematical studies. Although from the early 1880s they played a much reduced role in shaping the research interests of the top wranglers, the coaches continued to act as gatekeepers to mathematical knowledge in the university well into the first decade of the twentieth century. The altered relationship between coaches and the revived system of public tuition is best understood as a new division of pedagogical labor. According to this division, it was the coach's job to structure and discipline undergraduate mathematical studies, and to provide students with a good all-round knowledge of pure and applied mathematics upon which college lecturers of any speciality could build. For the majority of students, the coach remained the tutor who taught the art of problem solving and prepared them for the competitive ordeal of the Senate House examination.

The changing relationship between private and public teaching in the 1880s and 1890s was emphasized in the entries on the Mathematical Tripos published in new editions of the *Student's Guide.* The most able freshmen were now assured that "the rearrangements of College Lectures . . . and the establishment of systems of Inter-Collegiate Lectures" had rendered "the assistance of the private Tutor less a matter of necessity than [had] hitherto been the case." But for those who had not been trained under a wrangler schoolmaster, it remained the case that the aid of a good coach "must be regarded as a matter of necessity."[94] The continued relationship between coaching and undergraduate success was also made starkly visible in the at-

---

94. *The Student's Guide to the University of Cambridge,* 5th ed. (Cambridge: Deighton Bell, 1893), 20.

tempts made by women to compete in the Tripos examination during the last quarter of the century. Women were not awarded degrees in Victorian and Edwardian Cambridge, but, from the early 1870s, they were permitted to sit the Tripos examination and to receive an informal ranking relative to their male peers.[95] The published Tripos lists reveal that women were rarely placed among the top wranglers, but the occasional exceptions to this rule suggest that access to good coaching was an important and recognized determinant of examination success. The first woman to achieve a high place among the wranglers was Charlotte Scott (8W 1880), whose outstanding abilities in mathematics were recognized before she entered the university. Scott's tutors at Girton accordingly invited Routh to take her as a private pupil. When Routh was "unwillingly obliged to decline" on the grounds of his "numerous engagements," Scott was coached by Ernest Temperley (5W 1871) and George Walker (2W 1879).[96] Neither of these men was an especially notable coach, but Walker had been second wrangler the previous year and Temperley was one of the examiners in 1880. Scott was thus far better prepared for the examination than the majority of women students would have been. Her place among the upper ten wranglers caused a sensation when the order of merit was announced in the Senate House and was considered an important victory by "the advocates of higher education for women."[97]

Scott's achievement was eclipsed a decade later when, as we saw in chapter 4, Philippa Fawcett was placed "above the senior wrangler" in the Tripos of 1890. The daughter of Henry Fawcett (7W 1856), a fellow of Trinity Hall and enthusiastic university reformer, and Millicent Fawcett, an active supporter of women's suffrage and higher education, Philippa was coached in mathematics from the age of fifteen by G. B. Atkinson (12W 1856). This kind of pre-Cambridge coaching by a wrangler was, as we have seen, crucial to Tripos success in the last quarter of the century, and Atkinson, an experienced tutor and lecturer at Trinity Hall, predicted that Fawcett would eventually be placed among the top wranglers.[98] Well aware of the importance of a good private tutor, Fawcett's Cambridge mentors arranged for her to be tutored by one of the most successful coaches of the period, E. W. Hobson (SW 1878), and even persuaded the recently retired Routh to polish her prepara-

---

95. Gould 1993, 8. From 1882 the women's relative rankings were published in the *Reporter*.
96. *The Graphic* 21 (1880): 157. Routh presumably assumed that he would have to teach Scott alone.
97. Ibid. See also the *Times*, 4 February 1880, and *The Girton Review*, Michaelmas Term 1927, 10.
98. Gould 1993, 17.

tion for the Senate House.[99] Following her spectacular success, Fawcett invited Hobson to dine at a celebratory dinner held in her honor at Newnham College. Proposing a toast to Hobson before her jubilant supporters, Fawcett observed modestly that "in these matters the person 'coaching' was everything, the person 'coached' nothing."[100] The "immense applause" with which Hobson was then received suggests that the women were well aware that without access to the leading coaches they would remain unable to compete on equal terms with the men.

During the early years of the twentieth century the remaining work of the coaches was gradually taken over by college tutors and supervisors. The expansion of intercollegiate and university lectures at all levels through the 1880s and 1890s meant that, by 1900, it had become unnecessary for coaches either to lecture students or even to provide them with manuscripts covering the mathematical methods they were required to master. The prime job of the coach now was to ensure that students were attending an appropriate range of courses and that they understood what they were being taught. This was accomplished by setting numerous Tripos problems on the mathematical subject to hand and going over the solutions during the allotted coaching hour. It was now only on those occasions when a student had evidently failed to grasp a point made by a lecturer that a coach would recapitulate the theory in question in his own way. This curtailment of responsibility made it virtually impossible for a private tutor to dominate undergraduate training to the extent that Hopkins, Routh, and Webb had done. By 1900, even Webb's pupils were regularly being beaten by students of more amateur coaches (Appendix A1). Following Webb's retirement in 1902, the new leading coach, Robert Herman (SW 1882), never obtained the dominance of undergraduate training enjoyed by his predecessors (fig. 5.9). Described by one of his most eminent pupils as a man with a "genius for teaching" whose "manipulative skill was the envy of his colleagues," Herman combined private teaching with extremely successful courses of college lectures.[101] But, unlike previous leading coaches, Herman drew his private pupils almost entirely from his own college, Trinity, an act which appears to have enabled the college to claim thirteen of the last fourteen senior and second wranglers (Appendix A1).

---

99. *The Graphic*, 21 June 1890. Routh had ceased taking new male pupils in 1886 and so felt he had time to coach a female pupil individually.
100. *Cambridge Express*, 14 June 1890, 6.
101. Neville 1928, 237.

FIGURE 5.9. Robert Herman around the time (1886) he was appointed to a lectureship in mathematics at Trinity College. According to one of his most distinguished pupils, E. H. Neville (2W 1909): "To describe Herman's power as a teacher is difficult only because a weight of superlatives may stifle conviction. . . . Herman was the rare combination, conscientious and inspiring. The care with which his courses were planned has been discovered often by pupils who, beginning their own careers with every intention of being original, have found themselves reverting before long in subject after subject to the outlines which he traced" (Neville 1928, 238). Photograph AA.3.26, Wren Library, Trinity College. (By permission of the Master and Fellows of Trinity College, Cambridge.)

The combination of public lecturing and private, in-college tutoring practiced by Herman pointed the way to the teaching system that would dominate in twentieth-century Cambridge. The pressures that had led to the division of the Tripos examination in 1882 also led to the abandonment of the order of merit in 1909. By the turn of the century, the common Part I examination, in which no specialization was permitted, had become a recognized obstacle to the development of separate streams of pure and applied mathematics in undergraduate studies.[102] Ambitious students had little in-

---

102. The difficulties facing the Mathematical Tripos are discussed at length in the *Cambridge University Reporter*, 5 December 1899, 276–88.

centive to attend advanced courses all the while they knew that a broad knowledge of pure and applied subjects was the best route to Tripos success. The only way to make specialist knowledge pay in the examination would be to allow choice in the more advanced papers—as had been done in Part II since the 1880s—but this would undermine the credibility of that most famous and respected of Cambridge institutions, the order of mathematical merit. If students effectively sat different examinations, they could not be convincingly ranked individually according to differences of a few marks. Many senior members of the university, most notably Edward Routh, bitterly opposed the abandonment of the order of merit. They felt that an initial ranking of students by a single examination was the best way of preserving the breadth of undergraduate mathematical studies and of ensuring that students worked hard.[103]

By the turn of the century these arguments were rapidly losing ground on two fronts. First, many younger mathematicians in Cambridge were in favor of undergraduate specialization in order to enable pure mathematics to emerge as separate discipline. Second, students were abandoning the study of mathematics for other Triposes at an alarming rate.[104] Freshmen now realized that unless they could solve problems quickly and expertly enough to appear in the elite "upper ten" wranglers, they would automatically be branded as second rate, a situation that was considered intolerable by students and proponents of pure mathematics alike. With a virtual collapse in student numbers at the turn of the century, it was finally agreed that undergraduate specialization would be permitted and that the order of merit would be abolished.[105] From 1909, the Mathematical Tripos came into line with the other Triposes, successful honors candidates being placed alphabetically in one of three classes.[106] The abandonment of the order of merit precipitated the rapid demise of what remained of the coaching system. With undergraduate mathematicians able to pursue quite different courses of study, it became virtually

---

103. Ibid., 277–78.

104. *Cambridge University Reporter*, 14 May 1906, 873–90, at 873. This discussion gave a thorough airing to the heated opinion on both sides concerning the "crisis" now facing Cambridge mathematical studies.

105. The public debate surrounding the proposed abandonment of the order of merit can be followed in the *Times*; see for example 1 February 1907, 5. The final announcement of the order of merit in the *Times* is reproduced in fig. 4.5.

106. The decision to abandon the old Mathematical Tripos was made by a vote of all members of the university in February 1907. For the public announcement of the result, see the *Times*, 4 February 1907, 8. The events leading up to the vote and its aftermath are discussed briefly in Hassé 1951, 156.

impossible for one man to act as a common teacher. The job of directing undergraduate mathematical studies was now taken over by specially designated college tutors while that of meeting students individually—or in small groups—to solve problems and to discuss difficulties was delegated to "supervisors" who were also appointed and paid by the student's college.[107]

By the end of the first decade of the twentieth century, the teaching methods pioneered and practiced by the coaches throughout the Victorian and Edwardian periods had been redistributed among college and university teachers. Despite the continued emphasis on small-class and face-to-face teaching, however, the new mathematical lecturers and supervisors placed less emphasis on problem solving and the physical application of mathematical methods than had the coaches. As I noted in chapter 4, the rise of interest in pure mathematics from the late 1880s undermined the relationship between mathematics, physics, and such physically intuitive notions as muscular force, fostering instead an interest in the mathematical methods themselves and rigorous proof. The abrupt demise of what remained of the coaching system between 1907 and 1909 naturally gave college and university teachers much more influence over undergraduates, and, as we might expect, the interests and expertise of young graduates altered accordingly. As early as 1912, W. W. Rouse Ball observed that the "destruction of many of the distinctive features of the former [Tripos] scheme" had brought about a "curious alteration in the popular subjects." Students were rapidly losing interest in "those branches of applied mathematics once so generally studied," turning instead "to subjects like the Theories of Functions and Groups" (1912, 322). The Mathematical Tripos was, in H. R. Hassé's words, under "new management" (1951, 157), and, as we shall see in chapter 8, the distinctly physical and problem-solving approach to mathematics faded with the training system upon which it had been built.

107. Miller 1961, 103–5.

~ 6 ~
Making Sense of Maxwell's *Treatise on Electricity and Magnetism* in Mid-Victorian Cambridge

> In science one of the chief characteristics of a University, and the surest sign of its vigour, is the existence of schools of thought and research.... In physics ... Cambridge is well in the front rank ... [and] I take, as especially characteristic, the work which is being done in electricity under the influence of Maxwell's theory.
>
> J. H. POYNTING (3W 1876), 1883 [1]

6.1. Electromagnetism, Pedagogy, and Research

The publication of James Clerk Maxwell's *Treatise on Electricity and Magnetism* (hereafter *Treatise*) in March 1873 marked the culmination of almost twenty years of work by Maxwell on theoretical and experimental electricity. Since graduating in 1854, Maxwell had devoted much of his research effort to recasting Michael Faraday's novel conception of electric and magnetic lines of force in the form of a new, mathematical, field theory of electromagnetism, based on a dynamical ether and the principle of the conservation of energy.[2] During the 1860s, Maxwell had also served as a member of the British Association Committee on electrical standards, an experience that familiarized him with the instruments and techniques of experimental electricity, the definition and measurement of fundamental electrical units, and the theory and operation of the electric telegraph.[3] But, despite Maxwell's participation in these important projects, by the late 1860s his highly mathematical and conceptually difficult papers on electromagnetic theory had generated little informed response even from the handful of professional mathematical physicists capable of following his work.[4] The decision by the

1. Poynting 1920, 558.
2. On the development of Maxwell's electromagnetic field theory, see Siegel 1985 and 1991.
3. On Maxwell's work on electrical standards, see Schaffer 1995.
4. J. J. Thomson (1936, 100 –1) noted that, even in the early 1870s, Maxwell's scientific work "was known to very few" and his reputation "based mainly on his work on the kinetic theory of gases."

Board of Mathematical Studies in Cambridge in the mid 1860s to reintroduce electromagnetic theory to the Senate House examination from 1873 provided Maxwell with a timely opportunity to address this problem. He wrote the *Treatise* largely as an advanced textbook on the mathematical theory, instrumentation, and experimental foundations of electricity and magnetism, and included an account of his new field-theoretic approach. By the time the book was published, Maxwell had also been appointed to a new professorship of experimental physics in Cambridge, a job that expressly required him both to offer public lectures on electromagnetic theory to students of the Mathematical Tripos and to direct experimental research at the magnificent new Cavendish Laboratory to be opened in 1874.[5]

By the mid 1870s, then, Maxwell would appear to have been ideally placed to inaugurate a major program of research based on the new electromagnetic theory he had pioneered. The annual graduation of around a dozen brilliant wranglers, some of whom were keen to undertake experimental research at the Cavendish Laboratory under Maxwell's direction, provided a uniquely skilled workforce that might have explored both mathematical and experimental problems raised in the *Treatise*. Moreover, within a few years of Maxwell's untimely death in 1879, Cambridge had produced what would remain the only institutionally based research school in electromagnetic field theory in Victorian and Edwardian Britain. Yet to view Maxwell as the founder of this school would be mistaken in several important respects. First, Maxwell's public lectures on electromagnetism never attracted a large audience in the university and were singularly ineffective in communicating novel aspects of his work to Cambridge mathematicians. Their main effect in the early 1870s was to encourage a few young wranglers to read the *Treatise*, but, from the middle years of the decade, Maxwell's audience dwindled to one or two occasional students and no prospective wranglers at all. Second, even those wranglers who were prompted to study the *Treatise* in the early 1870s tended to avoid novel sections of the book in favor of those on familiar topics such as the calculation of electrostatic potentials. It was almost seven years before any Cambridge-trained mathematician gained sufficient mastery of the book to write a paper developing original aspects of Maxwell's work. Finally, neither Maxwell nor any of the students working with him at the Cavendish Laboratory were engaged in theoretical research on electromagnetic field theory. Only two of his research students, R. T. Glazebrook and

---

5. Maxwell's work for the British Association Committee and his professorial appointment in Cambridge are discussed in Sviedrys 1976 and Schaffer 1992.

J. H. Poynting, subsequently published important papers in this area, but their work owed virtually nothing to their personal relationship with Maxwell. Indeed, by the early twentieth century, when Maxwell's equations were recognized as the most important product of mid-Victorian physics, two of Maxwell's Cavendish students felt it necessary to defend the fact that even work on optical refraction carried out under Maxwell's supervision had not been approached theoretically in terms of Maxwell's electromagnetic theory of light.[6]

In this chapter, I use the Cambridge reception of Maxwell's *Treatise* as a vehicle for exploring the relationship between training and research in the university in the 1870s and early 1880s. The *Treatise* is an especially appropriate choice of book for this purpose, as the loose affiliation of wranglers working on electromagnetic field theory in late Victorian and Edwardian Cambridge constituted the first and most important research school in mathematical physics in the university. A number of historians of science have discussed the achievements of individual British "Maxwellians" in the 1880s and 1890s, yet none has emphasized the relatively enormous scale of the enterprise in Cambridge nor explained how the school was formed.[7] It will be the burden of this chapter to show that a major prerequisite to the emergence of the school was the incorporation of the *Treatise* in the pedagogical economy of undergraduate and graduate studies. Maxwell's book was certainly read by many individuals throughout Britain during the 1870s—two of whom, George Francis FitzGerald (Dublin) and Oliver Heaviside (London) made important theoretical contributions to the development of its contents—but only in Cambridge in this period was the *Treatise* used to train a substantial number of students in electromagnetic field theory.[8] I shall argue that it was this collective interpretation and local propagation of the *Treatise*'s contents that established and sustained Cambridge as by far the most important center in Britain for the development of Maxwell's work. But, as we shall see, the process by which the book was incorporated within various training regimes in the university was by no means straightforward.

Maxwell wrote the *Treatise* for a broad audience, and, to this end, he tried as far as was possible to keep advanced mathematical methods, novel physical theory, and electrical apparatus in respectively separate chapters. This separation meant that, at least in principle, an electrical engineer could read

---

6. Schuster 1910, 28; Glazebrook 1931, 137–38.
7. The two most important studies of British Maxwellians are Buchwald 1985 and Hunt 1991a.
8. FitzGerald's and Heaviside's work is discussed in Buchwald 1985, Nahin 1988, Hunt 1991a, and Yavetz 1995.

about a piece of electrical apparatus without running into higher mathematics, while a prospective wrangler could study the application of harmonic analysis to problems in electrostatics without running into unfamiliar electrical instruments.[9] For Maxwell himself, however, who was fully at home in any area of theoretical and experimental electricity, this division would have seemed somewhat artificial. He almost certainly hoped that his book would help to usher in a new era in university physics in which undergraduates would feel equally at home with higher mathematics, advanced physical theory, and electrical instruments. But in Cambridge in the early 1870s there were few if any teachers, apart from Maxwell himself, who possessed the skills required to deal effectively with such a broad range of material. Far from being reunited in undergraduate training, the three strands of electrical studies mentioned above were torn further apart as the *Treatise*'s contents were fragmented among the three major sites at which physics was taught—the coaching room, the intercollegiate classroom, and the Cavendish Laboratory (fig. 6.1). Teachers at these sites offered respectively different accounts of the book's contents, each emphasizing those aspects that suited their technical competencies and pedagogical responsibilities. The coherent field of theoretical and experimental study envisaged by Maxwell was thus fragmented through pedagogical expediency into three separate projects in, respectively, applied mathematics, physical theory, and electrical metrology.

A related problem to be overcome before the *Treatise* could be used effectively as a textbook on mathematical electrical theory concerned the difficulties inherent in translating what might be termed Maxwell's *personal knowledge* into one or more collectively comprehensible disciplines that could be taught to undergraduates and young wranglers in a matter of months. Much of Maxwell's presentation in the *Treatise* relied upon the physical intuition and case-by-case problem solutions he had developed through long years of experience. And, as I noted above, Maxwell did not engage in the kind of coaching or small-class teaching in Cambridge through which he might have nurtured similar physical and technical sensibilities in a new generation of wranglers. It was the coaches, intercollegiate lecturers, Tripos examiners, and, to some extent, the students themselves, who were left in the mid 1870s to provide a consistent and comprehensible account of the theory contained in Maxwell's book. In the following sections I discuss the production of this account, paying special attention to the different readings

9. This division often broke down in two ways. Both the theoretical analysis of the operation of many instruments and Maxwell's technical development of electromagnetic field theory were necessarily highly mathematical. See, for example, *Treatise*, vol. 2, chaps. 8 and 15.

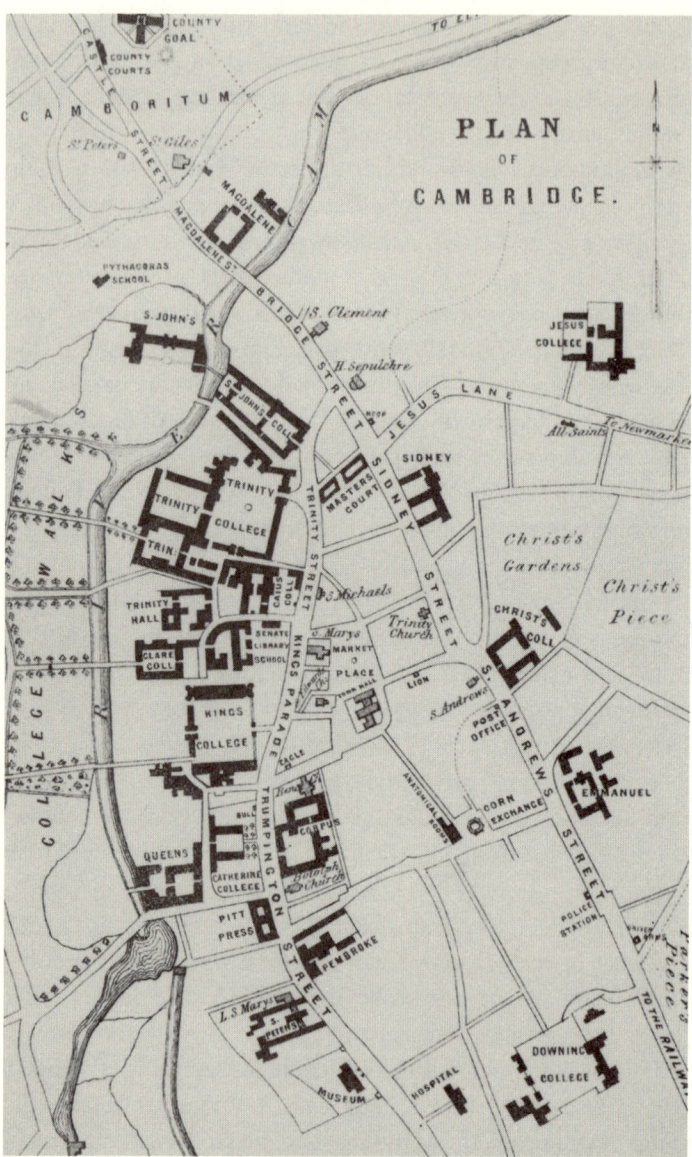

FIGURE 6.1. The pedagogical geography of mid-Victorian Cambridge. The Senate House is at the top of King's Parade, above the word "SENATE." Routh's coaching rooms were on Trumpington Street in the northeast corner of Peterhouse College (marked "S.PETERS") Webb's coaching rooms were in the southern extremity of the most westerly part of St. John's College. Herman's coaching room and Niven's lecture room were in Trinity College. The Cavendish Laboratory, begun three years after this map was published, was built on the land to the west of Corpus Christi College (marked "CORPUS"). The walk from the County Goal [sic] to Downing College is about a mile. From *The New Cambridge Guide*, 1868. (By permission of the Master and Fellows of St. John's College, Cambridge.)

of the *Treatise* provided at different teaching sites. I shall argue that the collective understanding of electromagnetic field theory that emerged in Cambridge circa 1880 was shaped by a combination of the problem-solving approach discussed in earlier chapters of this study and discussions at the intercollegiate lectures on the *Treatise* held at Trinity College by W. D. Niven. It was Niven's students who made the most important Cambridge contributions to Maxwell's theory, but, as we shall see, their work was also reliant upon techniques from other branches of mathematical physics taught at other sites in the university. I shall illustrate these latter points in this chapter through the early work of the two most important Cambridge contributors to field theory in the early 1880s, J. H. Poynting and J. J. Thomson. I begin however with an overview of the powerful pedagogical considerations that prompted Maxwell to write the *Treatise* and which shaped its initial Cambridge reception.

6.2. A Senate-House Treatise on Electricity

We saw in chapter 5 that one of the changes called for by the reformers of the 1860s was the reintroduction of electricity and magnetism to undergraduate mathematical studies. These changes were intended to make the study of mathematics at Cambridge more relevant to Britain's industrial and commercial interests, but the reformers, led by George Airy, were acutely aware that the provision of appropriate teaching materials was likely to prove a major obstacle to their proposals. In a letter to the Vice Chancellor in May 1866, Airy warned that the "real difficulties, and they are not light ones, would probably be found in providing Examiners and Books" (1896, 269). This might sound a peculiar state of affairs given that the two greatest exponents of the new subjects, William Thomson and James Clerk Maxwell, were both Cambridge graduates, but both had left the university shortly after graduating and neither had written an easily accessible account of his researches. It was in an effort to address the problem of examiners that the younger of the two men, Maxwell, was persuaded to become involved in the reform of physics teaching in his old university.

In 1865 the Board of Mathematical Studies began to strengthen mathematical physics in undergraduate studies by recommending that spherical harmonics and the shape of the earth be reintroduced as examinable topics from 1867.[10] To ensure that these and other physical subjects were properly

---

10. Glaisher 1886–87, 20. The introduction of these topics was prompted by Thomson's work in the early 1860s on the rigidity of the earth. See Kushner 1993, 199.

examined, the board coopted Maxwell, who had recently resigned his post as Professor of Natural Philosophy at King's College, London, as an examiner for the Triposes of 1866 and 1867. On the first of these occasions, Maxwell confined his problems to traditional Tripos subjects, but, early in 1866, as he contemplated how best to examine the new topics the following year, he realized that the absence of a suitable textbook on the mathematical theory of the shape of the earth placed severe limits on what he could reasonably expect even the most able questionists to know. In April 1866 he wrote to P. G. Tait, urging him to expedite publication of his and Thomson's *Natural Philosophy* (1867), which contained extensive coverage of these topics, so that he could "tickle the questionists next Jan[uary] with the Scotch School in a lawful manner."[11] Maxwell pointed out to Tait that until the book appeared he could only set long-winded questions with "ten lines of explanation or problem upon problem" in which students were required to derive results that, ideally, they ought already to have known. Maxwell's hopes were soon realized as the *Natural Philosophy* quickly became an important reference work in Cambridge. The top wranglers were referred to the book by their coaches as a definitive introduction to vibration theory, hydrodynamics, the shape of the earth, potential theory, harmonic analysis, and, most importantly for the establishment of mathematical physics in Cambridge, Lagrangian and Hamiltonian dynamics, and the principle of the conservation of energy.[12]

It was early in 1867 that the Board of Mathematical Studies decided to recommend the reintroduction of electricity and magnetism to the Senate House examination, a decision formally proposed in May of that year and approved as part of new Tripos regulations in June 1868.[13] Maxwell certainly took an active role in drafting the new scheme, and, towards the end of 1867, he also decided to write a major book on the theory and practice of electricity.[14] The first reference to this project occurs in a letter from Tait to Maxwell of 27 November 1867, in which Tait remarked that he was "delighted to hear" that Maxwell was "going to do a Senate-House Treatise on Electricity."[15] The timing of Maxwell's decision, Tait's epithet for the book, and, as we shall see, the structure of the finished volumes leave little doubt that Maxwell

---

11. Maxwell 1990–95 (hereafter referred to as *Scientific Letters*), 2:277. Tait replied that the book would "be out certainly in May" (278).
12. Smith and Wise 1989, 348–49.
13. Glaisher 1886–87, 22.
14. Niven 1890, iv.
15. Maxwell, *Scientific Letters,* 2:323, note 2.

conceived the project in large measure as an attempt to provide a textbook in electromagnetic theory for students of the Mathematical Tripos. He was acutely aware that good textbooks were vital to the successful introduction of new topics, even when, as in the case of the shape of the earth, a topic fell within a traditional field of study. The successful introduction of physical topics whose conceptual foundations and experimental practice were virtually unknown to university teachers would therefore be especially reliant upon the provision of a definitive teaching text.

Maxwell's worries about just how alien the new topics might seem when they first appeared on examination papers were heightened when he tried to manufacture an easy problem in electricity and thermodynamics for the Tripos of 1869.[16] Writing to Thomson for advice a few weeks before the examination, he remarked:

> If it is mathematically difficult it will be no fault in the eyes of the Cambridge men, though I would prefer it easy myself. . . . I think it is important to insert the wedge by the thin end and to "hold the eel of science by the tail." Great mental inertia will be called into play if the new ideas are not fitted on to the old in a continuous manner. (*Scientific Letters*, 2:464)

Maxwell was clearly aware that the application of new physical principles would not come easily to teachers and students steeped in the traditional methods of optics, hydrodynamics, and celestial mechanics. The best strategy for forestalling this difficultly, in anticipation of the full-scale introduction of the new topics in 1873, would be to write a comprehensive treatise that made the physical foundations and mathematical methods of theoretical electricity as similar as possible to those of traditional Tripos subjects.

The *Treatise* was written between 1868 and 1872, and published in late March 1873. It was a long, complicated, and difficult book, in two volumes, whose contents reflected Maxwell's broad knowledge of mathematical, theoretical, and experimental electricity. The book is divided into four major parts—Electrostatics, Electrokinematics, Magnetism, and Electromagnetism—written more or less in the sequence in which they appear in the published text.[17] It is significant that when Maxwell began writing the book in the early months of 1868, his initial concern was not, as one might have expected, to develop his field-theoretic approach to electromagnetism, but to

---

16. Maxwell was also appointed an examiner for the Triposes of 1869, 1870, and 1873.
17. Harman 1990–95a, 2, 26.

provide a detailed account of the application of potential theory (including the necessary mathematical methods not covered by Thomson and Tait in the *Natural Philosophy*) to the solution of difficult problems in electrostatics (*Scientific Letters*, 2:346–47). Maxwell correctly anticipated that it was material of this kind that would fit most naturally into undergraduate mathematical studies at Cambridge. The disadvantage of this pedagogical expediency was that the study of electrical potential, which formed the bulk of the first volume, required little or no reference to the novel physical aspects of Maxwell's own theory. Most of the examples he reproduced had originally been obtained using an imponderable-fluid theory of electricity in which electrical potential was propagated instantaneously between bodies in a manner analogous to gravitational potential. Maxwell did his best to play down such physical hypotheses, which he considered "entirely alien" to his own approach, but gave little indication of how the results were to be physically reinterpreted according to field theory. He simply noted that while from a physical perspective electrical action could be treated "either as the result of the state of the intervening medium, or as the result of a direct action between the electrified bodies at a distance," the distinction was of little importance from a computational perspective as "the two hypotheses [were] mathematically equivalent" (*Treatise*, x, 62–63). Maxwell's occasional remarks concerning the novelty of his theory could therefore be ignored by students who confined their studies to selected sections of the first volume and whose primary concern was the solution of specific problems using advanced mathematical methods.

Another aspect of the *Treatise* that positively discouraged mathematics students from straying beyond the sections on potential theory was the author's frequent discussion of electrical units, instruments, and experiments. Maxwell's work for the British Association Committee on electrical standards had familiarized him with a wide range of electrical instruments and had given him considerable expertise in experimental technique. This work was of great practical importance to the development of electrotechnology, but, for Maxwell, it was also of considerable theoretical significance. From the early 1860s until the late 1880s, the most important experimentally testable aspect of Maxwell's new electromagnetic theory of light was the numerical identity it required between the ratio of electromagnetic and electrostatic units (dubbed "v" by Maxwell) and the velocity of light. These numbers had long been known to have the same dimensions and to be approximately equal in magnitude, but to put the matter beyond doubt it would be necessary to establish electrical standards that supported the identity within the limits of experimental error and that were internationally agreed.[18] Maxwell devoted a good deal of work to securing these conditions during the 1860s, and his efforts were reflected in numerous sections of the *Treatise*. This metrological dimension to the book was so pronounced, in fact, that the wranglers who collaborated with Maxwell on the standards project at the Cavendish in the 1870s saw the *Treatise* more as a book about electrical measurement than as the harbinger of a new electromagnetic theory.[19] But, for the average Cambridge mathematics tutor, with no laboratory or industrial experience in electricity, passages in the book on instruments and experiments were hard going and appeared irrelevant to the needs of undergraduate mathematicians.

It was only in the last of the four major sections of the *Treatise* that Maxwell embarked on a detailed technical discussion of his dynamical theory of the electromagnetic field, but even here he had relatively little to say about the *physical* interpretation of electrical phenomena according to the dynamical approach. In contrast to earlier action-at-a-distance theories of electromagnetism, Maxwell's theory attributed all electromagnetic effects to changing physical states in an all-pervading ether, and he considered it a major advantage of this approach that he was able "to bring electrical phenomena within the province of dynamics." To this end, Maxwell, following Thomson and Tait, introduced his discussion of electromagnetism by outlining a form of mathematical analysis that allowed him to derive equations describing electromagnetic effects from the dynamical properties of the ether without explicit reference to its actual physical structure.[20] He needed to assume only that the ether could store energy in kinetic and potential form, that electromagnetic processes could be accounted for in terms of the flow of energy through the ether, and that during such processes the total energy was conserved. Maxwell also showed that the new equations he had developed to describe all electromagnetic phenomena predicted that electrical

---

18. See Schaffer 1995. The other areas in which the theory was to some extent experimentally testable concerned magneto-optic rotation and the prediction that the dielectric capacities of transparent media were equal to the square of their refractive indices. See *Treatise*, vol. 2, chaps. 20 and 21.

19. George Chrystal (1882, 238) claimed that Maxwell's book was "in the strictest sense a *Treatise* on Electrical *Measurement*" which should be understood as a continuation of the work of the British Association Committee.

20. *Treatise*, vol. 2, arts. 553–84 (quotation on p. 199). For a discussion of this aspect of Maxwell's work, see Buchwald 1985, 20–40.

and magnetic forces were transmitted through the ether at the speed of light and, famously, that light itself was a transverse electromagnetic wave in the ether. Remarkable as they were, these results were achieved at the cost of abandoning the traditional notion of electricity as some form of imponderable fluid. On Maxwell's showing, an electric field was due to a "displacement" (a kind of strain) in the ether, and an electric charge was generated by a discontinuity in this displacement at the surface of conductors (due to the fact that the ether could not sustain a displacement in regions where conductors were present). The fundamental process of electrical conduction, which in fluid theories was easy to conceptualize, was now rendered the most obscure of all common electrical phenomena from a physical perspective.[21] The most direct account of how charge and conduction were to be understood physically in Maxwell's new scheme was actually contained in a couple of short passages in the first volume of the *Treatise*, but, as I have already noted, these remarks could safely be ignored by an aspiring wrangler as he followed Maxwell's mathematical account of electrostatics, magnetostatics, and current electricity. A Cambridge reader who dipped into the last brilliant part of the *Treatise* would therefore have found it extremely difficult to make significant headway without entering fully into the mathematical detail of the argument and being prepared to tolerate only the sketchiest physical explanation of the electrical processes being described.

There were two final aspects of the *Treatise* that made it yet more inaccessible to the average reader: first, many of the results were obtained case-by-case in Maxwell's idiosyncratic style of physical reasoning, rather than by systematic deduction from general equations; and, second, the book contained numerous errors of varying degree of seriousness. These problems sprang partly from the fragmented state of the subject Maxwell was reviewing, and, no doubt, from his inexperience as a textbook writer and want of a reliable proof reader; but they also reflected Maxwell's style of work. Although an original geometer and outstanding physical theorist, Maxwell's shortcomings as an analyst were frequently remarked upon by his contemporaries. When Maxwell was an undergraduate, his coach, Hopkins, had famously commented that although it appeared "impossible for Maxwell to think incorrectly on physical subjects," in his analysis he was "far more deficient." Richard Glazebrook, who was taught by Maxwell at the Cavendish, recalled this comment in summing up the frustration he had experi-

---

21. Buchwald 1985, 28–34.

enced in trying to read the *Treatise* in the mid 1870s: "How many who have struggled through the *Electricity and Magnetism* have realized the truth of the remark about the correctness of his physical intuitions and the deficiency at times of his analysis!"[22] And, as we saw in chapter 3, George Darwin and others had made some extremely scathing remarks concerning Maxwell's style as an analyst and habit of jotting important calculations on bits of paper kept crumpled up in his pocket. In developing the mathematical theory of electricity and magnetism in the *Treatise,* Maxwell had made a number of errors, and for students with only a tenuous grasp of the physical concepts of basic electromagnetic theory and the specific techniques being employed to solve some problems, it was extremely difficult to discriminate between cases where Maxwell had made an error and cases where they simply failed to follow the physical or mathematical reasoning.[23]

The status of the *Treatise* as an undergraduate teaching text was, then, highly ambiguous. Although Maxwell had shown that electromagnetic theory could be grounded in dynamical principles and was therefore a suitable topic for undergraduate study, the sections most likely to be read made little reference to the new theory. Furthermore, those sections in which the new theory was expounded were conceptually difficult, contained misleading errors, and were in any case likely to be entirely ignored. There is also a sense in which the *Treatise* ought not to be regarded as a "textbook" at all. I noted in chapter 3 that most of the mathematical texts used by students in early nineteenth-century Cambridge were not "textbooks" in the twentieth century sense of the term because they were not written from a specifically pedagogical perspective. Early nineteenth-century authors generally wrote for a mathematically literate reader and expounded the subject only in so far as was necessary to discuss their own original contributions. Maxwell's book does not entirely fit this description, as he offered a fairly comprehensive coverage of electrical science, but in most other respects the book was written more in the style of a treatise than of a textbook. Thus he began by surveying (without elementary introduction) the many advanced mathematical

22. Glazebrook 1896, 32. Horace Lamb (1931, 145) made some similarly disparaging remarks concerning Maxwell's inability to handle the "details of minute calculation" both at the blackboard and in print.

23. A large book like the *Treatise* was bound to contain some typographical errors, but complaints of the kind cited above (and in section 6.3) were never made against Thomson and Tait's *Natural Philosophy* or Rayleigh's *Theory of Sound.* Rayleigh's book was considered a model of clarity, though, as his preface reveals, the proofs were read by H. M. Taylor (3W 1865) who removed many "errors and obscurities."

techniques employed in calculating the electrical potential around charged bodies, frequently including, without comment, his own previously unpublished theoretical and mathematical contributions to the subject. Likewise he sometimes discussed electrical machines and instruments, assuming that his readers would, like him, be equally at home with Green's functions, electrical images, telegraph cables, and Wheatstone's Bridge. Yet, in the early 1870s, only a handful of readers in the world, let alone in Cambridge, could have followed such a mixed and idiosyncratic narrative; and virtually no one could make physical sense of Maxwell's new dynamical theory of the electromagnetic field.[24]

For the purposes of understanding the pedagogical reception of Maxwell's work in Cambridge it is also important to keep in mind that the three distinct strands of electrical science referred to in the introduction to this chapter implied very different skills on the part of the prospective teacher. The first is the summary of techniques for calculating electrostatic and magnetostatic potentials and the electrical current in simple networks of conductors. This aspect of the book was relatively straightforward for mathematical coaches to teach as it required no knowledge of Maxwell's new field theory, very little knowledge of experimental electricity, and deployed mathematical techniques familiar from other areas of physical science. The second strand is the exposition of Maxwell's new theory given in the final quarter of the book. This aspect was very difficult for all Cambridge mathematicians to understand and, during the 1870s and early 1880s, was not taught by mathematical coaches but discussed in W. D. Niven's intercollegiate lectures at Trinity College. The third strand concerns the experimental and metrological content of the *Treatise*. This aspect of the book could neither be taught without practical demonstrations nor actively pursued by students unless they were able to develop skills in the hands-on manipulation of electrical instruments and apparatus. It was on this aspect that Maxwell focused his teaching at the Cavendish Laboratory and in which his first graduate students undertook research. Each of these three major aspects of the *Treatise* was discussed and taught, therefore, by different personnel at different sites within the university, but, as we shall see, a student trained at all three sites could acquire something akin to Maxwell's breadth of knowledge of the theory and practice of electrical science.

---

24. On the early reception of the *Treatise* beyond Cambridge, see Buchwald 1985 and Hunt 1991a.

6.3. "A Very Hard Nut to Crack"

> The difficulties experienced by a good mathematician in Maxwell's treatise arise more from what it omits than from what it contains. The difficulty lies in following Maxwell's train of thought, and in seeing what exactly it is he is trying to prove.
>
> CHARLES CHREE (6W 1883)[25]

One of the points I wish strongly to emphasize in this chapter is that making sense of a theoretically novel and technically difficult book such as Maxwell's *Treatise* was a collective enterprise in Cambridge and one that relied heavily on the whole pedagogical economy of undergraduate mathematical training. The years spent by Maxwell mastering and refashioning electromagnetic theory had led him to take for granted both the purpose and inherent difficulties of his project as well as the idiosyncratic techniques by which he solved specific problems. As Chree noted in the epigraph above, it was the absence not only of those steps in an argument that seemed too self-evident to Maxwell to be worthy of inclusion but sometimes of a clear statement of the point of the argument itself that made the *Treatise* so difficult for other mathematicians to follow.[26] We have seen that the technical expertise of Cambridge mathematicians was generated and preserved through a pedagogical economy based on small-class teaching, long hours of carefully supervised study, and the use of locally written textbooks and past examination questions. When it came to an original book like the *Treatise*, the only person to whom a baffled reader could turn for enlightenment was the author himself. But, as we shall shortly see, enquiries of this kind sometimes resulted in a two-way discussion over the meaning of the text. Even Maxwell's public lectures were ineffectual in overcoming these interpretational problems. As I argued in chapter 5, the transmission of highly technical knowledge was more effectively accomplished through interactive encounters between a teacher and a small group of pupils, supported by the supervised study of graded exercises and difficult problems, than through formal lectures to large classes. In this section I illustrate these points by describing some of the earliest and largely unaided attempts made by young Cambridge mathematicians to make sense of Maxwell's *Treatise*.

25. Chree 1908, 537.
26. Even P. G. Tait, an outstanding mathematical physicist who had corresponded extensively with Maxwell on the *Treatise* contents, noted in the margin of one of the proof sheets that while he found "nothing obscure" he found "a great deal that is hard" (JCM-CUL IV/2).

When the *Treatise* was published in March 1873, the only lectures on electromagnetic theory being held in Cambridge were those offered by Maxwell himself. Following his appointment as the first Professor of Experimental Physics in Cambridge in 1871, one of Maxwell's first duties was to deliver annual courses of public lectures on heat, electricity, and magnetism that would prepare students to answer questions on these topics when they appeared in the Tripos examination of 1873. Maxwell duly held public lectures on electricity and magnetism, beginning in the Lent and Easter terms of 1872 respectively, and, for the first couple of years, they appear to have been moderately well attended. In the Lent term of 1872 he had twenty-six students, more than two-thirds being wranglers from the years 1869–75, including the senior and second wranglers from both 1872 and 1873.[27] The latter two students, Harding and Nanson (the first to sit the examination under the new regulations), may well have derived some benefit from the lectures as Maxwell's mark books for the Tripos of 1873 reveal that both men gave almost faultless answers that year to a bookwork-and-rider question on electrostatics.[28] It is more likely however that it was either Harding and Nanson's coach, Edward Routh, or James Stuart (3W 1866) who taught them to apply potential theory to electrostatics.[29]

Whether or not Maxwell's lectures equipped students to answer Tripos problems on electrostatics remains an open question, but it is clear that they were of little help to Nanson when, in the autumn of 1873, he attempted to work his way through the *Treatise*. Evidently frustrated by his inability to follow the argument in several sections, this outstanding second wrangler wrote to Maxwell, drawing his attention to several errors of mathematical reasoning and to a number of passages where the meaning was opaque. Commenting on the section in which Maxwell explained how the method of electrical images was used to find the distribution of charge on intersecting planes and spheres, Nanson noted:

---

27. The students who attended Maxwell's lectures in 1872–73 are listed in JCM-CUL, V n/2 (back of notebook).

28. Maxwell was appointed to the new post of "additional examiner" (making five examiners in total) for the 1873 Tripos with special responsibility for problems in physical subjects. Similar appointments were made each year from 1873 to 1882. Maxwell allocated 43/75 marks for the bookwork and rider respectively, of which Harding scored 43/70 and Nanson 43/75. See *Cambridge University Examination Papers*, 2 January 1873, ix; and JCM-CUL, V k/8.

29. Stuart taught an intercollegiate lecture course on potential theory and harmonic analysis at Trinity College which included some discussion of electricity and magnetism. See *Cambridge University Reporter*, 7 December 1870, 144.

> Pardon me for remarking that the whole of Art[icle] 165 is very hard to make out; and I speak from knowledge of the experience of others in reading the same passage—this is the first case of successive images in the book and the principle of the method is not sufficiently explained. I have asked several men who have read the passage and not one of them could tell me *how* or why the images would determine the electrification.

Nanson went on to draw attention to incorrect definitions, the use of inconsistent nomenclature, significant mathematical and typographical errors, inaccurate or inappropriate citations, and a lack of consistency in what a mathematically literate British reader might reasonably be expected to know. In conclusion, Nanson begged Maxwell's pardon if he had "made any mistakes," and added, in mitigation, that the *Treatise* was "considered by all I know to be a very hard nut to crack."[30] These comments confirm that even the most able wranglers had great difficulty following the application of advanced mathematical methods to new and unfamiliar physical problems, even when no novel physical theory was involved.

The introduction of new and difficult physical concepts provided further problems for the student, as is illustrated in a letter to Maxwell by another of Routh's pupils, George Chrystal (3W 1875), who read the *Treatise* during the summer vacation of 1874. Routh would almost certainly by this point have introduced Chrystal to the theory of electrical potentials covered in the first volume of the *Treatise*, but when Chrystal strayed into the less familiar territory of the second volume he quickly ran into difficulties. He was troubled to find that an expression derived by Maxwell (for the force acting on an inductively magnetized body in a magnetic field) did not agree with the expression given elsewhere by William Thomson following a similar derivation. Unable to resolve the problem himself, Chrystal wrote to Maxwell asking for clarification. In his reply, Maxwell acknowledged that it was he, rather than Thomson, who had given the wrong expression for the force acting on the body, but he insisted that the error was no "mere slip." It sprang rather from the conceptual difficulties inherent in calculating the components of the total magnetic energy due respectively to the original magnetic field, the magnetic body introduced into the field, and the interaction between the two—though Maxwell did concede that such errors had taught him "not to be miserly in using symbols."[31]

30. Nanson to Maxwell, 5 December 1873 (JCM-CUL).
31. Chrystal to Maxwell, 7 July 1874; Maxwell to Chrystal, 22 July 1874 (JCM-CUL).

These examples illustrate the physical and mathematical difficulties experienced by young mathematicians reading Maxwell's book, but they also highlight the interactive relationship that quickly emerged between the *Treatise* and its Cambridge readers. Many of these enquiries prompted Maxwell to amend, rearrange, simplify, and in some cases entirely rewrite sections of the first nine chapters while preparing the second edition of the *Treatise* for publication. When the new edition appeared in 1882, most of the errors pointed out by Nanson had been corrected, and Maxwell's detailed response to Chrystal had become the basis of a whole new appendix explaining how the energy of induced magnetic fields was correctly calculated.[32] The *Treatise* was becoming, in effect, a collective product whose contents were based partly on Maxwell's original design but increasingly on the interactive struggles of other Cambridge mathematicians to understand and clarify his work.

The active exposition and interpretation of the *Treatise*'s contents was also pursued in Cambridge, as we might expect, through the manufacture of Tripos problems. In 1873, Maxwell himself acted as the "additional examiner" responsible for questions in the new physical subjects, but in subsequent years other outstanding Cambridge-trained mathematicians assumed responsibility for setting questions in electromagnetic theory.[33] In 1876 a complete collection of solutions to the previous year's examination was published specifically in order to provide model answers to problems in thermodynamics and, especially, electricity and magnetism. The additional examiner on this occasion was P. G. Tait—with whom Maxwell had corresponded extensively during the writing of the *Treatise*—who emphasized that complete solutions (rather than mere sketches) to "questions on the higher subjects" had been given in all cases where it was felt that "a reference to the Text-books in ordinary use was not sufficient" (Greenhill 1876, vii–viii). In many cases these were questions based on topics covered in the first volume of Maxwell's *Treatise,* and, in every case, the examiner showed in detail how the general theory expounded was correctly applied to solve a specific electrical problem.

The changing content of Tripos questions on electricity and magnetism over the 1870s and early 1880s also traces the gradual mastery of the *Treatise* by Cambridge mathematicians. In the mid 1870s the problems set were based

---

32. *Treatise,* 2:284. The nine chapters revised by Maxwell constituted about a quarter of the *Treatise.* For a review of the second edition, see Chrystal 1882.

33. The additional examiners for 1874–82 were William Thomson, P. G. Tait, Lord Rayleigh, H. W. Watson, Charles Niven, John Hopkinson, W. D. Niven, A. G. Greenhill, and G. H. Darwin.

mainly on electro- and magnetostatics, the few questions on current electricity being confined to the theoretical operation of Wheatstone's Bridge and its application to the location of faults in telegraph cables. The solution of these questions required a sound knowledge of potential theory and some familiarity with electrical measurement and current electricity, but required no knowledge at all of Maxwell's new electromagnetic field equations. Only in the latter 1870s did problems begin to appear that were clearly intended to be tackled by the direct application of mathematical theory unique to Maxwell's work. Yet, despite the fact that many of these questions were very technically demanding, they invariably tested the student's ability to apply Maxwell's equations to straightforward problems in electromagnetic induction rather than to display a grasp of their physical meaning and novel content.[34] It was not until the early 1880s that examiners began to refer explicitly to such peculiarly Maxwellian notions as "stress in the medium" and to expect students to demonstrate a mathematical understanding of such genuinely novel aspects of Maxwell's work as the electromagnetic theory of light.[35] Bearing in mind the close relationship that existed in Cambridge between advanced Tripos questions and the research interests of examiners, it is clear that the assimilation of successive sections of the *Treatise* within the university was a protracted process that took the best part of a decade to complete.

In the following two sections I shall show how this process was accomplished by contributions from two very different teaching sites in the university—the coaching room and the intercollegiate classroom—but before embarking upon this discussion it will be helpful to clarify Maxwell's role as a teacher of electromagnetic theory in Cambridge during the 1870s. Maxwell's status as author of the *Treatise* and Professor of Experimental Physics at Cambridge naturally leads one to suppose that he played a major role in the local interpretation of his new electromagnetic theory and in training the group of Cambridge researchers who took his work forward in the 1880s and 1890s. But, surprising as it may sound, Maxwell's presence in the university was almost entirely irrelevant to both of these developments. Although he lectured every year from 1871 until his death in 1879, and directed experimental work at the Cavendish Laboratory from 1874, Maxwell singu-

---

34. Compare the problems in Greenhill 1876 with those in Glaisher 1879. For an early question requiring a knowledge Maxwell's field equations, see Glaisher 1879, 217.

35. For the first references to "stress in the medium" and the electromagnetic theory of light, see respectively: *Cambridge Examination Papers*, 23 January 1880, morning paper, question x; and 9 January 1884, morning paper, question 11D.

larly failed to attract a sizable audience to his undergraduate lectures and showed no interest in encouraging able graduate students to undertake research on his new electromagnetic theory. This seeming failure on Maxwell's part is only apparent, however, and can be explained in the following way. First, the experience of men like Nanson and Chrystal makes it clear that Maxwell's introductory lectures on electricity and magnetism in the early 1870s were not particularly effective in teaching electromagnetic theory. As we have seen, of the eight prospective wranglers who attended his lectures in 1872, only two subsequently answered a straightforward question on electrostatics. The reason for this was that impersonal public lectures, unsupported by private tutors or college supervisors, were not an effective means of training undergraduates in the technical details of a new and difficult mathematical theory.

Second, apart from the fact that Maxwell was not a very engaging lecturer, the most able and ambitious mathematics students of this period were accustomed to being trained by private tutors. There was therefore little incentive to attend the lectures of a new professor, especially given that he would be discussing conceptually difficult and relatively minor subjects with which, prior to the publication of the *Treatise*, most coaches could offer very little support.[36] This point is borne out by the fact that just eight of more than a hundred prospective candidates for the Tripos of 1873 are listed as attending Maxwell's lectures in 1872.[37] Only the most enthusiastic and able students could spare the time and effort for such marginal studies, perhaps calculating that they would steal a lead on their rivals by tackling subjects the coaches could not teach. Third, once the Cavendish Laboratory opened in 1874 and Maxwell could teach using practical demonstrations in a purpose-built lecture theater, he chose to keep his presentation of mathematical theory to a relatively elementary level, focusing instead on physical principles and experiments. In order to understand this decision, it must be remembered that for Maxwell, as the newly appointed professor of *experimental* physics, the first task was to popularize his professed discipline, a task that was better accomplished by introducing students to otherwise alien instru-

36. Another useful introduction to these topics, Thomson's collected papers on electrostatics and magnetism, had appeared in 1872, but not in time to be of use for the Tripos of January 1873. See the discussion of the examination in the *Cambridge University Reporter,* 20 May 1873, 60–61.

37. Of just eleven (of over a hundred) students who attempted the electrostatics question, only four scored any marks at all for the rider (JCM-CUL, V k/8).

ments, experiments, and techniques of measurement than by discussing the intricacies of an extremely technical and conceptually difficult mathematical theory.[38]

This last point explains why attendance at Maxwell's lectures collapsed completely in the mid 1870s and, at least in part, why he was unsuccessful at generating research interest in his new electromagnetic theory. When it became clear that aspiring wranglers would not be required to take a practical examination and that the best coaches were rapidly mastering the basics of electromagnetic theory (including the schematic representation of simple instruments), the few undergraduate mathematics students who had been attending Maxwell's classes abandoned them entirely.[39] What is actually more surprising than Maxwell's quite reasonable decision not to discuss advanced aspects of electromagnetic field theory in his public lectures is his apparent lack of interest in discussing these issues with his graduate students at the Cavendish Laboratory. Yet here too Maxwell's actions are perfectly explicable. Maxwell saw it as his primary task at the Cavendish to teach his students— mainly successful wranglers—how to design, build, and carry out experiments. Moreover, some of these students did make an indirect contribution to the establishment of Maxwell's theory by assisting him in the accurate determination of the unit of electrical resistance.[40] The other likely reason that Maxwell did not encourage theoretical research in electromagnetism is that he almost certainly believed that there were no outstanding problems in this area that were likely to be tackled successfully by a recent graduate. The most obvious direction for research would have been an attempt to provide a fully electromagnetic account of optical reflection and refraction, yet Maxwell had himself tried and failed to make progress along these lines in the 1860s.[41] In summary, Maxwell's accomplishment during the 1870s was to establish the Cavendish Laboratory as a center for teaching and research in experi-

---

38. The elementary nature of Maxwell's Cavendish lectures is discussed in Lamb 1931, 144, and Newall 1910, 106. On the content of the lectures, see Maxwell 1881, v. The only firsthand record of the lectures is a collection of notes taken by J. A. Fleming in the Lent and Eater terms of 1879. These notes confirm that the lectures were elementary and that Maxwell's only concession to his own electromagnetic theory was a brief discussion of electric displacement. See JAF-CUL, Add 8083.

39. By the academic year 1878–79 Maxwell's audience was reduced to just one or two occasional students and no prospective wranglers at all (Fleming 1934, 64).

40. On Maxwell's wrangler collaborators at the Cavendish Laboratory, see Schaffer 1995.

41. Harman 1995, 10–11. Richard Glazebrook (1931, 137–38) noted that students at the Cavendish continued to approach optical effects in terms of "Elastic Solid theory" throughout the 1870s.

mental physics, but his direct contribution to the teaching and advancement of the new electromagnetic theory outlined in the *Treatise* was negligible.

6.4. Electricity in the Coaching Room

> There is a substantial substratum of truth in the remark once made to the writer that it would have been an immense improvement to Maxwell's "Electricity" to have been written by Routh.
>
> CHARLES CHREE (6W 1883) [42]

The site at which the vast majority of Cambridge undergraduates were formally introduced to technical aspects of the *Treatise* during the 1870s and 1880s was the coaching room. Although electricity and magnetism were first formally examined in 1873, it appears that the coaches made little or no attempt to prepare their pupils in the new subjects for this examination.[43] Prior to the publication of the *Treatise*, the coaches, like the students, had little to guide them beyond Maxwell's lectures, and no idea at all what form Tripos questions in these topics would actually take.[44] As soon as the *Treatise* appeared, Cambridge's leading coach, Edward Routh, set to work to master the sections he thought most relevant to undergraduate studies and to incorporate them in his coaching regime. Routh's thorough courses already covered most of the important mathematical methods used in the *Treatise*, and within months of reading the book he had written a new course in which examples from electrostatics, magnetostatics, and current electricity were used as illustrations of these methods. At the beginning of February 1874 one of Routh's most brilliant pupils, Joseph Ward (SW 1876), noted in his diary that he was "reading Electricity with Routh this term (using Maxwell's book)" and that Routh had recommended that he and his peers attend Maxwell's first course of lectures on experimental electricity at the Cavendish, as it would "be as well to see some experiments on the subject."[45] These unique remarks confirm the speed with which Routh incorporated the new subjects into his third-year coaching schedule, and also highlight his awareness that examination questions in these subjects were likely to refer to instruments

---

42. Chree 1908, 537.
43. G. H. Darwin's undergraduate notes reveal that Routh was certainly not teaching any electricity or magnetism in the mid 1860s (GHD-CUL).
44. The poor performance in electricity and magnetism by the majority of candidates in 1873 was attributed to the lack of suitable textbooks (*Cambridge University Reporter*, 3 May 1873, 42–43).
45. JTW-SJC, 4 February 1874. Marner (1994, Appendix C) provides further evidence that Routh's extant lecture notes on electricity and magnetism were written during 1873–74.

and practical devices with which he could not familiarize his pupils. I shall return shortly to the role of practical demonstrations in undergraduate mathematical studies, but it will be helpful first to look a little more closely at what Routh actually taught.

Routh's lecture notes reveal that he divided electricity and magnetism into three general areas. First, following roughly the order of exposition in the *Treatise* (from which selected sections were used as a textbook), Routh ensured that his students were thoroughly familiar with the fundamental definitions and principles of electrostatics, current electricity, magnetostatics, and electromagnetism. Second, he used numerous problems in electrostatics to exemplify the application of the mathematical methods of potential theory and harmonic analysis. These problems, which generally involved the relationship between electric charge, electric potential, and the surface geometry of conductors, were the ones that fitted most naturally into the existing Tripos syllabus. Third, and perhaps most surprisingly, Routh gave a detailed account of the theory and practical application of current electricity to telegraphy. Having discussed the electromotive force of a battery, the properties of insulators and conductors, the leakage of charge, Ohm's law, Joule's law of heating, and the practical measurement of electrical resistance (using Wheatstone's Bridge), he went on to explain the theoretical operation of submarine-telegraph cables and a range of methods for calculating the position of faults in such cables.[46]

Routh's teaching of electricity and magnetism was almost certainly the most thorough given by any coach in Cambridge and further illustrates the thoroughness and attention to detail that made him the most sought after and successful mathematics tutor in the university. It is nevertheless a very striking aspect of Routh's teaching that he introduced these new subjects, at least implicitly, in the form of an action-at-a-distance theory of electricity. This is nicely illustrated by the fact that Routh began his teaching of electromagnetic theory by establishing the inverse square law of attraction and repulsion for electrostatic force (fig. 6.2), an exercise calculated to drive home the formal mathematical analogy between gravitational and electrostatic phenomena. Routh in fact made no reference at all in his lecture notes to the field-theoretic approach adopted by Maxwell in the latter sections of the *Treatise,* nor did he discuss the electromagnetic theory of light. And, as we shall see, Routh's directions to his pupils concerning which sections of the *Treatise* were to be read and which were to be ignored almost certainly

46. EJR-PC, vol. 4, 484.

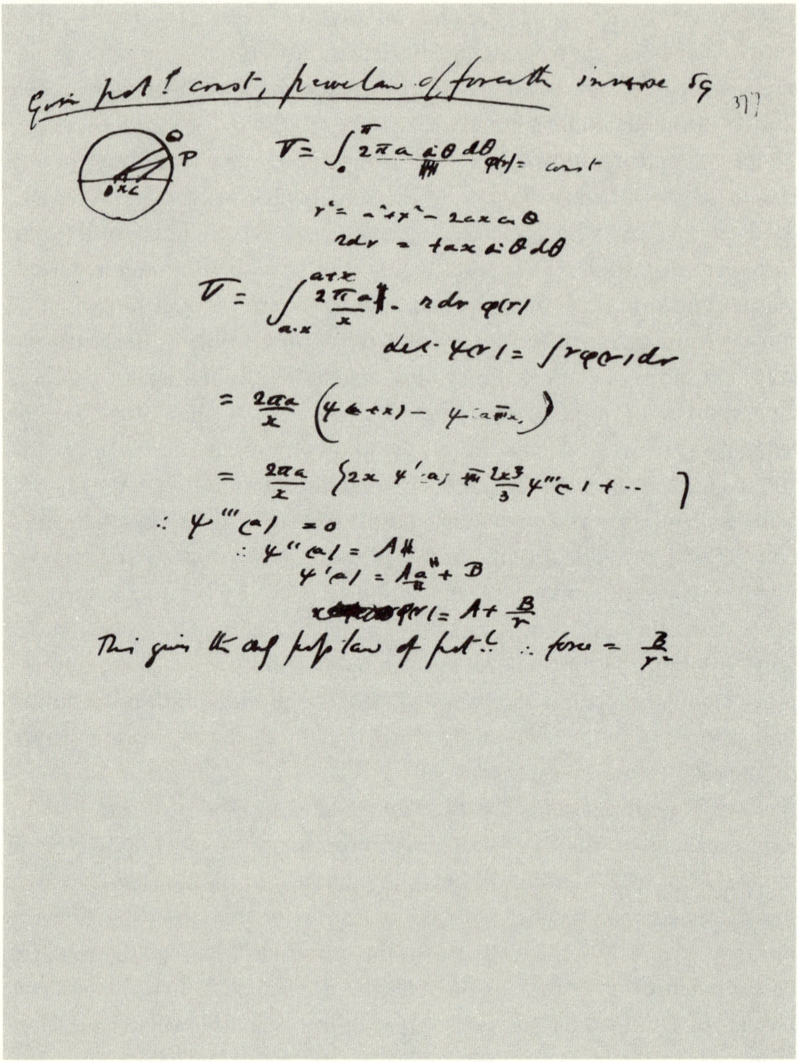

FIGURE 6.2. The first page of Routh's lecture notes on electrostatics. The page begins: "Given potential constant, prove law of force the inverse square." Routh proceeds to show that if the potential produced inside a conducting sphere by a uniform change on its surface is itself uniform (as experiment suggests it is), then the force acting between electrostatic charges must be inversely proportional to the square of the distance between them. This proof suggests a direct analogy between electrostatic force and the force of gravity (which also acts according to an inverse square law). In the nineteenth century the force of gravity was believed to act instantaneously across empty space, and this analogy would have encouraged Routh's students to believe that electrostatic force acted in a similar way. From EJR-PCC, Pet 731 B11, Volume IV (p. 377). (By permission of the Master and Fellows of Peterhouse College, Cambridge.)

reinforced this same conservative interpretation of electromagnetic theory on the book itself.

The complex relationship that existed in the mid 1870s between coaching, the contents of the *Treatise,* Maxwell's lectures at the Cavendish Laboratory, and the questions being set by Tripos examiners is succinctly illustrated by some reminiscences of the Tripos of 1876. The additional examiner for that year, Lord Rayleigh, had been disappointed as an undergraduate not to receive any training in experimental physics, and he strongly supported Maxwell's attempts at the Cavendish to introduce the subject to students of the Mathematical Tripos. In an effort to make attendance at Maxwell's lectures pay in the examination, Rayleigh set a question on the use of Wheatstone's Bridge, an instrument widely used by telegraphic engineers and in physical laboratories to measure electrical resistance. The bookwork part of the question required the student to explain both the principle of the instrument and the way it should be used to locate the position of a fault in the insulation of a cable submerged in a tank of water (both ends of the cable being accessible).[47] The content of this question is particularly interesting because the required technique was neither discussed in any of the books likely to have been read by Cambridge undergraduates nor taught even by Routh. The likely explanation is that Rayleigh had deliberately set a question that could only be answered by students who had witnessed Maxwell's demonstrations at the Cavendish. Maxwell certainly covered Wheatstone's Bridge and the electric telegraph in his lectures, and almost certainly had the apparatus to demonstrate exactly the problem Rayleigh had set.[48] This explanation is further substantiated by the rider to the question. Rayleigh required the student to derive an expression that would correct the error in the measured position of a electrical fault introduced by the finite resistance of the cable insulation.[49] The special approximation required is not discussed in any contemporary textbook and must therefore have been discussed by Maxwell in connection with a practical demonstration.

The deep impression made by Rayleigh's question on one of the candidates, Richard Glazebrook, illustrates both the students' general lack of fa-

---

47. Rayleigh 1899–1920 (hereafter, *Scientific Papers*), 1:282, question ix.
48. The apparatus Maxwell hoped to buy for the Laboratory is listed in Maxwell, *Scientific Letters,* 2:868–75.
49. Even a perfectly manufactured cable would leak charge through its insulating sheath so that the measured position of a fault would appear closer to the center of the cable than it actually was.

miliarity with common electrical apparatus and the coach's role in shaping a student's reading of the *Treatise*. Glazebrook, who subsequently worked with Maxwell at the Cavendish, was one of the few top wranglers tutored by Thomas Dale (3W 1862), a minor coach at Trinity, who, unlike Routh, had neither required his students to read the relevant chapter of the *Treatise* nor recommended that they attend Maxwell's Cavendish lectures. On two occasions when Glazebrook was invited to reminisce about Cambridge physics in the 1870s, he recalled Rayleigh's question and his own irritation at not being able to attempt a mathematically simple problem. Half a century after his examination he lamented:

> I had read carefully much of Maxwell's "Treatise on Electricity and Magnetism," published in 1873, but, alas, had paid no attention to chapter xi on "The Measurement of Electrical Resistance." My coach had originally marked it with a large O—omit. It is true he had corrected this later, but I had no idea of what was meant by Wheatstone bridge. My annoyance was great when, on my first visit to the Cavendish, Maxwell himself explained it and I realised the simplicity of the question I had passed over.[50]

Glazebrook's comments reveal both his lasting irritation at having dropped marks for a mathematically simple question and the way a coach would dictate which sections of Maxwell's book a student should read or ignore. In another account of the same events, Glazebrook alluded to the difficulties experienced by teachers and students alike in coming to terms with the new physical subjects and, on this occasion, made an even more explicit reference to Dale's role in shaping his understanding of Maxwell's work:

> My copy of Maxwell's *Electricity* was bought, I think, during the Long Vacation of 1874. I have it still, a few Chapters or sections of Volume 1 marked with Dale's R[ead], others that we should now think most interesting with an O[mit]. Volume 2 I hardly touched until after the Tripos when, in the summer of 1876, I went with Dale to Buttermere and we read much of it together. (Glazebrook 1931, 132–33)

The chapters of the first volume recommended by Dale were almost certainly those from Part 1 in which the study of electrostatics could be viewed

---

50. Glazebrook 1926, 53. The story is first referred to in Glazebrook 1896, 77. The practice of marking chapters "R" and "O" ("Read" and "Omit") went back at least until the early 1840s; see Bristed 1852, 1:301.

as a vehicle for the introduction and illustration of such advanced mathematical topics as Green's theorem and functions, inversion, harmonic analysis, confocal surfaces, and conjugate functions. A coach flicking through the first volume of the *Treatise* might reasonably have anticipated that it would be to these sections that an examiner would naturally turn when manufacturing mathematically difficult Tripos problems. Subsequent chapters of the first volume, dealing with the conduction of electricity in conductors and dielectrics, would have appeared pedestrian from a mathematical point of view and, in any case, contained much discussion of physical theory and practical electrical measurement that would have seemed obscure to the average coach. Only with the benefit of hindsight, as Glazebrook noted, was it these sections that seemed the most interesting. Moreover, it was the second volume of the *Treatise*, almost totally ignored by Glazebrook, which contained Maxwell's dynamical theory of the electromagnetic field, his equations of electromagnetism, and the electromagnetic theory of light.

Dale's selective reading of the *Treatise* was probably typical of that offered by other mathematics coaches and makes an interesting comparison with Routh's approach. Routh's use of Maxwell's book was, broadly speaking, very similar to Dale's, but he was clearly more systematic in his teaching and had a more informed view of what was likely to be set in the examination. A question on Wheatstone's Bridge had been set in 1873, for example, while questions requiring a knowledge of some sections of the second volume appeared in 1875. Routh had noted these questions and made sure that his students could tackle any similar ones set in the future.[51] But even Routh's thoroughness as a teacher could not guarantee that all of his pupils entered the examination room with an equally firm grasp of every topic. The examination script of another of his pupils, George Pitt (10W 1876), shows that although Pitt, unlike Glazebrook, could correctly sketch the physical layout of Wheatstone's Bridge (fig. 6.3), he could not remember the basic electrical principles that would have enabled him, in two or three lines of elementary algebra, to derive the balance condition.

Pitt's failure also makes a revealing contrast with the brilliant performance of the senior wrangler, Joseph Ward, on another of Rayleigh's electrical problems.[52] This, the most difficult question set by Rayleigh, concerned the flow of electricity through a three-dimensional mass of conducting

51. *Cambridge University Examination Papers* (Mathematical Tripos), 17 January 1873, afternoon, question viii. The question is solved in EJR-PC, vol. 4, 451–52.
52. Rayleigh, *Scientific Papers*, 1:286, question x.

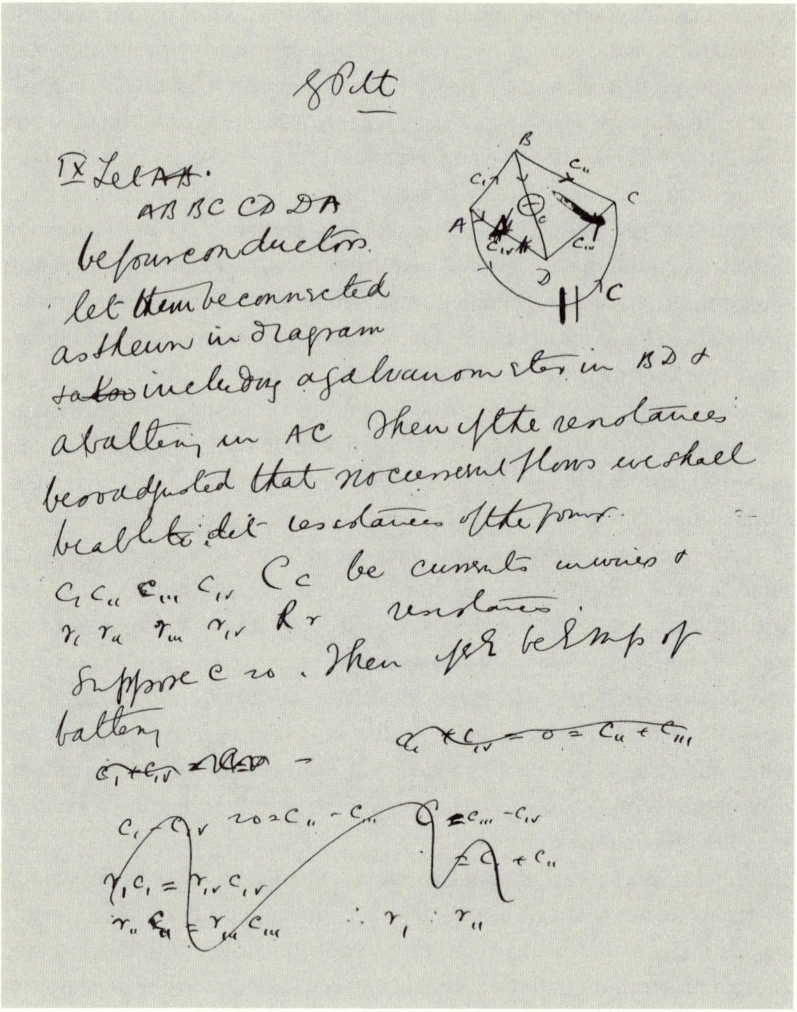

FIGURE 6.3. George Pitt's attempt to answer the following bookwork section of Rayleigh's question:

> ix. Explain the Wheatstone's Bridge method of measuring resistances. How is it applied to find the position of a fault in an otherwise well insulated cable immersed in tank of water, both ends being accessible?

To answer the question the student had to show that when no current flows through BD (on Pitt's diagram) the four resistances must be related as follows: $r_i/r_{ii} = r_{iv}/r_{iii}$. Pitt seems to have forgotten that when no current flows through BD, the electrical potential at B and D must be the same and the respective currents through r_i and r_{iv} must be the same as through r_{ii} and r_{iii}. This means that, by Ohm's Law, $c_i r_i = c_{iv} r_{iv}$ and $c_i r_{ii} = c_{iv} r_{iii}$, from which the required result follows at once by dividing one equation by the other. From a mathematical perspective this result is trivial and the inability of men like Pitt and Glazebrook to find the solution points to their unfamiliarity with electrical laws and instruments. From RC-PLRL, unpublished calculations, verso. (By permission of the Phillips Laboratory Research Library.)

material due to a general distribution of electrical potential. The candidate was told to assume that the potential was zero over the (x, y) plane at the z-origin, and required to derive a given expression (as a power series in z) describing the variation of the potential close to the plane. This part of the question was straightforward provided the student was familiar with the chapter on "Resistance and Conductivity in Three Dimensions" in the first volume of the *Treatise* and so realized that the required expression would be a solution of Laplace's equation.[53] Ward worked through this part of the problem (fig. 6.4) with the speed and confidence one might expect of a senior wrangler trained by Routh, displaying a firm grasp of the electrical principles and mathematical methods required. This page of Ward's script beautifully illustrates the general point that the vast majority of wranglers were much more comfortable tackling a mathematically difficult but physically straightforward problem than with one that required physical reasoning or a knowledge of practical instruments and experimental methods.

The second part of Rayleigh's final problem, for which Ward's attempted solution seems unfortunately not to have survived, is interesting for a different reason. This part of the problem required the student to derive an approximate expression for the conductivity of the thin slice of material between the zero-potential (x, y) plane and a neighboring equipotential surface.[54] The required solution was, in fact, a special case of a more general method developed by Rayleigh for tackling certain physical problems in potential theory using a special form of mathematical approximation. Rayleigh had corresponded with Maxwell on this topic in May 1870, and Maxwell had subsequently included some of Rayleigh's results in a chapter in the *Treatise* on resistance and conductivity in three dimensions (*Scientific Letters*, 2:545–49). Familiarity with this chapter, together with the result obtained in the first part of the question, would have enabled the student to solve the problem, provided he knew and recognized the method of mathematical approximation required. At the time Rayleigh was manufacturing this Tripos problem, he was writing an original paper on the use of his method to solve

---

53. *Treatise*, vol. 1, arts. 297–304.

54. The second part of Rayleigh's question (see fig. 6.4) read: "[Show that] if a neighbouring equipotential surface, $z = z_1$, coincide with the a plane parallel to xy, except over a certain finite region where there is slight deviation, the conductivity between the equipotential surfaces $z = z_1$, and $z = 0$, is expressed approximately by

$$\iint \left[1 + \frac{1}{3}\left(\frac{dz_1}{dx}\right)^2 + \frac{1}{3}\left(\frac{dz_1}{dy}\right)^2 \right] \frac{dx\,dy}{z_1}$$

the area of integration including the whole of the above-mentioned region."

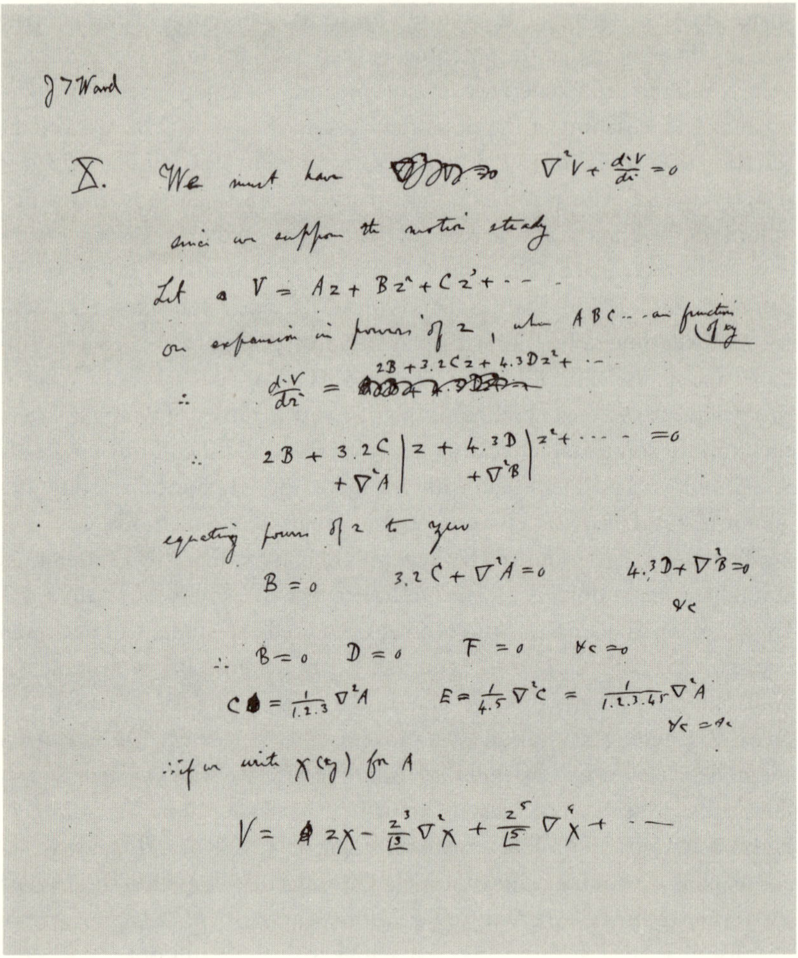

J T Ward

X. We must have ~~...~~ $\nabla^2 V + \frac{d^2V}{dz^2} = 0$

since we suppose the motion steady

Let $V = Az + Bz^2 + Cz^3 + \ldots$

on expansion in powers of z, where $A\,B\,C \ldots$ are functions of xy

$$\therefore \quad \frac{d^2V}{dz^2} = 2B + 3.2Cz + 4.3Dz^2 + \ldots$$

$$\therefore \quad 2B + \left.\begin{matrix} 3.2C \\ +\nabla^2 A \end{matrix}\right| z + \left.\begin{matrix} 4.3D \\ +\nabla^2 B \end{matrix}\right| z^2 + \ldots = 0$$

equating powers of z to zero

$$B = 0 \qquad 3.2C + \nabla^2 A = 0 \qquad 4.3D + \nabla^2 B = 0 \quad \&c$$

$$\therefore \quad B = 0 \quad D = 0 \quad F = 0 \quad \&c = 0$$

$$C = \frac{1}{1.2.3}\nabla^2 A \qquad E = \frac{1}{4.5}\nabla^2 C = \frac{1}{1.2.3.4.5}\nabla^4 A \quad \&c = \&c$$

$\therefore$ if we write $\chi(xy)$ for A

$$V = z\chi - \frac{z^3}{\underline{|3}}\nabla^2\chi + \frac{z^5}{\underline{|5}}\nabla^4\chi + \ldots$$

FIGURE 6.4. Joseph Ward's (SW 1876) attempt at the following part of Rayleigh's question on current flow in a conducting mass:

> x. Show that if in a uniform mass conducting electricity the potential be zero over the plane xy, its value at neighbouring points out of the plane will be
>
> $$z\chi - \frac{z^3}{1.2.3}\nabla^2\chi + \frac{z^5}{5!}\nabla^4\chi - \ldots$$
>
> where χ is a function of x and y, and $\nabla^2 = d^2/dx^2 + d^2/dy^2$.

Ward's script makes a sharp comparison with Pitt's in fig. 6.3. Where Pitt made little headway with a mathematically simple problem requiring a knowledge of instruments, Ward rapidly obtains the required result to a technically far more difficult problem requiring a knowledge of series solutions to Laplace's equation (written here in modified form). As Robert Webb later told his pupils (chapter 5), aspiring wranglers "cannot ever know too much about the solutions of $\nabla^2(V) = 0$." Note that Ward nevertheless makes and corrects a number of slips as he works his way towards the solution. From RC-PLRL, unpublished calculations, verso. (By permission of the Phillips Laboratory Research Library.)

problems in electrical conduction. When the paper appeared in the *Proceedings of the London Mathematical Society* a couple of months after the Tripos, Rayleigh gave the required result and pointed out that it had already been announced as a Tripos question (Rayleigh, *Scientific Papers*, 1:276). This example shows once again how advanced Tripos problems required students not merely to answer cunningly contrived though otherwise meaningless questions but, as we saw in chapter 5, to display the ability, albeit guided by the examiner, to solve a well-posed research problem.

Rayleigh's practical questions of 1876 represented the last attempt by an examiner to make success as a wrangler dependent upon attending lectures at the Cavendish Laboratory. Within a year or two it had become clear that mathematics students could grasp the principles of, say, Wheatstone's Bridge and its application to practical electrical measurements without actually witnessing the instrument in action. By the latter 1870s examiners had ceased to set problems on practical electro-technology, turning instead to more mathematically demanding questions in electromagnetic induction that required a knowledge of Maxwell's new field-theoretic equations. As coaches like Routh mastered the basic technical content of the first three quarters of the *Treatise* and learned to solve the kind of questions that examiners were likely to set, electricity and magnetism were effectively incorporated within the standard repertoire of undergraduate mathematical studies. Indeed, Routh in particular became a powerful gatekeeper to the mathematical methods necessary to make sense of Maxwell's book.

This point was nicely illustrated in 1883 when Michael Pupin, a mathematics student from Columbia College in New York, arrived in Cambridge in the hope of meeting Maxwell and learning his new electromagnetic theory. Pupin had the good fortune to make the acquaintance of Maxwell's former friend and colleague, W. D. Niven, who not only informed the embarrassed Pupin that Maxwell had died in 1879 but, having examined him in mathematics, offered him the following advice:

> Niven pointed out that a prospective physicist who wished to master some day Maxwell's new electrical theory must first master a good part of the mathematical work prescribed for students preparing for the Cambridge mathematical tripos examinations.
>
> "Doctor Routh could fix you up in a quicker time than anybody," said Niven with a smile, and then he added cautiously, "that is, if Routh consents to your joining his private classes, and if you can keep up the pace of the youngsters who are under his training." (Pupin 1923, 174)

FIGURE 6.5. Routh's team for the Mathematical Tripos of 1880. Routh had an annual photograph taken (towards the end of Michaelmas term) of the group of his pupils about the take the Mathematical Tripos. This one includes two of Cambridge's most important electromagnetic theoreticians of the 1880s and 1890s, J. J. Thomson and Joseph Larmor. Seven of Routh's pupils were in the top ten wranglers in 1880 (Appendix A1), six of whom are identified by name in the picture. Back row (standing): J. W. Welsford (7W), J. Larmor (SW), J. Marshall (22SO), Unknown, J. J. Thomson (2W), E. J. C. Morton (1SO), F. F. Daldy (16W). Second row (seated on chairs): T. Woodcock (23W), H. Cox (4W), Unknown, Edward Routh, P. T. Wrigley (33W), J. C. Watt (11W). Front row (seated on ground): Unknown, A. McIntosh (6W), W. B. Allcock (3W). Thomson, *Recollections and Reflections*, 1936. (By permission of the Syndics of Cambridge University Library.)

As we saw in chapter 4, Pupin entered fully into the spirit of Cambridge undergraduate life, working hard at mathematics in the mornings and evenings while rowing competitively in the college boat in the afternoons. After two terms of intense training in Routh's advanced classes, Pupin not only found himself able to "handle the mathematics of Maxwell's theory of electricity with considerable ease" (1923, 197) but was able to follow the professorial lectures in electromagnetic theory and optics given respectively by Rayleigh and George Stokes.

Despite Routh's brilliance at explicating some sections of Maxwell's *Treatise*, he offered no account of the new field theory discussed in the last quarter of the book. Routh certainly covered most of the mathematical methods employed in the latter sections, but these were taught in connection with other, more traditional Tripos subjects, such as celestial mechanics and the shape of the earth. The real significance of Routh's teaching to the study of electromagnetic theory in Cambridge was that it gave able undergraduates a thorough grounding in the principles and application of basic electromagnetic theory and familiarized them with the higher mathematical methods and dynamical concepts they would need if and when they ventured into the final sections of the book (fig. 6.5). It was, no doubt, the apparent ease and confidence with which Routh covered this material that encouraged students like Chree to suggest (see epigraph to this section) that the *Treatise* would have been a great deal easier to follow had it been written by Routh rather than by Maxwell. What Chree and his peers failed to take into account was that, in addition to his thorough mastery of Cambridge mathematics and unique ability as a teacher, Routh taught only those parts of the *Treatise* that were consonant with the wider content of undergraduate mathematical studies and drew upon the collective expertise of those students and examiners who had worked hard to find consistent and stable meaning in the earlier sections of Maxwell's text. Routh's pupils certainly received a solid training upon which more specialist lecturers could easily build, but his pupils would have acquired little or no sense that the latter sections of the *Treatise* developed a profoundly original approach to electromagnetic theory.

6.5. Teaching the *Treatise* at Trinity College

The site at which the novel, field-theoretic aspects of Maxwell's work were first taught and discussed in Cambridge was W. D. Niven's intercollegiate class in electricity and magnetism at Trinity College. We saw in chapter 5 that the reform of public teaching began in the late 1860s with the establishment of several intercollegiate courses which could be attended by any member of the university for a small fee. It will be recalled that one of the main purposes of these courses was to provide a forum at which a lecturer with expert knowledge in a particular field of mathematics or mathematical physics could attract a sizable class of advanced undergraduates and young wranglers in which productive intellectual exchange could take place. In the wake of Maxwell's move to the Cavendish Laboratory and poor student performance in the new physical subjects in the Tripos of 1874, the first intercollegiate courses in the mathematical theory of electro- and magnetostatics were held by James Stuart (Trinity College) and George Chrystal (Corpus Christi College) in 1875 and 1876 respectively.

In 1876, Stuart's course was taken over by W. D. Niven (3W 1866), who continued until 1882 to give an annual two-term lecture series based on

FIGURE 6.6. W. D. Niven in mid life. In addition to being a close acquaintance of Maxwell's during the second half of the 1870s, Niven not only edited Maxwell's collected scientific papers and the second edition of the *Treatise*, but made one of the first attempts (certainly the first in Cambridge) to clarify and develop technical arguments from the latter. Niven's greatest contribution to the development of electromagnetic theory in Cambridge was to use his rare knowledge of the *Treatise*'s contents to train a whole generation of "Maxwellians" during the period 1876–82. FA.I.72, Wren Library, Trinity College. (By permission of the Master and Fellows of Trinity College, Cambridge.)

Maxwell's *Treatise* (fig. 6.6).[55] A contemporary of Stuart's at Trinity (the two men were bracketed third wrangler in 1866), Niven had left Cambridge in 1867 to teach mathematical physics in London. Despite his absence from the university, Niven assisted with the introduction of the new physical subjects by acting three times as a Tripos examiner, and, in 1874, he was invited to re-

[55]. Thomson 1936, 42; Rayleigh 1942, 8. Thomson recalled that Niven's lectures were "on mathematical physics, mainly on Maxwell's treatise on *Electricity and Magnetism* which had then lately been published." The only extant notes from Niven's course are those taken in Michaelmas Term 1879 by J. A. Fleming. These show that Niven taught directly from the *Treatise* (JAF-UCL, 34).

turn to his old college as a mathematics lecturer.[56] Whether Niven was appointed specifically to teach the mathematical theory of electricity and magnetism is unclear—he appears to have spent all or part of his first year working on electrical experiments at the Cavendish Laboratory—but he certainly struck up a close friendship with Maxwell and became heir to his scientific writings following the latter's untimely death in 1879.[57] Niven completed the revisions for the second edition of the *Treatise* and edited the two posthumously published volumes of Maxwell's collected scientific papers.[58]

Niven's lectures were the first in Cambridge to treat the second volume of the *Treatise* as an advanced textbook on electromagnetic theory, and they soon made Trinity College by far the most important site in the university for the discussion of mathematically difficult and physically obscure aspects of Maxwell's work.[59] One of Niven's most distinguished students, Joseph Larmor (SW 1880), recalled not only that Niven's "class-room was a focus for [Maxwell's] theory, in which congregated practically all the mathematicians of the University," but also that Niven's influence as a teacher of the theory was so pervasive that "in later years he could count *nearly all* the active developers of electrical science on the mathematical side as his friends and former pupils" (Larmor 1917, xxxix, my italics).[60] As Larmor also noted, the great strength of Niven's teaching was not simply that he offered a systematic treatment of the most technically demanding sections of the book, but that he drew attention to, and attempted to clarify, the novel physical hypotheses upon which the book was based at a time "when they were largely misunderstood or not understood at all elsewhere" (xxxix).

An equally distinguished contemporary of Larmor's, J. J. Thomson (2W 1880), who attended the lectures during the academic year 1877–78, corroborates Larmor's assessment of Niven's teaching and provides further insight into his style. According to Thomson, Niven was "not a fluent lecturer nor was his meaning always clear," but he was a great enthusiast for Maxwell's "views" and managed to "impart his enthusiasm to the class." Thomson and

---

56. Niven was a Tripos examiner in 1871, 1873, and 1874.

57. Thomson et al. 1910, 331; Glazebrook 1896, 78.

58. Larmor 1917, xxxix. Niven's publications reveal no special interest in electromagnetic theory prior to his return to Cambridge.

59. Thomson et al. 1931, 21. Forsyth recalled that the "range" of the *Treatise* as taught by Niven "seemed to have little in common with the electricity of the blackboard and the examination paper" (1935, 163). The "blackboard" referred to was presumably Routh's.

60. The "active developers" referred to were those trained in Cambridge. There were two important non-Cambridge developers in Britain (George Francis FitzGerald and Oliver Heaviside) and many more in Continental Europe.

his peers recognized Niven's ability to inspire enthusiasm for Maxwell's work as his most important quality as a teacher, and one which, when combined with his apparently enigmatic attempts to explicate Maxwell's theory, was extremely productive. As Thomson noted:

> if we could not quite understand what he said about certain points, we were sure that these were important and that we must in some way or other get to understand them. This set us thinking about them and reading and re-reading Maxwell's book, which itself was not always clear. This was an excellent education and we got a much better grip of the subject, and greater interest in it, than we should have got if the question had seemed so clear to us in the lecture that we need not think further about it. (Thomson 1936, 42–43)

The reminiscences of A. R. Forsyth (SW 1881) add further weight to Thomson's claim that Niven's enthusiasm more than compensated for the alleged obscurity of his explanations. Forsyth saw Niven as "an interpreter of Maxwell's work," and although in his opinion Niven did not "possess a gift of clear and connected exposition, either in calculation or in explanation," he readily acknowledged that Niven "knew his subject well" and provided "an excellent stimulus to independent study" (Forsyth 1935, 164; Whittaker 1942–44, 210). At the end of an hour's lecture, Forsyth recalled, the student whose primary interest was mathematics found himself with "a collection of brief sentences, remarks, symbols and occasional diagrams" which were "far from comprehensible." Yet, convinced by Niven of the importance of Maxwell's work, Forsyth, together with the two other outstanding mathematicians of his year, R. S. Heath (2W 1881) and A. E. Steinthal (3W 1881), adjourned to Forsyth's rooms after each lecture and, with the help of the *Treatise* and other books from the library, "spent two or three hours in making a coherent account of the substance of Niven's lecture."[61]

Whether Niven really was as poor a lecturer as his students claimed him to be is difficult to judge, but there can be little doubt that the problems he experienced in trying to explicate the *Treatise* sprang as much from the inherent difficulty of his subject matter as from his shortcomings as a teacher. We have already seen that Maxwell's first readers in Cambridge found even the most familiar passages in his book difficult to follow. Younger students such as Larmor, Thomson, and Forsyth had the benefit of Routh's clear exposition of earlier sections of the *Treatise*, but they were still critical of

---

61. Rayleigh 1942, 8. Rayleigh reveals that this mathematical summary of Maxwell's work proved extremely useful to J. J. Thomson who "asked Forsyth for the loan of their joint notes" while revising for the Tripos of 1880.

Maxwell's presentation of the more novel aspects of his work in the latter sections. J. J. Thomson noted that the reason Maxwell's work initially had "few adherents" was that "his presentation of his theory [was] in some features exceedingly obscure." Indeed, in Thomson's opinion, Maxwell's attempt in the *Treatise* to explain the physical basis of his theory was one of "the most difficult chapters in the literature of Physics." Maxwell's analogy between "electric displacement" in an all-pervading electromagnetic "medium" and "displacement in an elastic solid under stress" left many of his readers extremely confused concerning the extent to which these analogical references were to be taken literally. As Thomson also pointed out, Maxwell used a range of sometimes contradictory physical analogies in the *Treatise*, often drawing, without reference, upon heuristic models he had employed in earlier papers and subsequently discarded.[62] The coupling of this confusing collection of physical and dynamical images with the forbidding array of advanced mathematical methods employed by Maxwell in deriving and applying his new equations made the last third of the *Treatise* very difficult reading indeed.

Niven's lectures helped Cambridge students to manage these difficulties in three ways. First, Niven's classes were clearly effective at convincing Cambridge mathematicians of the importance of Maxwell's new theory and in inspiring them to investigate it through the pages of the *Treatise*. As we have seen, his weekly lectures sent his students back again and again to search for meaning in Maxwell's enigmatic equations and prose. Second, Niven's classroom provided a meeting place where those engaged in this enterprise could share their insights and difficulties. This collective activity enabled Cambridge mathematicians to pool their skills in puzzling out opaque passages and difficult derivations, and, most importantly, to discriminate with confidence among problems that could be solved by adopting a particular interpretation, those due to Maxwell's errors, and those due to the fact that some aspects of the theory had simply been left unfinished. Finally, Niven's actual exegesis of the last quarter of the *Treatise*, opaque as it may in places have been, nevertheless impressed a particular reading of the book upon his class. Unlike Routh, Niven discussed the fact that, according to Maxwell's theory, electromagnetic effects were due to the flow of electric and magnetic energy in the space surrounding charged bodies, and he showed his students how

---

62. Thomson 1931, 35–36. For Maxwell's attempt to give a succinct account of the physical basis of his theory, see *Treatise*, vol. 1, arts. 59–60.

Maxwell's equations were correctly applied (albeit using approximations) to solve a range of standard problems in electromagnetism.

We cannot know for certain exactly what kinds of issue were discussed in Niven's lectures, but we can make some reasonable inferences based on two sources of indirect evidence. The first is the early research publications by Niven and his Cambridge peers. These publications are the subject of the next section, and I shall therefore defer discussion of the evidence they provide until then. The second source is the notes appended by Niven to the second edition of the *Treatise*. Niven explained in his preface to this edition that he had felt it appropriate to make "the insertion here and there of a step in the mathematical reasoning" and to add a "few foot-notes on parts of the subject which [his] own experience or that of pupils attending [his] classes shewed to require further elucidation" (xiii). These steps and notes provide a useful guide to the sections with which Niven and his students had difficulties and reveal two interesting characteristics of Niven's reading of the *Treatise*.

The first concerns the specific corrections that Niven made to the text. When Maxwell died in 1879 he had revised only the first nine of the twenty-three chapters in the book. Niven's editing of the remaining chapters was relatively light, but he did correct numerous errors and amend various derivations along the lines recommended by Nanson and Chrystal.[63] This work confirms that the many difficulties encountered and surmounted by early Cambridge readers of the *Treatise* had become common currency in Niven's lectures. The second characteristic revealed by Niven's notes is that his technical interest in the *Treatise* focused mainly on the advanced mathematical methods employed by Maxwell to find the potential distribution surrounding charged conductors, and on the application of the general equations of the electromagnetic field to the solution of problems in current electricity.[64] These were the topics on which Niven offered points of clarification and alternative derivations, and it is reasonable to infer that it was on these issues that advanced discussion in his classroom was focused. Conversely, Niven had no points of explanation, clarification, or alternative derivations to add either to the sections on the physical and dynamical foundations of the theory or to those on the electromagnetic theory of light. Those who studied the *Treatise* in Niven's classroom would therefore have gained the impression that Maxwell's theory was primarily a collection of equations and math-

---

63. George Chrystal (1882, 239) noted that only those "perfectly familiar" with the *Treatise* would appreciate the "conscientious labour" expended by Niven in preparing the second edition.

64. See, for example, Niven's notes to articles 160, 200, 603, 659, 701, and 705.

ematical techniques that could be applied to produce exact solutions to standard electrical problems. I shall comment further on this issue in the next section.

The combined efforts of Routh and Niven to teach the mathematical theory of electricity and magnetism gradually paid off through the latter 1870s. Where the examiners in 1873 and 1874 had been disappointed both by the general lack of interest in the new subjects and by the very poor quality of many of the answers submitted, the Board of Mathematical Studies reported in 1877 that "the work done by the best men in the higher physical subjects was very satisfactory."[65] By 1881, the additional examiner in the physical subjects, Alfred Greenhill, could report that the candidates had shown "a sound appreciation of principles and the power of applying them." In his capacity as president of that year's Board of Mathematical Studies, Niven, also commented with evident satisfaction that mathematics students now "devoted a great part of their time . . . formerly occupied by the study of Astronomy, including the Lunar and Planetary Theories, Figure of the Earth and Precession and Nutation," to the study of "Heat, Electricity and Magnetism." This was for Niven an extremely welcome change because the "progressive nature" of the latter subjects afforded in his opinion a better test of the "mathematical ability of the candidates."[66] Niven meant by this that it was relatively easy to produce new results—of the kind discussed above—in the new subjects and that such results could be set as difficult problems to Tripos candidates. Learning the progressive subject of electromagnetism with Routh and Niven, and mastering difficult Tripos problems based on the contents of the *Treatise,* also prepared the outstanding wranglers of the late 1870s to make their own contribution to the subject after graduation.

The attempts by Niven and his peers and students to develop the contents of the *Treatise* are discussed in the next two sections, but, in anticipation of this discussion, it will be helpful here to say something about the breadth of undergraduate mathematical studies in Cambridge during the 1870s. I noted in chapter 5 that an important characteristic of wrangler research was the deployment of mathematical techniques developed in one branch of physical science to the solution of problems in another branch. The uses of this approach are amply illustrated in the following sections, but it should be pointed out that one book above all others, Rayleigh's *Theory of Sound,* provided an exceptionally fruitful source of new physical imagery and mathe-

65. Quoted in Wilson 1982, 339.
66. *Cambridge University Reporter,* 7 June 1881, 640.

matical methods to several wranglers developing electromagnetic theory in the late 1870s and 1880s. Rather like the approach to the science of electromagnetism adopted by Maxwell in the *Treatise,* the *Theory of Sound* offered a general overview of recent work in the mathematical theory of acoustics as well as a systematic statement and development of Rayleigh's own research in this and related subjects.[67] Rayleigh's book was actually a much more ambitious and far-reaching study than the modest title implied. As we saw in chapter 5, it was the opinion of at least one mid-twentieth-century reviewer that there was virtually no vibrating system that could not be tackled successfully using the techniques developed by Rayleigh in the first ten chapters.

Similar sentiments were expressed by Rayleigh's contemporaries. When George Airy received a complimentary copy in May 1877, he wrote to the author expressing his gratitude and remarking that the technical discussion of "non-soniferous vibrations" was "worked out with a depth of mathematics applicable to even more complicated subjects." As Airy politely and accurately concluded, the book "almost merits a title of wider meaning." Edward Routh was equally impressed when he received a copy from the publishers in June 1877. He wrote at once to Rayleigh informing him that he expected to "learn much" from this "much needed" book. The outstanding coach also informed his brilliant ex-pupil that he intended to use the *Theory of Sound* as a "text-book" in the forthcoming term and that he and his advanced students would "go much further into the subject now than [they] ever did before."[68] The examiners' solutions to the problems and riders for the Tripos of 1878 suggest that Routh was as good as his word. The examiners for that year based several of the most difficult questions on results and techniques discussed in the *Theory of Sound* and made frequent reference to the book as an authoritative source on advanced mathematical methods. The sources referred to in the model answers for 1878 reveal that three books—Thomson and Tait's *Natural Philosophy,* Maxwell's *Treatise,* and Rayleigh's *Theory of Sound*—were now firmly established as the most important reference works for mathematics students studying the higher physical subjects.[69]

---

67. A method developed by Rayleigh for estimating the physical characteristics of resonators was borrowed by Maxwell who "translated it into electrical language for [his] book" (Maxwell, *Scientific Letters,* 2:605). Maxwell also introduced some of the mathematical methods developed in Rayleigh's early papers on acoustics in his Cambridge lectures on heat and electricity in 1873. See Strutt 1924, 81. G. H. Darwin heard Maxwell lecture on Rayleigh's "dissipation function" and elaborated the method in a notebook. See Maxwell, *Scientific Letters,* 2:941.

68. Strutt 1924, 83–84.

69. Glaisher 1879, passim.

All three of these books built their analysis of physical systems on the concept of the conservation of energy and on the application of one or more forms of generalized dynamics. There was, therefore, considerable overlap between the mathematical methods and physical principles adopted in each book, both Maxwell and Rayleigh acknowledging their debt in this respect to Thomson and Tait's *Natural Philosophy*.[70] During the late 1870s and early 1880s it was Rayleigh's mathematical analysis of vibrating systems, and especially of the energy radiated into the surrounding medium, that was to stimulate powerful new insights into Maxwell's electromagnetic theory.

6.6. The First Generation of Cambridge Maxwellians

The importance of W. D. Niven's classroom as a center for discussion of Maxwell's *Treatise* was due only in part to his role as a teacher. Niven was also one of three men, all of whom were undergraduates and fellows of Trinity College, who were responsible for the first original contributions to Maxwell's work in Cambridge. The two other members of the group were W. D. Niven's brother, Charles Niven (SW 1867), and Horace Lamb (2W 1872). I shall refer to these three men collectively as the first generation of Cambridge Maxwellians. The two Niven brothers probably first became interested in Maxwell's work as undergraduates in the mid 1860s. This was the time when the content of the Mathematical Tripos was beginning to be reformed, and W. D. Niven recalled that Maxwell's Tripos questions in 1866 and 1867 (the years in which he and his brother respectively sat the Tripos) "infused new life into the examination" (Niven 1890, iv, xvii).

It also appears that a rare interest in Maxwell's current publications in electromagnetic theory was being expressed at Trinity College at this time. Rayleigh recalled that he had been urged by W. D. Niven's contemporary, James Stuart, to read Maxwell's brilliant paper, "A Dynamical Theory of the Electromagnetic Field," shortly after it appeared in 1865.[71] Stuart was later the moving force behind the introduction of intercollegiate lectures in Cambridge and taught potential theory (including electricity and magnetism) at Trinity until W. D. Niven began his course in 1876.[72] It may be of considerable significance that Stuart and the Niven brothers were all Scotsmen who

---

70. *Treatise*, vol. 2, chap. 5; *Theory of Sound*, vol. 1, chap. 4. Rayleigh's theoretical discussion also drew heavily upon Routh's original works in dynamics.
71. Strutt 1924, 38.
72. *Cambridge University Reporter*, 7 December 1870, 144–45; Stuart 1912, 152.

shared Maxwell's background in Scottish natural philosophy. Growing up in Edinburgh in the 1840s, Maxwell's youthful interests in geometry, natural philosophy, and experimental science had been nurtured by visits to the Royal Society of Edinburgh and the optical laboratory of William Nicol, and by attendance at J. D. Forbes's undergraduate lectures in natural philosophy at Edinburgh University. Stuart and the Nivens took their first degrees respectively at the universities of St. Andrews and Aberdeen, the former being taught natural philosophy at St. Andrews by Forbes.[73] Maxwell himself also spent several weeks each year at Trinity College in connection with his duties as an examiner, and his personal presence seems further to have heightened local interest in his work in the latter 1860s.[74] Horace Lamb (2W 1872), the youngest member of the trio, was an undergraduate at Trinity College during this period and attended Maxwell's first professorial lectures during the academic year 1871–72. Lamb probably got to know W. D. Niven while both men were on the Trinity College staff in 1875, the older man quite possibly attending Lamb's brilliant course of intercollegiate lectures on hydrodynamics.[75] In this section I discuss the interpretation of and contributions to Maxwell's theory made by the first generation of Cambridge Maxwellians and then comment of the significance of their strong association with Trinity College.

It was in 1876, shortly after he had begun teaching from the *Treatise,* that W. D. Niven wrote what was in effect the first research paper by a Cambridge mathematician based on Maxwell's work. Niven's paper, like several of Nanson's queries to Maxwell, was concerned with the clarification of the mathematical methods used by Maxwell to solve problems in electrostatics. In the *Treatise,* Maxwell had employed the method of electrical images to find the distribution of electrical charge on two concentric conducting spheres due to a point charge between them. Niven's contribution was to show that this calculation could be recast into more familiar analytical form by deducing the required expressions as secondary solutions to Laplace's equation.[76] This first attempt to extend Maxwell's work displayed a characteristic common to

---

73. Harman 1990–95a, 2–7; Stuart 1912, 134–35; Larmor 1917, xxxviii; Macdonald 1923, xxvii. It may also be significant that Maxwell had taught natural philosophy at Aberdeen until 1860.

74. Strutt 1924, 44; Glazebrook 1896, 60–61.

75. J .J. Thomson et al. 1931, 142; Glazebrook 1932–35, 376–77; JCM-CUL, V n/2. Niven praised Lamb's abilities as a teacher in a letter of reference for the professorship of mathematics at Adelaide in 1875. Two of Lamb's examiners, Ferrers and R. K. Miller, also testified to his originality as a mathematician on the strength of answers he submitted in the Tripos of 1872.

76. Niven 1877, 65–66. The paper was read to the London Mathematical Society on 14 December 1876.

several papers on electromagnetism later published by wranglers. Despite the fact that his calculations were centrally concerned with charge distribution, Niven made no reference whatsoever to the fact that, according to Maxwell, charge was not a subtle fluid but a discontinuity in electrical displacement in the ether at the surface of a conductor.[77]

Given that Niven was at this time teaching from the *Treatise* and in regular contact with Maxwell, it seems highly improbable that he was ignorant of this fact, but, as far as he was concerned, his paper was primarily a mathematical exercise that required no discussion of the physical nature of charge. In this respect, Niven, an ex-pupil of Routh, formulated his research problems from the *Treatise* in much the same way that Routh himself might have done. Niven in fact understood Maxwell's original handling of mathematical methods in the *Treatise* to be as important a contribution to mathematical physics as was his new electromagnetic theory. Even in 1890, in his editor's preface to Maxwell's collected scientific papers, Niven claimed that the *Treatise* was "also remarkable for the handling of the mathematical details no less than for the exposition of physical principles" and that it was "enriched incidentally by chapters of much originality on mathematical subjects touched on in the course of the work" (1890, xxix). Apart from a handful of papers similar to the one described above, Niven devoted most of his research effort to investigating the mathematical properties of spherical harmonics using techniques first introduced by Maxwell.[78]

This highly mathematical approach to Maxwell's work is also illustrated in the research contributions of the other Cambridge Maxwellians. One of the first attempts actually to apply Maxwell's new equations was made by W. D. Niven's brother, Charles Niven, who had left Trinity College shortly after taking his degree to become professor of mathematics at Queen's College, Cork. Charles was also a keen reader of Maxwell's book, and, in 1879, he wrote a paper on the flow of currents generated in conducting sheets and spherical shells by a "varying external magnetic system" (C. Niven 1881, 307). Niven's paper was written as an illustration of how Maxwell's equations could be employed to provide an exact solution to a problem in induced currents by taking into account the new displacement current in the medium surrounding electrified bodies. The introduction to this paper also reveals that the Niven brothers were now offering a specific account of what

---

77. For other examples of this style see Rayleigh, *Scientific Papers*, 1: 272–76 [1876], Hicks 1878, Hill 1879 and 1880, Frost 1880, Niven 1880–81a and 1880–81b, and Gallop 1886.

78. Larmor 1917, xl. Niven developed a method, introduced by Maxwell, of defining the general spherical harmonic in terms of its poles.

they understood as the most significant mathematical and physical peculiarities of Maxwell's theory.[79] Charles opened his paper with the following summary of these peculiarities:

> The energy is supposed to be seated everywhere in the surrounding medium, and the free electricity is the convergence of a vector quantity termed the electric displacement. The total current, to which electromagnetic phenomena are due, is compounded of the current of conduction and the time-variations of the electric displacement. (1881, 308)

It was almost certainly one or more of these assumptions which, as Larmor remarked, readers of Maxwell's work who did not attend W. D. Niven's lectures found extremely difficult to grasp. In the balance of Charles Niven's paper, he reworked the problem of the currents generated in infinite conducting sheets and spherical shells in a more general and more systematic way than had Maxwell in the *Treatise*. The geometry of the conductors (planes and spheres) was chosen deliberately to make the problem soluble using standard mathematical methods, the main difficulty being the solution of Maxwell's equations using appropriate approximations and boundary conditions.[80]

That Charles Niven's application of Maxwell's equations typified the understanding of Maxwell's work in Cambridge in the mid 1870s is underlined by the fact that Horace Lamb independently undertook a remarkably similar investigation at around the same time. Lamb had left Trinity College in 1875 to become Professor of Mathematics at the University of Adelaide, but, even here, he remained in close contact with the Trinity network. One of his two closest friends in Australia was none other than Maxwell's young critic, Edward Nanson, now Professor of Mathematics at the University of Melbourne.[81] In words that echoed Niven's views on Maxwell's theory, Lamb wrote that the role assigned by Maxwell to "dielectric media in the propagation of electromagnetic effects" was so "remarkable" that it seemed worthwhile to "attack some problem in which all the details of the electrical processes could be submitted to calculation." Working in ignorance of Niven's

---

79. The Niven brothers were certainly in close contact on these matters circa 1880, as William warmly thanked Charles in the Preface to the second edition of the *Treatise* for his help in preparing the edition.

80. Edmund Whittaker (2W 1895) pointed out that the geometry of the conductors selected in early investigations of Maxwell's work was chosen mainly to facilitate ease of calculation (Whittaker 1910a, 344).

81. Glazebrook 1932–35, 378.

research, Lamb too tackled the problem of the currents generated in a conducting sphere "by electric or magnetic operations outside it" (Lamb 1883, 519). Lamb reproduced several of the results already derived by Niven— though by a different analytical route—and also drew attention to an effect that Niven had missed. Lamb was able to show that for "any ordinary metallic conductor" upon which existed a "non-uniform electrification of the surface," the "currents by which the redistribution of the superficial electrification [was] effected [were] confined to a very thin film" (547). This phenomenon, subsequently known as the "skin effect," was also predicted by action-at-a-distance theories of electromagnetism, but Lamb was the first to show how it was correctly calculated using Maxwell's equations.[82] Calculations of this kind lent additional support to Maxwell's work by showing that, despite its profound conceptual novelty, field theory readily yielded even the more subtle effects predicted by its Continental rivals.

The above papers provide considerable insight into the way the first generation of Cambridge Maxwellians understood the *Treatise* in the late 1870s. They were relatively little concerned either with the dynamical foundations of Maxwell's theory or with such novelties as the electromagnetic theory of light. It was the direct application of the new equations to problems in electrostatics and, especially, electrical currents that they identified as the obvious avenue of research. The latter problems were in the main soluble by both field and action-at-a-distance theories of electromagnetism, but the conceptual and mathematical approaches required by these respective theories were very different. According to action-at-a-distance theory, it will be recalled, electromagnetic phenomena occurred in conductors and dielectrics, and not at all in the space between these material bodies. In field theory, by contrast, electromagnetic effects were due almost entirely to the flow of energy in the ether, conductors acting merely to guide and partially to dissipate this energy. The solution of problems using the field-theoretic approach therefore required very careful consideration of the electromagnetic action and boundary conditions applicable at or near the surface of conductors.[83] In adopting this approach men like Lamb and the Nivens were both displaying a typically Cambridge understanding of research as a problem-solving exercise and following Maxwell's lead concerning the kinds of problem that should be tackled. The derivation of the general equations of

---

82. The effect was also found by Oliver Heaviside in 1885. See Nahin 1988, 142–43.
83. The field-theoretic and action-at-a-distance approaches are contrasted in Buchwald 1994, 333–39.

the electromagnetic field in the second volume of the *Treatise* was followed by a chapter on "current-sheets" in which Maxwell provided piecemeal solutions to the problems subsequently solved more systematically starting with Maxwell's equations by the first generation of Cambridge Maxwellians.[84] Finding the form of these currents for various initial conditions and conductor geometries was exactly the kind of problem-solving exercise that wranglers were ideally equipped to handle. In generating these solutions, Niven and Lamb very often drew heavily upon the methods of physical analogy and mathematical homology mentioned above.

Charles Niven, for example, worked simultaneously on problems in electrical conduction and the diffusion of heat in solids, and drew attention to the mathematically similar expressions obtained in both areas.[85] Lamb's work too relied on the application of mathematical similarities between the different physical problems in which he was interested. One reason Lamb's important paper on conduction appeared almost three years after Niven's paper was that in attempting to solve Maxwell's equations for the case of induced currents in conducting planes and spheres, Lamb developed "a certain system of formulae" (1883, 519) which he found to be applicable to problems in elasticity and hydrodynamics. He accordingly wrote two papers on these topics before returning to his research on Maxwell's equations. Furthermore, in working out the problems in elasticity and hydrodynamics he also ascertained the reliability of various approximations which he subsequently employed in solving Maxwell's equations for the case of the spherical conductor. This homological application of mathematical methods is further illustrated in a second paper by Lamb in which he extended his analysis of induced currents to include cylindrical conductors. In this paper, part of which had been anticipated by Rayleigh in 1882, Lamb drew on several sections of the *Theory of Sound*, especially those in which Rayleigh had shown how the vibrations of a membrane and of a gas inside spherical cavity could be expressed in terms of Bessel's functions.[86]

This focus on the solution of specific problems rather than the general mathematical properties of Maxwell's equations helps to explain a further peculiarity of Niven and Lamb's research on oscillatory discharge that is of considerable relevance to the next section. It perhaps seems strange, with

---

84. *Treatise*, vol. 2, chap. 12.
85. C. Niven 1881, 308; Macdonald 1923, xvii.
86. Lamb 1883, 525; Glazebrook 1932–35, 382; Lamb 1884, 140, 142, 147; Rayleigh, *Scientific Papers*, 2:128–29.

the benefit of hindsight, that neither of these extremely able mathematicians considered or stumbled upon the fact that some solutions to the very equations they were investigating predicted that rapidly oscillating currents would generate electromagnetic waves. The experimental detection of these waves some six years later by the German experimenter, Heinrich Hertz, would be heralded in Britain as providing spectacular confirmation of Maxwell's theory.[87] There are however two very good reasons why Cambridge men were unlikely even to have considered this possibility before 1883. First, although in the *Treatise* Maxwell famously claimed that light was an electromagnetic wave in the ether, he did not discuss the existence of such waves beyond the visible spectrum and certainly did not suggest that electromagnetic waves of any frequency could be generated by electrical currents. Niven and Lamb had no reason to suppose a priori that the production of such waves was a physical possibility and did not therefore seek such solutions to their equations.[88] Second, the solutions that these men *did* actively seek were precisely those for which the oscillating conduction and displacement currents were the most long lived—hoping that they would be detectable—and this led them, unwittingly, to make approximations that ignored the loss of electromagnetic energy by radiative effects.[89] A serendipitous discovery of the solutions predicting radiation was thereby rendered highly improbable.

The work of the first generation of Cambridge Maxwellians raises several issues concerning the Cambridge reception of Maxwell's *Treatise*. Bearing in mind that Lamb and the Niven brothers were unique among their Cambridge peers in contributing to the mathematical development of Maxwell's new electromagnetic theory, we are bound to wonder why all three of these men were closely associated with Trinity College. It is not possible to give a definitive answer to this question, but the circumstantial evidence provides at least a partial explanation. I noted above that there was an unusual interest in Maxwell's work on electromagnetism at Trinity in the second half of the 1860s and that members of the college (including Maxwell) were in the forefront of the reforms that culminated in the reintroduction of electro-

---

87. On the British reception of Hertz's work, see Hunt 1991, 158–62, and Buchwald 1994, 333–39.

88. W. M. Hicks claimed in 1910 that he had tried to detect "the velocity of propagation of electromagnetic waves" at the Cavendish Laboratory in 1874. What he had actually tried to measure was the finite rate of propagation of electromagnetic induction, probably using an experiment designed for the same purpose by Michael Faraday. See S. P. Thomson 1910, 19, and Simpson 1966, 425–26.

89. A similar point is made in Buchwald 1994, 335.

magnetic theory to undergraduate studies in 1873. The intercollegiate lectures in electricity and magnetism offered first by Stuart and then by W. D. Niven also provide evidence of Trinity's commitment to teaching these most important of the new physical subjects.[90] The close personal relationship that developed between Maxwell and W. D. Niven from the time of Niven's return to Cambridge may also have been a significant factor. Niven was a great enthusiast for Maxwell's work and provided a common link between his brother Charles and his fellow lecturer at Trinity, Horace Lamb. The importance of this link may well be reflected in the notable similarities between the studies undertaken by Charles Niven and Lamb in the late 1870s. As we have seen, both men saw the application of Maxwell's equations to the exact solution of problems in current flow as the obvious way of illustrating and advancing the power of field theory.

What is more certain is that W. D. Niven's lectures passed on his interpretation of Maxwell's theory to the next generation of Cambridge Maxwellians. And, if we make the reasonable assumption that this interpretation was the one embodied in the papers discussed in this section, we can also draw the following conclusion regarding the collective understanding of Maxwell's work prevalent in Cambridge circa 1880. The interpretation of field theory imparted to students was something akin to the summary (quoted above) given at the beginning of Charles Niven's paper on oscillatory discharge. The most significant peculiarity of this account is that the important physical concepts were referred directly to mathematical variables in Maxwell's equations. Thus electric charge is defined as the "convergence of a vector quantity termed the electric displacement," while electric displacement itself is defined only in terms of the electromagnetic effects due to its "time-variations." These examples show how Cambridge students were trained to conceptualize such quantities as "displacement" in mathematical rather than physical terms, and why they would have understood the general equations of the electromagnetic field as the ultimate foundation of Maxwell's theory. It is a corollary of this conclusion that while Niven's students were certainly aware of the physical significance of the displacement current in Maxwell's work, they recognized its key properties in mathematical rather than physical terms. For these students the advancement of Maxwell's field theory depended not on the clarification of its physical or dynamical foundations but on the application of the field equations to the

---

90. On the wider role of Trinity College in the reform of science teaching in mid-Victorian Cambridge, see Geison 1978, 94–115.

solution of new problems, especially those whose solutions could be verified experimentally. The account of Maxwell's theory that emerges from the work of the first generation of Cambridge Maxwellians thus accords very well with that inferred in the previous section from W. D. Niven's editorial work of the second edition of the *Treatise*.

6.7. The Second Generation of Cambridge Maxwellians

The important difference from a pedagogical perspective between the first and second generations of Cambridge Maxwellians lies in the form in which they were introduced to Maxwell's work. Where members of the first generation investigated the *Treatise* for themselves, having at each stage to decide which aspects of the book were the most important and innovative, the second generation—the principal members of which were J. H. Poynting, Richard Glazebrook, Joseph Larmor, and J. J. Thomson—received a systematic introduction to Maxwell's work in the coaching room and in W. D. Niven's lectures. I have already discussed several ways in which Routh's and Niven's teaching shaped the second generation's understanding of the *Treatise*'s contents, but two additional features are worthy of special note in connection with some of the research the younger men subsequently undertook. First, the second generation acquired from Niven's lectures a clear sense that it was Maxwell's equations, as set out in Part IV of the *Treatise*, that constituted the truly original content of the book. It was therefore upon these equations and their consequences that the research effort of the second generation was focused. Second, the informed and critical introduction to the *Treatise* experienced by Niven's students would have given them a definite sense not only of the mathematical principles of Maxwell's theory but also of some of its limitations.

The latter sense would have been further heightened by the contrast between the different readings of the *Treatise* offered respectively by Routh and Niven. Most undergraduates were introduced to electromagnetism by Routh and so assumed an action-at-a-distance theory of electrostatics and, quite probably, a fluid-flow theory of electrical conduction. During Niven's lectures these same students would have been told that "free electricity" was not an imponderable fluid but the "convergence of a vector quantity," and this would at once have raised the question of how electrical conduction was explained in Maxwell's scheme. But, as we have seen, the *Treatise* contained only a sketchy and incomplete account of conduction according to the principles of field theory.[91] Furthermore, although Maxwell gave expressions in the *Treatise* for the total electric and magnetic energy stored at each point in space due to charged and conducting bodies, he offered no account of how this energy moved around during conduction. We cannot now tell whether these issues were discussed explicitly in Niven's classroom, but Poynting was prepared to claim in 1883 that if one started with Maxwell's theory one was led "naturally" to pose the problem: "How does the energy about an electric current pass from point to point—that is, by what paths and according to what law does it travel from the part of the circuit where it is first recognisable as electric and magnetic to the parts where it is changed into heat or other forms?" This was an issue that might well have troubled students who had studied with Routh and Niven, but, as Poynting also noted, prior to his work in 1883 there existed no "distinct theory" of energy flow, most students of the *Treatise* holding the "somewhat vague opinion that in some way [the energy had] been carried along the conductor by the current" (Poynting 1920, 192).

In this section I examine the way Cambridge pedagogy helped to shape the early contributions to electromagnetic theory made by Poynting and J. J. Thomson, the two most prominent Cambridge Maxwellians of the early 1880s. The part played by pedagogy in each case is also informatively different. As we shall see, Thomson drew systematically upon the resources of all three sites at which electromagnetism was taught in Cambridge in order to fashion himself as the university's leading expert on Maxwell's theory. Poynting on the other hand devoted much of his early career to the accurate experimental determination of the mean density of the earth. His route to research in electromagnetic theory was far less calculated than was Thomson's. At the time Poynting derived the eponymous energy-flow theorem he was working in relative intellectual isolation in Birmingham. I shall argue that Poynting's sudden and dramatic intervention in the development of field theory was nevertheless a product of his undergraduate training in Cambridge and of a subsequent period as a research fellow at Trinity College.

J. J. Thomson (2W 1880) was in many respects the ideal type of a high wrangler of the late 1870s. Prior to arriving in Cambridge he studied mathematics, engineering, and physics at Owens College, Manchester, where he was taught by no less than three Cambridge-trained mathematicians. Unlike students prepared for Cambridge entrance in the public schools, Thomson

91. The difficulty of understanding conduction according to field theory is discussed in Buchwald 1985, 28–34.

also took courses in experimental physics with Balfour Stewart at Owens College and even attended a course of introductory lectures on Maxwell's *Treatise* given by Arthur Schuster in 1876.[92] As a Cambridge undergraduate, Thomson was, as we have seen, coached by Routh, and took Niven's intercollegiate courses on electricity and magnetism. Immediately after graduation, Thomson embarked simultaneously upon three distinct research projects, each of which drew heavily upon the resources of one or more of the main sites at which physics was taught in Cambridge. In order to display these pedagogical connections I shall give an overview of Thomson's early work, beginning with his experimental investigations at the Cavendish Laboratory.

Working under Rayleigh's direction, Thomson's first experiments in Cambridge were aimed at detecting the magnetic effects which, according to Maxwell, were generated by a displacement current in a dielectric.[93] But, unable to produce "sufficiently definite" results, Thomson was encouraged by Rayleigh to pursue more mundane experiments on the electrostatic capacity of induction coils. Once this work was satisfactorily completed, Rayleigh further encouraged Thomson to undertake some far more demanding and protracted research on the determination of the ratio ("v") between the electrostatic and electromagnetic units of electric charge.[94] Maxwell's theory required v to be numerically equal to the velocity of light and, as I noted above, the establishment of this identity with the limits of experimental accuracy was important to the theory's credibility. Using an experimental arrangement based closely on a design given by Maxwell in the *Treatise*, Thomson succeeded in producing a value for v that was the nearest to date to the accepted value of the velocity of light.[95] The completion of these experiments early in 1883 marked the end of Thomson's first period of work at the Cavendish.

---

92. Thomson 1936, 12–33; Thomson et al. 1910, 76–78. Thomson was taught elementary mathematics by A. T. Bentley (40W 1864), advanced mathematics by Thomas Barker (SW 1862), and engineering by Osborne Reynolds (7W 1867). Thomson was "raised to such a pitch of enthusiasm" by Schuster's lectures on the *Treatise* that he copied out the whole of Maxwell's "On Physical Lines of Force" in longhand. See Thomson et al. 1931, 34.

93. Thomson had visited the Laboratory "now and then" as an undergraduate to visit Schuster and Poynting, both of whom were undertaking experiments there in the late 1870s.

94. Rayleigh probably suggested this work in the light of his recent redetermination of the standard of the unit of electrical resistance (upon which the calculation of v depends). Thomson and W. D. Niven had both contributed to this project. See Rayleigh, *Scientific Papers*, 2:13–14, 44.

95. Thomson 1881b and 1884a. This work is discussed in Thomson 1936, 97, and Rayleigh 1942, 17–18.

The significance of this work to our present concerns is twofold. First, working in close collaboration with Rayleigh, Thomson had used the resources of the Cavendish to master many of the instruments and practical operations that were crucial to the enterprise of exact measurement in experimental electricity. Acquiring these skills not only enabled him to deal more confidently with those passages in the *Treatise* that assumed a knowledge of practical electricity but also established him at the heart of the Cavendish's long-standing project to support Maxwell's theory via electrical metrology. Thomson was in fact the first person at the Laboratory directly to measure the value of v, an accomplishment that certainly played a role in his appointment as Rayleigh's successor at the Cavendish in 1884.[96] Second, it is noteworthy that it was Thomson, a student of Niven, who was the first Cambridge experimenter to try to detect a novel physical phenomenon predicted in the *Treatise*. As we have seen, displacement currents and their effect on electrical discharge were quite familiar to W. D. Niven's students, and it was in late January 1880, just as Thomson was about to begin his experiments, that Charles Niven presented his paper on electrical discharge to the Royal Society. It will be recalled that Niven's paper began by drawing attention to the novel role of displacement currents in Maxwell's theory, and Thomson's first investigation was very likely an attempt indirectly to detect these currents experimentally. In this case it was the Cambridge mastery of Maxwell's equations that prompted a particular line of experiment, but, as we shall shortly see, the same mastery would shortly inhibit a potentially more fruitful line of investigation.

Thomson's first two research projects in mathematical physics drew heavily upon his training with Routh and Niven. The most ambitious of these concerned the application of Lagrangian dynamics to a range of problems in physics and chemistry, a project he had begun to formulate while still a student at Owens College.[97] Despite its Mancunian origins, the mathematical methods that brought this work to fruition were taken partly from Maxwell's *Treatise* and mainly from Routh's Adams-Prize essay of 1876 in which he introduced the "modified Lagrangian function."[98] On the strength of this highly mathematical work, Thomson was elected to a fellowship at Trinity College in the summer of 1880, his immediate future in Cambridge now being assured. The other mathematical project tackled by Thomson in

---

96. Rayleigh 1942, 20, 415.
97. Thomson 1936, 21, 75–77. The technical details of this project are discussed in Topper 1971.
98. Routh 1877; Topper 1971, 400–4. Thomson's dissertation was published in expanded form as Thomson 1888 and contains numerous references to Maxwell and Routh's work.

the early months of 1880 concerned the electromagnetic theory of light. The main purpose of Thomson's contribution was to generalize Maxwell's equations of electromagnetic wave propagation in order to derive two new expressions that could be compared either with established results from optical theory or directly with experimental data. The first derivation concerned the reflection and refraction of a ray of light at the surface of a transparent medium, the very problem that Maxwell had tried and failed to solve in the 1860s.[99] But where Maxwell had attempted to develop a complicated dynamical account of the interaction of electromagnetic waves with transparent media, Thomson assumed only that the rapidly oscillating electromagnetic fields of a light ray were subject to the same boundary conditions at the surface of a dielectric as were static electric and magnetic fields. This assumption enabled Thomson to calculate the relative intensities of the reflected and refracted portions of the ray, and to show that the results he obtained were consistent with those derived from other optical theories. The second derivation dealt with the convection of light as it passed through a moving transparent medium. Thomson showed that during this process the velocity of the ray would be increased by half the velocity of the medium, a result that was consistent with the limited experimental data then available.[100]

Thomson's paper constituted the first attempt by a Cambridge man to develop the electromagnetic theory of light, and, at the time of writing, Thomson believed his paper to the first of its kind anywhere in the world.[101] Despite the novelty of its subject matter, Thomson's investigation shows two connections with the teaching and research of the first generation of Cambridge Maxwellians. First, it displays great familiarity with the *Treatise*'s contents. Although the boundary conditions employed by Thomson are stated in the book, they are given in separate chapters and in connection only with electrostatic and magnetostatic fields (nor are they flagged as especially important results).[102] The suggestion that these conditions might also apply to rapidly oscillating fields at the surface of transparent media is exactly the kind of information that a member of the first generation might have passed on to a keen student like Thomson.[103] Second, like Niven and Lamb's work

99. Thomson 1880, 288–90.
100. Ibid., 290–91.
101. Ibid., 284. Unaware of Lorentz 1934–39, 1:1–192 [1875], and FitzGerald 1902, 45–73 [1880], Thomson wrote that Maxwell's papers "appear to constitute the literature on the subject."
102. *Treatise*, arts. 83, 427.
103. There is no clear discussion in the *Treatise* of the boundary conditions to be satisfied at the interface between dielectrics or at the surface of conductors. Niven (1881) and Lamb (1883) both discussed these conditions for the case of conductors.

on electrical discharge, Thomson's analysis of the interaction between light and transparent dielectrics proceeded more by the mathematical manipulation of Maxwell's equations and the application of specific boundary conditions than by consideration of the dynamical interaction between ether and matter. It was in fact precisely by ignoring such mechanical imagery in Maxwell's work that Thomson was able to make rapid progress with the electromagnetic theory of optical reflection and refraction.

In the summer of 1880, having secured his fellowship at Trinity College, Thomson embarked on a second and ultimately more important study of electromagnetic theory (Thomson 1881a). Prompted by some recently published experiments on cathode rays in which it was assumed that the rays were tiny charged particles moving with very high velocity, Thomson used Maxwell's equations to investigate the magnetic field generated by the motion of a charged sphere and the force that would act on the sphere when it moved through an external magnetic field. An especially important result derived by Thomson in this paper was an expression for the additional mass that a body would appear to have when charged (1881a, 234). This extra mass originates in the energy that has to be supplied to an accelerating charged body to generate the magnetic field produced by the moving charge. For our present purposes, however, it is the style of Thomson's analysis that is of most relevance, a style that is succinctly illustrated by an analytical step in the first derivation in the paper. A moving charged conductor will, according to Maxwell's theory, produce a changing electrical displacement at each point in the medium surrounding the conductor. This change constitutes a displacement current, and Thomson began his paper by calculating the total magnetic vector potential generated at each point in space by the current. Having derived the required expression, he noticed that it did not satisfy a mathematical condition generally deployed in Maxwell's theory (1881a, 232).[104] Thomson solved this problem by simply adding a term to one component of the vector so that the condition was satisfied. He then proceeded with his analysis without pausing to consider the physical meaning or implications of the term he had added.

This peculiarity of style was pointed out a few months later by the Irish physicist, George Francis FitzGerald, who had been making a careful and critical study of the *Treatise* since the mid 1870s.[105] In a short response to

---

104. Thomson, following Maxwell, required the divergence of the magnetic vector potential he had calculated to be zero (which it was not). This point is discussed in Buchwald 1985, 270.

105. On FitzGerald's early career, see Hunt 1991, chap. 1. Thomson's derivation and FitzGerald's response is discussed in Buchwald 1985, 269–76, and Hunt 1991, 187–88.

Thomson's paper, FitzGerald (1902, 102–7) offered a physical interpretation of Thomson's mathematical analysis and showed what physical assumptions had to be made in order consistently to derive an expression for the magnetic vector potential that satisfied the required condition. The fact that Thomson had either failed to consider these issues or believed them to be unimportant points yet again to the primacy of Maxwell's equations in Thomson's work. Thomson's goal was to tackle a problem that seemed to be straightforwardly soluble (if mathematically complicated) by the direct application of Maxwell's equations, his main guide throughout the calculation being the requirement that Maxwell's equations remain satisfied at each step in the analysis. When this requirement broke down, Thomson clearly thought it appropriate to proceed by making the simplest possible mathematical adjustment to the troublesome expression such that Maxwell's equations would be satisfied. This very literal interpretation and application of Maxwell's equations was quite consistent both with the problem-solving training Thomson had received as an undergraduate and with the application of Maxwell's equations made by the first generation of Cambridge Maxwellians.

This interpretation was also one that permeated much of Thomson's early work on Maxwell's theory and which was passed on to the next generation of wranglers via Tripos problems. An excellent illustration of these points is afforded by Thomson's 1880 derivation of the convection of light passing through a moving transparent medium (mentioned above). To make the calculation Thomson assumed (i) that the electric field of a light ray would polarize a moving dielectric, (ii) that the moving polarization constituted an electric current and generated an additional potential, and (iii) that the velocity of the light ray would depend on the total potential. The calculation of the light velocity inside the moving dielectric was then obtained by the analytically complicated but otherwise straightforward manipulation of Maxwell's equations. To FitzGerald, who relied more heavily upon physical reasoning, Thomson's result appeared implausible because, if generalized, it seemed to lead to paradoxical conclusions. Despite the fact that he could find no specific faults in Thomson's calculations, FitzGerald concluded that they must be based on an incorrect interpretation of Maxwell's theory.[106] In Cambridge, by contrast, this and other results obtained by Thomson were believed to follow quite straightforwardly from Maxwell's equations and were soon being set as advanced Tripos problems. Thus in 1886, candidates

---

106. These conclusions are briefly discussed in Hunt 1991, 203–4.

for Part III of the Mathematical Tripos were required to reproduce not only Thomson's derivation of the drag coefficient but a related effect associated with the rotation of a dielectric (fig. 6.7).[107]

One final example of the relationship between Thomson's early research and that of the first generation of Cambridge Maxwellians occurs in an important paper he wrote in 1884 on the propagation of electromagnetic waves. In order to grasp the significance of this paper we must briefly turn again to the work of FitzGerald. In 1879, FitzGerald heard the young experimenter, Oliver Lodge, make the striking suggestion that electromagnetic waves were generated by rapidly alternating currents and that it might therefore be possible to generate them artificially. Like most other readers of the *Treatise*, FitzGerald had not considered this possibility, and his calculations soon convinced him that Maxwell's theory ruled out the production of such waves.[108] FitzGerald was led to this mistaken conclusion by taking some of Maxwell's statements in the *Treatise* too literally, and, in 1882, a passage in Rayleigh's *Theory of Sound* led him to tackle the problem afresh. In his book, Rayleigh had discussed an equation analogous to the one derived by FitzGerald for the electromagnetic disturbances surrounding an oscillating current, but, unlike FitzGerald, Rayleigh showed that some solutions to the equation represented the propagation of energy in the form of progressive waves. Having reworked his calculations in the light of Rayleigh's findings, FitzGerald concluded that changing electric currents necessarily radiated electromagnetic energy and that, contrary to his earlier pronouncement, electromagnetic waves must be ubiquitous.[109] These far-reaching claims, together with a proposed method for the practical generation of electromagnetic radiation, were made public by FitzGerald at the British Association meeting in September 1883 and published shortly thereafter.[110]

It was almost certainly FitzGerald's announcement that prompted Thomson to take up the theoretical question of the generation, propagation, and detection of electromagnetic waves. Prior to the British Association meeting, no one in Cambridge had considered such matters, and it is almost inconceivable that Thomson would not very soon have heard or read about

---

107. The question was not set by Thomson (who was not an examiner that year) but by J. W. L. Glaisher (2W 1871).

108. Hunt 1991, 33–34.

109. The main sources for FitzGerald's calculation are *Treatise*, vol. 2, chap. 11, and *Theory of Sound*, vol. 2, arts. 276–80.

110. FitzGerald 1902, 128–29 (13 October 1883).

> MATHEMATICAL TRIPOS. PART III. 179
>
> MONDAY, *January* 4, 1886. 9—12.
>
> GROUP D.
>
> 1. SHEW that, if a bar of isotropic material expands under tension, it cools during the process unless heat is supplied from the outside, and *vice versâ*.
>
> Shew also that, if we take into account changes of temperature, the velocity of propagation of sound along the rod is, in both cases, less than if these are neglected, and the diminution is proportional to the absolute temperature of the rod.
>
> 4. In any electromagnetic field establish the equations
>
> $$P = cq - br - \frac{dF}{dt} - \frac{d\psi}{dx},$$
>
> $$Q = ar - cp - \frac{dG}{dt} - \frac{d\psi}{dy},$$
>
> $$R = bp - aq - \frac{dH}{dt} - \frac{d\psi}{dz},$$
>
> where P, Q, R are the components of the electromotive force, a, b, c those of magnetic induction, F, G, H of the vector potential, p, q, r of the velocity of the medium, and ψ is the electric potential.
>
> On the assumptions of the electro-magnetic theory of light, prove the following results, in the case of a beam of plane polarized light travelling parallel to the axis of x.
>
> (i) If the medium is moving parallel to the axis of x with uniform velocity, the velocity of propagation of the light is increased or diminished by half this velocity, according as the medium and the light are moving in the same or opposite directions.
>
> (ii) If the medium is rotating uniformly round the axis of x, the plane of polarization turns approximately as if it were rigidly connected with the medium.

FIGURE 6.7. J. W. L. Glaisher's (2W 1871) 1886 Tripos question (question 4) based on Thomson's new results in the electromagnetic theory of light. The bookwork is taken from Article 598 of the second volume of Maxwell's *Treatise*. The two riders, (i) and (ii), are drawn respectively from Thomson (1880) and Thomson (1885). The equations for P, Q, R derived in the bookwork were Thomson's starting point for results required in the riders. Tripos problems thus linked standard derivations in the *Treatise* to important new results in electromagnetic theory. I have included question 1 to give a flavor of the other subjects covered on this paper. Cambridge University Examination Papers, Mathematical Tripos, Part III, Monday, 4 January 1886, 9–12 (Group D). (By permission of the Syndics of Cambridge University Library.)

FitzGerald's paper.[111] Thomson would at once have realized that the generation and detection of electromagnetic radiation would constitute a new and important test of Maxwell's theory. Moreover, like Lodge and Fitz-

---

111. Thomson probably attended the meeting, as he was elected an annual member of the Association in 1883 and invited to prepare a "Report on Electrical Theories." Rayleigh certainly attended and would surely have mentioned FitzGerald's paper to his Cambridge colleagues. *British Association Report*, 1883, lxvi.

Gerald, Thomson would soon have worked out that the only practicable way of demonstrating the existence of these tiny oscillating fields would be by generating standing waves in a suitable detector. In order to find out how much radiation would be emitted by a practically realizable oscillating electrical system, Thomson investigated the problem from a theoretical perspective. Within a few months of FitzGerald's announcement, Thomson had calculated exactly how much energy would be radiated by currents oscillating on the surface of a hollow conducting sphere.

The production of surface currents by external electromagnetic disturbances was, as we have seen, a problem that had already been investigated in some detail by the first generation of Cambridge Maxwellians. Thomson drew heavily upon these earlier investigations in formulating the problem, but, unlike his older peers, he focused on solutions to the general equations that represented cases in which the initial potential distribution over the surface of the sphere altered (typically vanished) extremely rapidly (Thomson 1884b, 201).[112] He showed that in these circumstances a significant fraction of the energy of the induced surface currents would be propagated into the surrounding medium. At the end of a series of long and complicated analytical manipulations Thomson arrived at two important results. He first showed that if the initiating source was inside the sphere, the standing oscillations set up would be wholly confined within the sphere and would, at least in principle, be sufficiently powerful and long-lived to be detectable. Thomson acknowledged that the rapidly oscillating nature of these effects would make them hard to detect, but concluded that their existence nevertheless offered "a promising way of testing Maxwell's theory" (1884b, 209). It is not known whether Thomson actually tried to detect these effects, but even if he had detected them it would not have constituted an especially impressive verification of Maxwell's theory.[113] The energy due to the standing waves inside the sphere was not propagated over a significant distance and would, in practice, have been very hard to distinguish from well-known inductive effects.

---

112. FitzGerald's original paper (1902, 122–26) analyzed radiation from a current loop. Thomson followed his Cambridge peers in choosing the mathematically convenient geometry of a sphere, even though it was a very poor radiator of electromagnetic energy. The sphere offers a convenient geometry in these calculations because in spherical polar coordinates the radial distance from the origin 'r' can often be treated as a constant (thus reducing the equations to two variables).

113. A letter from FitzGerald to Thomson a few months later reveals that Thomson was searching for a device capable of detecting these standing waves. See Rayleigh 1942, 22.

Thomson's second result was far more important. Here he showed that the electromagnetic energy due to oscillating surface currents initiated by a disturbance *outside* the sphere would "be dissipated into space in the course of a few vibrations, so that we could not expect any measurable result in this case" (1884b, 210). In other words, the rapid discharge of a large conducting sphere *would* emit electromagnetic radiation, but in the form of a brief pulse rather than a coherent train of waves. This piece of theoretical work goes a considerable way towards explaining why no one at the Cavendish Laboratory sought to generate and detect electromagnetic waves in the mid 1880s. Thomson's early papers had established him in Cambridge as the leading exponent of Maxwell's new electromagnetic theory: it was Thomson who was invited in 1883 to prepare a "Report on Electrical Theories" for the British Association; he who replaced Niven as a college lecturer at Trinity in 1882; and he who was appointed in June 1884 to one of the five new University Lectureships in Cambridge with special responsibility for the teaching of electromagnetic theory.[114] Six months later, Thomson was elected Rayleigh's successor as Professor of Experimental Physics and Director of the Cavendish Laboratory. But, as we have just seen, less than a year before Thomson took up the latter appointment, he had demonstrated to his own satisfaction that practical arrangements of spherical conductors could not radiate energy into space with sufficient quantity, coherence, or duration to render it detectable.[115]

Thomson's early career provides an excellent illustration of the way a student could draw upon the pedagogical resources available in late Victorian Cambridge to try to establish himself as a noteworthy researcher. The very speed with which he produced his first papers on Lagrangian analysis and Maxwell's electromagnetic theory is in itself an indication of the profound debt he owned to the teaching of Routh and Niven respectively. Where men like Heaviside and FitzGerald had taken years of private study to obtain a sufficient grasp of Maxwell's theory to begin offering their own original contributions, Thomson wrote two important papers on electromagnetic theory within months of graduating while working simultaneously on several other projects. His work with Rayleigh at the Cavendish Laboratory also helped to equip Thomson with that rare combination of mathematical and

---

114. Rayleigh 1942, 14; *Cambridge University Reporter*, 3 June 1884, 791. The establishment of these lectureships was discussed in chapter 5.

115. Thomson continued to try to devise experiments to detect the existence of displacement currents in dielectrics. See Thomson et al. 1910, 135.

experimental skills peculiar to Maxwell himself. Moreover, despite the novelty of the conclusions reached in his early research papers, the style of Thomson's analysis clearly displayed the hallmarks of his problem-based training and the research undertaken by the first generation of Cambridge Maxwellians.

A final aspect of Thomson's early career that is worthy of comment is why he, more than any other member of Niven's class, chose to devote his early research efforts to the investigation of novel aspects of Maxwell's theory. Thomson's rare enthusiasm for Maxwell's work can partly be traced to his early encounter with the *Treatise* at Owen's College, but the confidence and technical skill with which he attacked specific projects probably owes more to his close personal acquaintance with W. D. Niven. We have already seen that Thomson's careful reading and understanding of the *Treatise* owed a good deal to Niven's lectures, but Niven also took a close personal interest in the young Thomson's career. Their relationship was in fact so close that in a referee's report to the Royal Society on one of Thomson's early papers, Niven felt obliged to inform the Secretary not only that Thomson was a "great friend" but that of all the pupils who had "passed through [his] hands at Cambridge," Thomson was "the one in whom [he took] the greatest interest."[116] Thomson for his part described Niven as one of the "kindest friends [he] ever had" and recalled how from his freshman days at Trinity College he "went very often to [Niven's] rooms" and regularly accompanied him on long constitutional walks.[117] Thomson does not record the precise topics of their conversations during these frequent meetings, but it is hard to believe that they did not regularly discuss their common interest in Maxwell's work and the ways in which it might be developed. As Niven remarked in the report quoted above, he had "a good deal of experience of Mr Thomson's work" and it might well be that Niven acted as a sounding board for Thomson's ideas and offered regular advice and encouragement based on his extremely thorough knowledge of Maxwell's *Treatise*.

The origin and development of J. H. Poynting's work on the flow of energy in an electromagnetic field shows similarly intriguing connections both with Cambridge pedagogy and with the analogical use of techniques from Rayleigh's *Theory of Sound*. As I noted in chapter 1, Poynting entirely effaced these connections from his published account, implying instead that the energy-flow theorem had emerged simply by reflection upon problems inher-

---

116. W. D. Niven to George Stokes, 2 October 1883, Royal Society, RR.9.191.

117. Thomson 1936, 43. W. D. Niven played a major role in all of Thomson's professional appointments up to and including his election as Rayleigh's successor.

ent in Maxwell's theory. In order to clarify this point it is important to make the following distinction: the problem of providing a clear and quantitative account of how energy was conveyed around an electric circuit according to Maxwell's theory might well have been one that concerned a number of careful readers of the *Treatise*, but finding a satisfactory solution to the problem was a very different matter. I shall suggest that Poynting was led to his famous theorem by consideration of problems beyond Maxwell's *Treatise* and that his dramatic intervention in electromagnetic theory was made possible by the pedagogical resources upon which he could draw. In order to develop this suggestion I begin with a brief review of Poynting's career as an undergraduate and young researcher. Like Thomson, Poynting studied at Owens College, Manchester, where he too received an excellent grounding in mathematics from Cambridge-trained teachers.[118] Unlike his slightly younger peer, however, Poynting completed his undergraduate studies too early to attend either Schuster's lectures on the *Treatise* in Manchester or those given by Niven in Cambridge. When Poynting graduated in 1876, his knowledge of Maxwell's electromagnetic theory would have been confined almost entirely to what he had learned through coaching with Routh.[119] And, since the *Theory of Sound* did not appear until 1877, Poynting's knowledge of acoustics and vibration would also have come from Routh together with relevant sections of Thomson and Tait's *Natural Philosophy*.

After graduation, Poynting returned to Owens College as a demonstrator in physics in Balfour Stewart's physical laboratory, a position he held until 1878. During this period Poynting showed no special interest in advanced electromagnetic theory but began a major series of extremely delicate experiments designed to provide an accurate determination of the mean density of the earth. In 1878 he was elected to a fellowship at Trinity College, whereupon he returned to Cambridge to continue his experimental research under Maxwell at the Cavendish Laboratory. Poynting remained in Cambridge until 1880, when he was appointed to the new Professorship of Physics at Mason College, Birmingham. This appointment led him to leave Cambridge for good, and, as one of his obituarists recorded, to throw himself "wholeheartedly into the arduous duties connected with the starting of a new University College." Over the next few years, Poynting's energies were absorbed in preparing a large number of lectures and demonstrations, and in establishing a new physical laboratory in Birmingham (Poynting 1920, xvi).

118. Poynting 1920, xv. Poynting too was taught by Thomas Barker (SW 1862).
119. Poynting's first paper (1920, 165–67) was a simplified version of Routh's proof (fig. 6.2) of the inverse-square law of electric attraction.

It was Poynting's work as a teacher at Mason College that seems to have prompted the specific line of investigation that led eventually to his important contribution to electromagnetic theory. Poynting took his duties as a teacher extremely seriously, and he devoted a good deal of effort to contriving clear elementary proofs and demonstrations of basic physical principles to show to his classes. If these pedagogical devices were especially successful, he would sometimes publish them for the benefit of other teachers. In a paper of this kind read to the Birmingham Philosophical Society at the beginning of November 1883, Poynting gave an elementary derivation of the velocity of propagation of sound by consideration of the rate at which longitudinal waves transmit energy through the conducting medium (Poynting 1920, 299–303). The method was based upon the fact that the energy produced in the medium by an advancing plane wave can be calculated in two ways. First, it can be assumed that the sum of the potential and kinetic energies in a given volume immediately behind the wavefront will be transported in a given time into an equivalent, contiguous volume by the advancing wave. Second, the energy delivered to the medium by the wave can be calculated in terms of the work done by the advancing wavefront. By equating the expressions obtained in these calculations, Poynting was able to show how the wave velocity could be found in terms of the physical properties of the medium. In a concluding section of the paper he demonstrated that the same technique could be applied to derive the velocity of transverse waves—the kind of waves of which light was believed to be composed. Although the contents of this paper were relatively elementary, they are extremely important to our present concerns for the following two reasons: first, they reveal that it was the above derivations that piqued Poynting's interest in the relationship between the energy and velocity of waves; and, second, the citations in the paper show that the method employed by Poynting was "merely an application" of a technique developed by Rayleigh in the *Theory of Sound* (1920, 300; cf. *Theory of Sound*, 1:475–80).

The significance of these points lies in the fact that within weeks of completing the paper, Poynting had written his now celebrated work on the flow of energy in the electromagnetic field.[120] As we have seen, the introduction to the latter work implies that Poynting's investigation was prompted wholly by reflection upon difficulties inherent in Maxwell's electromagnetic theory, but a subsequent letter to Rayleigh confirms a quite different course of events.

---

120. Poynting's paper on the energy-flow theorem was received by the Royal Society on 17 December 1883.

Hoping to get his paper published in the prestigious *Philosophical Transactions of the Royal Society,* Poynting sent the finished paper to Rayleigh, asking if he would be prepared to "communicate" it to the society. In the covering letter Poynting also made the following revealing comment on the origins of the enclosed paper: "I had read your note in the Theory of Sound and it was I believe my starting point. I tried to get a simple proof of the velocity of sound and of transverse vibrations by a consideration of the flow of energy *and an attempt to apply the method to Maxwell led to the law for the transfer of electro-magnetic energy.*"[121] These comments confirm that the work upon which Poynting's fame would eventually rest almost certainly emerged in roughly the following way. Having found that his "simple proof" was applicable to the kind of waves (transverse) of which light was believed to be composed, Poynting next considered whether the proof would work if one began with Maxwell's electromagnetic theory of light rather than purely mechanical considerations.[122] In order to explore this possibility he needed both an expression for the total energy associated with a light wave in a given volume of space and an independent expression for the work done by the optical wavefront as it advanced through the medium. The attempt to obtain these expressions quickly led Poynting to the heart of the problem of energy flow in the electromagnetic field.

The first of the required expressions could be obtained fairly straightforwardly from article 792 in the *Treatise,* in which Maxwell showed how to calculate the total energy in a given volume due to the fields associated with an electromagnetic wave. Finding an electromagnetic equivalent of the second expression was rather more difficult. Maxwell's theory posited no electromagnetic analogues to the mechanical concepts used by Poynting in his derivation of the work done by the advancing wavefront. Poynting would therefore have had to try to derive the required expression by an alternative route, and, judging by the published version of his paper, he proceeded as follows. He began with Maxwell's volume integrals representing the total electromagnetic energy in a given region of space. Poynting differentiated this expression to reveal how the energy in the region would alter as the electric and magnetic fields changed with time. He then used several substitu-

---

121. Poynting to Rayleigh, 14 December 1883 (RC-ICL); my italics. Parts of this letter are reproduced in Strutt 1924, but are wrongly dated 14 December 1882.

122. The suggestion that Poynting's derivation of the energy-flow theorem emerged from his work on acoustics is made in Noakes 1992, 17–18. Joseph Larmor was the only contemporary of Poynting to remark upon the mechanical and optical foundations of the theorem. See Poynting 1920, x–xi.

tions (using Maxwell's field equations) and partial integration to split the changing energy into several easily identifiable volume components and one surface component. Having equated the sum of the volume components with the surface component, Poynting (1920, 180) offered the following interpretation of the equation: the volume components represented the change with time of the total electromagnetic energy inside the region, while the surface component represented the flow of energy in or out of the region necessary to preserve a balance (fig. 1.1). It was the integrand of the surface integral—representing the energy flow per second per unit area at a point in space due to the electric and magnetic forces acting at that point—that Poynting had sought in order to complete his derivation of the velocity of light. This was also the expression that would become known as the Poynting vector (1920, 181).[123]

Poynting must quickly have realized that the expression he had derived was of far greater significance than was a mere derivation of the velocity of light. It showed that the energy of an electromagnetic field always flowed in a direction perpendicular to the lines of electric and magnetic force of which the field was composed, a result that Poynting referred to as the "energy-flow theorem." The fact that the mathematical expression of the theorem (the Poynting vector) made it easy to calculate both the direction and magnitude of energy flow in any electromagnetic field also meant that Poynting had a powerful new tool with which to investigate electrical processes that remained obscure in Maxwell's scheme. The single most important clarification of this kind concerned the process of electrical conduction. As Poynting remarked in the introduction to his paper, a conduction current had hitherto been regarded as "something travelling along the conductor" while the associated energy in the space surrounding the conductor was "supposed to be conveyed thither by the conductor through the current" (1920, 175). Yet according to Maxwell's theory, the energy associated with conduction ought, as Charles Niven had remarked, to be "seated everywhere in the surrounding medium" rather than in the wire itself. The Poynting vector showed that this was indeed the case and that the process of conduction was properly interpreted as one in which field energy flowed radially into the wire, where, by an unexplained mechanism, it was converted into heat. Poynting went on to explain several other common electromagnetic processes in a

123. Poynting showed that if $\mathbf{E}$ and $\mathbf{H}$ are vectors representing respectively the electric and magnetic field strengths at a point in space, the direction and magnitude of energy flow ($\mathbf{P}$) at that point is given by their vector product ($\mathbf{P} = \mathbf{E} \times \mathbf{H}$). Poynting did not use vector notation in his original derivation (see fig. 1.1).

similar way, concluding with a derivation of the velocity of light along the lines of his original paper on the velocity of sound waves. Thus the problem from which Poynting's research had begun was tucked away without comment at the end of the paper as a relatively minor corollary of the more general theorem. Equally misleading to the historian trying to piece together the origins of the energy-flow theorem is Poynting's claim in the introduction that it was the theorem that had been the initial object of his inquiry.

My account of the origin of the Poynting vector both explains a number of previously puzzling aspects of Poynting's research and raises a number of new questions concerning his knowledge of Maxwell's *Treatise* and Rayleigh's *Theory of Sound*. The fact that Poynting's published papers failed to mention the relationship between his respective researches in acoustic and electromagnetic waves has led historians to assume that he must have been wrestling for some time with the problem of electromagnetic energy prior to the publication of his paper. Yet this assumption neither explains why Poynting chose to work on this particular problem nor how or why the work came to fruition in mid-December 1883. As we have seen, Poynting did not study the relevant sections of Maxwell or Rayleigh's books as an undergraduate, had specialized in experimental rather than mathematical research after graduation, made no visible contribution whatsoever to electromagnetic field theory prior to 1883, and was occupied from 1880 with physics teaching and the establishment of a new physical laboratory in Birmingham. The course of events outlined above explains why Poynting was exploring the problem of energy flow in the electromagnetic field in the autumn of 1883, but provides no account either of his evident familiarity with Rayleigh's *Theory of Sound* or of his ability to deal so effectively with the most technically demanding and conceptually obscure aspects of Maxwell's *Treatise*. In order to give a plausible account of these competencies we must return to the Tripos examination Poynting sat in 1876 and his time as a research student in Cambridge between 1878 and 1880.

Although the *Theory of Sound* did not appear until after Poynting had graduated, he was introduced as an undergraduate to some of the most important problems that Rayleigh was currently trying to solve. It will be recalled that Rayleigh was the "additional examiner" for the Tripos of 1876 and that he was at this time in the process of completing his book. When he began setting problems towards the end of 1875, he naturally turned to his current research for inspiration. Several of his questions were based on acoustics and the theory of vibration, and one of the most difficult required the candidates to derive an expression for the energy radiated into the atmosphere by a point source of sound.[124] Rayleigh reproduced this original derivation in chapter 14 of the *Theory of Sound* (noting its announcement in the Tripos of 1876), the very chapter that would shortly lead FitzGerald to derive an analogous expression for the energy emitted by an oscillating electrical current. In attempting to solve Rayleigh's problems in 1876, Poynting was therefore struggling to reproduce techniques and results that were not only central to Rayleigh's approach to vibration theory but which would shortly bridge the gulf between acoustic and electromagnetic radiation.

What is equally noteworthy is that at the end of the chapter in which Rayleigh provided the solution to the Tripos problem just mentioned, he offered a method for deriving a relationship between the work done by sources of sound in a given volume of a vibrating medium and the energy transmitted by the medium across the volume's surface. Even more striking is that fact that the starting point and mathematical steps in the derivation are precisely the ones employed by Poynting in arriving at his energy-flow theorem.[125] It is also likely that Poynting had studied this passage. We know that he had read at least parts of the *Theory of Sound* by 1883, but there is reason to suppose that he began studying the book well before that. When Poynting returned to Cambridge in 1878, Rayleigh's work was being widely studied and discussed in the university, not least because it had just become required reading for aspiring wranglers. Given the status of problems on the annual Tripos papers, Poynting might well have obtained a copy in 1877 in order to study Rayleigh's solutions to the questions that had helped to decide his fate in the Senate House. An enquiry of this kind would have led Poynting directly to the crucial passages in the book.

Rayleigh's Tripos problems also confirm that the prospective wranglers of 1876 were not expected to have any knowledge of Maxwell's field-theoretic approach to electromagnetism. Poynting's knowledge of this subject almost certainly derived from Niven's intercollegiate lectures at Trinity College. It will be recalled that although Poynting returned to Cambridge in 1878 to work with Maxwell at the Cavendish Laboratory, he was also a fellow of Trinity College and friend of J. J. Thomson (whom he had taught in Manchester). At the Cavendish he would also have been reacquainted with Richard Glazebrook, who had been an undergraduate contemporary of Poynting at Trin-

---

124. Rayleigh, *Scientific Papers*, 1:280, question viii.
125. Rayleigh, *Theory of Sound*, vol. 2, art. 295. The analogical origin of Pointing's derivation is also supported by Thomson's contemporary observation that the theorem was "open to question" from a purely electromagnetic point of view (Thomson 1886, 151).

ity and who was now a keen reader of the *Treatise*. Like the vast majority of Cambridge mathematicians with an interest in novel aspects of Maxwell's field theory, Thomson and Glazebrook both attended Niven's lectures, and it is extremely likely that Poynting would have done the same.[126]

What is certain is that by the early 1880s Poynting recognized the existence in Cambridge of a powerful school of research in electromagnetic theory. At the beginning of October 1883, just a few weeks before he embarked upon his ground-breaking research, Poynting gave a lecture in Birmingham on university training. In the lecture he argued (see epigraph to this chapter) that Cambridge University was especially notable for the school of research in electromagnetism it had cultivated under the influence of Maxwell's theory. A prerequisite to research of this kind, Poynting added, was teachers who were "no mere connecting links between investigators and learners" but rather men who were either "working themselves to extend the bounds of knowledge," or else were "so well acquainted with the methods of research" that they were able to "give sympathy and encouragement to those of their students who are strong enough to question nature for themselves." The "school" to which Poynting referred can only have been that principally constituted by the Nivens, Rayleigh, Glazebrook, and Thomson, while the teacher in question must have been W. D. Niven.[127] The only person in this group not to have been taught by Niven was Rayleigh, but, as he recalled in 1882, it was his arrival in Cambridge as Maxwell's successor in 1880 that stimulated his interest in electromagnetic theory. Rayleigh recalled that he found Trinity College and the Cavendish Laboratory "so saturated with [electricity]" that he thought it "best to make it his particular study" (Campbell and Garnett 1882, 358).

The above reconstruction of Poynting's early education and research career offers an explanation not only of why he was so familiar with the content of Maxwell and Rayleigh's important books but also of how, working in relative isolation in Birmingham, he was able so rapidly to turn his pedagogical derivation of the velocity of sound into a profound paper on the flow of energy in the electromagnetic field. As we have seen, there is no evidence that Poynting was especially interested in research problems in electromag-

---

126. A highly mathematical paper on the induction of coils shows that late in 1879 Poynting was certainly studying the sections of vol. 2 of the *Treatise* discussed in Niven's classroom (Poynting 1920, 170–74).

127. Poynting 1920, 559. Poynting explicitly differentiated in this passage between the research school in electromagnetic theory and the work being undertaken at the Cavendish Laboratory.

netic theory prior to the summer of 1883, and both his route to the problem of energy flow as well as his initial strategy for tackling the problem almost certainly came from Rayleigh's *Theory of Sound*. It was thanks to his familiarity with Maxwell's equations and the interpretation placed on them by the Cambridge Maxwellians that Poynting was able so rapidly to derive the energy-flow theorem, to recognize its importance to electromagnetic theory, and to apply it to several important cases. The fact that Poynting so readily identified and built upon what would, for most electricians, have been a highly counterintuitive view of electrical conduction leaves little doubt that he was aware that the phenomenon remained unexplained in the Maxwellian scheme and knew what kind of account would constitute an adequate explanation. The speed and enthusiasm with which the theorem was accepted by other Maxwellians further indicate that these were shared sensibilities, yet ones that no one before Poynting had fully articulated or addressed directly.[128] It is my contention that Poynting acquired these sensibilities through Niven's lectures and conversations with his Cambridge colleagues, and that, prior to the autumn of 1883, he was no more concerned with the problem of energy flow in the electromagnetic field than was any other Maxwellian.[129]

6.8. A Research School in Electromagnetic Field Theory

The above discussion highlights the several and somewhat complicated senses in which Cambridge can usefully be described as the institutional home of a research school in electromagnetic field theory in the early 1880s. Consider first Maxwell's role in founding the school. One of the chief characteristics attributed by historians of science to successful research schools is the presence of a productive, energetic, and charismatic leader.[130] Maxwell certainly fulfilled the leadership role in his capacity as director of experimental physics at the Cavendish Laboratory, but his contribution to the investigation of more theoretical aspects of the *Treatise* was far more indirect. His most important functions in this respect were to support the reform of the Tripos, to act on several occasions as an examiner, to write the *Treatise* as an advanced textbook, and, by his lectures and increasing stature in the university through the 1870s, to encourage interest in the *Treatise*'s

---

128. On the reception of Poynting's energy-flow theorem, see Buchwald 1985, 47–49, and Hunt 1991, 113–14.

129. Indeed, he was probably less concerned than were men like Thomson and Glazebrook, who had undertaken research on Maxwell's theory.

130. Geison 1993, 234.

contents. There is nevertheless no evidence that he either assisted Cambridge mathematicians in the interpretation of novel aspects of field theory or suggested potentially fruitful avenues for further research.

The key skills required to build a research school around the *Treatise* in Cambridge circa 1880 were actually distributed among many individuals and were coordinated through the local training system. Although Maxwell's emergent reputation played a part in encouraging Cambridge men to study his works, the task of making sense of the mathematical and theoretical content of the *Treatise* fell to the combined efforts of those recent graduates who first struggled through the book, the first generation of Cambridge Maxwellians who sought to apply the new equations to the solution of specific electrical problems, and teachers such as Routh and Niven whose job it was to prepare able undergraduates to answer difficult questions in the Tripos examination. It was this collective interpretation and teaching of Maxwell's work that, by the early 1880s, had produced a generation of mathematicians who understood that Maxwell's equations constituted the innovative part of his theory and knew how to manipulate the equations to solve problems in electromagnetic induction. This knowledge also enabled wranglers like Thomson and Poynting to make fundamental contributions to electromagnetic field theory in the early 1880s. From a technical perspective, therefore, the leaders of the nascent research school in the late 1870s were the examiners, the leading coaches—especially Routh—and, most importantly of all, W. D. Niven.

It should also be kept in mind that the technical resources employed by both the first and second generations of Cambridge Maxwellians in building on the *Treatise*'s contents went beyond the mathematical and physical theory contained in the book itself. As we have seen, the first generation drew heavily upon analogical arguments from heat theory and hydrodynamics, while the second found Rayleigh's *Theory of Sound* especially productive. Another example of this kind that might easily have been discussed alongside Thomson and Poynting's work in the previous section is Glazebrook's use of a hydrodynamical model of the electromagnetic field in 1881 to explain magneto-optic rotation and the recently announced Hall effect.[131] Where Glazebrook's familiarity with relevant sections of the *Treatise* derived mainly from Niven's lectures, his thorough knowledge of hydrodynamics originated in Horace Lamb's intercollegiate course on that subject held at

---

131. The Hall effect concerned the electrical effect of a magnetic field on a current-carrying conductor. The effect and the technical content of Glazebrook's paper are discussed in detail in Buchwald 1985, 84–95, 111–17.

Trinity College in 1874–75.[132] The existence of a research school in field theory in Cambridge circa 1880 was thus a product not just of specialized teaching in electromagnetic theory but also of a much broader training in the mathematical methods and physical principles of several branches of mathematical physics. It is noteworthy in this context that, unlike some of their non-Cambridge peers, the men I defined for convenience as members of the first and second generations of "Cambridge Maxwellians" did not use such a collective term to delineate their work.[133] As far as they were concerned, electromagnetic theory was but one of a much wider range of physical sciences which shared a common foundation in dynamics.

It important to keep in mind, however, that despite the productivity of the second generation of Cambridge Maxwellians, it remains the case that the rich resources of Cambridge pedagogy did not provide exclusive access either to the meaning of the *Treatise*'s contents or to the ways in which they could most effectively be developed. Neither FitzGerald nor Heaviside, for example, was taught from the *Treatise*, yet they each managed over a period of years to make sense of the most novel sections the book and to make original contributions that were of comparable significance to those made by their Cambridge contemporaries: it was FitzGerald who first developed an electromagnetic theory of optical reflection and refraction, showed that changing currents generated electromagnetic waves, and criticized Thomson's derivation of the field due to a moving charge; Heaviside, working completely outside the academic world, independently derived (by a different route) an equivalent expression to the Poynting vector and also corrected errors in Thomson's early work.[134] My purpose here is not to investigate the circumstances that enabled these men to develop the *Treatise* but is rather to use their achievements to highlight two important characteristics of the Cambridge school of Maxwellians.

The first concerns the significance of training and collaboration to research in electromagnetic theory in Britain in the early 1880s. As we have seen, the reintroduction of electricity and magnetism to undergraduate studies in Cambridge took place as part of a wider program of reform intended to promote research in progressive and commercially important areas of

132. Glazebrook 1932–35, 376–77. Glazebrook was recommended to attend Lamb's lectures by his coach.

133. On FitzGerald, Lodge, and Heaviside's collective sense of constituting a group of "Maxwellians," see Hunt 1991a, chap. 8.

134. Nahin 1988, 115–19; Hunt 1991a, 121. The other contributions made by FitzGerald and Heaviside in the 1880s are discussed in Buchwald 1985, Nahin 1988, Hunt 1991a, and Yavetz 1995.

mathematical physics. It was impossible to predict in advance how successful these reforms would be in generating new knowledge, but the work of men like Poynting, Thomson, and Glazebrook nevertheless emerged from a training program intended, among other things, to make original investigations in electromagnetic theory as likely as possible. The contributions of FitzGerald and Heaviside, by contrast, were to a large extent the fortuitous products of the personal interests, ambitions, and abilities of the individuals concerned.[135] Moreover, Heaviside taught himself those skills of paper-based calculation that he felt necessary for his own research. His methods were idiosyncratic and frequently frowned upon by academic mathematicians, but they turned out to be as powerful in electromagnetic theory as those Cambridge pedagogy generated in wranglers.[136] What is striking when one compares the accomplishments of individuals such as FitzGerald and Heaviside with those of individual members of the Cambridge school is their approximate parity. Given the enormous disparity in the pedagogical resources available to Cambridge and non-Cambridge Maxwellians respectively in the 1870s, one might have expected a comparable disparity in their respective contributions to electromagnetic theory in the early 1880s. What the similarity suggests is that the training system in Cambridge was not especially successful in generating new research problems from the *Treatise*, but was successful in producing a relatively large community of mathematicians who, as we saw in the previous section, were capable of applying Maxwell's equations to relevant research problems as and when they arose. Both Thomson and Glazebrook wrote major theoretical studies in response to novel developments in electromagnetism (experimental and theoretical), while Poynting's solution to the energy-flow problem arose unexpectedly from his work on the velocity of waves.

This brings us to a second and related characteristic that is also highlighted by comparison with the work of FitzGerald and Heaviside. Both of these men communicated their research through formal publications, and neither personally trained a new generation of investigators capable of continuing their work. In Cambridge, by contrast, the training and examination system provided an accumulative cycle that preserved and propagated not only the accomplishments of Cambridge Maxwellians but also those of

---

135. FitzGerald's mathematical training at Trinity College, Dublin, had much in common with that at Cambridge.

136. Heaviside's academic isolation must be qualified. Many of the books from which he taught himself mathematics and mathematical physics were products of the very Cambridge system discussed in this study. See Hunt 1991a, 105, 122–23, and 1991b; and Nahin 1988, chap. 10.

non-Cambridge researchers that seemed important to Cambridge coaches, lecturers, and examiners. It is noteworthy that even the most unorthodox of Heaviside's productive mathematical methods were eventually systematized, taught, and published by Cambridge mathematicians.[137] Cambridge was the only site in Britain at which new knowledge in the mathematical theories of electricity and magnetism was passed on directly to subsequent generations, and, as we shall see, by the late 1890s it had become the only British center where advanced research on these subjects was being actively pursued.

The importance of the research school in Cambridge during the late Victorian and Edwardian periods almost certainly lies, therefore, more in the conservative nature of the local training system than in direct interaction between senior members of the university interested in electromagnetic theory. From the late 1880s, the key non-Cambridge Maxwellians in Britain (FitzGerald, Lodge, and Heaviside) seem to have enjoyed a far more productive professional relationship through correspondence than did their Cambridge contemporaries through face-to-face interaction.[138] Indeed, the two greatest exponents of Maxwellian electromagnetic theory in late Victorian Cambridge, Larmor and Thomson, seem rarely, if ever, to have discussed their common intellectual interests once they left Niven's classroom.[139] Productive intellectual communication in Cambridge seems to have taken place mainly at the level of coaching and intercollegiate lectures, and through advanced examination questions based on recent research. It was mainly through this formal training mechanism that graduate researchers passed on not only their own work but also their interpretation of what appeared to be important research in electromagnetism carried out beyond Cambridge. This interaction between conservative pedagogy in Cambridge and new knowledge produced at sites outside of Cambridge will form a major theme in the remaining chapters of this study.

137. Nahin 1988, chaps. 10 and 13.
138. Hunt 1991, chap. 8.
139. Larmor certainly had much closer and more productive contact with the non-Cambridge Maxwellians than with his Cambridge colleagues in the 1890s. See chapter 7.

~ 7 ~
Joseph Larmor, the Electronic Theory of Matter, and the Principle of Relativity

> [I]t is now natural to make the conjecture, commonly spoken of as the Lorentz-FitzGerald hypothesis, that the system S when set in motion with a velocity u assumes the configuration of the system S', this latter being a [contracted] configuration of equilibrium for the moving system. Indeed, if we suppose all forces to be electrical in origin, this view is more than a conjecture; it becomes inevitable.
>
> JAMES JEANS, 1911[1]

7.1. Pedagogy and Theory Change

In the final three chapters of this study I introduce a new line of enquiry to my exploration of Cambridge mathematical culture. Until now I have discussed mathematical physics in Cambridge largely as an isolated and self-contained enterprise. My primary aim has been to chart the emergence of a new economy of undergraduate and graduate mathematical studies and to show how this economy both shaped the research efforts of the most able wranglers and transmitted their accomplishments to subsequent generations of students. The latter two issues will remain central to the discussion that follows, but I shall henceforth be concerned much less with pedagogical innovation than with the interaction between Cambridge pedagogy and mathematical knowledge generated at other sites. This shift of focus actually began in the previous chapter, where we saw how Maxwell's electromagnetic field theory, developed in London and Scotland, was gradually and painstakingly made the foundation of Cambridge's first major research program in mathematical physics. There I emphasized the interpretive work required to transform the *Treatise*'s contents into a teachable theory, but it should be kept in mind that this transformation was in some respects a comparatively straightforward exercise for Cambridge mathematicians. Maxwell was a graduate of

---

1. Jeans 1908, 2d ed., 577.

the Mathematical Tripos and so shared a common sense of mathematical purpose and technique with his Cambridge peers and students. As we saw, many of his most able students were actually trained by his most able undergraduate peer, Edward Routh. Maxwell also prepared the ground for the introduction of electromagnetism in Cambridge by acting five times as a Tripos examiner, and by holding lectures and publishing a Senate-House treatise on the subject. Last but not least, his very presence in the university helped to convince teachers and students alike that the *Treatise* was an important book that would repay careful study.

But what of innovative work in mathematical physics that either had no Cambridge connection at all, or, worse still, was viewed in Cambridge as irrelevant, incompatible, or hostile to an extant tradition within the university? In these cases there would be no local advocates to champion the cause, let alone to hold lectures, write textbooks, or set canonical Tripos problems. As we saw in chapter 5, an important component in the propagation of technical expertise was face-to-face interaction with a skilled practitioner, but in Cambridge it was invariably high wranglers who were appointed to senior college and university positions in mathematically based disciplines. Given the conservative nature of such a cyclical training system, it is important to enquire in what circumstances and by what routes mathematical knowledge generated beyond Cambridge sometimes found its way into graduate and undergraduate studies. This question is addressed in these last three chapters by examining the interaction between Joseph Larmor's theory of the electrodynamics of moving bodies and Albert Einstein's special and general theories of relativity. In Cambridge during the period 1905 to 1920, Einstein's work was, by turns, ignored, reinterpreted, rejected, and, finally, accepted and taught to undergraduates. This series of events is fascinating as an historical case study in its own right, but it also illustrates the doubly conservative role training systems can play, initially by resisting and then by preserving and propagating a new theory. In the latter sense these concluding chapters raise a much broader range of issues concerning the mechanisms by which major new theories in mathematical physics are assessed and disseminated.

In the following sections I prepare the ground for this discussion by tracing the development of Larmor's electrodynamics during the 1890s as well as the process by which it was mastered, modified, and applied by his students during the first two decades of the twentieth century. I shall shortly explain why a firm grasp of these events is crucial to understanding the Cambridge reception of relativity theory, but it will be helpful first to dispel a common misconception concerning the state of Cambridge mathematical physics in the late nineteenth and early twentieth centuries. There is a widespread view, repeated or insinuated in many textbooks on the theory of relativity, that physics circa 1900 faced a serious theoretical crisis in the form of the failure by theoreticians to provide a satisfactory explanation for the null result of the Michelson-Morley ether-drift experiment.[2] In these accounts it is Einstein who averted the crisis in 1905 by proposing the so-called "special theory" of relativity. This theory not only explained Michelson and Morley's null result by making the velocity of light a universal constant but rendered the notion of an electromagnetic ether superfluous to physics. Closely related to the above view is the claim by some historians and philosophers of science that mathematical physicists in Cambridge were unreasonably slow to accept Einstein's theory on account of their strong commitment to the ether concept. Cambridge mathematicians are judged to have behaved *unreasonably* in this respect because they had already resorted to such ad hoc devices as the Lorentz-FitzGerald contraction hypothesis in an attempt to preserve an ether-based electrodynamics in the face of disconfirming experimental evidence.[3] Failure to abandon this line of argument even after Einstein had offered what is now regarded as the correct explanation of the null result seems therefore to confirm the proposition that Cambridge mathematical physics had become excessively insular and introspective during the early years of the twentieth century.

What emerges from these remarks is that the historiography of Cambridge physics circa 1900 derives in large measure from our understanding of the early history of relativity theory. It is clear with the benefit of hindsight that what is now known as Einstein's special theory of relativity did indeed point the way forward in physics after 1905, but we ought not to *assume* this knowledge in trying to understand how the theory was initially assessed, accepted, or rejected by Einstein's contemporaries. To do so is to prejudge precisely what we are seeking to understand and to explain. These issues are properly the subject of the next two chapters—in which they are explored in greater depth—but there is one aspect of them that is relevant to our immediate concerns. The perception that Michelson and Morley's null result could not legitimately be explained by the Lorentz-FitzGerald contraction

---

2. See, for example, Bergman 1942, 26–28, Dixon 1978, 10, Lawden 1985, 14, and D'Inverno 1992, 16.

3. Hunt suggests that belief in the Lorentz-FitzGerald contraction hypothesis by Cambridge mathematicians represented a "strange contortion" (1986, 134). Cantor and Hodge gloss the same event as an attempt to "save the phenomena, or more exactly, save the theory"(1981, 52). The depiction of the contraction hypothesis as a classic ad hoc hypothesis is discussed in Hunt 1988.

hypothesis originates in the belief that the proper explanation was that given by Einstein in 1905. On this showing the historical problem is one of explaining why Cambridge mathematicians stubbornly refused to accept the correct theoretical explanation of a troublesome experimental result. There are two problems with this approach. First, as Darrigol (1996, 310) has pointed out, in the immediate wake of Einstein's first publication on relativity there was no conclusive experimental evidence by which to decide between his theory of the electrodynamics of moving bodies and those proposed by his immediate predecessors and contemporaries. Any judgment regarding the validity of Einstein's theory would therefore have had to be made on other grounds. Second, Cambridge-trained mathematical physicists believed the contraction hypothesis to emerge naturally from Larmor's theory of the electrodynamics of moving bodies and so saw no reason to search for any other explanation of Michelson and Morley's null result. Since this latter point is an important one that is touched upon many times in this chapter and the next, I shall briefly sketch the history of the experiment and its subsequent theoretical interpretation.

We saw in chapter 6 that Maxwell's field theory identified light as an electromagnetic phenomenon and attributed electromagnetic effects to physical states in an all-pervading ether. It was Maxwell himself who suggested in 1878 that it might be possible to detect the earth's orbital motion through the ether by comparing the times taken by a light beam to travel with and against the earth's motion.[4] In the early 1880s, the American experimenter Albert Michelson proposed to try a modified form of this experiment using a device that would come to be known as an interferometer. Michelson's plan was to split a light beam into two perpendicular rays which would then be made to traverse equal distances (along perpendicular arms of the interferometer) before being reflected back to recombine and form an interference pattern. If the rays happened to travel respectively parallel and perpendicular to the earth's direction of motion through the ether, they ought to take different times to make the round trip, a difference that would reveal itself through shifts in the interference pattern when the interferometer was rotated. In a series of experimental trials undertaken between 1881 and 1887, those from 1885 being carried out in collaboration with Edward Morley, Michelson invariably obtained a null result; the earth's motion through the ether did not effect the journey time of light rays in the expected manner.[5] One way of ex-

4. Swenson 1972, 57.
5. Ibid., 54–97.

plaining this result was to assume that the ether was dragged along, or "convected," at the earth's surface, but this assumption did not appeal to the leading theoreticians of electromagnetism of the time.[6] Instead, FitzGerald and the Dutch mathematical physicist H. A. Lorentz suggested independently in 1889 and 1892 respectively that matter (and hence one arm of the interferometer) contracted minutely in its direction of motion through the ether, thereby concealing an effect that would otherwise be detectable.[7] The suggestion came to be known as the Lorentz-FitzGerald contraction hypothesis.[8]

When presented in the above fashion, the contraction hypothesis does indeed appear ad hoc; that is, introduced with little or no theoretical foundation simply to explain away an otherwise troublesome result. But when viewed in proper historical context the situation appears rather different. As Bruce Hunt has shown, FitzGerald actually had some theoretical grounds for believing that the contraction he proposed might really occur.[9] For example, he was aware that the strength of the electric field surrounding a moving charge should be reduced in the charge's direction of motion through the ether by precisely the factor required by the contraction hypothesis. If one were prepared to assume that intermolecular forces were wholly electrical in origin, then it followed that moving matter would *have* to contract in order to preserve the electrical equilibrium between its molecules. Despite this speculative justification, the Maxwellian community in Cambridge in the early 1890s considered FitzGerald's hypothesis "the brilliant baseless guess of an Irish genius."[10] As far as they were concerned, the contraction of moving matter was an untenable idea because a plausible electromagnetic theory of matter had not been proposed. Only after 1897, when Larmor published a completely electronic theory of matter that embodied the contraction hypothesis, did Cambridge mathematicians begin to accept that Lorentz and FitzGerald had offered the correct explanation of Michelson and Morley's null result. From a Cambridge perspective, then, the result did not constitute a serious problem in physics after the late 1890s. Indeed, it was commonly believed, as the epigraph above reveals, that if one accepted Larmor's electron-based electrodynamics, the contraction of moving matter became inevitable.

---

6. Larmor's reasons for rejecting the convected-ether explanation are discussed in Warwick 1991, 35.
7. The contraction factor was $\sqrt{(1 - v^2/c^2)}$, where v is the matter's velocity through the ether and c the velocity of light.
8. Hunt 1991a, 192–93.
9. Ibid., 185–97.
10. Glazebrook 1928, 287–88.

It should now be clear why a sound grasp of Larmor's work and its Cambridge dissemination are fundamental to our understanding of the reception of relativity theory in the university. Throughout this study I have emphasized the importance of training in shaping the technical skills and kinds of argument that mathematical physicists at a given site will typically take for granted. It is this training that shapes their sense of what physical and mathematical principles are admissible and how they are properly employed to generate meaningful solutions to physical problems. In order to explain the Cambridge response to Einstein's special theory of relativity, it is therefore necessary to understand how the theory appeared to mathematical physicists trained and working in the Cambridge tradition. In the balance of this chapter I prepare the ground for this analysis in the following three stages. I first show that the development of Larmor's electrodynamics was not a rearguard action designed merely to accommodate awkward experimental facts, but an active research program that emerged from his Cambridge training and the Maxwellian tradition within the university. Second, I show how the building of a research school around his completed electrodynamics at the turn of the century was accomplished through the pedagogical mechanisms discussed in earlier chapters. It was through examination questions, lectures, and textbooks that many wranglers trained circa 1900 learned to accept the central tenets of Larmor's theory and to apply its mathematical methods to tackle problems. Finally, I briefly illustrate the use made of Larmor's theory by one of his students engaged in research on electron physics circa 1910. As we shall see, this example clearly displays the taken-for-granted status of Larmor's work in Cambridge at a time when his students continued to regard Einstein's fully relativistic approach to electrodynamics, insofar as they were aware of it, as inferior to their own.

This is also an appropriate point at which to reiterate the cautionary remark made in the introductory chapter concerning the level of technical knowledge expected of the reader. Thus far I have sought to *explain* rather than to *assume* the kind of expertise possessed by mathematical physicists, but, from this point forward, I shall to some extent assume that the reader is a product of the kind of training system whose history we have been exploring. This turn in the narrative is not only unavoidable if we are fully to grasp the technical issues that shaped the Cambridge reception of relativity, but is in some senses appropriate. In order to convey an accurate idea of the differences in approach that separated physicists such as Einstein from his Cambridge contemporaries, it is necessary to assume that the reader is already familiar with a range of technical terms and possesses the necessary expertise in physics and mathematics. Moreover, the extreme difficulty of elucidating such differences for those who have not studied the relevant technical disciplines exemplifies the power of training to generate highly specialized technical sensibilities. This is not to say that the narrative will be completely impenetrable to the general reader. Most of the conclusions drawn are not couched in explicitly technical terms, and the gist of the argument, if not the evidence, should be comprehensible even without the technical detail. There is nevertheless one dimension to this chapter that will surely have more impact upon physicist than non-physicist readers. Larmor's everyday use of concepts long discredited in post-Einsteinian physics will perhaps seem to the former strange, amusing, or simply absurd. Yet, as we shall see, the constructive power of training to shape the physicist's understanding of the world is made more starkly visible when what we regard as falsity rather than truth is being propagated.

7.2. Joseph Larmor's Training and Early Career

From intimate association with [Larmor] I can testify to his extraordinary quickness in going direct to the heart of a mathematical question, alike in the case of known theory and of new problems.

J. D. EVERETT, 1884[11]

Joseph Larmor was by far the most important contributor to the development of mathematical electromagnetic theory in Cambridge during the 1890s. An almost exact contemporary of J. J. Thomson (fig. 6.5), Larmor too conformed closely to the ideal wrangler-type of the late 1870s. Born in Country Antrim, Ireland, in 1857, Larmor was educated at the Royal Belfast Academical Institution and Queen's College, Belfast.[12] At the Academical Institution he was taught by the schoolmaster-wrangler, Randal Nixon (21W 1864), before going on to obtain the "highest honours" in mathematics and natural philosophy at Queen's College.[13] As an undergraduate Larmor was taught mathematics by a graduate of Trinity College, Dublin, but the examiner of his Bachelor and Master of Arts examinations was none other than

---

11. Everett was professor of Mathematics at Queen's College, Galway. The letter is one of a number of testimonials written when Larmor applied for the Chair of Mathematics at Queen's College, Galway, in 1880 and the Chair of Mathematics at University College London in 1884. Larmor Box (JL-SJC).

12. Larmor's biography is taken from Eddington 1942–44, unless otherwise stated.

13. *The Cambridge Review*, 4 February 1880, 5. Queen's College generally employed wranglers to teach mathematics, but the premature death of George Slesser (SW 1858) in 1862 brought the tradition to an end (Moody 1959, 2:603–4).

Charles Niven, then professor of mathematics at Queen's College, Cork.[14] Larmor's precocious ability in mathematics took him next to St. John's College, Cambridge, where, supported by a series of mathematical scholarships, he was coached by Routh and attended W. D. Niven's intercollegiate course on electricity and magnetism. Routh subsequently wrote in a testimonial that, having known Larmor "intimately" and overseen his work "day by day," he was prepared to "certify that [Larmor's] mathematical talent [was] of no ordinary kind." As several other of his Cambridge teachers willingly attested, even by Cambridge standards Larmor "possessed in an eminent degree the faculty of rapidly and correctly thinking out the answers to the most difficult questions."[15] His acquisition of the senior wranglership in 1880 not only prompted great celebrations in Belfast (fig. 7.1) but also won him a fellowship at St. John's College and the professorship of natural philosophy at Queen's College, Galway.[16] In 1885 Larmor returned to St. John's College as a lecturer in mathematics, a position he retained until 1901 when he succeeded George Stokes as Lucasian Professor of Mathematics.

Larmor was (by the definition given in chapter 6) a member of the second generation of Cambridge Maxwellians, but in the decade that followed his graduation he wrote only one substantial paper on the mathematical theory of electromagnetism. This paper, written in Galway in 1883, is nevertheless of some interest, as it provides important insight into Larmor's early understanding of Maxwell's theory as well as his method of approaching original investigations. The subject of the paper was electromagnetic induction in "conducting sheets and solid bodies," a topic initiated by Maxwell in the *Treatise* and, as we saw in chapter 6, one that was taken up by the first generation of Cambridge Maxwellians. It appears in fact to have been Charles Niven's 1881 paper on this topic that inspired Larmor's investigation. Larmor began by acknowledging that Niven had already given "a complete mathematical solution of Maxwell's equations" for induction in conducting planes and spheres, but noted that the very generality of Niven's solutions made them extremely complicated from a mathematical perspective. Larmor's aim was to derive concise and complete solutions to a number of relatively simple

---

14. Niven's account of Larmor's brilliant performance in mathematical examinations is given in his 1880 testimonial. Larmor Box, JL-SJC.

15. Almost all of the testimonials from Larmor's Cambridge teachers comment on his extraordinary problem-solving ability. The citations are from letters by Routh and Ernest Temperley (5W 1871), respectively; the latter was the Senior Moderator in 1880. Larmor Box, JL-SJC.

16. The scale of the celebrations in Belfast was due to the fact that the previous year's senior wrangler was also educated at the Academical Institution and Queen's College.

FIGURE 7.1. This "torchlight procession" held in Larmor's honor in Belfast on 12 February 1880 was a mark of the importance associated with the senior wranglership throughout Britain. The procession, which attracted "large crowds of delighted spectators," was described as follows by a local reporter: "The students, arrayed in fantastic costumes, and each bearing a torch, left the College in procession, and having passing through the principal streets of the city, marched to the residence of Mrs Larmor, where hearty cheers were given for her talented son. On the way back to the College a number of rockets were discharged, and in front of the building a bonfire constructed of tar-barrels was burnt" (*The Graphic*, 6 March 1880, 243). (By permission of the Syndics of Cambridge University Library.)

cases of induced currents (not specifically solved by Niven), concentrating, as far as possible, on "physical quantities" rather than difficult "mathematical transformations." [17] The first interesting characteristic of Larmor's paper, then, is that it was very much a product of his Cambridge training. The topic was one taught and developed by the Niven brothers, while the problem-solving style was typical of a young wrangler.

The second interesting aspect of the paper concerns the kind of solution Larmor applied to a particular class of problem. In the fifth section of the paper, he considered the current generated in a conductor by motion between the conductor and a magnetic field. Larmor showed that, provided the conductor formed a closed circuit, the magnitude of the current generated depended only on the relative motion between the circuit and the field. It

---

17. Larmor 1929, 1:8. The mathematical simplicity of Larmor's solutions was only relative, however, as he still made extensive use of advanced harmonic analysis.

therefore made no difference in tackling problems of this kind "whether the relative motion of the conductor and the field be ascribed to the conductor or to the magnetic field, or to both conjointly" (1929, 1:17). This result was not original to Larmor, as its proof had been given by Maxwell in the *Treatise* and had even been set as a Tripos problem in 1878.[18] The original point Larmor sought to establish was that one could exploit the result to solve specific problems by adopting whichever set of coordinate axes was the most convenient for the problem in hand.

Larmor next showed how the same technique could be applied to open circuits. In this case, he noted, one could expect only approximate results because, according to Maxwell, the potential generated in an open circuit by a changing magnetic field depended partly on the frame of reference adopted by the observer. Larmor demonstrated that these frame-dependent effects were nevertheless so small that for all practical purposes the technique in question could safely be applied to open circuits.[19] He then showed how the calculation of the currents generated in moving conductors by magnetic fields could be greatly simplified by reducing the conductors to rest and solving the "corresponding relative problem, where motion across the lines of force is replaced by a variation of the field itself" (1929, 1:19). These results are striking in two respects: first, they reveal that Larmor's earliest investigations of Maxwell's theory were centrally concerned with the electromagnetic effects of relative motion between fields and conductors; and, second, they show that Larmor was able to exploit these effects to generate relatively simple solutions to otherwise technically complicated problems. These were techniques that, as we shall shortly see, he would employ to far greater effect in electromagnetic theory in the 1890s.

A final and related point that is usefully emphasized at this juncture is Larmor's belief in the existence of an absolute frame of reference in the form of the electromagnetic ether. I noted above that, according to Maxwell, the electrostatic potential ascribed to an open circuit depended partly upon the frame of reference in which the potential was calculated.[20] This claim troubled Larmor because Maxwell had not explained to which frame of ref-

---

18. Maxwell, *Treatise*, vol. 2, arts. 600–601; Glaisher 1879, 217–20. Larmor's copy of this collection of Tripos problems is held in the Scientific Periodicals Library in Cambridge.

19. Larmor showed that these effects would be "second order," that is, proportional to the second power of the ratio (v/c) where v is the velocity of the system through the ether and c the velocity of light.

20. *Treatise*, vol. 2, arts. 600–601. Maxwell noted only that the value of the potential would be different according to the frame of reference chosen. He did not suggest that the ether provided an absolute frame of reference.

erence the *true* value of the electrostatic potential was properly ascribed. A possible solution to the problem (though not one adopted by Maxwell) was to assume that the true value was the one that would be measured in a frame of reference that was at absolute rest. But how was such a frame to be defined? The notion of absolute rest was alien to accepted mechanics since, as Larmor himself remarked, there existed "no criterion of absolute rest at all as far as matter [was] concerned." Larmor's solution was simply to assert that if mechanics provided no frame of absolute rest, then one was inevitably led to the conclusion that "the true value of [the electrostatic potential] is that derived from axes fixed with reference to some system or medium which is the seat of the electromagnetic action" (1929, 1:18). In other words, Larmor believed that electromagnetic effects were due to physical processes in some underlying medium (the electromagnetic ether) and that the true value of the electrostatic potential was the one that would be calculated in the frame of reference that was stationary with respect to this medium. This belief in the existence of an absolute frame of electromagnetic reference was also one that would play an extremely important role in the development of Larmor's electromagnetic theory in the 1890s.

7.3. The Development of the Electronic Theory of Matter (ETM)

With the exception of the paper just discussed, most of Larmor's original investigations during the 1880s were concerned with the mathematical methods and physical applications of analytical dynamics. Foremost among the mathematical methods was the Principle of Least Action, a technique that Larmor believed to represent the clearest, most compact, and most general means of expressing any physical problem. Larmor was convinced that this form of expression best revealed the formal mathematical connections that existed between the "different departments" of mathematical physics, and so facilitated the analogical solution of a wide range of problems (1929, 1:31). It was Larmor's combined interests in dynamics, electrodynamics, and the Principle of Least Action that led him in the early 1890s to begin searching for a dynamical foundation for electromagnetic phenomena. This line of research was initially inspired by a paper of FitzGerald's that Larmor studied while preparing a report on magneto-optics for the British Association.[21] FitzGerald had noticed that a remarkable formal analogy existed between the expression given by James MacCullagh for the mechanical energy stored

---

21. Hunt 1991a, 212.

in his rotationally elastic ether and those given by Maxwell for the energy stored in the electromagnetic field. By replacing the mechanical symbols in MacCullagh's theory with appropriate electromagnetic symbols and applying the Principle of Least Action to the resulting Lagrangian, FitzGerald was able to follow MacCullagh's analysis and obtain an electromagnetic theory of the propagation, reflection, and refraction of light. Through his work on the analytical dynamics of magneto-optic rotation, and through reading FitzGerald's paper, Larmor became convinced that MacCullagh's ether could provide a common dynamical foundation for Maxwell's synthesis of electromagnetic and luminiferous phenomena.

The fruits of Larmor's research were published by the Royal Society as "A Dynamical Theory of the Electric and Luminiferous Medium" (hereafter "Dynamical Theory") in three instalments (with various appendices) between 1894 and 1897, but during this period his theory changed considerably. The first instalment came in for some powerful criticism from FitzGerald, who acted as a referee for the Royal Society. During a period of intense correspondence between the two men in the spring and summer of 1894, FitzGerald persuaded Larmor to introduce the concept of "discrete electric nuclie," or "electrons," into his theory.[22] But the introduction of the electron did far more than solve the immediate problems that troubled Larmor's theory; over the following three years it also had a profound effect upon his understanding of the relation between the electromagnetic ether and gross matter. As we saw in chapter 6, in Maxwell's scheme an electric current was thought of not as a material flow of electrical substance but as a breaking down of electric displacement which, by some unknown mechanism, converted the electromagnetic energy stored in the ether into heat energy in the conductor. Prior to the introduction of the electron, the electromagnetic ether and gross matter were largely distinct concepts, the mechanism of their interaction seldom being discussed. Likewise the effects produced by the motion of a charged body was something of a side issue, not perceived as having any real relevance to the much more central problem of the nature of the conduction current. But with the introduction of the electron as the fundamental carrier of a natural unit of electric charge, the situation changed considerably.

The nature of the magnetic field generated by a moving charge was first considered by J. J. Thomson in 1881, and, as we saw in chapter 6, he drew attention to the "extra mass" that such a body would appear to have due to the

---

22. The development of electromagnetic theory during the 1890s is discussed in detail in Buchwald 1985, 133–73, Hunt 1991a, chap. 9, and Warwick 1991 and 1995a.

energy of self-induction stored magnetically in the field. With the introduction of the electron this work assumed much greater significance. According to Larmor's fully developed theory, the universe consisted of a sea of ether populated solely by positive and negative electrons. These electrons could be thought of mechanically as point centers of radial strain in the ether and the sole constituents of ponderable matter. This view of the universe partially solved the problem of the relationship between ether and matter by reducing all matter to moveable discontinuities in the ether. Furthermore, if all matter was constructed solely out of electrons, then perhaps its inertial mass was nothing more than the electromagnetic mass of the electrons as they are accelerated with respect to the ether. Thomson's "extra mass" now became the *only* mass, and the mechanics of the universe was reducible to electrodynamics. This is theory that I shall refer to as the Electronic Theory of Matter (hereafter ETM).[23]

The ideal of reducing mechanics to electrodynamics gained credence with a number of European physicists towards the end of the nineteenth century, and has been characterized by McCormmach (1970) as the *Electromagnetic View of Nature*. Like Larmor, who was actually the first to suggest that the entire mass of the electron might be electromagnetic in origin, these physicists imagined a universe made only of ether and electrons, and whose dynamics was governed solely by the equations of electromagnetism. But Larmor's ETM differed in one very important aspect from the continental Electromagnetic View of Nature. Unlike some of his European contemporaries, Larmor continued to believe that the equations of electromagnetism were not themselves fundamental but were ultimately to be derived from the dynamical properties of the underlying ethereal medium by the application of the Principle of Least Action. It is important to keep in mind, however, that during the 1890s Larmor increasingly tackled problems in the electrodynamics of moving bodies by direct application of the equations of electromagnetism, and, in practice, concerned himself relatively little with their dynamical foundation.

A further consequence of Larmor's introduction of the electron was that the previously minor problem of the electrodynamics of moving bodies was brought to center stage. If electrical conduction and associated electromagnetic effects were due solely to the motion of electrons, and if matter was itself composed exclusively of electrons, then virtually every problem, both in

---

23. The development of Larmor's electronic theory of matter is discussed in detail in Warwick 1991.

electrodynamics and matter theory, became a problem in the electrodynamics of moving bodies. Indeed, these two previously distinct realms of physical theory became virtually inseparable. Such well-known effects as the electric polarization and magnetization of matter—which previously had been ascribed to changes in the dynamical properties of the ether somehow brought about by the presence of matter—could now be explained in terms of the electronic micro-structure of matter. Moreover, by 1895 Larmor had shown that if the universe consisted of nothing but positive and negative electrons whose motions were governed solely by the equations of electrodynamics (that is, if the ETM was correct) then moving matter would contract minutely in its direction of motion through the ether.[24] As I noted above, a similar suggestion by FitzGerald in 1887 had been derided in Britain as theoretically untenable, but following the advent of Larmor's ETM it quickly acquired a new status.[25]

The accommodation of the contraction hypothesis within Larmor's theory is of such importance to the discussion that follows that I shall briefly explain the two lines of investigation by which it occurred. The first concerns the phenomenon of light convection. When a beam of light passes into a moving, transparent medium, part of the medium's velocity is imparted to the beam. Larmor constructed a theoretical model of this effect and used it to investigate the behavior of the rays in Michelson and Morley's interferometer. He found he could explain the null result if he employed new space-time transformations to relate the electromagnetic variables of the stationary-ether frame of reference to those measured in the rest-frame of the moving interferometer. Larmor also found that Maxwell's equations were invariant to the second order under these new space-time transformations provided that new electromagnetic vectors were introduced in the moving system.[26] This important finding was eventually explained by Larmor in the following way. He claimed that the electromagnetic fields measured by observers in moving systems (on the earth's surface, for example) were not the physically real fields that would be measured by observers who were stationary in the ether. Rather, they were subsidiary fields generated partly by the earth's motion through the ether. It just happened that these subsidiary fields also satisfied Maxwell's equations. Larmor noted that the new space-time transforma-

---

24. On the Michelson-Morely experiment and its relation to British electromagnetic theory in the late Victorian period, see Hunt 1991a, 185–97, and Warwick 1995a, 302–22.

25. Warwick 1991, 48–56.

26. Ibid., 83–91. Maxwell's equations are invariant to *all* orders under these transformations, but Larmor did not notice this until around 1904. This is discussed further below.

tions had a strange property; they seemed to imply that moving electrical systems contracted in precisely the manner suggested by FitzGerald.[27]

The second line of research that led Larmor to his explanation of Michelson and Morley's null result concerned the potential surrounding a moving electron. By 1897, Larmor was convinced that positive and negative electrons were the sole constituents of gross matter. In that year, he derived a general expression for the total electromagnetic potential surrounding a moving electron. This expression indicated that the potential surfaces surrounding the electron would contract by exactly the amount predicted by FitzGerald (and by the new space-time transformations). In order to remain in electromagnetic equilibrium, a stable group of electrons moving through the ether would therefore have to move closer together by the same amount. Since Larmor believed gross matter to be solely electronic in construction, it seemed reasonable to him to suppose that moving matter must also contract by this amount at the macroscopic level. Larmor now had a physical explanation for the contraction implied by the new space-time transformations. A very important point to note in this context is that, during 1897, the relationship between the ETM and the Michelson-Morley experiment changed dramatically. From this point on, Larmor claimed that the null result was actually *predicted* by his theory and so provided important experimental evidence that matter was electrically constructed and that his theory was correct.[28]

Towards the end of 1897 Larmor collected and clarified the most important results of the new electrodynamics for his 1899 Adams Prize essay "The Theory of the Aberration of Light." In this essay he gave the first systematic account of the ETM including the following new space-time transformations. If (x, y, z, t) $(\mathbf{E}, \mathbf{B})$ and (x_1, y_1, z_1, t_1) $(\mathbf{E}_1, \mathbf{B}_1)$ represent respectively the space-time and electromagnetic-field variables measured in the stationary ether frame of reference and a frame of reference moving through the ether with velocity v, then these variables are related as follows:

$$x_1 = \beta(x - vt), \quad y_1 = y, \quad z_1 = z, \quad t_1 = \beta\left(t - \frac{vx}{c^2}\right) \quad (1)$$

$$\mathbf{E}_1 = (E_{x1}, E_{y1}, E_{z1}) = \beta[\beta^{-1}E_x, (E_y - vB_z), (E_z + vB_y)] \quad (2)$$

$$\mathbf{B}_1 = (B_{x1}, B_{y1}, B_{z1}) = \beta\left[\beta^{-1}B_x, \left(B_y + \frac{v}{c^2}E_z\right), \left(B_z - \frac{v}{c^2}E_y\right)\right] \quad (3)$$

27. Ibid., 59.
28. So strong was Larmor's reliance on the null result that when it was called into question, circa 1900, he sought new experiments to prove that motion through the ether could not be measured to the second order. See Warwick 1995a.

where $\beta = (1 - v^2/c^2)^{-1/2}$ and c is the velocity of light. These were the transformations later given independently by Lorentz and by Einstein, equations (1) now being known as the "Lorentz transformations."[29] It was a revised and expanded version of this essay that appeared in 1900 as Larmor's best-known and most widely read publication, *Aether and Matter* (1900).[30]

Larmor was the first mathematical physicist to employ the above transformations, but his interpretation of their physical meaning was very different from that subsequently given by Einstein.[31] The transformations were used by Larmor to "compare a system shrunk in the FitzGerald-Lorentz manner and convected through the ether, with the same system unshrunk and *at rest*."[32] This understanding is difficult for the reader trained in modern physics to grasp, for two reasons. First, following Einstein, we are used to thinking of motion as a purely relative quantity, so that any inertial frame can legitimately be considered to be *at rest*. Second, the modern reader likewise recognizes no privileged description of the electromagnetic fields surrounding a charged system — only different accounts given by observers in different states of uniform relative motion. For Larmor, however, electromagnetic fields were the observable manifestations of real physical states of a dynamical ether. He accordingly used the above transformations in the following way: he first imagined an electromagnetic system *at rest* in the ether and surrounded by the electric and magnetic fields ($\mathbf{E}, \mathbf{B}$); he then employed the transformations (1), (2), and (3) to calculate the new fields ($\mathbf{E}_1, \mathbf{B}_1$) that would surround the *same system* (as measured in the stationary ether frame) if it were moving through the ether with the uniform velocity v (with the moving earth, for example). Larmor referred to systems related in this way as "correlated systems."[33]

The fields ($\mathbf{E}_1, \mathbf{B}_1$) were not of course those that would be measured by an observer in the rest frame of the moving system ($\mathbf{E}_r, \mathbf{B}_r$). The moving observer might assume ($\mathbf{E}_r, \mathbf{B}_r$) to be real fields, but they actually owed their existence to the following three effects upon ($\mathbf{E}_1, \mathbf{B}_1$): first, the motion of the observer through the ether, which had the effect of "mixing up" the real

---

29. Lorentz and Einstein introduced these transformations in 1904 and 1905 respectively. See Lorentz 1934–39, 5:172–97, and Einstein, *Collected Papers,* vol. 2, doc. 23, sections 3 and 6. The transformations (1) were first referred to as the "Lorentz transformations" by the French mathematician, Henri Poincaré, in 1905 (Miller 1981, 80).

30. The transformations (1), (2), and (3) actually preserve the form of Maxwell's equations to all orders of v/c, but Larmor did not notice this property until after the publication of *Aether and Matter.*

31. On Einstein's interpretation of these equations, see Miller 1981, chaps. 6 and 7.

32. Quoted in Warwick 1991, 62.

33. Ibid., 63.

electric and magnetic fields so that part of the electric field appeared as if it were a magnetic field and vice versa; second, the physical contraction of the moving matter which led the moving observer to adopt a new scale of measurement; and third, the new standard of time measurement unknowingly adopted by the moving observer. It was for Larmor a mere peculiarity of the fields (E_r, B_r) that they too satisfied Maxwell's equations (at least to the second order) and so concealed the earth's proper motion through the ether. Only by referring back to calculations made according to coordinates that were stationary in the ether could the real fields be calculated.

From a mathematical perspective, Larmor's application of the new transformations was similar to a range of analytical and geometrical mapping techniques routinely employed by wranglers in order to solve problems in various branches of mathematical physics. For example, by transforming the known solution of a problem in electrostatics, it was often possible to find the solution to another previously unsolved problem of a similar nature.[34] Maxwell had discussed numerous examples of this kind in the first volume of the *Treatise,* and Routh's lecture notes reveal that he taught such methods to his students. Routh also employed these methods to solve problems in mechanics and even referred to dynamically related bodies as "correlated" bodies (1868, 169–83). Larmor made the method applicable to moving electrical systems, thereby simplifying calculations that previously had been extremely complicated.[35]

From a physical perspective, the stationary ether frame of reference remained for Larmor a truly privileged frame, compared to which moving bodies actually became shorter and moving clocks actually ran more slowly. Conversely, a system that was stationary in the ether would have to appear *longer* to an observer moving through the ether. That Larmor really did think of his system in this way is highlighted by the fact that he sometimes referred not to a "contraction" of moving matter, but an "elongation" of the system when brought to rest; and, on at least one occasion, he even referred to the FitzGerald-Lorentz "elongation hypothesis."[36] Larmor made little explicit comment on the physical meaning of his new nonstandard space-time transformations, presumably because he did not regard them as problematic. The transformation $x_1 = \beta (x - vt)$ expressed the contraction of moving

---

34. This technique is illustrated below and in chapter 8.

35. An example of how Larmor's transformation technique could dramatically simplify an otherwise complicated calculation in electrodynamics is discussed in Warwick 1991, 55.

36. Larmor 1929, 2:41, and 1900, 155. For Larmor's reference to the "elongation" hypothesis, see Larmor to Lodge, 24 October 1901 (OJL-UCL).

matter, but this was to be expected in the ETM. The transformation $t_1 = \beta$ $(t - vx/c^2)$ received no specific physical interpretation but would, according to Larmor, be "unrecognizable" because time in any frame was "isotropic." By this Larmor appears to have meant that an observer in a moving frame (as on the earth's surface) would be bound to employ clocks made of electrons and that these clocks would naturally record t_1. But even if clocks really did run more slowly in a moving system, this local time could in theory be related back to the "real" time measured in the stationary ether frame. We should therefore think of Larmor's transformations as being *in* space and time rather than *of* space and time.

Finally, we must briefly consider why Larmor thought it not only possible but desirable that the earth's motion through the ether would eventually be detected experimentally. Larmor's mathematical formulation of the ETM did not rule out the possibility that the earth's motion would remain undetectable to all orders of v/c. However, from a physical perspective this possibility required one also to assume that positive and negative electrons could legitimately be treated as point sources in the ether. The negative result of the Michelson-Morley and other experiments designed to detect the earth's motion indicated that this assumption was empirically admissible up to the second order.[37] But if the electron constituted a finite structure in the ether, as Larmor believed it might, then it was possible that matter would not contract in the predicted way when electrons were brought into close proximity with one another. It would then be possible not only to detect the earth's proper motion through the ether at higher orders—and hence to calculate the real electromagnetic fields in the stationary ether—but also to learn something about the electron's size and structure.[38] These considerations led Larmor always to approximate his own calculations to the second order, since he believed them to be empirically unfounded at higher orders.[39] It was this habit of approximating calculations that prevented him from noticing until after the publication of *Aether and Matter* that Maxwell's equations actually transformed exactly under the Lorentz transformations.[40]

---

37. On Larmor's attempt to find alternatives to the Michelson-Morley experiment to justify the contraction hypothesis, see Warwick 1995a.

38. Larmor 1900, 190. Larmor stated that his theory required the "explicit postulate" that the orbital electrons of which matter was constituted be separated by 1,000 times their "structural diameters" (Larmor 1936–39, 318).

39. Larmor 1900, 190, 276.

40. This property was pointed out to Larmor circa 1904, but he considered it unimportant as it was not supported (though not contradicted) by the available experimental evidence. This point is discussed further in chapter 8.

Larmor's new electrodynamics, based on the electron and incorporating new space-time and electromagnetic transformations, marked an important break, both conceptually and geographically, with the British Maxwellian tradition of the 1880s and early 1890s. From a theoretical perspective, the ETM transcended the work of Thomson, Poynting, and others by positing the electron as the natural origin and unit of positive and negative electric charge and by offering a new unification of optics, electrodynamics, and matter theory. The ETM embodied what would shortly become known as the "Lorentz transformations" and relied heavily upon the null result of the Michelson-Morley experiment to provide empirical support for its central claims that matter was electrically constituted, that all mass was electromagnetic in origin, and that matter contracted minutely in its direction of motion through the ether. Larmor's theory also provided the theoretical tools necessary to tackle a wide range of problems in the electrodynamics of moving bodies, an issue to which I return below. From a geographical perspective, the emergence of the ETM was accompanied by an effective localization of British research in mathematical electromagnetic theory in Cambridge. We saw in chapter 6 that during the 1880s several non-Cambridge men, most notably FitzGerald and Heaviside, mastered Maxwell's *Treatise* and made extremely important contributions to its theoretical interpretation. As I also emphasized in that chapter, the most important theoretical results achieved by these men were not passed on to students in Dublin or London, but to Cambridge undergraduates preparing for the Mathematical Tripos. Only the latter institution produced a steady stream of Maxwellians during the 1880s and 1890s, the emergence of the ETM around 1900 further isolating Cambridge as the lone center in Britain for research in mathematical electromagnetic theory. Larmor's extremely technical papers on the subject found few, if any, competent British readers beyond Cambridge, but, as we shall shortly see, the ETM was effectively transmitted to a new generation of wranglers during the first decade of the twentieth century.

The British localization of electromagnetic theory within Cambridge in the 1890s was counterbalanced during the first decade of the twentieth century by an increasing internationalization of the discipline. Growing emphasis on the formal-mathematical structure of electron-based electromagnetic theory, visible not only in Larmor's work but also in that of several of his European contemporaries, made it increasingly easy for students trained by Larmor to see themselves as engaged in a common project with Continental mathematical physicists. Where Larmor himself had sharply distinguished his work from that of Lorentz, Max Abraham, and, especially, Henri

Poincaré, the next generation of wranglers had few qualms about drawing freely on their work. It is also likely that Larmor's effective withdrawal from research on the ETM after 1900 positively encouraged his students to look elsewhere for inspiration in developing the electrodynamics he had helped to pioneer. Following his appointments as Secretary of the Royal Society in 1901 and Lucasian Professor of Mathematics in 1903, Larmor devoted much of his time to writing semi-popular lectures and to academic administration. Thereafter he published just a handful of short papers clarifying his earlier work and made only occasional contributions to ongoing debates in other branches of mathematical physics.[41] Larmor's final attempt during World War I to defend the ETM as a viable alternative to the general theory of relativity is discussed in chapter 9.

7.4. Cambridge Pedagogy and the Establishment of the ETM

When Larmor's *Aether and Matter* was published in 1900, the physics community was just beginning to accept that the tiny charged particles, or "corpuscles," identified experimentally by J. J. Thomson and others in the late 1890s, were both the fundamental carriers of negative electric charge and subatomic particles common to all atoms.[42] Not surprisingly, British proponents of the ETM immediately went further and identified the corpuscles with Larmor's massless "electrons."[43] But while what was sometimes called simply "electron theory" was quickly and generally accepted by virtually all physicists during the first four years of the twentieth century, it would be wrong to infer that all physicists in Britain, or even in Cambridge, subscribed to the ETM. In order to adopt electron theory, it was necessary to believe only that electrons were the fundamental carriers of negative electricity, and that they were one of the constituent parts of the atom. This meant that certain physical and chemical properties of atoms and of the radiation they emitted might be explicable in terms of these tiny particles and their orbital motions within the atom. Most British physicists did in fact believe that the whole mass of the electron was electromagnetic in origin, but this was not equivalent to the belief that atoms were made *solely* of positive and negative

---

41. Larmor's most significant work in electromagnetic theory in this period is discussed in Warwick 1995a.

42. See, for example, Falconer 1987. Falconer discusses the diverse views expressed in Britain on the nature and importance of Thomson's corpuscles after 1897, and the way in which they gradually became "electrons."

43. Falconer 1987, 271.

electrons and that *all* of the mass of the atom (and hence of matter) was really electromagnetic mass.[44]

The ETM achieved its most widespread popularity in Britain around 1904, but soon after this a number of experimenters began to doubt the utility of the assumption that the whole mass of the atom was electromagnetic in origin. For one thing, the predicted positive electron had not been detected and, for another, ongoing research in experimental atomic physics appeared to indicate that the positive part of the atom was quite different in nature from the negative electron.[45] The mathematical details of Larmor's ETM were in any case of little relevance to the vast majority of British experimenters. They were not trained in higher mathematics and employed their own conceptual models of the atom in their experimental work. Conversely, the vast majority of wranglers had little or no experience in experimental physics and learned and researched electromagnetic theory as a fundamentally mathematical discipline. During the first two decades of the twentieth century, then, both commitment to the principles of the ETM as well as mastery of its mathematical methods were confined to graduates of the Mathematical Tripos. In this section I discuss in turn the three main channels through which the ETM was taught in Cambridge: examination questions, new textbooks, and Larmor's college and professorial lectures.

Examination questions based on the mathematical methods of Larmor's electrodynamics of moving bodies began to appear in Part II of the Mathematical Tripos just a year after the publication of *Aether and Matter*. In 1901, H. M. Macdonald (4W 1889) set a question (fig. 7.2) requiring the student to derive the electrical potential and forces surrounding a charged conductor moving rapidly through the ether.[46] These expressions had been derived by Larmor in Part III of the "Dynamical Theory" in 1897 and were reproduced in *Aether and Matter*.[47] In 1902, Larmor himself set a similar question (fig. 7.2), but on this occasion he also asked the student to investigate the relationship between the potential surrounding a system of moving conductors and that surrounding the *same system* at rest in the ether and elongated by a factor of β.[48] This was, as I explained in the previous section, the

---

44. Many physicists believed into the 1930s that the whole of the electron's mass was electromagnetic in origin. See, for example, Kaye and Laby 1936.

45. Warwick 1989, 210–25.

46. Macdonald had adopted many of Larmor's techniques, including the new space-time transformations, in his own work on electrodynamics. Macdonald's work is discussed in Warwick 1989, 113–16 and 1995a, 321.

47. Larmor 1900, 11. Larmor's original derivation is discussed in Warwick 1991, 57, 81–83.

48. Where β is defined as in equations (1), (2), and (3) above.

> MATHEMATICAL TRIPOS. PART II. 739
>
> SATURDAY, *June* 1, 1901. 2—5.
>
> DIVISIONS VI, VIII.
>
> 1. Explain a theory of the alterations necessary in the equations of electrodynamics to make them applicable to a region of space in which there is matter in motion, and state the optical relations which have to be satisfied when the matter is transparent.
>
> 2. A conductor is moving through the aether with uniform velocity v in the direction of the axis of x; shew that the components of the electric force at any point in the aether are given by
>
> $$-(V^2-v^2)\frac{\partial\phi}{\partial x},\quad -V^2\frac{\partial\phi}{\partial y},\quad -V^2\frac{\partial\phi}{\partial z},$$
>
> where V is the velocity of radiation and ϕ satisfies the differential equation
>
> $$(V^2-v^2)\frac{\partial^2\phi}{\partial x^2}+V^2\frac{\partial^2\phi}{\partial y^2}+V^2\frac{\partial^2\phi}{\partial z^2}=0.$$
>
> If the conductor is a sphere of radius a, prove that $\phi=A\eta$, where
>
> $$V^2a^2\tanh^2\eta+(V^2-v^2)(y^2+z^2)\sinh^2\eta=v^2a^2,$$
>
> and determine A in terms of the charge on the sphere.

FIGURE 7.2. Mathematical Tripos questions on Larmor's electrodynamics of moving bodies set in 1901 and 1902 by H. M. Macdonald (4W 1889) and Larmor respectively. Questions of this kind helped to legitimize the concept of "motion through the aether" as well as the contraction (or "elongation") hypothesis. Today they serve to remind the historian of the range of advanced mathematical and problem-solving skills within which these hypotheses acquire meaning. After the examination such questions formed an archive of exercises from which subsequent generations of students could hone their technical skills. I have included question 1 on each paper to give a flavor of the other subjects covered. Cambridge University Examination Papers, Mathematical Tripos, Part II, Saturday, 1 June 1901, 2–5, Divisions VI, VIII, q. 2; and Friday, 30 May 1902, 2–5, Division VIII, q. 6. (By permission of the Syndics of Cambridge University Library.)

technique of "correlated systems" that had led Larmor to the contraction hypothesis and hence to his explanation of the null result obtained in Michelson and Morley's ether drift experiment. The final part of Larmor's question was also related indirectly to the contraction hypothesis. The student was required to show that the potential of an asymmetrical conductor was a function of its orientation with respect to its direction of motion through the ether, the implication of this result being that an electro-mechanical torque would act on the conductor to bring it to the orientation of minimum potential. In 1902, Larmor had not only demonstrated to his own satisfaction that no such a torque existed but had used this demonstration to provide further evidence that moving matter contracted by a factor of $\sqrt{(1-v^2/c^2)}$ in its direction of motion through the ether.[49] We must as-

---

49. Larmor had argued that if the torque existed it would be possible to build a perpetual motion machine using the earth's motion through the ether. The contraction of the moving

> 746 MATHEMATICAL TRIPOS. PART II.
>
> FRIDAY, May 30, 1902. 2—5.
>
> DIVISIONS V, VIII.
>
> 1. The principal moments of inertia of the earth being A, A, C, let θ_0 be the inclination of the equator to the principal plane of the earth's motion; ϕ_0 the angle between an axis A fixed in the earth and the descending node of the equator on the principal plane; n the angular velocity of the earth's rotation; ϵ the epoch of ϕ_0, so that
>
> $$\phi_0 = -\int_0^t \frac{C-A}{A} n \, dt + \epsilon;$$
>
> and L, M the perturbing couples acting on the earth about the axis A and an axis 90° west of A in the equator; prove by the method of variation of parameters that the equations of motion of the earth are
>
> $$k \frac{d\theta_0}{dt} = -L \cos \theta_0 \sin \phi_0 - M \cos \theta_0 \cos \phi_0,$$
>
> $$k \sin \theta_0 \frac{d\epsilon}{dt} = -L \cos \phi_0 + M \sin \phi_0,$$
>
> where k is a certain constant.
>
> How do you explain the fact that in the unperturbed motion ϕ_0 diminishes, whereas the earth has a positive rotation?
>
> 6. Prove that, in the case of a system of electrified bodies travelling through the aether with uniform velocity, the electric force (i.e. the force acting on electric charge) is derived from a potential.
>
> Shew that in the case of a system of charged conductors, their several potentials are ϵ times as great as those of a system of correlative conductors at rest, of the same geometrical form except that the system is elongated in the ratio $\epsilon^{\frac{1}{2}}$ in the direction of the motion, where $\epsilon = (1 - v^2/c^2)^{-1}$, v being the velocity of convection and c that of radiation.
>
> Shew, by consideration of an ellipsoid, that an elongated charged conductor moving at high speed through the aether tends to set itself lengthways to the direction of motion.

sume therefore that a student who had attended Larmor's lectures and studied his most recent publications would have been aware—and perhaps have noted in his answer—that no electro-mechanical interaction actually occurred (at least to the second order of v/c) between moving charged bodies and the ether, and that this furnished important evidence for the contraction hypothesis (and hence the ETM).

Tripos questions of the kind just discussed are significant in the development of electrodynamics in early twentieth-century Cambridge in two respects. First, it was in large measure through questions of this kind that the generation of Cambridge mathematicians trained in the early 1900s developed their skill in the mathematical manipulation of such theoretical entities as electric and magnetic fields. Their commitment to Larmor's electro-

---

conductor rendered its potential independent of its orientation with respect to its motion through the ether. This argument is examined in detail in Warwick 1995a.

dynamics thus emerged not simply from a dogmatic belief in the existence of an electromagnetic ether, but as part of a protracted process of learning to conceptualize and to solve difficult problems in the electrodynamics of moving bodies using Larmor's conceptual apparatus and mathematical methods. This route to mastering Larmor's work, as preparation for Part II of the Mathematical Tripos, also helps to explain the difference between Larmor's own attitude to the ETM and that adopted by most of his students. Wranglers such as Larmor, trained in the late 1870s, preserved something of Maxwell's sense of the physical and dynamical reality of the electromagnetic ether. It is important to keep in mind that Larmor's route to the electron and the new electrodynamics of moving bodies had begun in an attempt to find a common dynamical foundation for electromagnetic and luminiferous processes, and that, along the way, he had often found it necessary to consider the physical micro-structure of the ether.[50] For Larmor's students from 1900 the situation was very different. From an ontological perspective they needed to know only that the universe consisted of positive and negative electrons in a sea of ether, the application of the ETM then following as a largely mathematical exercise based on Maxwell's equations and Larmor's electrodynamics of moving bodies. As we shall see, these students were much less concerned with the physical and dynamical nature of the ether than they were with the application of electron theory and the electrodynamics of moving bodies to specific physical problems.

The second, and closely related, way in which Tripos questions can inform us about the wrangler approach to electrodynamics after 1900 concerns the form of many of the questions set. Students were often asked to *demonstrate* the fundamental properties of moving electrical systems according to the ETM but were not asked to comment on the order of v/c to which such demonstrations should be limited. Thus although the second question above requires candidates to prove that moving and stationary systems are "correlated" in the manner claimed by Larmor, it does not invite comment on Larmor's implicit caveat that such results were justified empirically only to the second order. This omission not only points to the increasingly mathematical way in which even Larmor taught electrodynamics after 1900 but helps to explain why wranglers readily conflated Larmor and Lorentz's work on electrodynamics. When Lorentz demonstrated in 1904 that equations (1), (2), and (3) above actually transformed Maxwell's equa-

---

50. Larmor's views on ether structure, culminating in the introduction of the electron, are discussed in Buchwald 1985, 141–73, and Hunt 1991a, chap. 9.

tions exactly to *all* orders of v/c, most young wranglers understood the demonstration as a straightforward mathematical extension of Larmor's work. They believed Lorentz to have shown that the earth's motion through the ether was completely undetectable by electromagnetic means and were prepared to follow their European colleagues in referring to this property as constitutive of a "principle of relativity."[51] Unlike Einstein's axiomatic use of the term in 1905, however, wranglers understood the principle merely to describe the covariance of Maxwell's equations under the Lorentz transformations.[52] And, as we shall see in chapter 8, they also believed that the principle could hold true only if matter was constructed electronically and all mass was electromagnetic in origin.

Another important vehicle for the propagation of the fundamental principles of the ETM in early twentieth-century Cambridge was textbooks on electromagnetic theory. The first and by far the most influential book of this kind was James Jeans's (2W 1898) *Mathematical Theory of Electricity and Magnetism* (hereafter *Mathematical Theory*), published in 1908. Jeans was, as I noted in chapter 5, a typical late Victorian wrangler; he was prepared by a wrangler-schoolmaster at Merchant Taylors' School for a mathematical scholarship at Trinity College and then coached to the second wranglership in just two years by R. R. Webb and R. A. Herman. In 1900 Jeans was placed in the first division of Part II of the Mathematical Tripos, and, in 1901, he was awarded a Smith's Prize for an essay on the "distribution of molecular energy."[53] Jeans left Cambridge in 1905 to take up a professorship in applied mathematics at Princeton University, his Princeton lectures on electricity and magnetism forming the basis for the *Mathematical Theory*.[54] In the preface to the first edition, Jeans explained that there was (by 1907) "a certain well defined range of Electromagnetic Theory, which every student of [mathematical] physics may be expected to have covered," and that while Maxwell had written the *Treatise* for the "fully equipped mathematician," his book

---

51. In 1904 Poincaré first described Lorentz's demonstration that Maxwell's equations transformed exactly under the "Lorentz transformations," as constituting a "principle of relativity" (Miller 1981, 79). This point is discussed further in chapter 8.

52. Larmor would not have gone along with this definition, as he believed the transformations to be empirically justified only to the second order of v/c. Lorentz evaded the question of what would happen when electrons were forced into close proximity by assuming that they too were subject to the FitzGerald-Lorentz contraction (Miller 1981, 67–75).

53. Barrow-Green 1999, 310.

54. Milne 1952, 14–15. Jeans also took the Cambridge style of mathematical training to Princeton. By 1907 his students gently mocked him with the couplet, "Here's to Jimmy Hopwood Jeans. He tries to make us Math-machines."

had been written "especially for the student, and for the physicist of limited mathematical attainments" (Jeans 1908, v).

Jeans's remark that the book was suited to the mathematician of "limited" attainment needs to be treated with some caution. As Charles Chree (6W 1883) accurately wrote in a review for *Nature*, the *Mathematical Theory* provided "a good book of reference to the physicist of superior mathematical attainment," but was "best adapted" to the needs of the student "preparing for mathematical examinations, such as the Cambridge Tripos" (Chree 1908). In addition to reproducing many of the most important theoretical results given by Maxwell in the *Treatise*, Jeans's book also covered the higher mathematical methods required by the student. The *Mathematical Theory* thus united many of the technical skills that, in the 1880s, would have been taught by Niven and Routh. Other reviews of the first and subsequent editions of the book show that it was in general extremely well received, partly because there existed a real need for a comprehensible textbook covering mathematical electromagnetic theory. From 1908 until at least the late 1920s, Jeans's *Mathematical Theory* remained the standard student reference work on this important subject.[55]

Approximately the first three quarters of the *Mathematical Theory* consisted in a lengthy summary of Maxwell's account of electro- and magnetostatics, current electricity, and electromagnetic induction in which the advanced mathematical methods employed by Maxwell were explained in terms that would be comprehensible to an aspiring wrangler (fig. 7.3). In these and subsequent sections of the book, Jeans also drew upon the accomplishments of the Cambridge Maxwellians by including literally hundreds of examples from "the usual Cambridge examination papers." He noted that these examples were extremely important as they not only provided problems for the mathematical student to tackle as exercises but formed "a sort of compendium of results for the physicist, shewing what types of problem admit to exact mathematical solution" (1908, v). As we saw in chapters 3 and 5, these end-of-chapter collections of problems—by this time standard in textbooks on mathematical physics—formed an important and cumulative link between the original investigations of Cambridge mathematicians, the ideals of competitive examination, and the regimes of

---

55. The reviewer for the *Philosophical Magazine* wrote, "undoubtedly it appears to us to be a book which is destined to take a leading place amongst expositions of electromagnetic theory" (Anon. 1908a). The reviewer for *The Electrician* also commented upon the need for such a book (Anon. 1908b). The lasting popularity of the book is reflected in a review of the second edition published in *The Electrician* (Anon. 1912).

HARMONIC POTENTIALS.

257. We are now in a position to apply the results obtained to problems of electrostatics.

Consider first a sphere having a surface density of electricity S_n. The potential at any internal point P is

$$V_P = \iint \frac{S_n dS}{PQ} = \iint \frac{S_n dS}{\sqrt{a^2 - 2ar\cos\theta + r^2}}$$

$$= \iint \frac{S_n}{a}\left(1 + \frac{r}{a}P_1(\cos\theta) + \frac{r^2}{a^2}P_2(\cos\theta) + \ldots\right) dS$$

$$= \frac{4\pi}{2n+1} a^2 \frac{r^n}{a^{n+1}} (S_n)_{\cos\theta=1}, \text{ by the theorems of §§ 237 and 255,}$$

$$= \frac{4\pi}{2n+1} \frac{r^n S_n}{a^{n-1}} \quad\ldots\ldots\ldots\ldots\ldots\ldots\ldots\ldots\ldots\ldots\ldots(173),$$

this expression being evaluated at P.

Similarly the potential at any external point P is

$$V_P = \frac{4\pi a^{n+2} S_n}{(2n+1) r^{n+1}}.$$

These potentials are obviously solutions of Laplace's equation, and it is easy to verify that they correspond to the given surface density, for

$$\left(\frac{\partial V}{\partial r}\right)_{\text{outside}} - \left(\frac{\partial V}{\partial r}\right)_{\text{inside}} = 4\pi S_n.$$

This gives us the fundamental property of harmonics, on which their application to potential-problems depends: *A distribution of surface density S_n on a sphere gives rise to a potential which at every point is proportional to S_n.*

258. The density of the most general surface distribution can, by the theorem of § 240, be expressed as a sum of surface harmonics, say

$$\sigma = S_0 + S_1 + S_2 + \ldots,$$

in which S_0 is of course simply a constant. The potential, by the results of the last section, is

$$V = 4\pi a \left\{ S_0 + \frac{S_1}{3}\left(\frac{r}{a}\right) + \frac{S_2}{5}\left(\frac{r}{a}\right)^2 + \ldots \right\} \text{ at an internal point } \ldots(174),$$

$$V = 4\pi a \left\{ S_0\left(\frac{a}{r}\right) + \frac{S_1}{3}\left(\frac{a}{r}\right)^2 + \frac{S_2}{5}\left(\frac{a}{r}\right)^3 + \ldots \right\} \text{ at an external point } \ldots(175).$$

FIGURE 7.3. These two sections from Jeans's *Mathematical Theory* form a link between eighteen pages on the solution of Laplace's equation [$\nabla^2 V = 0$] in terms of spherical harmonics (see also fig. 8.4 for the beginning of that section) and numerous examples showing how the solutions were used to tackle problems in electrostatics. Jeans essentially offers a more formal and succinct account of the presentation of the subject given by his own coach, Robert Webb (chapter 5) in the late 1890s. Jeans, *Mathematical Theory*, 1908. (By permission of the Syndics of Cambridge University Library.)

supervised rehearsal by which undergraduate mathematicians mastered their craft.

In the last quarter of the *Mathematical Theory,* Jeans grounded the fundamentals of electrodynamics on the post-1900 concept of the electron and fully embraced the ETM. Towards the end of the book he summed up his discussion of electromagnetic mass in the following way:

> We may say that experiment makes it certain that a large part of the mass of the electron is electromagnetic in its origin, and makes it probable that the whole mass is of this nature. If so, it is natural to suppose further that the mass of the carriers of positive electricity must also be electromagnetic. Thus we may conjecture with a good deal of probability, that all mass is of electromagnetic nature, so that all kinetic energy of material bodies can be explained as the electromagnetic energy of displacement currents—currents in the ether. (Jeans 1908, 512)

The experiments to which Jeans referred were those undertaken by the German experimenters Walter Kaufmann, Adolf Bestelmeyer, and Alfred Bucherer in order to try to ascertain which of several models of the electron was the correct one.[56] By 1908, most European physicists believed it likely that the whole mass of the electron was electromagnetic in origin, but Jeans's evident commitment to the ETM required him to go further and to claim that these results made it probable that *all* mass was really electromagnetic mass. Jeans commented on the debate over which model of the electron was best supported by the experimental data, and it was during this discussion that he made passing reference to Einstein's work. According to Jeans, Einstein, along with Abraham, Bucherer, and Lorentz, had proposed a "hypothesis" concerning the distribution of charge on a moving electron. Jeans gave no indication that he considered Einstein's hypothesis any more important or radical than those proposed by the other three men, let alone that it might be part of a completely new way of looking at electromagnetic theory or even a new theory of space and time.[57] Jeans concluded, in any case, that the most recent experimental evidence favored the hypothesis put forward by Max Abraham and Bucherer (Jeans 1908, 511).

By the time Jeans prepared the second edition of the *Mathematical Theory* (1911), he had returned to Cambridge to take up the Stokes Lectureship

---

56. For discussion of these experiments and their relevance to various theories of the electron, see Goldberg 1970–71, Cushing 1981, and Miller 1981, chap. 12.

57. Einstein was not in fact concerned with the distribution of charge on the electron and published no "hypothesis" regarding its distribution.

in mathematics. In this edition the latter section of the book had been virtually rewritten, and Jeans did not repeat his earlier claim that the experiments of Kaufmann and Bucherer made it probable that all mass was electromagnetic in origin. By 1911, several experimental physicists had openly attacked the ETM on the grounds that it was overly speculative and irrelevant to the practice of experimental physics.[58] Jeans had therefore to exercise considerable care in discussing the experimental foundations of the theory. He nevertheless retained his faith in the ETM, turning now to a discussion of the Michelson-Morley experiment and the FitzGerald-Lorentz contraction hypothesis in order to substantiate his arguments. In a new section of the electrodynamics of moving bodies, Jeans followed Larmor in discussing the "correlation" of moving and stationary electrical systems, remarking at one point that it was now "natural" to adopt the "Lorentz-FitzGerald hypothesis, that the system S when set in motion with a velocity u assumes the configuration of the system S'," this latter being "a [contracted] configuration of equilibrium for the moving system" (Jeans 1908, 2d ed. 1911, 577). He pointed out that it would not be possible to measure this contraction with a measuring rod because it too would be subject to the same contraction, but, he added, "as we shall now see, optical methods are available where material means fail, and enable us to obtain proof of the shrinkage." As we should by now expect, Jeans went on to describe the Michelson-Morley experiment and to claim that the null result obtained thereby "amounts almost to positive proof of the Lorentz-Fitzgerald contraction hypotheses" (577). It is important to note that the order of argument adopted by Jeans was the same as that used by Larmor: the contraction hypothesis followed quantitatively from the ETM, the null result of Michelson and Morley's experiment confirming this result experimentally. Thus the result was not presented as offering a major challenge to electromagnetic theory (which could only be circumvented by invoking the contraction hypothesis) but rather as confirming evidence that a theoretically predicted effect did in fact occur. The null results of additional experiments designed to detect an electro-mechanical interaction between moving charged bodies and the ether were similarly referred to as providing further evidence that the contraction of moving matter was a fact.

As in the first edition of the *Mathematical Theory*, Einstein was briefly mentioned, this time in connection with the problem of the correlation of

---

58. See, for example, Soddy 1907. Soddy argued that the study of experimental radioactivity had "furnished ... the most damaging evidence against the possibility of an electronic constitution of matter."

electromagnetic forces between moving and stationary electrical systems. Having demonstrated that the electrostatic equipotentials surrounding a charged body contract as the body moves through the ether, Jeans remarked that "Lorentz, to whom the development of this set of ideas is mainly due, and Einstein have shown how the theorem may be extended to cover electromagnetic as well as electrostatic forces" (1908, 2d ed. 1911, 577). He did not make any distinction between the work of Lorentz and Einstein—nor, by implication, between that of Larmor and Einstein—and did not believe that Einstein's work was in any sense incompatible with that of Lorentz. In fact, at the very end of the book Jeans discussed the "Lorentz deformable electron," and drew three conclusions from the latest experimental results (obtained by Bucherer) on the velocity dependence of the mass of an electron. These conclusions were: first, that the results confirmed Lorentz's hypothesis that the electron contracted in the direction of its motion (but did not expand laterally as suggested by Bucherer); second, that they provided further confirmation of the Lorentz-FitzGerald contraction hypothesis, of which the concept of the contracting electron was an extension; and third, that they further confirmed that the mass of the electron was of purely electromagnetic origin (579). These arguments, all of which repeated in the third edition of the *Mathematical Theory* (1915), clearly indicate that Jeans neither perceived Einstein's work as representing a new theory of electromagnetic action based on two clearly defined postulates, nor that the application of these postulates might require a major change in the way physicists understood the concepts of space and time. In short, there was no mention of any new theory of relativity.[59]

Jeans's *Mathematical Theory* was by far the most popular introductory textbook in the study of advanced mathematical, electromagnetic theory, and, although used throughout Britain, it was particularly popular with students of the Cambridge Mathematical Tripos—for whom it had almost certainly been written. It is therefore quite reasonable to suppose that the views expressed by Jeans in the book were commonly held by graduates of the Mathematical Tripos during the period 1900–19. That this was indeed the case is made even more probable by the fact that it was Joseph Larmor who taught electromagnetic theory to these students. Larmor's college and professorial lectures constitute the third and most direct route by which the ETM was propagated in Cambridge during the first decade of the twentieth

---

59. The first account of the "Theory of Relativity" appeared in the fourth edition of *Mathematical Theory*, published in 1920.

century. According to the only firsthand account of these lectures, Larmor's style of presentation was very similar to that of his own teacher in the late 1870s, W. D. Niven. Larmor apparently adopted a "generality of view" that seemed "slow and inconclusive" to the "average student"; but, to those who were "prepared to follow with attention," the lectures were "full of stimulus, sometimes by their very incompleteness provoking the mind to wrestling and questioning."[60] As with Niven's lectures, it appears to have been Larmor's references to aspects of electromagnetic theory that remained obscure or inconsistent that prompted his most able students to undertake research in this area.

There is no firsthand account of the actual content of Larmor's lectures, but we can obtain an indirect sense of the material he covered from a textbook, *The Theory of Electricity,* completed in 1918 by one of his students, G. H. Livens (4W 1909). This book was partly intended to supersede Jeans's *Mathematical Theory* and was based explicitly on Larmor's lectures for the academic year 1909–10.[61] In the preface to his book, Livens explained that his treatment of electromagnetic theory was based on "the original Faraday-Maxwell theory, generalized and extended to the case of moving systems by Sir Joseph Larmor." According to Livens, this form of electromagnetic theory had "been almost entirely abandoned" in recent textbooks on the subject, but remained the only approach that was "completely satisfactory from the point of view of mathematical and physical consistency" (1918, v).[62] On the important question of the origin of inertial mass Livens followed Jeans's cautious optimism in stating that although no "decisive answer" had yet been reached, it was "generally held" that there existed "no material mass in the ordinary sense of the word in the positive particles" and that "the material mass of any ordinary piece of matter is nothing other than the inertia mass of the electric charges involved in it" (1918, 645). Writing in 1917, Livens could not consider his treatment of electromagnetic theory complete without at least some mention of the use of the principle of relativity as an axiom in physics. By this time most British physicists were aware that Einstein used the principle in this way, and, at the very end of his book, Livens accordingly devoted a short section to the discussion of "Relativity." Even in 1917, how-

60. See Ebenezer Cunningham's (SW 1902) *DNB* entry on Larmor.
61. Livens 1918, v. Livens book failed to supercede Jeans's *Mathematical Theory.* Livens would have attended the lecturers in preparation for Part II of the Mathematical Tripos in June 1910 (in which he was placed in the second division of the first class).
62. Livens no doubt had in mind the many textbooks on electromagnetic theory then appearing for use by engineers and students of experimental physics.

ever, Livens did not especially associate relativity with the name of Einstein, nor with any well-defined theory of relativity. In fact, he began the section by noting that the discussion that followed would be simplified by assuming that "the whole electrodynamic properties of matter can be explained on the basis of a stationary aether and electrons" (1918, 656)—in other words, by assuming the ETM.

The aspect of Liven's account of electromagnetic theory that is of most relevance to our present concerns is his evident belief that all the "facts" of electromagnetic theory were accounted for by the ETM and that it was this theory that held the greatest promise for the future progress and unification of physics. For him, as for most other graduates of the Cambridge Mathematical Tripos who were interested in electromagnetic theory during this period, the "principle" or "theorem" of relativity was simply an extension of the ETM due to Lorentz's discovery that Maxwell's equations transformed exactly under the Lorentz transformations. Here, as I noted above, they parted company with Larmor himself, who continued to believe that it might be possible to measure motion through the ether at higher orders (and that this was an essentially empirical question). But the younger generation of wranglers was prepared either to consider the electron for all practical purposes as a mathematical point, or else to follow Lorentz in assuming that, whatever its structure, the electron too would be subject to the FitzGerald contraction; either way, the Lorentz transformations became exact to all orders, even for closely packed electrons.

The establishment of the basic tenets of the ETM among wranglers interested in electromagnetic theory during the first decade of the twentieth century was accomplished through the pedagogical devices discussed in earlier chapters of this study. Coaches such as Webb and Herman continued to drill students in the advanced mathematical methods common to most branches of mathematical physics, while professorial lectures, advanced Tripos problems, and new textbooks propagated the fundamental principles and analytical techniques proper to Larmor's theory. The existence of a well-defined school of electromagnetic theoreticians in early twentieth-century Cambridge was also recognized by contemporaries both within and beyond the university. In 1908, the *Electrician*'s reviewer of Jeans's *Mathematical Theory* noted the pressing need for such a book and expressed surprise that "some distinguished member of the *Cambridge school* [had] not written a similar work before" (Anon. 1908b; my italics). Likewise in 1912, when W. W. Rouse Ball prepared a short history of mathematics in the university for the International Congress of Mathematicians, he entitled his talk "The Cambridge School of Mathematics." What is especially interesting about Rouse Ball's remarks is that the work he most readily identified as constituting a mathematical school in Cambridge was that associated with the continuation of Maxwell's electromagnetic field theory. According to Rouse Ball, this branch of Cambridge mathematics was "still making history" and, he concluded, was not only the most striking achievement of the school in the last century but the one with which Newton, "were [he] here now," would wish "to be particularly associated" (1912, 321–22). As a wrangler from the early 1870s who had participated in the reform of undergraduate and graduate teaching over the subsequent two decades, Rouse Ball was especially pleased to note that the existence of "numerous investigators" working on common research projects had been made possible by "teaching and inspiration" that was "directly due to the professors" (322). It had been major goal of those reforms (see chapter 5) to stimulate research by bringing professors and college lecturers into closer contact with the most able young wranglers, and, in Rouse Ball's opinion, the active development of Maxwell's field theory that had taken place in Cambridge over the previous forty years was ample evidence that the goal had been reached.

Further evidence for contemporary awareness of a well-defined Cambridge school of electromagnetic theory occurs in the form of negative comments on the school's narrow and local focus. An anonymous review of Livens's *Theory of Electricity* published in the *Philosophical Magazine* in 1919 acknowledged it to be a useful "handbook on the writings of the famous Cambridge School of mathematical physicists," but criticized the author for sticking too closely to the "original work of Sir Joseph [Larmor] and his school" (Anon. 1919, 199). This reviewer felt that Livens ought to have taken more account of the latest experimental work on x-rays and the conduction of electricity through gases, and noted that the coverage of "relativity" was "very meagre indeed." This reference to relativity almost certainly referred to Einstein's general relativistic theory of gravitation, which, for reasons discussed in chapter 9, was by this time casting doubt on the competence of the ETM correctly to explain a number of cosmological phenomena. The anonymous reviewer argued that both the empirical success of general relativity and the attention it was receiving from "many of the best mathematical physicists in Europe" made it worthy of "more serious exposition or criticism" (199). By 1919, the school of electromagnetic theory built upon Larmor's writings and developed by his students was so firmly established in Cambridge that it had, in the opinion of at least one reviewer, become insular and indifferent to relevant theoretical developments taking place beyond the university. It is nevertheless important to note that comments of this kind did not appear until after World War I, and that, as we shall also see in chapter 9, by the time they were published plans were already afoot to include the special and general theories of relativity in the Cambridge curriculum.

7.5. Putting the ETM to Work

Having outlined the origin and rise of the Cambridge-based research school in the ETM from around 1900, I conclude this chapter by looking briefly at an example of the theory being applied by one of Larmor's students to tackle a problem in the electrodynamics of moving bodies. I noted above that commitment to the ETM was not simply a matter of belief in the existence of an electromagnetic ether, but emerged in the process of learning to apply a series of specific principles and analytical techniques taken from Larmor's work. The example below not only conveys a powerful sense of how naturally Larmor's students applied these techniques but illustrates their differences both from their older Cambridge colleagues and from Einstein. One of the most important of the techniques just referred to was the application of equations (1), (2), and (3) above to correlate a system moving through the ether with the equivalent elongated system, at rest. For Larmor and his students these equations were demonstrably valid only if matter were composed entirely of electrons whose mass was completely electromagnetic in origin. When older Cambridge mathematicians, trained before the publication of *Aether and Matter,* made electromagnetic calculations that flouted these conditions, Larmor's students were quick to criticize their error and to offer what they saw as the correct analysis in terms of the ETM. Their application of equations (1), (2), and (3) was also superficially similar to uses made of the same equations by Einstein. It was therefore difficult, as we shall see in more detail in chapter 8, for Cambridge mathematicians to appreciate any practical advantages to adopting the Einsteinian approach.

The example in question concerns an attempt by Livens in 1911 to analyze the accelerated motion of a perfectly conducting electrified sphere. Wranglers like Livens, whose primary interest was physics rather than mathematics, regarded Larmor's theory of the electrodynamics of moving bodies not as a subject that required development in its own right, but as a serviceable tool for tackling problems in such areas as x-rays and atomic theory.[63] An

---

63. The work of wranglers who tried to develop Larmor's theory per se is discussed in chapter 8.

excellent example of this kind was the challenge of finding an expression for the motion of an accelerated (or decelerated) charged sphere that took account of the mechanical reaction on the sphere produced by the radiation it emitted. The significance of this calculation lay in the fact that it would provide a powerful theoretical tool for investigating the process by which x-rays were generated when high-speed electrons (conceived as tiny charged spheres) struck the anode inside a cathode-ray tube.[64] The nature and origin of x-rays was a subject of considerable debate in Britain circa 1907, theoretical interest in the topic being piqued in Cambridge at this time by the announcement of the following title for the Adams Prize of 1909: "The radiation from electronic systems or ions in accelerated motion and the mechanical reaction on their motion which arises from it."[65] Livens was too young to enter an essay in this competition—he did not graduate until June 1909—but the prize did indirectly prompt his investigation of the issues in question.

Although the Adams Prize was eventually won by G. A. Schott, it was one of his competitors, G. W. Walker (4W 1897), who was the first Cambridge man to publish a paper on the dynamics of a charged sphere.[66] One of the main purposes of Walker's paper was to determine the sphere's initial acceleration without adopting the so-called "quasi-stationary" hypothesis. This hypothesis amounted to the assumption that the instantaneous electromagnetic mass of an accelerating charge was the same as the mass of an equivalent charge in a state of uniform motion with the same velocity. This was a reasonable and generally accepted assumption provided that the sphere's acceleration was small.[67] In order to avoid the quasi-stationary hypothesis and produce a more exact solution to the problem, Walker found it necessary to ascribe a small, non-electromagnetic, "rest mass" to the sphere. This step rendered Walker's analysis inconsistent with the ETM.[68] Walker's paper was at once attacked by two of Larmor's students in the pages of the *Philosophi-*

---

64. Wheaton 1983, 135–39.

65. *Cambridge University Reporter,* 12 March 1907, 645.

66. Schott 1912; Walker 1910. Of two other entries deemed of almost equal merit to the winner, one was certainly by G. F. C. Searle. The title and submission date of Walker (1910) make it virtually certain that he was the other runner up. See *Cambridge University Reporter,* 11 May 1909, 845; Searle to Heaviside, 10 May 1909 (OJH-IEE).

67. For a contemporary discussion of the "quasi-stationary hypothesis," see Cunningham 1914, 147–48.

68. The form of equations Walker employed became unstable when the velocity of the electron went to zero unless this extra, non-electromagnetic, mass was included.

cal Magazine, first by J. W. Nicholson (10W 1904), who took him to task for assuming the existence of non-electromagnetic mass, and then, early in 1911, by Livens, who similarly attacked the paper and also offered an alternative solution to the problem that was fully compatible with the ETM.[69] One of the most interesting aspects of Livens's analysis is that he appears at first sight to be adopting something akin to a modern relativistic approach to the problem. But, by attending carefully to the meaning he ascribed both to the principle of relativity and to the technical detail of his argument, we shall see that this is not the case.

Livens began by employing Maxwell's equations to investigate the motion of a charged sphere accelerating from rest in the ether (Maxwell's equations definitely applying in this case). In this part of the paper he established that the complicated terms describing the radiative back-reaction on the sphere due to its initial acceleration from rest were only transitory and that the electromagnetic mass of the sphere would quickly settle down to its generally accepted value.[70] Livens then turned to the case of a charged sphere that began its acceleration (or deceleration) from an initial velocity "v" through the ether.[71] In order to solve this problem, Livens turned at once to Larmor's transformation technique and, referring to the appropriate passage in *Aether and Matter,* remarked that according to Larmor, "if we refer the phenomena to a set of axes moving with a velocity v, the fundamental equations of the theory, the Maxwell equations referred to moving axes, assume exactly the same form as they had originally referred to fixed axes" (1911, 644). It is here that Livens's calculation appears very like a relativistic analysis, since he claimed that the solution to a problem formulated in one frame of reference must have an identical solution when similarly formulated in a second frame moving uniformly with respect to the first. It was for this reason that Larmor's students were prepared to refer to this property of electrical systems as constituting a "principle of relativity." But, unlike Einstein, they believed that the principle had to be proved analytically rather than assumed as an axiom of physics.

Livens's procedure from this point forward illustrates very clearly the difference between the Einsteinian and Larmorian interpretations of equa-

---

69. Nicholson's work is discussed further below and in chapter 8.

70. Livens 1911, 641–44. That is $m = 2e^2/3ac^2$ (where e is the charge on the sphere, a its radius and c the velocity of light).

71. Ibid., 644. This calculation is particularly relevant to the case of an electron striking the anode of a discharge tube because the tube would be moving with the earth's velocity through the ether.

tions (1), (2), and (3). Livens now considered two identical electrical systems, one at rest in the ether and the other moving through it with a constant velocity v. He then pointed out that the electromagnetic states of these identical systems could be correlated in the following way: insofar as the electromagnetic vectors (**E**, **B**) of the stationary system could be expressed (using Maxwell's equations) as functions of (x, y, z, t), so the correlated states of the moving system (**E**$_1$, **B**$_1$)—as measured in the moving frame—would be expressed as the same functions of (x$_1$, y$_1$, z$_1$, t$_1$).[72] In other words, the solution he had found for the case of the sphere accelerating from rest in the ether would apply equally to the sphere beginning with a velocity v, provided that it was written in terms of (**E**$_1$, **B**$_1$) expressed as a function of (x$_1$, y$_1$, z$_1$, t$_1$).

Livens now took a final step in his calculation that further betrayed his debt to Larmor's work. The variables (**E**$_1$, **B**$_1$) and (x$_1$, y$_1$, z$_1$, t$_1$) were those that would be measured by an observer moving with velocity v through the ether. In order now to find the "true" value of the electromagnetic variables associated with the accelerating sphere he would have to calculate their value as measured in the stationary-ether frame of reference. Livens showed that the actual electromagnetic mass (M$_x$, M$_y$, M$_z$) of a sphere moving through the ether with velocity v was related to that measured by an observer moving with the system (m$_x$, m$_y$, m$_z$) by the expression (M$_x$, M$_y$, M$_z$) = (β^3m$_x$, βm$_y$, βm$_z$).[73] Thus the components of electromagnetic mass (M$_e$) of a uniformly accelerated charged sphere, starting from any constant velocity less than that of light, are given by the expression:

$$M_e = \frac{2e^2}{3ac^2}(\beta^3, \beta, \beta) \qquad (4)$$

which is the same value as is obtained when the quasi-stationary hypothesis is employed.[74] Livens had succeeded in demonstrating the validity of the quasi-stationary hypothesis in this calculation by a routine application of Larmor's transformation techniques and without assuming the existence of any non-electromagnetic mass. That Livens was following Larmor throughout the calculation is underlined by his claim that the real electromagnetic fields surrounding a moving electrical system can be found only by referring back to the stationary ether frame of reference.

---

72. Ibid., 645. Where the electromagnetic, space, and time variables are related by equations (1), (2), and (3) above.

73. A more detailed analysis of this calculation is given in Warwick 1989, 131–34.

74. 1911, 645. The value of v in β is the sphere's absolute motion through the ether. Lorentz's version of this calculation using the quasi-stationary hypothesis is discussed in Miller 1981, 73–75.

The differences between Livens and Walker concerning the proper way to tackle the problem of the initially accelerated motion of a conducting sphere illustrate both the theoretical core and the generational divide associated with Larmor's electrodynamics. Walker had graduated before the publication of the final instalment of Larmor's "Dynamical Theory" and had thereafter worked for seven years at the Cavendish Laboratory.[75] Those working at the Cavendish under J. J. Thomson during the period 1897–1904 regarded the negatively charged, subatomic particles discovered by him in 1897 as "corpuscles," and were loth to identify them with Larmor's massless electrons.[76] Walker therefore had no qualms about attributing a small non-electromagnetic mass to his charged sphere. But, as we have seen, when he published his analysis, he was immediately challenged by the two of Larmor's students most responsible for applying the ETM to problems in electron dynamics. These young wranglers believed the ETM to point the way to a fully electronic theory of the atom and appreciated that the theoretical consistency of the project required the assumption that the atom was made solely of positive and negative electrons whose mass was entirely electromagnetic in origin.

Larmor's students continued to develop the ETM well into the second decade of the twentieth century. For example, in 1915 Nicholson read a paper before the Physical Society in which he stoutly defended the notion that all mass was electromagnetic in origin. We have already seen that Nicholson attacked Walker for assuming the existence of non-electromagnetic mass, and, in 1910, he published two papers (1910a and 1910b) showing how the stability of Walker's equations could be retained without this assumption. In his 1915 paper, Nicholson applied the ETM to stellar spectroscopy and the explanation of the periodic properties of the elements.[77] His main purpose was to challenge Ernest Rutherford's planetary model of the atom—in which the hydrogen atom consisted of an electron orbiting a positively charged particle—by suggesting that the nucleus should be regarded as "an aggregate of [positive] nuclei and electrons whose total charge amounts to $+e$" (1915, 217). Nicholson hoped this model would enable the ETM to accommodate the atomic phenomena associated with experimental radioactivity, and he made

---

75. Thomson et al. 1910, 334. Both Schott and Searle likewise trained before 1900, worked at the Cavendish Laboratory, and did not adopt Larmor's methods.

76. By 1909, Cavendish workers generally used the generic term "electron" for negatively charged, subatomic particles, but it was of little concern to them whether electrons had a small non-electronic mass. Warwick 1993.

77. For a detailed discussion of Nicholson's atomic theory, see McCormmach 1966–67.

a start by proposing an electronic theory of isotope formation (1915, 221–24).[78] In order to justify the arguments deployed in the paper, Nicholson claimed that it was now "generally believed" that the electron's mass was "purely of electromagnetic origin" and that positive electricity was also discrete and existed in "portions equal to e or multiples of e." He readily acknowledged that the mass of a positive particle was much greater than that of an electron but insisted that the properties of the two fundamental particles were otherwise so similar that it was reasonable to assume that the mass of both was purely electromagnetic and subject to the same formula (1915, 217).

The work of men like Livens and Nicholson clearly illustrates that Cambridge electromagnetic theory was perceived by its practitioners neither to be in a state of decline in the first fifteen years of the twentieth century nor to require a special explanation of Michelson and Morley's null result. Thanks to their Cambridge training, Larmor's students assumed the ETM to provide both a sound framework for the reduction of physics to electrodynamics and a series of practical techniques for tackling outstanding problems in the micro-physical properties of matter, the origin of electromagnetic radiation, and the structure of the atom. Recognition of this state of affairs not only helps to explain the lasting appeal of the ETM but also enables us to begin to understand why most Cambridge mathematicians took little interest in Einstein's work on the electrodynamics of moving bodies between 1905 and 1919. When Einstein showed in 1905 that equations (1), (2), and (3) followed from the principle of relativity and the principle of the constancy of the light velocity (treated axiomatically), men like Livens and Nicholson (insofar as they were aware of Einstein's work at all) assumed that he had merely rederived, by "philosophical" considerations, relationships that were already well known and regularly used in Cambridge electrodynamics.[79] They believed, moreover, that the new principles were empirically unfounded and led to no new experimentally testable results. In Nicholson's opinion, as we shall see in chapter 8, Einstein's relativistic approach to electrodynamics entailed serious physical contradictions that did not arise in the ETM.

---

78. Nicholson argued that atomic nuclei contained negative electrons and smaller positive particles. He suggested that isotopes were formed by the emission of alpha particles whose constituents acquired greater electronic mass when ejected from the nucleus. He called this difference the "mutual mass." Nicholson's atomic theory is discussed in McCormmach 1966–67, but this paper ignores Nicholson's commitment to, and deployment of, the ETM.

79. The "philosophical principle of relativity" is discussed in Livens 1918, 661.

7.6. Training, Truth, and Rationality

> What can it mean to say that nothing makes people do or believe things which are rational or correct? Why in that case does the behavior take place at all?
> DAVID BLOOR, *KNOWLEDGE AND SOCIAL IMAGERY* [80]

The continued development of an ether-based ETM in Cambridge over the first two decades of the twentieth century effectively highlights the constructive power of training systems to shape scientists' beliefs and technical competencies. One reason that training has been little studied in the context of the contemporary mathematical sciences is that it is generally regarded as a passive process that merely passes on to a new generation that which is taken for granted by more experienced practitioners. How and why students slowly master and accept a body of established knowledge are not seen as problems that require explanation. As David Bloor (1976, 10–11) pointed out more than twenty years ago, causal explanations of intellectual activity, especially if they are explanations of a social or cultural kind, tend to be evoked only when the activity in question is understood to have been irrational or erroneous. All the while students correctly learn and reproduce knowledge that their teachers regard as self-evidently true, nothing more needs to be said.[81] Moreover, my description of the process by which Larmor's ETM was propagated in Cambridge after 1900 might appear to some readers to exemplify this explanatory strategy. Since the ETM is now believed not only to be a false theory but one devised largely to preserve the concept of an electromagnetic ether, one might conclude that what I have produced is a detailed explanation of why Cambridge-trained mathematical physicists continued to cling to it long after Einstein had proposed a better theory.

This is not the conclusion I wish to draw. What I have sought to demonstrate in this chapter is that adherence to the ETM in Cambridge after 1900 was not just a product of dogmatic belief in the ether's existence and of hostility to an alternative theory that dismissed the ether as superfluous. As far as Larmor's students were concerned, his work stood as a comprehensive and progressive addition to a research tradition in Cambridge that stretched back to Maxwell himself. Their commitment was not simply to the notion of an ether, but to a sophisticated conceptual structure and range of practical calculating techniques that were gradually acquired through years of coach-

---

80. Bloor 1976, 10.
81. Ibid., 11. Bloor's remarks refer to research rather than training, but they are applicable in both cases.

ing and problem solving. As they acquired these skills, the ether became an ontological reality that lent meaning both to the idea of an ultimate reference system and to the application of dynamical concepts to electromagnetic theory. In order to abandon the ETM in favor of a new theory, they would have had to give up not just the ether concept, but also a whole research programme in the form of hard-learned skills of calculation and successfully accomplished problem solutions.

The reason we are inclined in this case to point to the Cambridge training system as the active *cause* of their beliefs is that the knowledge being imparted is now judged to be false. In this and similar cases we cannot explain the intellectual activity in question by implicit reference to its truth content, and so, as Bloor suggests, we are inclined to look for another explanation. Yet the knowledge content of *any* technical training system is potentially vulnerable to criticism of this kind. Physics students are routinely taught theories that are known to be false (if nevertheless useful), but they master and apply this knowledge as readily and uncritically as they would knowledge that is currently regarded to be true.[82] And even theories regarded today as true may well come to be seen as false fifty or a hundred years hence. What I wish to call into question is not the veracity of modern physics, but the usefulness of such categories as truth and falsity in explaining how students master technical disciplines. Much of the earlier part of this study was devoted to tracing the historical development of the pedagogical techniques by which mathematical physics was taught in Cambridge, and, as we shall see in chapter 9, these resources worked equally well regardless of whether they were employed to teach the ETM or the general theory of relativity. The reason the constructive power of training is more visible in the former than the latter case is that it is only in those historical situations where falsity appears to us to pass for truth that we are apt to seek some kind of explanation.

The need for such an explanation becomes more pressing when the false beliefs persist well after their errors have been highlighted by the appearance of a better theory. Thus it might be objected to some of the points just made that the error made by Cambridge mathematicians lay not in their belief in an ether, but in their apparent unwillingness properly to investigate an

---

82. Physics students are initially taught the so-called classical theories of mechanics, electromagnetism, and thermodynamics, only to be taught later that these theories require drastic revision in the light of relativity theory and quantum mechanics. These revisions often render the conceptual foundation of the older theory (for example, Newton's theory of gravitation) completely false. My point is that we ought not to refer to the ultimate truth of the former theories in trying to understand how they are learned by physics students.

alternative theory once it had been proposed. The weakness in this objection lies in the assumption that they were in some sense in a position objectively to compare the two theories. A comparison of this kind would require them to be equally familiar with both theories and to possess agreed criteria by which a definitive choice could be made. There are however very good reasons why these conditions could not, except with the benefit of considerable hindsight, have been met. We have seen that mastering mathematical physics in the Cambridge style was a process that required a protracted period of guided study, problem solving, and, ideally, face-to-face interaction with a competent practitioner. These were the pedagogical resources that generated the wranglers' sense of what criteria a good theory had to fulfil and how it was properly applied to solve problems. It follows that a symmetrically firm and sympathetic grasp of Einstein's work would be virtually impossible to obtain simply by reading his published papers.[83] As we shall shortly see, Cambridge mathematicians interpreted and judged Einstein's work according to their own criteria. The persistence of the ETM after 1905 points, then, not to the narrow mindedness of its proponents, but to the inherent stability of institutionally based research programs. Physicists trained to interpret the physical world in a certain way generally resist or ignore alternative interpretations not because of dogma, but because acquiring a truly comparable sympathy and grasp of an alternative approach is a much more complicated and protracted process than we generally allow. It is to this issue that we now turn.

83. On the importance of personal interaction in learning the special theory of relativity, see Laub to Einstein, 2 February 1908, Einstein, *Collected Papers,* vol. 5, document 79.

∽ 8 ∾

Transforming the Field
The Cambridge Reception of Einstein's Special Theory of Relativity

> I am sorry that I have so long delayed to write to thank you for sending me . . . a copy of your paper on the principle of relativity. I have not been able so far to gain any really clear idea as to the principles involved or as to their meaning and those to whom I have spoken in England about the subject seem to have the same feeling.
>
> G. F. C. SEARLE TO EINSTEIN, MAY 1909 [1]

8.1. A Letter from Einstein

Sometime during the winter of 1908–9, G. F. C. Searle (28W 1887), a graduate of the Mathematical Tripos and Demonstrator in experimental physics at the Cavendish Laboratory, received a letter from Albert Einstein, then still a technical expert at the Swiss Patent Office in Bern. Einstein's letter contained a copy of a review article he had written in 1907 on the principle of relativity and some of its consequences.[2] Owing to illness and the pressure of work, it was some months before Searle found time to study the paper, but, when he finally did so, he made little headway. Writing to his friend Oliver Heaviside in early March 1909, Searle mentioned the paper and confided that he had "no idea" what the "principle [of relativity]" was. Two months later Searle wrote to Einstein himself thanking him for the paper but adding apologetically that neither he nor any of his Cambridge acquaintances could "gain any really clear idea as to the principles involved or as to their meaning." Searle could only suggest that Einstein "write a short account of the subject" for translation into English and publication in a British

---

1. Searle to Einstein, 20 May 1909; Einstein, *Collected Papers*, vol. 5, doc. 162.
2. Ibid. It is actually unclear which paper Einstein had enclosed, but the timing and Searle's references to "the principle" make it almost certain that it was "On the Principle of Relativity and the Conclusions Drawn from It" (Einstein, *Collected Papers*, vol. 2, doc. 47).

scientific journal, though he hastily added that he was "not an expert in German."[3]

There are several reasons why Searle's brief exchange with Einstein is an appropriate episode with which to open a discussion of the early Cambridge reception of relativity. Searle is the only British physicist to the best of my knowledge to have corresponded with Einstein on the subject before 1919. His receipt of a paper both *by* and *from* Einstein also makes him the only British physicist to have "received" relativity in a literal sense; that is, to have had it brought directly to his attention by the author as a new approach to electrodynamics that was worthy of study. Searle is thus a unique historical figure in Cambridge physics from the first decade of the twentieth century, in that we know he received and actually read Einstein's work, and from whom we have a brief, firsthand account of his immediate response. It is moreover the *nature* of this response that is of greatest significance to our present concerns. Given the radical and revolutionary status subsequently bestowed upon Einstein's work of 1905, one might assume that it made quite a stir among his peers at the time. Yet if Searle's remarks are anything to go by, it was greeted, at least in Britain, with a mixture of indifference and incomprehension.

It is tempting at first sight to explain Searle's bafflement by assuming him to have been an especially unreceptive reader of Einstein's work, but this assumption does not bear closer analysis. Searle was a regular visitor to Germany, had studied electrodynamics from German textbooks, was personally acquainted with several German physicists, and had recently prepared two extremely technical articles on the electrodynamics of moving bodies. On the strength of this work he was invited to present two papers at the 80th meeting of the German Naturforscherversammlung held in September 1908 in Cologne. While in Cologne he was the guest of Alfred Bucherer, who had just announced an experimental proof of the principle of relativity (a subject on which he also spoke at the Cologne meeting) and who had corresponded with Einstein on the matter. It is extremely likely, therefore, that Bucherer and Searle actually discussed Einstein's work, not least because it was Bucherer who subsequently urged Einstein to send Searle a copy of his review paper on relativity.[4] In short, Searle's knowledge of electrodynamics and familiarity with German physics probably made him much more re-

---

3. Ibid. Einstein wrote just such an article towards the end of 1909 for a French audience, but, as far as I am aware, it was never published in English (Einstein, *Collected Papers*, vol. 3, doc. 2).

4. The events that led Einstein to write to Searle are discussed in Warwick 1993, 13–18.

ceptive to Einstein's work than was the average British physicist. Why, then, did he find it so inaccessible and apparently unexciting?

One very specific way of answering this question is to point to the generational shift in Cambridge electrodynamics described in the previous chapter. Having prepared for the Mathematical Tripos in the mid 1880s, Searle's approach to electrodynamics derived from the teachings of J. J. Thomson (also his sometime collaborator) and the writings of Oliver Heaviside. Unlike wranglers trained after 1900, Searle never adopted the transformation techniques introduced by Larmor in the "Dynamical Theory" and *Aether and Matter*. In 1896 Larmor in fact blocked the publication of part of a paper by Searle on the grounds that he had not used Larmor's "well-known transformation" published the previous year but had worked out his results by "labourious and indirect processes."[5] One of Searle's younger colleagues at the Cavendish Laboratory noted in a similar vein in 1909 that Searle had "maintained the old traditions of Cambridge physics, handed down from Stokes, Maxwell, and Rayleigh, in publications dealing with purely mathematical physics."[6] Searle was evidently unaware or uninterested by the fact that, as we saw in chapter 7, Maxwell's equations were widely believed from 1904 to be valid to all orders of magnitude in any reference frame moving uniformly through the ether—a property sometimes referred to as constituting a "principle of relativity." He would therefore have regarded Einstein's axiomatic use of the relativity principle as completely arbitrary and failed to see how it was relevant to the kinds of computational technique he routinely employed himself.

A more general way of responding to the question posed above is to recognize Searle's inability to find meaning in Einstein's paper as exemplary of an important point made at the end of the previous chapter. There I noted that Cambridge physicists were not in a position directly to compare the respective merits of Larmorian and Einsteinian electrodynamics because such a comparison would require them, among other things, to have a roughly comparable mastery of both approaches. But, as we have seen, the rapid mastery of a physical theory was typically generated through a complex pedagogical economy so that Cambridge mathematical physicists were unlikely to gain much understanding of Einstein's work even by reading what he had composed as an expository paper on the principle of relativity and its

---

5. This episode is discussed in Warwick 1991, 55–56.
6. Thomson et al. 1910, 247. Searle's role at the Cavendish Laboratory is discussed in Warwick 1993, 17–18.

consequences.[7] Searle's case is nicely illustrative of this point in two respects. First, it provides a unique and stark example of a physicist who, having been sent a paper on relativity by Einstein, found it completely incomprehensible. Second, it provides strong evidence that the vast majority of Searle's Cambridge colleagues in both experimental and mathematical physics had either read and been similarly baffled by Einstein's work, or, as is far more likely, had simply ignored it as irrelevant to their own researches. As we saw in chapter 7, the authors of Cambridge textbooks on electrodynamics mentioned Einstein's work only in passing and continued well into the second decade of the twentieth century to conflate it with that of Lorentz.

There was however a tiny handful of Cambridge mathematicians who not only identified Einstein's work as important, but as early as 1907 had actually begun to use it in their own research. This would, at first sight, appear very seriously to undermine the point I have just made. If even one or two wranglers managed to grasp Einstein's meaning simply by reading his text, then it would appear possible that any wrangler could in principle have done so. Yet, as we shall see in this chapter, closer examination of these rare cases actually strengthens rather than weakens my argument. The reason Cambridge mathematicians would initially have found Einstein's work irrelevant or incomprehensible is that they would not have recognized (i) the problems to which Einstein's paper was supposed by him to constitute a solution, (ii) the legitimacy of the new principles he employed to effect the solution, and (iii) the practical techniques by which the principles were employed to tackle specific problems. They lacked, that is, precisely the carefully honed attitudes and skills that their coaches and lecturers had cultivated in them with respect to the local approach to electrodynamics. The really important point to note is that these very Cambridge skills could nevertheless make Einstein's work seem important and relevant to some Cambridge mathematicians, though typically not in ways that Einstein would himself have recognized as legitimate. When Einstein's work was read in Cambridge, it took on *new* meanings that can only be understood in terms of the local training and research practices.

In this chapter I illustrate this point with two examples. The first concerns a collaborative venture between Ebenezer Cunningham (SW 1902) and Harry Bateman (SW 1903) intended to extend what they understood as the "principle of relativity" to make it applicable to electrical systems in states

---

7. The paper Einstein sent to Searle began with an exposition of electrodynamics circa 1900 and Einstein's own recent contribution.

of non-uniform motion. This work has been understood by some historians as a contribution to the development of the special theory of relativity, but, as we shall see, this was not really the case. Cunningham and Bateman actually assumed Einstein's work to be a contribution to the Electronic Theory of Matter (ETM) and exploited it to develop the latter theory in new ways. The mathematical methods they employed also display the constructive role their Cambridge training played in shaping their research as well as their attitude to the notion of an electromagnetic ether. Cunningham's work in particular shows that while the ether remained an important ontological entity, its properties could be altered dramatically to suit the changing mathematical structure of the ETM. The second example focuses on J. W. Nicholson's attempt in 1912 to defend the Cambridge use of the principle of relativity (as analytically derivable from the ETM) against Einstein's use of the term as a fundamental postulate. Cambridge mathematicians were aware by this time that Einstein applied the principle in this way, but did not understand him to be proposing a fundamentally new theory and were not inclined to adopt his point of view. Once again, this reluctance did not spring primarily from Einstein's denial of the utility of the ether concept, but from Nicholson's belief that Einstein's axiomatic use of the relativity principle led to paradoxical conclusions that did not arise in the ETM.

The broader point I wish to address through these case studies is that the translation of new theories between distant sites is often a more complicated and fragile process than most reception studies allow. Circulating a radically new theory in the form of a paper publication provides no guarantee that it will be studied by the wider community of potential readers, and, even if it is studied, no guarantee that it will be understood or interpreted in the manner intended by the author. This lack of fixity provides another route by which to highlight the role of training in generating the background skills and assumptions through which a reader produces meaning from a text. It will also highlight a new sense in which pedagogical economies are conservative. We shall see that in addition to preserving indigenous traditions, they tend also to reinterpret and to resist alternative ones. In the case of relativity this meant first incorporating some aspects of Einstein's work in the ongoing project of the ETM and then rejecting its foundational principles as illegitimate. We shall also see that work on the ETM continued to flourish in Britain until the outbreak of World War I, and that its demise thereafter is to some extent attributable to pedagogical factors. I begin however with some general remarks on reception studies and Einstein's early publications on relativity. Although Einstein's work is not itself a principal focus of this chapter, some explanation of its content and presentation will both prepare the ground for the case studies that follow and clarify my approach to the international reception of relativity theory.

8.2. Reception Studies and the "Theory of Relativity"

Einstein's theory of relativity is widely regarded today as one of the greatest intellectual products of the twentieth century. It is fundamental to the practice of modern physics and has revolutionized our understanding of the concepts of space and time. As an intellectual achievement, its origin and development both warrant, and have received, close examination by historians of physics.[8] Yet it is not simply the monumental status of relativity theory that has generated the extensive literature devoted to its early international reception.[9] Also of importance to these reception studies is the fact that the founding principles of the so-called "special" theory of relativity can be traced to a single paper, "On the Electrodynamics of Moving Bodies," published by a single author, Albert Einstein, in September 1905.[10] In this paper Einstein not only clearly set out the foundations of the special theory of relativity but did so in a form that remains acceptable to physicists to this day. The identification of Einstein's paper as the well-defined harbinger of a new and radical physical theory has encouraged several historians to look at early responses to the theory in national context.[11] It has been argued that the range of these responses provides a useful means of comparing national styles of physics during the first decade of the twentieth century.[12]

There can be little doubt that the diverse uses made of Einstein's work in this period can throw light on the style of physics practiced at specific sites, but for our current purposes it is important to add the following qualification to the sense in which these uses are understood as responses to a "theory of relativity": although important innovations in physics can certainly be associated with Einstein's first paper on the subject, it is inappropriate

---

8. The literature on the early history of relativity is massive. For a useful overview and bibliography of studies to 1986, see Cassidy 1986. For an overview of more recent studies, see Darrigol 1996.

9. For studies of the international reception of relativity theory, see Warwick 1992, 625.

10. Einstein, *Collected Papers,* vol. 2, doc. 23. Einstein's unique authorship of the theory was first emphasized in Holton 1973, 165.

11. Had the theory been announced piecemeal over an extended period of time, and/or been constantly revised, it would have provided a much less well-defined object to be "received."

12. See Goldberg 1970, 90, and Glick 1987a, 381.

to depict these innovations as the *cause* of early commentaries on his work. Rather than viewing such commentaries as passive responses to a widely recognized theory of relativity, I shall try to provide an explanation of why a tiny handful of mathematical physicists in Cambridge identified Einstein's paper as relevant to their own research. When understood as an active process of reinterpretation, the Cambridge reception of Einstein's work becomes a far more powerful tool for revealing locally taken-for-granted beliefs and practices in the development of physical theory. Since this claim is central to much of the discussion that follows, I shall develop it a little further in this section.

The historian Stanley Goldberg, who undertook the first study of the early international reception of special relativity, began by noting the three features of Einstein's "innovative theory" that set it apart from other contemporary studies of electrodynamics.[13] According to Goldberg, the three features were: first, that the theory was derived from two postulates (the principle of relativity and the principle of the constancy of the velocity of light); second, that it was a *kinematic* rather than *dynamical* theory; and, third, that it was not dependent upon the existence of an electromagnetic ether.[14] For Goldberg, it is Einstein's clear expression of these features that establishes the latter as the unique author of the special theory of relativity. It is also this clarity of expression that makes the special theory of relativity such a seemingly appropriate topic for reception studies. Those who read Einstein's paper in the *Annalen der Physik*—then the leading German journal of theoretical physics—would surely have noticed the novel content and profound implications of the new theory. On Goldberg's reading, Einstein's paper seems almost calculated to raise a response.

Yet it is important to recognize that those far-reaching aspects of Einstein's paper that now seem so obvious to historians of physics were not so evident to Einstein's contemporaries circa 1905. The three key features identified by Goldberg are described by him as foundational to the "special theory of relativity," yet this theory took a decade to emerge as a well-defined entity. Einstein's 1905 paper was entitled "On the Electrodynamics of Moving Bodies," and nowhere in its pages did Einstein claim to offer, or even mention, a "the-

---

13. Goldberg followed Holton in identifying the innovative features unique to Einstein's work (Goldberg 1970, 91).

14. Ibid., 91. I agree with Goldberg that these were innovative features in the sense that Einstein himself would probably have identified them as crucial in 1905. For a technical exegesis of Einstein's paper, see Miller 1981.

ory of relativity." The latter epithet (Relativitätstheorie) was first used by one of Einstein's contemporaries in 1906 and was only systematically, and somewhat reluctantly, adopted by Einstein himself in 1911.[15] Prior to this adoption, Einstein usually spoke or wrote of the "principle of relativity" when he wished to refer broadly to the principles, arguments, and techniques introduced in his 1905 paper. It was only towards the end of 1915, after he had generalized the theory of relativity to include accelerated frames of reference and gravitation, that Einstein introduced the phrase "special theory" of relativity to distinguish his earlier and now restricted work from the new "general theory."[16] Most contemporary readers of his 1905 paper would not therefore have understood Einstein to be announcing a new and well-defined theory of space and time that would revolutionize physics, but as showing how certain problems in the electrodynamics of moving bodies could be tackled by assuming the principle of relativity and the principle of the constancy of the light velocity.[17]

The historical significance of these cautionary remarks becomes evident in three ways as soon as we begin to explore the Cambridge reception of Einstein's work. First, very few physicists in the university saw fit to publish any response to Einstein's new theory. We can note right away, therefore, that those who were prompted to respond identify themselves not as typical of their compatriots but rather as unusual. Second, as I have argued elsewhere, the meaning ascribed to Einstein's work by experimenters trained in the Cavendish Laboratory differed dramatically from that ascribed by graduates of the Mathematical Tripos. Any scheme for explaining the Cambridge reception of relativity must therefore be capable of differentiating between the intellectual resources available to different scientific subgroups within a single university.[18] Far from the nation state being a useful unit of analysis in receptions studies, it seems that, at least in this case, even a single academic institution is too large a unit from a geographical perspective. Third, of the handful of wranglers who did comment on Einstein's work, most failed to notice all three of the allegedly novel features identified by Goldberg. As we

---

15. Einstein, *Collected Papers*, vol. 2 (German), 254.
16. This point is discussed further in chapter 9.
17. The term "theory of relativity" was first used in Cambridge in connection with Einstein's work by H. R. Hassé (7W 1905) in April 1909. Hassé referred to the "Lorentz-Einstein theory of relativity" in order to indicate that these men assumed the truth of the principle of relativity without analytical proof. See Warwick 1989, 187–88.
18. The response by experimenters at the Cavendish Laboratory to Einstein's work is discussed in Warwick 1993.

saw in chapter 7, Jeans's expository remarks on the electrodynamics of moving bodies in the *Mathematical Theory* made no distinction between the work of Lorentz and Einstein.[19] And, as we saw above, Searle and his Cambridge acquaintances were unable to make any sense of Einstein's paper even when it was brought directly to their attention. Moreover, apparent oversights of this kind cannot generally be ascribed to professional incompetence or to cursory reading of the paper. As we shall see, Cambridge mathematicians attributed meaning to Einstein's work that accorded with their own research interests.

What emerges from my remarks so far is that despite the novel features of Einstein's 1905 paper, there is little meaningful sense in which early Cambridge commentaries can be construed as responses to the "special theory of relativity." Cambridge mathematicians simply did not recognize the innovative features which, according to Goldberg, were constitutive of the new theory. Furthermore, the term "response" is itself rather misleading because it seems to attribute Einstein's work with the active capacity to command the attention of the reader. I stated above that it was not simply the high status of relativity theory in modern physics that prompted the extensive literature on its early reception, but that status has shaped the historiography of early twentieth-century physics in a way that impinges directly upon reception studies. Historians have projected the modern status of relativity theory back onto Einstein's work of 1905, so that its widespread importance to the physics of the first decade of the twentieth-century has become self-evident.[20] Within this relativity-centred historiography, it has become unnecessary to provide an historical explanation of why any of Einstein's contemporaries chose to comment on his early publications. Despite the extreme paucity of such comments, their existence is generally treated as unexceptional.[21] It was probably the title of Einstein's paper, "On the Electrodynamics of Moving Bodies," that determined its likely (small) readership, but, as in Searle's case, even those deeply interested in this topic would not necessarily have found its contents important or comprehensible.

Rather than taking it for granted that physicists should have been interested in Einstein's work, I shall offer a more symmetrical account by seeking

---

19. This conflation of Lorentz's and Einstein's work was not confined to British commentators. See Miller 1981, 227.

20. Joseph Illy (1981, 206) goes as far as to suggest that recent historical reconstructions have turned the history of physics into the history of relativity.

21. For some notable attempts to explain Planck's early interest in Einstein's work of relativity, see Goldberg 1976, Heilbron 1986, 28–32, and Pyenson 1985, 194–214.

to explain why a tiny handful of Cambridge mathematicians were prompted to make the remarks that they did. Put slightly differently, I shall claim that these commentators should be regarded not as mere passive respondents to the appearance of a new and novel theory but as active interpreters who, for whatever reason, identified the work of a then little-known physicist as relevant to their own research. The importance of this seemingly small shift of emphasis is considerable. By abandoning the relativity-centred historiography, we need no longer define the work of those who commentated on Einstein's paper in terms of relativity theory itself. The burden of historical explanation is thereby shifted from accounting for the failure of most commentators to understand what Einstein really meant, to accounting for what it was about the research interests of the commentators themselves that made Einstein's work seem relevant. As this last remark implies, I shall not attribute a unique or essential meaning to Einstein's text, and will not, therefore, speak of Cambridge commentators as having "misread" or "misunderstood" the author's true intentions.

Since the meaning of Einstein's text clearly changed according to the tradition within which it was being appropriated, the stability of a particular interpretation must depend upon elements beyond the text itself. I shall argue that individual interpretations (including Einstein's own) derive in large measure from the locally taken-for-granted research practices within which an interpretation holds currency. As I noted at the end of the previous chapter, the local traditions made up from such practices are usually evoked only insofar as they are perceived to act as obstacles to the adoption of the "proper" Einsteinian meaning.[22] I shall assume that the reconstruction of any interpretation must be couched in terms of these local practices and, indeed, is historically meaningless without them. An important implication of this assumption is that the successful transmission of a new physical theory is extremely difficult to achieve simply by the distribution of a text that allegedly embodies the theory. This problem is in fact one that we encountered in chapter 5 in a slightly different form. There we saw that the highly prized problem-solving skills of the wrangler were not easily transmitted beyond the university simply by the distribution of textbooks. The reproduction of these skills in students at new sites was most effectively accomplished by Cambridge-trained teachers who were capable of recreating the whole pedagogical economy from which such skills emerged. In the final two chapters of this study,

---

22. Goldberg argues that relativity theory could not be accepted in Britain "until it could be made consonant with the existing British ideas about the ether" (1970, 113).

the problem of the transmission of new knowledge in mathematical physics is approached not in terms of the translation of mathematical culture *from* Cambridge, but in terms of the translation (or failed translation) of Einstein's special and general theories of relativity *to* Cambridge. As in chapter 5, we shall see that paper resources alone (in the form of published articles) were not effective vectors in this process.

8.3. Ebenezer Cunningham's Training in Cambridge

Ebenezer Cunningham's research in electrodynamics has attracted little informed comment from historians of science. It has been variously characterized as an *anticipation* of relativity theory, a direct *contribution* to relativity theory, and, most recently, as an attempt to *modify* relativity theory in order to prevent it from undermining the concept of the electromagnetic ether.[23] Notice that in each case his work is defined not in terms of Cambridge mathematical physics, but in terms of the emergent theory of relativity to which his work is seen, retrospectively, as a response. Only one study, by Stanley Goldberg, makes any attempt to establish a link between Cunningham's approach to relativity theory and his Cambridge training. However, Goldberg (1984, 240) uses what he sees as the "rigid and narrow" education offered by the Mathematical Tripos as a means of explaining why Cunningham was unable to accept the theory in the form intended by Einstein. By avoiding this relativity-centered approach and paying careful attention to the research in which Cunningham was actively engaged when he began to exploit Einstein's paper, I shall show that his work was a direct contribution to Cambridge electrodynamics.

Cunningham went up to Cambridge in 1899, having won a major scholarship to read mathematics at St. John's College.[24] At St. John's he was coached for the Mathematical Tripos by R. R. Webb and was taught electromagnetic theory by Joseph Larmor. After becoming the last of Webb's students to win the coveted senior wranglership, Cunningham (fig. 8.1) went on to take Part II of the Mathematical Tripos the following year (1903). His postgradu-

---

23. Goldberg portrays Cunningham as an eclectic who tried to bring about a reconciliation between the "ether-denying" theory of relativity and "pure-ether" mechanics. See Goldberg 1970, 113–17, and 1984, 230. McCrea argues that Cunningham's early work shows that the "body" of physicists (including the British) had "already arrived at all the essentials of the special relativity without reliance upon Einstein" (1978, 124). Darrigol claims that Cunningham was "the English herald of Einstein's relativity" (1996, 288).

24. For an autobiographical account Cunningham's life, see Cunningham 1969. McCrea 1978 contains a list of Cunningham's publications.

FIGURE 8.1. Ebenezer Cunningham (aged 21) from a popular postcard depicting the senior wrangler of 1902. Cunningham was a deeply committed Congregationalist Christian and total abstainer from alcohol. Possessing a "fine physique with an erect bearing," he was also an outstanding oarsman who rowed competitively for St. John's College and kept up his sport well into middle age (McCrea 1978, 116–17). Photograph from the archive of St. John's College Library. (By permission of the Master and Fellows of St. John's College, Cambridge.)

ate training included attendance at Larmor's Part II lectures on electromagnetic theory, and, in preparation for the examination, he would have worked through problems of the kind illustrated in figure 7.2 (Larmor being one of the Part II examiners that year). Cunningham then began research in the theory of differential equations for which he was awarded a Smith's prize in 1904.[25] Alongside his mathematical research he also made a long and careful study of *Aether and Matter* in order to find out more about some of the issues touched upon in Larmor's lectures.[26]

25. Cunningham's Smith's Prize dissertation was entitled "On the Normal Series Satisfying Linear Differential Equations" (Barrow-Green 1999, 310).
26. It will be recalled that it was Cunningham who noted that the "incompleteness" of Larmor's lectures provoked the more able students to explore the issues for themselves (see section

The aspect of the ETM that had evidently caught Cunningham's attention was Larmor's use of what would later be called the Lorentz transformations—equations (1) of chapter 7. As an expert in the solution of differential equations, Cunningham investigated the transformation technique introduced by Larmor and soon made what was, for him, a very important discovery. As we saw in chapter 7, when Larmor had first employed the Lorentz transformations in *Aether and Matter* he had noted, erroneously, that they transformed Maxwell's equations only to the second order of v/c.[27] Cunningham quickly spotted that they actually transformed Maxwell's equations exactly, and at once told Larmor of this important finding.[28] Larmor responded by remarking that he had since noticed this property himself, but considered it unimportant because it was not possible to ascertain experimentally whether motion through the ether was actually detectable above second order.[29] The following summer (1904) Lorentz drew attention in print to the exact nature of the transformations—which thereafter bore his name—and incorporated them into his theory of the electrodynamics of moving bodies.[30] Cunningham left Cambridge in October of 1904 to take up a junior lectureship in mathematics at Liverpool University, where he remained for the next three years. Sometime during this period he came across Einstein's now famous paper "On the Electrodynamics of Moving Bodies" and began the research for which he is now remembered.[31] It was also during this period that he began to use the term "principle of relativity" but, as we shall shortly see, his understanding of the principle differed markedly from Einstein's understanding.

Cunningham's first explicit reference to the principle of relativity appeared in 1907 in a reply to a critique of Lorentz's electron theory by Max Abraham. Lorentz had made it a fundamental principle of his electron the-

---

7.4). Cunningham's mathematical research in this period mainly concerned the classification and solution of differential equations (see McCrea 1978). Cunningham recalls this period in an interview for the AHQP.

27. Warwick 1991, Appendix D. In this Appendix Larmor's reasons for reaching this erroneous conclusion are discussed in detail.

28. Larmor and Cunningham were both fellows at St. John's at this time and so would have come into contact daily.

29. This incident is recorded in Cunningham 1969, 29, and 1907a, 539.

30. Lorentz's article first appeared in English in July 1904. The equations were dubbed the "Lorentz transformations" by Poincaré in 1904 (Miller 1981, 80).

31. Cunningham recalled in his AHQP interview that Einstein's was an unknown name at this time and that he came upon the paper "quite by chance," probably in 1906. As will shortly become clear, however, Cunningham was clearly following most publications (including those in German) on the electrodynamics of moving bodies.

ory that moving electrons contracted by a factor of $\sqrt{(1 - v^2/c^2)}$ in their direction of motion through the ether.[32] Abraham, who supported a rigid electron theory, claimed that Lorentz's deformable electron was untenable because it led to an ambiguous definition of the electromagnetic momentum of the electron. The central point of Cunningham's reply to Abraham was that Abraham's own discussion was flawed because he had not made use of the principle of relativity in analyzing the electron's motion. In reworking Abraham's analysis, Cunningham showed that Lorentz's deformable electron was consistent with a unique definition of electromagnetic momentum and, crucially, revealed his own understanding of the principle of relativity.

Cunningham began his paper by explaining that it had been proved analytically that the Lorentz transformations transformed Maxwell's equations exactly, provided that the electric and magnetic fields measured by observers in states of relative motion were related in a certain well-defined way (equations [2] and [3] of chapter 7). He then defended Lorentz's deformable electron by asserting that:

> Lorentz's hypothesis of the reduction of the dimensions of a body when it moves relatively to an observer is reduced by this geometrical correspondence to the assumption that in the variables associated with the axes moving with it its shape is unaltered—an assumption suggested by the fact that the electromagnetic equations associated with those variables are independent of the motion through the ether, and by the attempt to form a purely electromagnetic theory of nature. (Cunningham 1907a, 540)

Cunningham's justification of Lorentz's deformable electron postulate thus contained two steps. He first claimed that it was unnecessary separately to postulate the contraction of moving electrons because the contraction followed as a geometrical consequence of the Lorentz transformations. If these transformations were applied, then an electron would appear spherical in its own rest frame but contracted by a factor of $\sqrt{(1 - v^2/c^2)}$ when moving with velocity v. This first step in Cunningham's argument appears compatible with Einstein's definition of the principle of relativity, but the second step reveals his differences with Einstein. The second step in Cunningham's argument was the assertion that the correctness of the Lorentz transformations had to be justified on theoretical grounds. They were justified mathematically because they (unlike the Galilean transformations) could preserve

---

32. On Lorentz's notion of a contracting electron, see Miller 1981, 72–73.

the form of Maxwell's electromagnetic field equations for all uniformly moving frames of reference; and they were justified physically because the contraction of moving matter implied by the transformations was predicted by the ETM. For Cunningham, therefore, the term "principle of relativity" was merely a shorthand statement of the fact that the Lorentz transformations provided a powerful mathematical tool for dealing with certain problems in the electrodynamics of moving bodies.

Cunningham's position on the status of the Lorentz transformations was thus more akin to Larmor's position than to Einstein's. For example, Cunningham noted that the "transformations in question" had been given by Einstein but added that they were, "in substance," the same as those given by Larmor in *Aether and Matter* (1907a, 547). Larmor and Cunningham nevertheless differed crucially over the precise physical status of the transformations: Cunningham was prepared to accept that the *exact* mathematical correlation between the electromagnetic variables of different systems implied by the transformations was a physical fact (so that motion through the ether could never be detected), whereas Larmor insisted on restricting the transformations to the limit of experimental accuracy. But despite this important difference with Larmor, Cunningham remained a long way from adopting Einstein's interpretation of the principle of relativity, and did not even recognize Einstein's second postulate (the principle of the constancy of the velocity of light).

This last point is clarified by Cunningham's comments during a brief dispute that followed the publication of his reply to Abraham. Towards the end of this reply, Cunningham had remarked, in passing, that the principle of relativity that he (Cunningham) had employed was, in essence, identical to a similar principle that had been stated elsewhere by Alfred Bucherer.[33] Cunningham (1907a, 544) claimed further that the employment of any such principle was bound to lead to the Lorentz transformations (which Bucherer had not used), because it was necessary to *explain* how the velocity of light could be the same for all observers in relative states of motion. This comment prompted a rapid response from Bucherer, who had come to no such conclusion. Bucherer replied that he was not aware of any "known fact of observation" that required one to consider the velocity of light to be the same for all observers (Bucherer 1908). Cunningham replied as follows:

33. Bucherer 1907. Bucherer attributed the principle of relativity to Lorentz and did not at this point mention Einstein. Bucherer first submitted the paper to the *Annalen der Physik*, but it was rejected by the editor Max Planck. This episode is discussed in Pyenson 1985, 201–2.

> I did not wish to assert that it was required by any known fact of observation but took it to be *involved in the statement of the principle*. . . . [I]f Maxwell's equations are assumed to hold when referred, as occasion requires, to various frames of reference moving relative to one another, the deduction cannot be escaped that the velocity of propagation of a spherical wave will be found to be exactly the same, whatever the frame of reference. (Cunningham 1908, 423; my italics)

As I noted above, Einstein grounded his kinematical arguments upon two independent principles, the second of which (the constancy of the light velocity) uncoupled his arguments from electromagnetic theory. Cunningham, by contrast, assumed the principle of relativity to be a statement of the fact that if Maxwell's equations were equally applicable in every frame of reference, then it followed that the velocity of light must itself appear the same in all frames of reference. Thus Einstein's second postulate was, for Cunningham, simply a consequence of the principle of relativity. Indeed, the principle of relativity itself was not really a fundamental postulate for Cunningham because, as we have seen, it rested ultimately upon the ETM and the mathematical properties of the Lorentz transformations.

8.4. Making Sense of Einstein

Having sketched Cunningham's interpretation of the principle of relativity, I now return to the issue of why he first identified Einstein's work as relevant to his own research interests. A complete answer to this question will have to wait until the following section in which I explain how Einstein's work partly prompted Cunningham's own original contribution to electrodynamics. In this section I prepare the ground for this discussion by suggesting why Cunningham's understanding of the principle of relativity would have led him, unlike the vast majority of his British colleagues, to take the trouble to read Einstein's paper, even though it was by a little-known author and written in German. We have seen that, by 1905, Cunningham was committed to the ETM and had discovered for himself that Maxwell's field equations transformed exactly under the Lorentz transformations. Furthermore, the exact nature of the transformations, the significance of which Larmor had denied, had subsequently been exploited by Lorentz in his electromagnetic theory of 1904. Following this last development, Cunningham took a keen interest in all publications on the electrodynamics of moving electrical systems and would therefore have been attracted by a short paper explicitly entitled "Zur

Elektrodynamik bewegter Körper" (On the Electrodynamics of Moving Bodies).[34]

There were, I suggest, two closely related aspects of Einstein's paper that would have caught Cunningham's interest. The first concerned Einstein's development of the mathematics of the transformation technique introduced by Larmor and Lorentz. Larmor, it will be recalled, had explained the failure to measure second-order effects of the earth's motion through the ether by introducing transformations that retained the form of Maxwell's equations up to that order. When Cunningham later discovered that the transformations given by Larmor actually preserved the form of Maxwell's equations to all orders he (unlike Larmor) took this to imply that no effects of the earth's motion would ever be detected experimentally. For Cunningham, Einstein's paper of 1905 was primarily an insightful discussion of the mathematical consequences of ascribing truly equivalent status to the space-time and electromagnetic variables of all uniformly moving reference systems. Einstein's analysis showed, for example, that if the space and time variables of all moving systems were granted equivalent status, then a new formula for the addition of velocities was implied. Furthermore, once the new velocity addition formula was adopted, the Lorentz transformations formed what would now be called a "mathematical group" (successive transformations being replaceable by a single transformation of the same form).[35] Einstein's analysis thus brought a pleasing mathematical symmetry to the electrodynamics of moving bodies. The observations made by all observers in states of relative motion could now be correlated without reference to any uniquely privileged ether frame of reference.[36] In Cunningham's terminology, Einstein had shown how to bring complete "geometrical correspondence" between moving frames of reference simply by following up the mathematical implications of Larmor's original transformations.[37]

The second important insight read into Einstein's paper by Cunningham was the demonstration that the mathematical symmetry now inherent in the

---

34. The references in Cunningham's papers reveal him to have been familiar with Anglo-German work on electrodynamics, but not with that of Poincaré.

35. Einstein, *Collected Papers*, vol. 2, doc. 23, section 5. Cunningham was not the only Cambridge electromagnetic theoretician to read Einstein's paper in this way. G. A. Schott (1907, 687) also noted that Einstein had given the "correct" expression for velocities measured in moving systems if the Lorentz transformations were adopted.

36. Einstein, *Collected Papers*, vol. 2, doc. 23, section 9.

37. Cunningham recalled that the significance of Einstein's paper struck him as soon as he began reading it. This also suggests that it was essentially the mathematical completion of Larmor's scheme the he took from the paper. See interview for the AHQP.

electrodynamics of moving bodies was actually a physical consequence of the ETM. For Cunningham, the discussion of kinematics and simultaneity with which Einstein began his paper amounted to the following: if the world were ultimately electromagnetic in structure, the only practical way of defining simultaneity between two spatially separated events would be by employing electromagnetic signals; and since, according to the ETM, electromagnetic radiation traveled with the same velocity in all directions for all observers, any observer could legitimately employ Einstein's definition of simultaneity. That this was actually Cunningham's view is fully confirmed by the following remarks with which he began his own major contribution to electrodynamics in 1910:

> The absence, as far as experiment can detect, of any phenomena arising from the earth's motion relative to the electromagnetic aether has been fully accounted for by Lorentz and Einstein, *provided the hypothesis of electromagnetism as the ultimate basis of matter be accepted,* so that the only available means of estimating distances between two points is the time of propagation of effects between the bodies, such propagation taking place in accordance with the equations of electron theory. (Cunningham 1910, 77; my italics)

Cunningham in fact required both of the above insights to make sense of Einstein's paper. As a Cambridge wrangler trained in the early years of the twentieth century, Cunningham would have been very impressed by the mathematical symmetry brought to the electrodynamics of moving bodies by Einstein's work. But the mathematics would have been of little physical significance for him, were it not for the fact that Maxwell's equations transformed exactly under the Lorentz transformations and the fact that the ETM offered a physical explanation (in any given frame of reference) of the contraction of moving matter. For Cunningham, Einstein had developed a very powerful mathematical technique in electromagnetic theory (which could, for convenience, be referred to as the "principle of relativity") as a consequence of the ETM. Furthermore, the strategy that Cunningham assumed Einstein to have followed in developing the principle of relativity (following up the mathematical implications of the ETM) suggested to him how he might further develop the principle in a similar manner.

8.5. The New Theorem of Relativity

During the academic year 1906–7, Cunningham was not primarily concerned with defending Lorentz's electron theory. He had, in fact, begun col-

FIGURE 8.2. Illustration depicting the two bracketed senior wranglers of 1903, Harry Bateman and P. E. Marrack, and "the only woman wrangler," Hilda Hudson. A student of Newnham College, Hudson was ranked unofficially as equal to the seventh male wrangler. *The Graphic*, 27 June 1903, 865. (By permission of the Syndics of Cambridge University Library.)

laborating with another Cambridge mathematician, Harry Bateman (SW 1904), in an attempt to establish mathematically what he would eventually call the "new theorem of relativity"; a theorem that would be valid for electrical systems in states of accelerated as well as states of uniform motion.[38] Bateman's Cambridge career was almost identical to Cunningham's. A student of Trinity College, he was coached by Robert Herman to the senior wranglership (fig. 8.2) of 1903, and awarded the first Smith's Prize in 1904.[39] After graduation, however, Bateman spent the two following academic years traveling in Europe and studying at Göttingen and Paris.[40] In the autumn of 1906, he joined Cunningham at Liverpool University, where he too had been appointed to a junior lectureship in mathematics. During the following academic year the two mathematicians found that they shared common interests in the solution of differential equations and certain aspects of transformation theory.[41] Through these shared interests, and Cunningham's understanding

38. Sanchez-Ron (1987a, 42) notes correctly that Bateman's approach to electrodynamics was geometrical, but claims, wrongly, that Cunningham and Bateman worked independently on the covariance of Maxwell's equations.

39. For biographical details of Bateman and a complete list of publications, see Erdélyi 1948. Bateman's Smith's Prize essay was entitled "The Solution of Differential Equations by Means of Definite Integrals" (Barrow-Green 1999, 310).

40. It is possible that Bateman heard Minkowski lecture in Göttingen, but this would have been before the latter developed his four-dimensional approach to relativity theory.

41. Bateman's early research in mathematics was also concerned with the solution of differential and integral equations.

of Einstein's paper, they began to work together to establish a new, more general, principle of relativity. The importance of Einstein's work for Cunningham and Bateman was, as we have seen, that it demonstrated how the principle of relativity could lend mathematical symmetry to the problem of the electrodynamics of moving bodies. It also suggested how a whole new range of problems in electromagnetic theory might be made accessible to mathematical solution by bringing together the principle of relativity and their common expertise in transformation theory.

During 1907, both Cunningham and Bateman became interested in the physical application of a mathematical technique known as *inversion* and its generalization in the form of the *conformal transformation*. Both men had been taught inversion as a geometrical technique that was chiefly of use in electrostatics. As the technique is central to Cunningham's early research in electrodynamics, I shall explain how the technique was employed by Cambridge wranglers. The geometrical interpretation of inversion in two dimensions, as described in Jeans's *Mathematical Theory*, is illustrated in figure 8.3. Consider a circle of radius K with center O. For any point P inside the circle, an *inverse* point P' can be generated outside of the circle by imposing the condition $OP.OP' = K^2$. In this transformation K is known as the *radius of inversion*. If the point P now moves to generate the line PQ, an inverse line $P'Q'$ will be generated outside the sphere, each inverse pair of points satisfying the relationship $OP.OP' = K^2$. A number of geometrical relationships between the inscribed figures and their inverses can be demonstrated, the most important of which for our present purposes being that if two lines inscribed inside the circle intersect in an angle θ, then their inverse images outside the circle will also intersect in angle θ. This property is actually definitive of a much wider group of transformations, known as the *conformal* transformations, of which inversion in a circle may be considered a simple, geometrical, example.

The physical importance of the technique of inversion in electrostatics is illustrated in figure 8.4, also taken from Jeans's *Mathematical Theory*. If the geometrical structure of a charged conductor (in this case two intersecting planes) and its surrounding field lines (not shown) is inverted in a sphere whose centre is at O, then a new conductor (in this case two intersecting spheres) with its field lines (not shown) is generated. That the field lines in the inverted image will continue to intersect everywhere at right angles (as they must) is guaranteed by the property, mentioned above, that angles between intersecting lines are preserved under inversion. By inversion it is

202 *Methods for the Solution of Special Problems* [CH. VIII

INVERSION.

226. The geometrical method of inversion may sometimes be used to deduce the solution of one problem from that of another problem of which the solution is already known.

Geometrical Theory.

227. Let O be any point which we shall call the centre of inversion, and

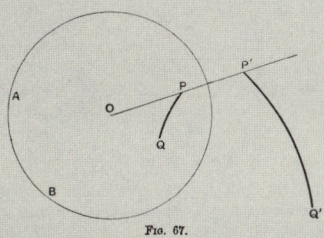

FIG. 67.

let AB be a sphere drawn about O with a radius K which we shall call the radius of inversion.

Corresponding to any point P we can find a second point P', the inverse to P in the sphere. These two points are on the same radius at distances from O such that $OP \cdot OP' = K^2$.

As P describes any surface $PQ\ldots$, P' will describe some other surface $P'Q'\ldots$, each point Q' on the second surface being the inverse of some point Q on the original surface. This second surface is said to be the *inverse* of the original surface, and the process of deducing the second surface from the first is described as *inverting* the first surface.

It is clear that if $P'Q'\ldots$ is the inverse of $PQ\ldots$, then the inverse of $P'Q'\ldots$ will be $PQ\ldots$.

If the polar equation of a surface referred to the centre of inversion as origin be $f(r, \theta, \phi) = 0$, then the equation of its inverse will be $f\left(\dfrac{K^2}{r}, \theta, \phi\right) = 0$. For the polar equation of the inverse surface is by definition $f(r', \theta, \phi) = 0$, where $rr' = K^2$ for all values of θ and ϕ.

FIGURE 8.3. Jeans's explanation of the geometrical theory of inversion. The property of inversion that makes it useful for tackling problems in electrostatics is that lines crossing at right angles in any geometrical pattern will continue to do so in the pattern's inverted image. If therefore a geometrical pattern represents a charged conductor and its associated lines of equipotential and electric force (which must cross at right angles), the inverted image of the pattern will represent a new conductor together with its lines of equipotential and force. Put a different way, if one pattern represents a solution to Laplace's equation, so will the inverted image. Jeans, *Mathematical Theory*, 1908. (By permission of the Syndics of Cambridge University Library.)

206   *Methods for the Solution of Special Problems* [CH. VIII

231. *Intersecting Planes.* As a more complicated example of inversion, let us invert the results obtained in § 212. We there shewed how to find

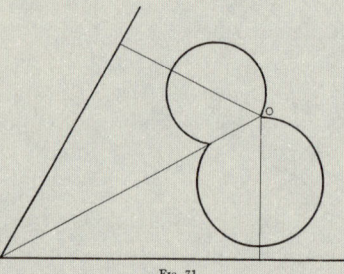

Fig. 71.

the distribution on two planes cutting at an angle $\frac{\pi}{n}$, when put to earth under the influence of a point charge anywhere in the acute angle between them. If we invert the solution we obtain the distribution on two spheres, cutting at an angle π/n, raised to a given potential. By a suitable choice of the radius and origin of inversion, we can give any radii we like to the two spheres.

If we take the radius of one to be infinite, we get the distribution on a plane with an excrescence in the form of a piece of a sphere: in the particular case of $n = 2$, this excrescence is hemispherical, and we obtain the distribution of electricity on a plane face with a hemispherical boss. This can, however, be obtained more directly by the method of § 219.

SPHERICAL HARMONICS.

232. The problem of finding the solution of any electrostatic problem is equivalent to that of finding a solution of Laplace's equation

$$\nabla^2 V = 0$$

throughout the space not occupied by conductors, such as shall satisfy certain conditions at the boundaries of this space—*i.e.* at infinity and on the surfaces of conductors. The theory of spherical harmonics attempts to provide a general solution of the equation $\nabla^2 V = 0$.

This is no convenient general solution in finite terms: we therefore examine solutions expressed as an infinite series. If each term of such a series is a solution of the equation, the sum of the series is necessarily a solution.

FIGURE 8.4. An example of the technique outlined in fig. 8.3. The illustration shows two planes intersecting in an acute angle π/n together with their inverted image, two intersecting spheres. The inverted image of the field lines (not shown) due to a point charge anywhere inside the acute angle between the planes will give the field lines surrounding the intersecting spheres (if they are electrically charged). Thus by knowing the solution to one problem (the field surrounding intersecting planes) the solution to a new problem (the field surrounding intersecting charged spheres) can be found. Note also the discussion of solutions to Laplace's equation by spherical harmonics that begins below this example. Jeans, *Mathematical Theory*, 1908. (By permission of the Syndics of Cambridge University Library.)

therefore possible to find the form of the electric field surrounding a complicated system of charges and conductors by inverting the known solution to a simpler case. This was precisely the kind of physical-mathematical problem that was employed in the Cambridge Mathematical Tripos to introduce students to advanced mathematical techniques. In 1895 a Tripos problem explicitly required the candidates to "Apply the method of conformal transformation to establish the correlation between the fields of two freely electrified cylinders."[42] As H. R. Hassé (7W 1905) later recalled, "to the average student[,] pure geometry at that time [1900 –1907] can have meant little more than the use of one or other of the processes of inversion, reciprocation or projection for the solution of problems."[43]

In February 1908 Cunningham read a paper before the London Mathematical Society (hereafter LMS) entitled "Conformal Representation and the Transformation of Laplace's Equation." Although the paper was never published, it is reasonable to infer that in it Cunningham investigated the possibility of generating solutions to Laplace's equation (a second order partial differential equation) by the transformation of known solutions.[44] In regions where the charge density is zero (outside of a charged sphere, for example), the distribution of electric potential (V) surrounding a charged conductor will always satisfy Laplace's equation

$$\frac{\partial^2 V}{\partial x^2} + \frac{\partial^2 V}{\partial y^2} + \frac{\partial^2 V}{\partial z^2} = 0$$

with appropriate boundary conditions applied. But, as we have just seen, it is possible to move from the solution of one electrostatic problem to another, and hence from one solution of Laplace's equation to another, by the process of inversion. This is the analytical analogue of the geometrical technique described above, and Cunningham was very likely concerned with the generality of the solutions that could be generated by this method.[45]

Bateman was also interested in the physical application of the conformal transformations. In October 1908 he submitted a paper to the LMS on

---

42. The problem was complicated by the cylinders having different oblique sections. This example is from Leathem 1906, 56. Jeans gave ninety-four examples of Tripos problems of this kind at the end of chapter 8 of the *Mathematical Theory*.

43. Hassé 1951, 154. On the fundamental status of geometry as the study of space in Cambridge mathematics during the nineteenth century, see Richards 1986.

44. A list of papers read at the February meeting is given in the *Proceedings of the London Mathematical Society* 6: viii.

45. Cunningham may have been unaware that Forsyth 1898 already addressed this topic, which perhaps explains why the paper was never published.

conformal transformation in a four-dimensional space and its application in geometrical optics (Bateman 1909). Bateman opened his paper by recounting the history of the application of the method of inversion to problems in electrostatics, explaining that any conformal transformation could be made up out of a series of inversions in suitably chosen spheres. Indeed, he pointed out that the demonstration of this mathematical property had actually been set as a question in the Mathematical Tripos he had sat in 1903.[46] He then showed that just as new solutions to Laplace's equation could be obtained by the conformal transformation of known solutions (Cunningham's problem), so new solutions of the four-dimensional equations

$$\frac{\partial^2 V}{\partial x^2} + \frac{\partial^2 V}{\partial y^2} + \frac{\partial^2 V}{\partial z^2} - \frac{1}{c^2}\frac{\partial^2 V}{\partial t^2} = 0$$

$$\left(\frac{\partial V}{\partial x}\right) + \left(\frac{\partial V}{\partial y}\right) + \left(\frac{\partial V}{\partial y}\right) - \frac{1}{c^2}\left(\frac{\partial V}{\partial t}\right) = 0$$

(which were fundamental to the electromagnetic theory of light) could be obtained by the four-dimensional conformal transformation of known solutions (1909, 72–78). Finally, Bateman applied his results to some problems in optics and mentioned, in passing, that the fundamental equations of electron theory could also be written in mathematically symmetrical form in four dimensions by employing the new dimension s (s = ict).[47] This paper represented Bateman's initial contribution to the collaborative venture he began with Cunningham during the academic year 1906–7. Together they sought new transformations (other than the Lorentz transformations) that would transform Maxwell's equations exactly.

It was Cunningham's understanding of the principle of relativity that formed his point of departure for the new research project. According to Cunningham, the principle of relativity expressed the fact that the electromagnetic field equations transformed exactly under the Lorentz transformations. The principle was extremely useful in electromagnetic theory because it sometimes enabled one to simplify otherwise awkward calculations in the electrodynamics of moving bodies by transforming from one reference frame to another. However, this technique could only be applied to systems that were in *uniform* states of relative motion; it could not be ap-

---

46. Bateman 1909, 71, footnote. See *Cambridge Tripos Papers*, Mathematics, Part I (1903), Paper 1, Questions 3 and 4.
47. Bateman 1909, 82–87. Where i is $\sqrt{-1}$.

plied to accelerated systems.[48] Cunningham now set out to find new transformations that would preserve the form of Maxwell's equations for the case of accelerated motion.

Cunningham's strategy in tackling this problem was to imagine that electromagnetic processes occurred in a four-dimensional space, so that electromagnetic fields that changed with time could be thought of as static geometrical forms, rather like three-dimensional electrostatic fields.[49] Each four-dimensional solution could then be inverted in a four-dimensional *hypersphere* of *pseudo-radius K* in order to produce a new solution. Central to Cunningham's paper was the demonstration that Maxwell's equations retained their form under these transformations. This effectively established that if the electromagnetic fields in one four-dimensional frame of reference were solutions of Maxwell's equations, then the inverted images of those fields would also be solutions of Maxwell's equations. Any one of these frames of reference could therefore represent a possible physical reality in the sense that an observer working in that frame would find that Maxwell's equations applied to all electromagnetic processes. Since the conformal group of transformations could be built up by repeated inversions (on this point Cunningham acknowledged Bateman's assistance), Cunningham concluded that this must be the most general group of transformations with this property.[50]

In order fully to appreciate the importance of this result, it must be remembered that for Cunningham the principle of relativity was valid only insofar as it had been analytically derived. Larmor and Lorentz had shown that the Lorentz transformations exactly correlated the fields of electrical systems in states of uniform motion, and that the ETM made these transfor-

---

48. For a technical discussion of why, according to Cambridge electrodynamics, the transformations were valid only for uniformly moving systems, see Warwick 1991, 84–88.

49. Cunningham cited Minkowski as the first to note that in a four-dimensional space (x, y, z, ict) the Lorentz transformations represented a finite rotation of the space around the $y = 0$, $z = 0$, axes, but Cunningham and Bateman were not following Minkowski. Minkowski's "Das Relativitätsprinzip" lecture of 1907 was not published until 1915, and his famous "Raum und Zeit" did not appear until 1909. Only his "Die Grundleichungen für die elektromagnetischen Vorgänge in bewegten Körpern" had reached Cambridge (May 1908) when Cunningham and Bateman submitted their papers. Cunningham and Bateman had begun their collaboration two years earlier. Bateman's four-dimensional, integral, representation of the electromagnetic equations was taken from the work of his Liverpool colleague, Richard Hargreaves (5W 1876), but it is quite likely that he had also read Minkowski's paper. See Minkowski 1908, 1909, and 1915, and Hargreaves 1908.

50. Cunningham 1910, 87–89. Nicholson referred to the physical interpretation of the four-dimensional representation of electromagnetic processes as a "fiction" which "in the mathematical sense, was effective after the manner that ordinarily electrical images in electrostatics are effective." See McLaren 1925, 37.

mations physically plausible. For Cunningham, Einstein had completed this work mathematically, but only for systems in uniform motion. Now Cunningham had shown, analytically, that there were other transformations that could preserve the form of Maxwell's equations between electrical systems in much more complicated states of relative motion. This was the foundation of what Cunningham called the "new theorem of relativity" (1910, 87).

This whole enterprise was grounded upon a commitment to the ETM. Just as this theory guaranteed the physical plausibility of the Lorentz transformations, so the new transformations could only represent possible physical systems because the world was ultimately electromagnetic in nature. Cunningham pointed out that the kinds of motion of electrical systems represented by the inversion of a simple system were a good deal more complicated than those considered by Einstein. Even in the most simple cases, a stationary system transformed into one that was expanding or contracting radially about a fixed point in a particular way, and, as with the electrostatic case, it was often difficult to decide to which physical problem any given transformation was the solution (1910, 93).[51] Cunningham had nevertheless succeeded in finding a new and more general group of transformations that preserved the form of the electromagnetic equations, and this for him was equivalent to extending the field of application of the principle, or "theorem," of relativity.

8.6. Re-Defining the Ether

> The principle of relativity then does not deny the existence of an aethereal medium; that is only the interpretation of an individual. What it does do is to emphasize the insufficiency of the existing conceptions of the aether, and to set up a criterion by which suggestions as to the nature of the aether may be examined.
>
> EBENEZER CUNNINGHAM, 1914[52]

In the discussion so far I have concentrated on the technical practice of Cunningham's physics and have not explicitly considered the status of the electromagnetic ether in his work. The reader will probably have noticed that despite his belief that motion through the ether could not be detected (at any order), Cunningham was not inclined to follow Einstein in abandoning the

---

51. For an example of how Cunningham and Bateman's work was used in electromagnetic theory, see Hassé 1913, 181–206.

52. Cunningham 1914, 410.

ether concept as "superfluous."[53] In this section I shall describe Cunningham's notion of the ether and suggest how it related to the technical practice of his research. The important point to notice is that the ontological status of Cunningham's ether reflected the changing mathematical practice inherent in his electrodynamics. The ether concept itself is therefore of limited utility in explaining Cunningham's reading of Einstein's work.

Cunningham's first published statement on the status of the ether occurred at about the time he was beginning his work on the new theorem of relativity. In 1907, several physicists had objected to the claim, made originally by Larmor, that magnetic effects might be attributable to a real physical flow of the (otherwise stagnant) ether.[54] The objection was made on the grounds that relative motion between the rapidly moving earth and the ether did not appear to give rise to any measurable magnetic effects. In a letter to *Nature* magazine, Cunningham pointed out that this objection was valid only if one assumed that relative motion between a moving observer and the ether was equivalent to a real flow of ether in the stationary-ether frame of reference. But, he continued, this was not the case because it was "not permissible to speak of the velocity of an observer relative to the aether as though the aether were a material medium given in advance" (1907b, 222).

Cunningham went on to explain that the ether was *defined* by its electromagnetic properties, and that one of these was that a uniform drift through the ether had no effect upon the observed phenomena. In order to justify this statement he quoted at length from Larmor's *Aether and Matter* and insisted that: "The aether is in fact, not a medium with an objective reality, but a mental image which is only unique under certain limitations.... *Two frames of reference imply two aethers;* so long as we restrict ourselves to a single frame, the objection to Larmor's scheme does not arise" (1907b, 222; my italics). Cunningham clearly saw himself as championing Larmor's theory in arguing that the ether was only a useful concept when referred to any given observer's frame of reference. His notion of a plurality of ethers (one for each inertial frame of reference) was quite new.[55] Following his discovery that

---

53. Einstein, *Collected Papers,* vol. 2, doc. 23, Introduction.

54. Larmor suggested in the early 1890s that a magnetic field probably represented some kind of real flow of the ether in the stationary ether frame, but this was not to be confused with motion *relative* to the ether, which would not generate such an effect. On Oliver Lodge's attempt to measure this effect, see Hunt 1986.

55. Larmor himself (1907, 269–70) immediately rejected Cunningham's claim by asserting that arguments based upon relativity must "lapse" because absolute motion could be measured in relation to the distant "quiescent" ether.

Larmor's transformations were exact, Cunningham must have concluded that the notion of a unique frame of reference (the stationary ether frame) had to be abandoned as meaningless. As he later remarked in discussing Larmor's contribution to electromagnetic theory:

> Every electromagnetic chain of phenomena is in a sense an electromagnetic clock or space-scale. . . . To grant the completeness of Larmor's scheme is therefore to admit the impossibility of determining physically a unique system of variables (x, y, z, t) which may be called absolute space and time variables; in other words it is to admit the impossibility of establishing the uniqueness of the aether. (Cunningham 1914, 44)

Cunningham nevertheless believed, as did many wranglers, that electromagnetic effects had, ultimately, to be attributed to a dynamical medium if they were to remain comprehensible to the human mind.[56] He solved this problem by redefining the ether in a way that was compatible with his mathematical description of electromagnetic processes. Each observer now associated an electromagnetic ether with his own inertial frame of reference and could therefore give a consistent dynamical account of electromagnetic processes as observed in that frame. The theorem of relativity (in the form of the Lorentz transformations) could then be employed to reveal how the electric and magnetic vectors of different ethers (observed in different frames of reference) were related. Cunningham's scheme thus retained the ontological purpose of the ether (to render electromagnetic processes dynamically comprehensible) while making it compatible with the new mathematical formalism.

The point I want to emphasize is that we should not try to explain this conceptual shift in the ether's status as somehow forced on Cunningham by his contact with Einstein's work. The shift was not designed to protect the ether from being undermined by the theory of relativity. As the above quotation reveals, Cunningham's notion of a plurality of ethers owed as much to his reading of Larmor's *Aether and Matter* as to his reading of Einstein's 1905 paper. Likewise the shift did not originate in an attempt to bring about a "reconciliation between the theory of relativity and the ether."[57] Cunningham neither recognized any theory of relativity in Einstein's work nor believed that Einstein had called for the ether concept to be abandoned.

---

56. On Cunningham's subsequent attempts to develop an ether whose physical properties were Lorentz invariant, see Warwick 1989, 192–96.

57. This remark is from Goldberg 1970, 117.

This last point might seem very surprising given the importance now associated with Einstein's apparent rejection of the ether concept in 1905. It should be remembered, however, that Einstein made just one reference to the ether in his first paper on relativity, in which he claimed only that the concept was "superfluous, inasmuch as in accordance with the concept to be developed here, no 'space at absolute rest' endowed with special properties will be introduced."[58] This brief remark would have been read by someone like Cunningham not as a flat denial of the ether's existence but as consistent with his own emergent view that the concept of a unique, stationary ether was becoming ontologically meaningless within the mathematical structure of electrodynamics.[59] It should also be noted that as it gradually became clear to Cambridge mathematicians over subsequent years that Einstein and some of his followers had abandoned the concept of the ether entirely, they tended to treat this as a metaphysical rather than physical stance. The reason they sometimes referred to Einstein's use of the relativity principle as a "philosophical" one is that they regarded him to be drawing conclusions that, while they might be logically correct, were based on premises that had not been shown to be true. Cambridge-trained mathematical physicists would in any case have found it hard to take seriously any claim that no form of ether existed. How in that case was the nature of electric and magnetic fields to be explained, and how did an electromagnetic wave convey energy through empty space? Cunningham wrestled with questions of this kind between 1905 and 1914, but they were generated more by his own work on the conformal transformations than by Einstein's work of 1905.

The shift in the ether's status brought about by Cunningham actually points to the fact that early twentieth-century British electromagnetic theory was not simply, or even primarily, a theory of the ether. Many of the mathematical techniques employed by wranglers originated in continuum mechanics, and the ether concept lent plausibility to the application of these techniques in electromagnetic theory. But by the early years of the twentieth century, the ether no longer acted as a heuristic guide to the development of electromagnetic theory in the way it had from the 1860s to the early 1890s. As we saw in the previous section, the ether played no direct role in Cunningham's research on the conformal transformation of Maxwell's equations. It was, in fact, the other way around; Cunningham's sense of the ontologi-

---

58. Einstein, *Collected Papers*, vol. 2, doc. 23, Introduction.

59. Einstein claimed himself in 1919 that the "hypothesis of ether in itself is not in conflict with the special theory of relativity" (Einstein 1922, 16).

cal reality of the ether derived from the mathematical techniques he had been taught in Cambridge. These techniques were, moreover, quite different from those employed by Einstein, so that even if Cunningham had (for whatever reason) been prepared to abandon the notion of the ether, he would still not have been a practitioner of relativity theory.[60]

8.7. J. W. Nicholson and the Rotating-Wheel Paradox

> The principle of relativity may be regarded from two points of view. In the first place, it may be postulated, as by Einstein and others, for the case of uniform translation, in the form of a general law underlying the general description of electromagnetic phenomena, quite apart from its actual analytical origin. Secondly, it may be derived, as originally by Larmor, from the result of an analytical transformation applied to the variables in a set of equations occurring in electromagnetic theory. When approached from this stand point, it is endowed with a mode of proof which is otherwise lacking.
>
> J. W. NICHOLSON (12W 1904)[61]

My second example of the interaction between Einstein's early work on relativity and Cambridge electrodynamics concerns a paper published by J. W. Nicholson in 1912 on the validity of the principle of relativity as applied to uniformly rotating electrical systems. Nicholson was, as we saw in chapter 7, an ardent supporter of the ETM and, thanks to his work on atomic theory, has remained one of Larmor's best-known students. And, as I emphasized in the previous chapter, Nicholson routinely used Larmor's transformation techniques in aspects of his own work that involved the electrodynamics of moving bodies. The epigraph above reveals that, at least by 1912, Nicholson was well aware that Einstein's interpretation of the relativity principle was very different from his own. In this section I show why Nicholson was prepared to claim not only that Einstein went too far in raising the principle to a postulate but also that the Cambridge interpretation was demonstrably superior to the one adopted by Einstein. As we shall also see, Nicholson further differentiated himself from Einstein by suggesting that it might be possible to measure the earth's absolute velocity of rotation with respect to the ether.

Before turning to this episode, it will be helpful to say a few words about the events that led Cambridge mathematicians to begin to notice and to comment on the fact that Einstein's approach to electrodynamics was significantly

---

60. Einstein used the axiomatic status of his two principles, rather than complicated mathematical analysis, to find new relationships between physical entities. The most remarkable example of this kind in 1905 was the expression $E = mc^2$. Einstein, *Collected Papers*, vol. 2, doc. 24.

61. Nicholson 1912, 820.

different from their own. One of the research problems explored between 1908 and 1912 by Cunningham, Bateman, and another young wrangler from St. John's College, H. R. Hassé (7W 1905), concerned the application of Larmor's transformations to the electromagnetic equations as applied to ponderable matter. In this case the equations had to include terms to account for the electrodynamic effects of material magnetization as well as electrical polarization, convection, and conduction. Thus Maxwell's equation expressing the magnetic-field vector **H** generated in the ether by changing electrical-displacement vector **D**:

$$curl H = \frac{1}{c} \frac{\partial D}{\partial t}$$

becomes for ponderable matter:

$$curl H = \frac{1}{c} \left[J + \frac{\partial D}{\partial t} + \rho v + curl(P \times v) \right]$$

where **J**, **P**, and **v** are vectors representing respectively the conduction current, the electric polarization, and the velocity of the matter with respect to the observer's ether-frame of reference. In order for the latter equation to preserve its form under the Lorentz transformations it is necessary to introduce new transformations for the quantities **J** and **P** of a similar form to equations (2) and (3) in chapter 7. The question confronted by the Cambridge-trained group circa 1908 was what form these transformations took and how they were correctly derived.[62]

The reason this issue highlighted the difference between Cambridge and German approaches to the relativity principle is that Hermann Minkowski, as well as Einstein in collaboration with Jacob Laub, had already published a solution to the problem.[63] What all three of the above mentioned wranglers noticed around 1909 was that the German authors had adopted a rather different procedure from their own. The Germans *assumed* the validity of the principle of relativity and simply applied the Lorentz transformations to find out how the electric and magnetic variables measured in different frames of reference must be related in order to satisfy the principle. The Cambridge men proceeded in a quite different way, by first deriving the required transformations from electromagnetic arguments using the ETM, and then show-

62. This issue is discussed in technical detail in Warwick 1989, 181–90.
63. Minkowski 1908, and Einstein and Laub, in Einstein, *Collected Papers*, vol. 2, docs. 51, 52, 53, and 54.

ing that they satisfied the Lorentz transformations. This procedure provided the necessary analytical demonstration that the relativity principle was valid in this case. It was thus in studying how Minkowski, Einstein, and Laub employed their definition of the relativity principle to tackle a specific problem that the Cambridge men began to appreciate the extent of their differences with their German peers.

Having recognized these differences, they had no hesitation in taking the Germans to task for failing to ground their work in the ETM. In April 1909, Hassé completed a paper in which he strongly criticized Einstein, Laub, and Minkowski for using what he referred to as "the Lorentz-Einstein *theory of relativity*" directly to derive relationships between electromagnetic variables as measured in different frames of reference.[64] This is to the best of my knowledge the first time a British physicist acknowledged the existence of a "theory of relativity" in print.[65] A year later Bateman made exactly the same point in noting that while the transformations that "leave the electrodynamical equations for ponderable bodies unaltered" were properly derived from the ETM, the German authors had "proceeded in the opposite way ... by *assuming* the transformations [were] Lorentzian transformations."[66] From this point on, Cambridge-trained mathematicians became increasingly aware of their fundamental differences with Einstein and his German followers concerning the status of the relativity principle, and it was against this background that Nicholson sought in 1912 to assert the superiority of the Cambridge interpretation.

The debate Nicholson entered concerned a paradox, propounded in 1909 by the German physicist Paul Ehrenfest, which ran as follows: when a material disc begins to spin, the circumference should be subject to the FitzGerald-Lorentz contraction by virtue of the tangential velocity of each of its elements; but the radii of the disc should not contract because there is no radial motion along their length. We are therefore faced with a seemingly paradoxical situation in which the circumference shrinks as the disc rotates faster, while the radius remains constant.[67] Two Cambridge physicists working at the Caven-

---

64. Hassé 1910, 192; my italics. Searle could not have consulted Hassé, as he had left Cambridge in 1908 to take up an assistant lectureship in mathematics in Liverpool.

65. Hassé's reference to the "Lorentz-Einstein" theory indicates that he had no clear conception at this time of what we would recognize as Einstein's special theory of relativity. Darrigol (1996, 311) shows that this conflation of Lorentz and Einstein's work was also commonplace in Europe.

66. Bateman 1910, 225; my italics. Similar sentiments are expressed in Cunningham 1912, 126.

67. Ehrenfest 1909, 918. Ehrenfest contrived the above paradox to challenge a relativistic definition of rigidity propounded by Max Born. For a discussion of this episode, see Miller 1981, 245–49.

dish Laboratory had offered a solution to this problem which they hoped would save the principle of relativity for the case of uniform rotation, but Nicholson proposed a quite different solution.[68] He thought the paradox both highlighted the danger inherent in introducing new principles to physics and exemplified the advantage of the Cambridge school's interpretation of the principle of relativity over that adopted by Einstein and his followers.

We have seen that Cambridge mathematical physicists trained after 1900 did not think of the principle of relativity as an axiom, but as an expression of the fact that the fundamental equations of the ETM transformed exactly under equations (1), (2), and (3) of chapter 7. It should be noted, however, that Larmor's demonstration of this property, even when extended to all orders of v/c, applied only to electronically constructed systems in states of *uniform* motion with respect to the ether. The analytical demonstration of the wrangler interpretation of the principle of relativity therefore placed definite physical and kinematic constraints upon the principle's realm of applicability. Nicholson pointed out that by raising the principle to an axiom, these constraints were likely to be ignored, and cited the paradox propounded by Ehrenfest as a case in point. He then demonstrated that the mathematical form of Maxwell's equations was not preserved beyond the second order when they were transformed into a frame of reference that rotated with a rotating electrical system (Nicholson 1912, 824). The ETM therefore predicted that the principle of relativity did not apply to rotating systems, and Nicholson noted accordingly that "it [was] not a matter of surprise that any attempt to extend the scope of the principle into this domain leads to immediate difficulties" (1912, 821).[69] This demonstration enabled Nicholson not only to assert the superiority of the Cambridge interpretation of the principle of relativity, but to claim that it might still be possible to measure the earth's motion through the ether using a more sensitive form of Michelson and Morley's interferometer. Having offered his solution to Ehrenfest's paradox, Nicholson concluded:

> The principle of relativity thus cannot be derived analytically for the case of rotation, except as a first order principle, and as we have seen, its extension

---

68. The Cavendish workers suggested that the disc must form a shallow bowl to allow for the circumferential contraction. This work is discussed in Warwick 1993, 18–21.

69. Nicholson did not consider Einstein to have applied the principle incorrectly. The offending paper was Born 1909. Born claimed that the principle of relativity could be applied kinematically to any infinitesimal element of a body whatever its state of motion.

beyond the problems contemplated by Einstein is not warranted. We must therefore, since the motion of the earth is not translation, expect second order velocity effects due to the motion of the earth in experiments like those of Michelson and Morley. (Nicholson 1912, 825)

Nicholson's remarks further highlight three important points with respect to the attitude of Cambridge-trained mathematicians to the legitimate use of the principle of relativity. First, Nicholson, like Livens, Cunningham, Bateman, and others, did not regard Einstein as the author of an essentially new theory.[70] Rather they understood him to have drawn attention to the consequences, both mathematical and philosophical, of raising an analytical result originating in the ETM to the status of a postulate. For Cambridge mathematicians this was an extremely dubious move since it was not only empirically unjustified, but severed the proper connection between the principle and the ETM, and, as far as they were concerned, led to no new empirically testable results. Indeed, on Nicholson's showing it led to results that could be demonstrated to be false by the proper application of the ETM. Second, there is a clear symmetry between the respective attitudes towards the principle of relativity expressed by Cunningham and Nicholson. Where the former extended the principle by finding new transformations that preserved the fundamental equations of the ETM, the latter resolved Ehrenfest's paradox by showing that no such transformations existed for the case of rotating systems. In both cases, the principle's legitimacy was assumed ultimately to rest on the analytical properties of Maxwell's equations. Finally, Nicholson was not in the least perturbed to discover that the ETM predicted that it should be possible to measure second-order effects of the earth's rotatory motion through the ether. Einstein ruled out effects of this kind as a matter of principle, but Nicholson saw a pleasing symmetry between the ETM and Newtonian mechanics—both theories now recognized rotation as the only form of absolute motion.[71]

8.8. Training and the decline of the ETM

The case studies above reveal how misleading it is to depict the work of men like Cunningham, Bateman, and Nicholson either as having contributed di-

---

70. For other examples of wranglers who adopted this view of the principle of relativity in their work, see H. R. Hassé (7W 1905) 1910, and S. B. McLaren (3W 1899) 1913.

71. Nicholson 1912, 825. Nicholson noted that the earth's relatively small angular velocity with respect to the stars meant that the sensitivity of Michelson and Morley's interferometer would have to be enhanced enormously before any effect would be detected.

rectly to the development of relativity theory or else as having rejected the theory because it denied the existence of the ether. They understood themselves to be engaged in the development of the ETM, and, in the case of Cunningham and Bateman, it was the attempt to broaden the scope of the principle of relativity (according to their definition) that lent meaning and value to Einstein's work on the electrodynamics of moving bodies. My detailed analysis of Cunningham's research goals and methods also reveals that this work sprang almost entirely from local concerns and techniques. Cunningham sought to extend the ETM by developing four-dimensional forms of problem-solving methods that were propagated in Cambridge through coaching, lectures, and Tripos questions. Einstein's work of 1905 seemed relevant to this project because it not only revealed and repaired an apparent mathematical asymmetry in Larmor's original formulation of the ETM, but suggested to Cunningham how the theory could be extended to cover electrical systems in states of non-uniform motion. It is also clear that Einstein's brief remarks concerning the status of the ether were not an important factor in Cunningham's appropriation of his work. Cunningham adopted those parts he found useful and then altered the ether's ontological status to make it consistent with the new mathematical structure of the ETM. Nicholson's work displays a similar lack of immediate concern with the ether. His worries about raising the relativity principle to the status of a postulate were expressed in terms of the computational ambiguities such a move might produce and made no reference to the existence or otherwise of an electromagnetic medium. It is by paying attention to the practical skills and resources employed by wranglers like Cunningham and Nicholson in developing their technical arguments that the constructive power of their Cambridge training is properly displayed.

The content of this chapter also highlights a need to reassess the historiography of Cambridge electrodynamics during the first fifteen years of the twentieth century. Far from recognizing a crisis in physics circa 1900, Cunningham and his contemporaries felt themselves to be building successfully on a Cambridge research program stretching back some thirty years. They did not regard Einstein's work of 1905 as a dramatic intervention in their discipline but as a contribution to a progressive project initiated by Larmor and Lorentz. Only from around 1909 did these young wranglers begin to appreciate that Einstein, Minkowski, and others regarded the relativity principle as a new postulate in physics that was independent of any specific physical theory, but, even then, they did not regard it as a serious threat to their own approach. What does require explanation within the continuous and progressive history of Cambridge electrodynamics sketched in this chapter is why the ETM apparently petered out around 1915. Unlike Maxwell and Larmor's work, the new electrodynamics never appeared on Tripos examination papers and found no new supporters among wranglers after 1909.[72] As we shall see in the next chapter, by 1917 the task of defending the ETM against challenges raised by Einstein's *general* theory of relativity was left to an older generation in the form of Larmor and Lodge.

There are several strands to the likely explanation of these problems, two of which point directly to pedagogical issues peculiar to Cambridge. First, it is a striking fact that all of the wranglers who worked on Larmor's ETM in the period 1904–15 studied for Part II of the Mathematical Tripos between the publication of *Aether and Matter* in 1900 and the reform of the Tripos in 1909.[73] This reform, as we saw at the end of chapter 5, dramatically altered the range of subjects and skills taught to undergraduates. With the abandonment of coaching in favor of lectures and small-group supervision, the long-standing emphasis on mathematical physics and the solution of problems gave way to greater emphasis on pure mathematics and the reproduction of examples. The wide choice of pure-mathematical subjects on offer in Parts I and II of the Mathematical Tripos seems also to have generated a wave of enthusiasm for the kind of research being undertaken by men like G. H. Hardy and J. E. Littlewood. This shift marginalized the mathematical tradition in which Larmor and his students had been trained. The most able students turned away from what the pure mathematicians referred to scoffingly as the "water, gas and electricity" subjects in favor of the theories of functions and groups.[74]

The second important factor in the nonappearance of the new electrodynamics in Cambridge was the lack of a competent teacher in the subject in the university. All of the men whose work has been discussed or mentioned in this chapter won research fellowships in Cambridge shortly after graduation, but every one of them left the university within two or three years to take up a junior teaching position at a provincial university.[75] In 1912 Cunningham became the only one of their number to return to Cambridge when he

---

72. Questions posed circa 1915 continued to be posed in terms of Larmor's methods. See *Cambridge University Examination Papers*, Mathematical Tripos, Part II, Schedule B, 9 June 1915, Question B12, and 6 June 1916, Question B12.

73. Livens exercised his right to take Part II of the Mathematical Tripos in 1910 according to the "old regulations." This was the final year this was possible.

74. Littlewood 1986, 82.

75. Cunningham, Bateman, and Hassé went to Liverpool, McLaren to Birmingham, Nicholson to Queen's College, Belfast, and Livens to Sheffield. Further details of their careers can be

took up a lectureship in mathematics at St. John's College. He also began writing two textbooks, *The Principle of Relativity* (1914) and *Relativity and the Electron Theory* (1915). These books set out the accomplishments of the Cambridge group over the previous decade, including Cunningham's most recent work on the multiple-ether concept.[76] He also reported some of the most important accomplishments of the German approach to relativity which he was beginning to explore through such textbooks as Max von Laue's *Das Relativitätsprinzip*.[77] Cunningham was now prepared to acknowledge the heuristic value of applying the principle of relativity even in situations where it had not been shown to be valid by strictly analytical means. It is quite possible that he would in time have established a school of Cambridge-style relativity and electrodynamics through his college lectures, textbooks, and the setting of Tripos questions, but he barely had time to settle into his new post before his own academic life and that of the university were severely disrupted by the outbreak of World War I.

This brings us to the question of why work on the ETM by Cambridge-trained researchers came to a halt around 1915. Part of answer to this question lies in the fact that, as we have just seen, there was no competent practitioner of the subject teaching in Cambridge until just before the outbreak of the war. Larmor's lectures remained the most advanced introduction to electrodynamics in Part II of the Mathematical Tripos, but he had little sympathy with the work of men like Cunningham and Bateman, and continued to teach the ETM much as he had in 1900. A factor of even greater and more immediate significance is that virtually all of those who had helped to develop the post-Larmorian ETM had moved on to other projects by 1915. Harry Bateman emigrated to the United States in 1910, where he eventually held chairs in mathematics, theoretical physics, and aeronautics at the California Institute of Technology. He continued to publish at a prodigious rate, but took no further research interest in the principle of relativity.[78] As a deeply committed, nonconformist Christian, Ebenezer Cunningham became a conscientious objector at the outbreak of World War I, refusing military service in 1915. He offered to work as a schoolmaster, but the local Tribunal ruled

---

found in McCrea 1978, Erdélyi 1948, Vint 1956, W. Wilson 1956, Nicholson 1917, and Spencer Jones 1951.

76. The Preface to Cunningham's 1914 book gratefully acknowledges the "great help" he had received from Hassé in its preparation.

77. von Laue 1911. Cunningham occasionally borrowed results from von Laue. See Warwick 1989, 199–206.

78. Erdélyi 1948.

that he must leave the academic world and work on the land growing food. After the war, Cunningham returned to St. John's College as a lecturer in mathematics but never again published a research article in mathematical physics.[79] The research activities of some of the wranglers who had made more minor contributions to the ETM were also diverted by the war effort: H. R. Hassé went to the Woolwich Arsenal to work on ballistics; J. W. Nicholson worked on radio-telegraphic communications and the production of mathematical tables; S. B. McLaren died of wounds received in action at Abbéville in August 1916.[80] Following the publication of his textbook, *The Theory of Electricity* in 1918, G. H. Livens devoted his professional life to mathematical education.[81]

It is clear, then, that the ETM rapidly lost currency in Cambridge after the first decade of the twentieth century because it was not passed on to a new generation of practitioners. But this does not explain why the generation trained between 1900 and 1910 abandoned this field of study by 1915. The likely explanation is that the ETM was essentially a highly mathematical enterprise that found little practical application in new areas of experimental atomic physics. As I noted above, most experimenters (even in Britain) were not prepared to accept that matter was composed solely of positive and negative electrons. After about 1906, the electromagnetic view of nature gradually lost ground to the special theory of relativity, especially in Europe. As we have seen, when Cunningham studied von Laue's 1911 textbook on the special theory of relativity, even he began to appreciate the theory's power to solve difficult problems without recourse to advanced mathematical methods. The Cambridge interpretation of the principle was neither mathematically simple nor easily applicable to any new area of physics because the practitioner was always required to demonstrate the principle's validity in a particular case by reference to the ETM. From around 1912 it was probably unclear how the ETM could usefully be developed further either as a mathematical structure or as means of explaining or predicting new experimental results.

The effective stalemate reached by the outbreak of World War I between the Cambridge and Einsteinian interpretations of the relativity principle was never explicitly resolved. The matter could not be decided experimentally and, as we have just seen, was not explored further by the younger generation of Cambridge mathematical physicists during World War I. At the end

79. McCrea 1978.
80. Vint 1956; W. Wilson 1956; Nicholson 1917.
81. Spencer Jones 1951.

of the war, Cunningham returned to his lectureship at St. John's College and began teaching an annual course for mathematicians and physicists based on his book *Relativity and the Electron Theory*.[82] This course, entitled "Electron theory," served for many years as an introduction to the application of what was now fast becoming known as the "special theory of relativity" to electrodynamics.[83] Cunningham taught relativity in his own style, continuing to offer mechanical interpretations of results in terms of multiple ethers and depicting Einstein's contribution essentially as a mathematical completion of work in electromagnetic theory due to Larmor and Lorentz (fig. 8.5).[84] He was nevertheless aware that he had lost touch with the wider development of the subject in Europe during World War I and made no attempt either to build upon his own ideas or to supervise research students in this area.[85] Growing awareness of Einstein's general theory of relativity in Britain from around 1916 made it clear that he had taken the debate over relativity beyond electrodynamics and electron theory into the realms of gravitation and cosmology. As we shall see in chapter 9, the general theory was unlikely to be conflated with alternative theories of gravitation, made a number of testable predictions that were not shared by the ETM, and found a powerful British advocate in the form of the Cambridge astronomer A. S. Eddington.

The failure by Cambridge mathematicians to adopt Einstein's interpretation of the relativity principle during the first two decades of the twentieth century also illustrates an additional sense in which technical training systems are inherently conservative. We saw in chapters 6 and 7 that Cambridge was a uniquely important center in Britain for clarifying, circulating, and propagating the methods of Maxwell's electromagnetic field theory and Larmor's ETM. It was this ability to accommodate and to preserve extremely specialised concepts and technical skills that I labeled as the first conservative property of a local pedagogical economy. In this chapter we have seen that students trained in the Cambridge tradition not only acquired, practiced, and furthered the indigenous technical culture, but successively ignored, reinterpreted, and rejected Einstein's early work in the same area. They did

---

82. Kragh 1990, 9, n. 23.

83. *Cambridge University Reporter*, 9 October 1919, 99. The Preface to the second edition (1920) of Cunningham's *Relativity and the Electron Theory* redefined its virtually unaltered contents as an account of the "special principle" of relativity.

84. A set of notes taken at Cunningham's lectures at St. John's by Llewellyn Thomas in 1923–24 is reproduced in the AHQP.

85. Paul Dirac was sent by Hassé to Cambridge in 1923 to work with Cunningham on relativity. Cunningham declined to supervise him, as he felt he had lost touch with recent developments (Kragh 1990, 6–7).

The Lorenz-Einstein transformation.

Suppose we wish to find the field of an electron moving along the x-axis with velocity v. It would be convenient to use $x' = x - vt$ instead of x, so as to be able to treat the phenomenon as stationary.

Actually we find it convenient to take
$$x' = \beta(x - vt)$$
$$y' = y$$
$$z' = z$$
$$t' = \beta(t - \tfrac{vx}{c^2})$$
where $\beta = \frac{1}{\sqrt{1-v^2/c^2}}$

as new independent variables.

The components of force on a stationary charge are $(E_x, E_y, E_z) \cdot q$.

Following Lorenz, on the moving charge they would be $(E_x, E_y - v\tfrac{H_z}{c}, E_z + v\tfrac{H_y}{c})$. This suggests a transformation, but again it is convenient to introduce β.

We put
$$E_x' = E_x$$
$$E_y' = \beta(E_y - \tfrac{v}{c}H_z)$$
$$E_z' = \beta(E_z + \tfrac{v}{c}H_y)$$
and likewise
$$H_x' = H_x$$
$$H_y' = \beta(H_y + \tfrac{v}{c}E_z)$$
$$H_z' = \beta(H_z - \tfrac{v}{c}E_y)$$

It is now found that the field equations for free space
$$\tfrac{1}{c}\tfrac{\partial E}{\partial t} = \operatorname{curl} H$$
$$0 = \operatorname{div} E$$
$$-\tfrac{1}{c}\tfrac{\partial H}{\partial t} = \operatorname{curl} E$$
$$0 = \operatorname{div} H$$

transform exactly into
$$\tfrac{1}{c}\tfrac{\partial E'}{\partial t'} = \operatorname{curl}' H'$$
$$0 = \operatorname{div}' E'$$
$$-\tfrac{1}{c}\tfrac{\partial H'}{\partial t'} = \operatorname{curl}' E'$$
$$0 = \operatorname{div}' H'$$

where $(\operatorname{curl}' H')_x = \tfrac{\partial H_z'}{\partial y'} - \tfrac{\partial H_y'}{\partial z'}$ etc.
$\operatorname{div}' E' = \tfrac{\partial E_x'}{\partial x'} + \tfrac{\partial E_y'}{\partial y'} + \tfrac{\partial E_z'}{\partial z'}$ etc.

Larmor showed in his book 'Ether and Matter', in the chapter on 'correlation of stationary and moving systems' that this was true to the second order. Lorenz showed this also about the same time.

But more still can be done; the transformation extends to the Lorenz scheme of equations.

Notice that if $\underline{v}$ has components (u_x, u_y, u_z), to be consistent with (1)
i.e. $\delta x' = \beta(\delta x - v \delta t)$
$\delta y' = \delta y$
$\delta z' = \delta z$
$\delta t' = \beta(\delta t - v \tfrac{\delta x}{c^2})$
we must have
$u_x' = \tfrac{u_x - v}{1 - v\tfrac{u_x}{c^2}}$
$u_y' = u_y/\beta(1 - v\tfrac{u_x}{c^2})$
$u_z' = u_z/\beta(1 - v\tfrac{u_x}{c^2})$

If we put also $\rho' = \beta\rho(1 - v\tfrac{u_x}{c^2})$
we find that the equations
$$\tfrac{\rho \underline{v}}{c} + \tfrac{1}{c}\tfrac{\partial E}{\partial t} = \operatorname{curl} H$$
$$\rho = \operatorname{div} E$$
$$-\tfrac{1}{c}\tfrac{\partial H}{\partial t} = \operatorname{curl} E$$
$$0 = \operatorname{div} H$$
transform exactly to
$$\tfrac{\rho' \underline{v}'}{c} + \tfrac{1}{c}\tfrac{\partial E'}{\partial t'} = \operatorname{curl}' H'$$
$$\rho' = \operatorname{div}' E'$$
$$-\tfrac{1}{c}\tfrac{\partial H'}{\partial t'} = \operatorname{curl}' E'$$
$$0 = \operatorname{div}' H'$$

This was the great achievement of Einstein's paper of 1905 in Annalen der Physik.

If we have one solution $\rho = \rho(x,y,z,t)$, $\underline{v} = \underline{v}(x,y,z,t)$, $E = E(x,y,z,t)$, $H = H(x,y,z,t)$ the equations in another
by transforming the equations $\rho' = \rho'(x',y',z',t')$ etc. back to ρ etc. x,y,z,t and obtain a new solution such that stationary charges in the first correspond to …

FIGURE 8.5. A page from the notes taken by L. H. Thomas at Cunningham's 1923–24 lecture course at St. John's College on "Electron Theory." Note that Cunningham presents "the great achievement of Einstein's paper of 1905" (four lines up from bottom of page) as showing how the transformations developed to second order by Larmor and Lorentz could be made exact. Thomas Collection, LTP-NCSU. (By permission of North Carolina State University Libraries.)

so, moreover, not through an unreasonable attachment to the notion of an electromagnetic ether, nor even after some kind of neutral comparison between the two approaches, but by finding (or failing to find) meaning in Einstein's papers in terms of the intellectual framework provided by their training. The second sense it which this economy was conservative, then, is that it generated new meanings in texts from other technical cultures and resisted approaches that seemed to be at variance with accepted local practice.

These latter characteristics are best understood as exemplary of my comments in chapters 3 and 5 concerning the rich pedagogical resources required to preserve locally generated mathematical knowledge and the fragility of such knowledge when conveyed in textbooks or other printed forms without the aid of a human teacher. Einstein's work was made meaningful and acceptable to Cunningham by his ongoing research in the ETM, and, when interpreted in that context, acquired new meanings that were consistent with an essentially Cambridge research project. That project was conserved by the unwitting replacement of meanings tacitly assumed by Einstein with those assumed by Cunningham due to his Cambridge training and research. An important point to note in this context is that what counts as "tacit" in a text can depend as much upon the skills of the reader as on those of the author. From Cunningham's perspective circa 1906, Einstein's seemingly clear statement that the two principles he adopted were to be regarded as independent postulates obviously has to be counted as tacit. Furthermore, when Cambridge-trained mathematicians began to realize around 1909 that Einstein's approach was actually rather different from their own, they still neither accepted nor fully understood the meaning of his work. These respective failures to *accept* and to *understand* the content of texts generated in alternative intellectual traditions occurred for different reasons which are usefully distinguished in my explanatory scheme.

We have seen that it was by comparing approaches to specific problems in electrodynamics that Cambridge mathematicians became aware that Einstein and his German followers attributed axiomatic status to the relativity principle. The Cambridge men could not accept this approach because they believed it to be empirically unjustified and because they thought that a proper electrodynamics had to be built analytically from the principles of the ETM. At this level, their reasons for rejecting Einstein's work were based on an overt conflict of theoretical principles that was empirically unresolvable. What they did not appreciate was what Einstein would have recognized as the conceptual holism and heuristic power of his theoretical approach, qualities that could only be grasped by exploring the theory more deeply and finding out how it was applied to tackle specific problems. Yet it is precisely this shift from the statement of theoretical principles to their application in the solution of problems that is most difficult to achieve without the guidance of a skilled practitioner. As we have seen, students forced by circumstance to study alone often believed they had grasped the meaning of an expository passage until they attempted to apply their putative knowledge to the solution of a few end-of-chapter problems.

This barrier to the transmission of technical knowledge as text assumes further significance when we consider its implications to the spread and assessment of new theories. Just as students find it difficult to master new subjects simply by memorizing laws, principles, and definitions, so practicing physicists can find it equally hard to acquire a working knowledge of a new theory simply by reading research publications that contain formal statements of its precepts and key conclusions. The examples discussed in this chapter can be understood to illustrate this point, Cunningham's work clearly displaying the changing phases in the interpretive process. He initially found Einstein's work meaningful without recognizing or adopting any of the principles that the author himself would have regarded as central to the emergent theory of relativity. A more protracted interaction with further research publications led Cunningham to identify some of the fundamental differences between Cambridge and Einsteinian approaches to the principle of relativity, at which point he rejected the latter as irreconcilable with his own technical practices and beliefs. Finally, when he began, around 1912, systematically to study an increasingly well-defined theory of relativity as expounded in such textbooks as von Laue's *Das Relativitätsprinzip,* he slowly acquired some aesthetic appreciation of the theory's structure and a better sense of its heuristic value in providing new results and simple solutions to otherwise complicated problems.

It would nevertheless be wrong on several counts to regard these events either as minimizing the problem of theory transfer via text, or as evidence that Cunningham had acquired a fully Einsteinian understanding of the theory of relativity. First, it should be noted that by 1914 Cunningham had been in contact with the work of Einstein and his German followers for at least eight years. His experience therefore points at the very least to the inefficiency of formal publication as a medium of theory transfer. Second, in studying German textbooks on relativity, Cunningham was taking advantage of works produced for explicitly pedagogical purposes. As we saw in chapter 5, books of this kind by no means guaranteed the reader a firm grasp of the theory they contained, but for someone like Cunningham they would have helped to clarify the differences between the Einsteinian and Cambridge approaches. Furthermore, Cunningham to some extent drew upon the exposition he found in these books in preparing his own accounts of relativity. When deriving the Lorentz transformations, for example, he began by introducing an operational definition of simultaneity in the stationary-ether frame of reference and then showing that such a definition was meaningless if motion through the ether could not be measured. He then introduced the "hypothesis" of the constancy of the velocity of light (though not the principle of relativity) as a means of defining simultaneity in any inertial frame of reference and showed that the variables in such frames had to be related via the Lorentz transformations (Cunningham 1915, 27–34).[86] Cunningham also adopted a more symmetrical view of the respective merits of Einsteinian and Cambridge approaches to the principle of relativity. While he still maintained that the principle had not been shown experimentally to be valid to all orders of magnitude, he also acknowledged that the ETM was likewise built on hypotheses (such as the electronic nature of positive electricity) for which there was little empirical evidence (1915, 47–54).

This brings us to the final reason why Cunningham ought not to be regarded as having mastered relativity theory in the Einsteinian sense. By 1915 he was certainly becoming familiar with, and to some extent sympathetic to, the Einsteinian approach, but he remained very uneasy about attributing truly axiomatic status to Einstein's two postulates. Furthermore, he made no attempt to contribute to the development of the Einsteinian position himself. In the final chapter of his 1915 book, Cunningham set out his own view on the necessity for preserving the ether concept in physics. He now argued that the mechanical properties of the ether had to be the same in every inertial frame of reference and that these properties had to be chosen so that they transformed exactly under the Lorentz transformations (1915, 87–94). As Cunningham himself pointed out, for him the principle of relativity now provided an important guide to the kinds of property that could legitimately be attributed to the ether (1915, 48). And, as we saw above, when he returned to Cambridge after World War I, he continued to teach this approach in his course on "Electron Theory." As the very title of his lectures and book sug-

---

86. It is actually unclear whether Cunningham meant the light-velocity "hypothesis" as a fundamental postulate or a generalization from experiment.

gests, Cunningham also continued to regard the principle of relativity as a theorem in electron theory rather than as foundational to a wholly new kinematics and mechanics.

Despite his familiarity with Einstein's work, then, Cunningham was by no means a practitioner of the Einsteinian approach. The tools and sensibilities with which he pursued his research were still those of the wrangler trained in the ETM. His contact with papers on relativity had lent him neither the conviction nor the technical mastery of the subject necessary to recognize what in Einstein's terms would have counted as a legitimate research problem. To have acquired this knowledge he would have had to interact much more closely with the community in which the theory of relativity was not only taken for granted but being actively developed. He would also have had to begin tackling problems using a properly Einsteinian approach, including those that could only be tackled in this way. Cunningham's understanding of Einstein's work remained on a different level. By 1915 he appreciated its logical structure but was not prepared further to pursue a theory that seemed in its extreme form to negate the ether's existence altogether and to marginalize the importance of the ETM. In this sense Einstein's relativity remained for Cunningham a collection of propositions that never constituted a familiar theoretical tool. Likewise Cunningham's lectures both before and after World War I did not constitute an introduction to the special theory of relativity in Einstein's sense, and no research school in relativity ever emerged from them. Had it done so, moreover, it would more properly be regarded as the continuation of a Cambridge project in electrodynamics than as the successful translation of a German one in the theory of relativity.

∼ 9 ∼

Through the Convex Looking Glass
A. S. Eddingtion and the Cambridge Reception of Einstein's General Theory of Relativity

> As long as there was no interruption, as long as each generation could hand over its method to the next, everything went well and the science flourished. But as soon as some external cause brought about an interruption in the oral tradition and only books remained, it became extremely difficult to assimilate the work of the great precursors and next to impossible to pass beyond it.
> "THE DECAY OF GREEK MATHEMATICS"[1]

9.1. From the Special to the General Theory of Relativity

An important conclusion to emerge towards the end of the previous chapter is that until at least the end of World War I, none of Cambridge's leading experts in electromagnetic theory had become a competent or enthusiastic practitioner of the Einsteinian approach to the principle of relativity. They had appropriated those aspects of Einstein's work that seemed compatible with the Electronic Theory of Matter (ETM) but saw no reason to explore his approach any further. Indeed, men like Nicholson believed that Einstein's attribution of axiomatic status to the relativity principle generated serious and possibly insurmountable theoretical difficulties in electromagnetic theory without offering any substantially new or testable results. It will also be recalled that I explained this stalemate between theoretical approaches in terms of the conservative properties of training systems. Those pedagogical resources that made the ETM seem so plausible and researchable to the generation of wranglers trained by Larmor offered no similarly detailed and compelling account of electrodynamics according to Einstein. In this chapter I explore the process by which this apparent impasse was broken between 1916 and 1920, and, as in previous chapters, I shall do so in such a way as to cast light both directly and indirectly upon the importance of training.

---

1. Van der Waerden 1954, chap. 8.

The change in attitude towards relativity that occurred in Cambridge during the above period was prompted by Einstein's publication of the *general theory of relativity* and by the powerful support the theory received in the university from the Plumian Professor of Astronomy, A. S. Eddington (SW 1904). It is Eddington's advocacy of the general theory of relativity that will form the main focus of this chapter, but, before turning to this subject, it will be helpful to say a little about the origin of Einstein's theory as well as its relationship to his earlier work on the principle of relativity. We saw in chapter 8 that Einstein's first applications of the relativity principle were concerned mainly with problems in electrodynamics and confined to non-accelerated reference systems. But towards the end of 1907, Einstein showed how the principle of relativity could be applied (at least approximately) to physical processes taking place in a reference frame that was in a state of uniform acceleration with respect to other inertial frames. This line of argument constituted his first steps towards a relativistic theory of gravitation. The key to forging a link between gravity and accelerated frames of reference was a new hypothesis that would later become known as the "equivalence principle." This principle asserted the "complete physical equivalence of a gravitational field and a corresponding acceleration of the reference system," and was heuristically useful because the latter was "to some extent accessible to theoretical treatment."[2] By analyzing the physical properties of a gravitational field as if they were due to an equivalent uniform acceleration of the reference system, Einstein investigated the effects of the field on clocks and electromagnetic processes. The most important results to emerge from this analysis were that the light emitted by atoms in a powerful gravitational field would appear to an observer outside the field to be minutely shifted towards the red end of the spectrum (the "gravitational red shift" phenomenon) and that the path of a ray of light passing close to a massive gravitating body would be minutely deflected.[3]

Einstein had little more to say on the subject of relativity until 1911, when he published a restatement of the equivalence principle, a clearer derivation of the light-deflection effect, and pointed out that this effect might be detectable by measuring the positions of stars close to the sun during a total solar eclipse.[4] Shortly after this he embarked upon what turned out to be

---

2. Einstein, *Collected Papers*, vol. 2, doc. 47, section 17. On the history of this principle, see Norton 1989a.
3. Ibid., sections 19, 20.
4. Ibid., vol. 3, doc. 23.

a four-year struggle to construct a completely new theory of gravitation. Building on the equivalence principle, Einstein now sought to formulate the laws of nature in terms of equations that would hold good in at least some classes of non-inertial frames of reference.[5] In the final form of this general theory of relativity, published in late November 1915, the gravitational field of a body was represented in any frame of reference by ten potential functions, the latter being linked via Einstein's "field equations" to the metrical properties of the space surrounding the body.[6] Within this theory the previously separate notions of inertia and gravitation were united in a single inertio-gravitational field. And, just as the special theory of relativity rendered the electric and magnetic components of an electromagnetic field relative entities (depending on the *unaccelerated* reference-frame chosen by the observer), so the general theory made the inertial and gravitational components of the inertio-gravitational field similarly relative entities (depending on the *accelerated* reference-frame chosen by the observer).[7] It was also a requirement of the general theory that gravitational effects travel with the velocity of light. Einstein now described his earlier work as constituting the "special" theory of relativity, to which the all-encompassing "general" theory reduced when gravitational effects could be ignored.[8]

It was a great accomplishment of the general theory not only that it satisfied Einstein's requirement that the laws of nature be expressed in a form that was independent of any specific coordinate system, but also that it explained a long-standing anomaly in the motion of the planet Mercury and made two further testable predictions.[9] These predictions were the gravitational red shift and bending of light referred to above, the quantitative value of the former being preserved, while that of the latter was approximately

---

5. Einstein's route to the general theory remains keenly disputed among historians of physics, but the details of the debate need not concern us here. For discussion of these issues see the editors' introductions to vols. 4 and 6 of Einstein, *Collected Papers*. For a more technical discussion, see Pais 1982, chaps. 9–14, and especially Norton 1989b and 1993.

6. Einstein's clearest contemporary account of the foundations of the new theory is given in section A of Einstein, *Collected Papers*, vol. 6, doc. 30.

7. These points are discussed in Stachel 1995, 289.

8. Einstein seems first to have designated his earlier work as constitutive of a "special" theory of relativity in November 1915. A year earlier he contrasted the new "general" theory of relativity with his "original" theory of relativity. See the introductions to Einstein, *Collected Papers*, vol. 6, docs. 9 and 21.

9. Einstein hoped as early as 1907 that a new relativistic theory of gravitation would explain the advance of 43" of arc per century in Mercury's perihelion. The search for this explanation was one factor that guided Einstein's path to the general theory. See the editors' introduction to document 14 in Einstein, *Collected Papers*, vol. 4.

doubled by the new field equations.[10] Despite these remarkable successes, it was a considerable drawback to the theory that it was conceptually difficult to grasp and relied upon advanced mathematical methods that were little known to the vast majority of physicists. From a conceptual perspective the theory required a firm grasp of four-dimensional, non-Euclidian geometry, while from a mathematical point of view it required considerable dexterity in the algebra and calculus of tensors. Einstein himself had taken some time to master these methods and had been helped in doing so by his mathematician friend and colleague at the Swiss Federal Institute of Technology, Marcel Grossmann.[11] It is a mark of Einstein's recognition that the majority of his contemporaries had little if any understanding of these methods that he often devoted whole sections of his published research papers to explaining the mathematical techniques he was about to employ.[12]

There are three aspects of the Cambridge perception of the general theory of relativity that need to be clarified before turning to Eddington's role in its British reception. First, despite the novelty and heuristic power of Einstein's work on the equivalence principle, it was ignored almost entirely in Britain until 1916. This was to a large extent a reflection of the wider lack of British interest in Einstein's work, but even men like Cunningham, Bateman, and Nicholson saw little reason to follow up his occasional remarks on gravitation. They were concerned mainly with electromagnetic theory in the form of the ETM and would have recognized no obvious connection with theories of gravity. Furthermore, they did not approve of attributing axiomatic status to the relativity principle, an assumption that was fundamental to its application to gravitation.[13] And, insofar as Cambridge mathematicians disapproved of making the relativity principle a postulate, they would a fortiori have disapproved of attributing similar status to the equivalence principle.[14] In short, there was very little knowledge in Britain of Einstein's work on the general theory prior to its publication in completed form at the end of 1915.

The second important point to keep in mind is that the general theory presented a far greater challenge to the ETM than had the special theory of

---

10. Einstein, *Collected Papers*, vol. 6, doc. 30, section 22.

11. The Einstein-Grossmann collaboration is discussed in the editors' introduction to document 13 in Einstein, *Collected Papers*, vol. 4.

12. See, for example, Einstein, *Collected Papers*, vol. 4, doc. 13, part II; and vol. 6, doc. 30, part B.

13. Einstein, *Collected Papers*, vol. 2, doc. 47, section 17.

14. As we saw in chapter 8, the relativity principle could to some extent be justified analytically by assuming the ETM (though not in the case of gravitation), but the equivalence principle would have appeared to be pure speculation.

relativity. The new theory of gravitation (which included the relativity principle) explained the anomalous motion of Mercury and made testable predictions regarding the effects of gravity on atomic spectra and the path of a ray of light. In this sense the respective Einsteinian and Cantabrigian interpretations of the relativity principle became accessible to comparative empirical test. Third, despite the interest that such a radical and testable theory was likely to generate, its conceptual novelty and mathematical difficulty meant that it was not going to be easy for Cambridge mathematicians to master its technicalities. As we saw in the case of Cunningham, even the special theory of relativity, which was far less demanding both conceptually and mathematically, had never been mastered and practiced in a fully Einsteinian sense in the university. Bearing in mind that the progress of World War I effectively ruled out any personal contact between British and Continental mathematicians, the translation of the general theory from Berlin to Cambridge was not going to be an easy process.

These observations take us directly to the heart of the issues that will be central to this chapter. Most people familiar with a little history of modern physics will be aware that Einstein's new theory achieved rapid international recognition in November 1919 following sensational reports in the press that it had been verified by Eddington. Earlier that year, Eddington had led one of two British expeditions to the South Atlantic to measure the deflection of starlight due to the sun during a total solar eclipse, measurements which he claimed to verify Einstein's prediction that the path of a ray of light was bent by a powerful gravitational field.[15] Eddington had also defended and popularized the general theory of relativity in Britain since the latter months of 1916, and, in 1918, had written and published the first technical account of the theory by a British scientist. Yet, in the light of the discussion of earlier chapters in this study, these familiar facts raise a number of obvious questions. Why did Eddington take a sympathetic interest in Einstein's work circa 1916? In terms of training, he was a typical member of the generation of wranglers discussed in the previous chapter, most of whom, as we shall shortly see, initially ignored or rejected the general theory on several grounds. How was Eddington apparently able to master the technicalities of the new theory so rapidly? He was unable to visit or attend the lectures of any of the German or Dutch masters of relativity, and, as we have seen, new theories do not travel easily in the form of published research papers alone. Why did

---

15. The background to Eddington's eclipse expedition and some of his reasons for undertaking it are discussed in Earman and Glymour 1980a.

Eddington set out to verify the new theory in 1919, and why, given their antipathy to Einstein's work, were his Cambridge peers apparently so ready to acknowledge that he had succeeded? Lastly, what pedagogical role did Eddington play in establishing relativity in Cambridge? Convincing his peers and students that Einstein's predictions had been verified was by no means the same thing as making them competent practitioners, let alone competent researchers, in the general theory of relativity.

These are questions that have been little addressed in the historical literature, mainly because their answers are tacitly assumed to be obvious.[16] As with Cunningham's apparent enthusiasm for Einstein's earlier work, it is simply assumed that the essential truth of relativity was self-evident to a few far-sighted men. But, as we have seen, Cunningham's work appears in a very different light when properly situated in its local context. Eddington's case is rather different, but there is equally much to be gained by applying the approach to reception studies outlined towards the beginning of the previous chapter. What is required are historical explanations of what attracted Eddington to Einstein's work, what resources he used to master this most conceptually novel and mathematically demanding of theories, and how he was able to popularize and propagate the theory among a new generation of Cambridge mathematicians during the 1920s and 1930s. In the following sections I offer answers to each of these questions and in doing so will attempt to cast further light on the process by which novel theories are successfully translated between geographically distant sites. Up until now my discussion of the Cambridge reception of relativity has focused on the conservative nature of the local culture of training and research, and on how that culture acted as an effective barrier to the arrival of relativity in the university. In this chapter I investigate the conditions that enabled the barrier to be overcome, even in circumstances in which there was no direct personal interaction between Cambridge mathematicians and their Continental peers. It is largely because the progress of World War I rendered such interaction impossible that the difficulties of communicating novel theoretical work in purely textual form are in this case made especially visible.

My analysis is broken into several stages. I begin by examining Eddington's mathematical training and early career as an astronomer in order to highlight those factors that made him unusually receptive to a new theory

---

16. Earman and Glymour (1980a) question why the British tried to verify the general theory, but conclude, in effect, that it was due to Eddington's enthusiasm for Einstein's work. The origin of this enthusiasm is not explained.

of gravitation. I then examine the relevance of World War I to the British reception of relativity and argue that it was initially much less significant than is usually supposed. It was only after May 1916, when Eddington was informed of the cosmological consequences of Einstein's work by the Dutch astronomer Willem de Sitter, that the war became an important factor in impeding the transmission of relativity to Cambridge. I next explore the gradual process by which Eddington, with de Sitter's help, took his first steps towards mastering the theory. As we shall see, this process was indirectly aided by pedagogical resources from Leiden and directly aided by Eddington's own efforts to write popular, semi-popular, and highly technical accounts of the theory for a British audience. At the same time that he was beginning to master relativity, Eddington also engaged in a debate with Oliver Lodge over the competence of the ETM to make the same astronomical predictions as those made by the general theory. I briefly discuss this debate in order to show its importance in shaping the view of British scientists towards the respective merits of general relativity and the ETM. Finally, I explore Eddington's role as a teacher of relativity in Cambridge in the interwar period. Having become a competent practitioner of the theory himself, he used the pedagogical apparatus of the Cambridge training system to pass on his own skills to several generations of graduate students. In doing so he not only made the university an active site for research in several aspects of general relativity, but also exploited the conservative properties of the training system to preserve the technical skills associated with the theory during a period when competent practitioners were few and far between.

9.2. Eddington and the Principle of Relativity

Like most successful wranglers of the late-Victorian and Edwardian eras, Eddington's formation as a mathematician began well before he entered Cambridge.[17] His outstanding ability in mathematics first emerged while he was a schoolboy in Weston-super-Mare, and, in 1898, he was awarded a Somerset County Scholarship. This award enabled him to enter Owens College, Manchester, at just fifteen years of age, where he was taught physics by Arthur Schuster and mathematics in the Cambridge style by Horace Lamb (2W 1872). In 1902 Eddington not only graduated top of his year with a first-class degree in physics but won a Natural Science scholarship to Trinity College, Cambridge, to begin preparation for the Mathematical Tripos. At Trinity

---

17. Eddington's early career is discussed in Douglas 1956.

MR. A. S. EDDINGTON (TRINITY)
Senior Wrangler.

Mr. Arthur Stanley Eddington, Senior Wrangler, is a son of the late Mr. A. H. Eddington, of Kendal. He was born at Kendal in 1882, and was educated at Brynmellyn School, Weston-Super-Mare, and Owen's College, Manchester. His private tutor was Mr. R. A. Herman. Our portrait is by A. H. Legg, Weston-Super-Mare.

FIGURE 9.1. Arthur Eddington, the 21-year-old senior wrangler of 1904. Eddington was the first second-year man to be placed unbracketed senior wrangler, a feat made possible in part by his pre-Cambridge training at Owens College by Arthur Schuster and Horace Lamb (2W 1872). *The Graphic*, 18 June 1904, 834. (By permission of the British Library.)

College he consolidated his emergent reputation as an outstanding applied mathematician by becoming the first second-year man to be placed unbracketed Senior Wrangler (fig. 9.1).[18] The following year, 1905, Eddington was placed in the top division of the first class in Part II of the Mathematical Tripos. In 1907 he was awarded a Smith's Prize for a dissertation on the "systematic motion of the stars."[19]

Eddington's mathematical education at Trinity College was, broadly speaking, typical of that offered at Cambridge in the early Edwardian period. There is however one peculiarity of his training that is of special relevance

18. Eddington was the only "second-year man" ever to become Senior Wrangler.
19. Barrow-Green 1999, 310.

to his later work on the general theory of relativity. The mathematical methods that lay at the heart of Einstein's technically forbidding work on gravitation were those of differential geometry, but, as I noted above, few Cambridge mathematicians, and even fewer physicists, could claim any real expertise in this area in the early twentieth century. A rare exception to this general rule was Eddington's coach, Robert Herman (see chapter 5), who was not only "devoted" to the study of differential geometry but was its "acknowledged master" in the university. As Herman's outstanding pupil and obituarist, E. H. Neville (2W 1909), later recalled, when Einstein's work placed differential geometry at the heart of cosmological theory, Herman "could have contributed more easily than anyone else in [England] to the flood of expositions." That he did not do so was due mainly to his "somewhat naive realism with regard to Euclidian space" (Neville 1928, 238), a prejudice that, as we shall shortly see, he shared with many of his Cambridge contemporaries. Despite his scepticism regarding the existence of non-Euclidian space, Herman nevertheless held an annual course at Trinity College on differential geometry which was attended by his pupils. As Eddington's near contemporary at Trinity, J. E. Littlewood (SW 1905), later confessed, both he and G. H. Hardy had subsequently used the notes they took in Herman's classes as the basis for their own lectures on the subject at Manchester and Oxford respectively.[20] Herman's most able pupils thus formed a select group in Cambridge who, at least from a purely mathematical perspective, were unusually well placed to make technical sense of Einstein's work.

Eddington's postgraduate career began with a brief and unsuccessful period of experimental research at the Cavendish Laboratory, during which he earned a living as a coach and assistant lecturer in mathematics at Trinity College.[21] He also began research in applied mathematics, hoping, no doubt, eventually to win a college fellowship. Eddington's first mathematical research appears to have focused on potential theory, with special reference to the astronomical effects of attributing a finite rate of propagation to gravitational effects.[22] Before this work came to fruition he was offered, in February 1906, the position of Chief Assistant at the Royal Greenwich Observatory.[23] Having taken up the post, Eddington began research on the systematic

---

20. Littlewood 1986, 83.

21. Douglas 1956, 11–14.

22. Douglas 1956, 9; Stachel 1986, 226; Plummer 1945, 115. This subject was one to which his teacher and mentor, E. T. Whittaker, had made major contributions. See Temple 1956.

23. The RGO was looking at this point for an Assistant with a sound knowledge of mathematical physics, rather than a standard knowledge of dynamical astronomy. Trinity fellows such

motion of the stars, a project that formed the basis of his first major publication and, in the form of a dissertation, won him both a Smith's Prize and a fellowship at Trinity College in 1907. During the next eight years he steadily built a considerable reputation as a mathematical astronomer and practical observer.[24] In 1913 Eddington was invited to return to Cambridge as Plumian Professor of Astronomy, a position he retained for the rest of his life.

One final episode from the Greenwich period of Eddington's career that is of particular significance to his subsequent work on the general theory of relativity is his participation in an expedition to observe a total eclipse of the sun in Brazil in 1912. The expedition was ultimately unsuccessful due to dense cloud, but one of the expedition's leaders, C. D. Perrine, hoped to photograph the eclipse in order to test Einstein's 1911 prediction that starlight grazing the sun's limb would be minutely deflected by its gravitational field. It is virtually certain, therefore, that by 1912 Eddington was aware not only of Einstein's prediction but also that it was accessible to experimental test in the manner proposed by Perrine.[25] Eddington's move into astronomy in 1906 was a very important factor in determining his subsequent enthusiasm for a relativistic theory of gravitation. Most professional astronomers were little concerned at this time with problems in the electrodynamics of moving bodies. When they first encountered the principle of relativity circa 1910, they understood its importance and genealogy in a rather different way from their contemporaries in mathematical physics. Since Eddington broadly shared the astronomer's view of these matters, we must explore it a little further before turning to his initial response to Einstein's work.

The earliest references to the principle of relativity by British astronomers were concerned not with the electrodynamics of moving bodies but with informing professional astronomers of the possible astronomical implications of the principle. The first such reference was made in 1910 by the astronomer H. C. Plummer in an article on the aberration of starlight.[26] Plummer noted that most astronomers were quite unfamiliar with the principle of relativity, but wished to bring it to their attention because "the developments of mod-

---

as E. T. Whittaker and W. W. Rouse Ball were instrumental in securing Eddington's appointment. See Douglas 1956, 13–14.

24. For an overview of Eddington's astronomical work in this period, see Douglas 1956, chap. 4.

25. Douglas 1956, 18–19; Stachel 1986, 227. Stachel notes that it might well have been on this expedition that Eddington first heard of Einstein's work and prediction.

26. Plummer (1875–1946) was an Oxford-trained mathematician who was an Assistant at the Oxford Observatory. In 1912 he became Professor of Astronomy at Trinity College Dublin and Astronomer Royal for Ireland.

ern physical theory concern[ed] the astronomer no less than the physicist" (Plummer 1910). According to Plummer, it was to the Dutch mathematical physicist, Lorentz, that the development of the principle of relativity was chiefly due, Einstein being mentioned only as having shown how the principle led to modifications of the standard expressions for the aberration of starlight and the Doppler effect. An article by Whittaker in a similar vein made no mention at all of Einstein, but cited Max Planck and Hermann Minkowski as having shown that Newton's laws had to be modified in the light of the principle of relativity.[27]

The most authoritative account in English of the astronomical importance of the principle of relativity was published by the Dutch astronomer Willem de Sitter. This work is of considerable importance because it was the first detailed exposition of the principle of relativity offered to British astronomers and affords considerable insight into the new genealogy attributed to relativity by astronomers. Although writing in 1911, well before the appearance of Einstein's general theory, de Sitter portrayed the importance of the relativity principle solely in terms of current debates over the nature of gravitation and its possibly finite rate of propagation.[28] He claimed that the principle had been fully established in the case of electromagnetic processes and argued that if, as some physicists had suggested, the principle was universally valid, then gravitational effects must also be propagated with the velocity of light. De Sitter attributed the development of the principle of relativity chiefly to his Leiden colleague, Lorentz, and discussed the subsequent attempts made by Lorentz and the French mathematician, Henri Poincaré, to formulate laws of planetary motion that were invariant under the Lorentz transformations (1911, 389). It was the astronomer's job, de Sitter insisted, to search for anomalies in planetary orbits (such as the advance of Mercury's perihelion) that would test not only the proposed new laws but also the scope of validity of the principle of relativity.

De Sitter did not mention Einstein's name in the paper, but, mindful of the likely attitude of some British readers towards the generalized principle of relativity, sought to draw a distinction between the proper work of the positional astronomer in testing the principle's validity and any debate over the

---

27. Whittaker 1910. Whittaker had left Trinity College in 1910 to become Astronomer Royal for Ireland, and moved to Edinburgh University as professor of mathematics in 1912. His account of the history of relativity appeared in Whittaker 1951–53, esp. 2:27–77, 144–96.

28. De Sitter 1911. At this point Einstein had shown only that it followed from the "equivalence principle" that the velocity and path of a light ray must change in the proximity of a gravitating body. He had not discussed the possibly finite rate of propagation of gravitational effects. Einstein, *Collected Papers*, vol. 2, doc. 47, vol. 3, doc. 23.

existence of the ether. Having noted that Plummer and Whittaker's otherwise admirable papers on relativity "made free use of the word 'aether,'" he remarked in a footnote:

> As there are many physicists nowadays who are inclined to abandon the aether altogether, it may be well to point out that the principle of relativity is essentially independent of the concept of the aether, and, indeed, is considered by some to lead to a negation of its existence. Astronomers have nothing to do with the aether, and it need not concern them whether it exists or not. (de Sitter 1911, 389)[29]

Many of the elite of British mathematical astronomy had been trained in the Cambridge Mathematical Tripos and would, by 1911, have been aware that advocates of the Einsteinian interpretation of the relativity principle denied the existence of the ether. De Sitter sought to diffuse any antipathy towards the principle within the British astronomical community by depicting its application and testing as purely mathematical and observational matters.

His strategy was well judged. In arguing that the professional loyalty of the astronomer ought to be to exact measurement rather than to squabbles over the existence of an electromagnetic medium, de Sitter was exploiting the fact that the ether actually played no practical role in the business of positional astronomy.[30] Neither the observational practice nor the mathematical theories of stellar motion and cosmology made reference to the ether concept.[31] Unlike the case of British electromagnetic theory, therefore, the ether could be abandoned, or simply ignored, in astronomy, without devaluing the professional practice of the discipline and without requiring any retraining on the part of the astronomers. Prior to 1915, the principle of relativity appeared to British astronomers simply as part of well-accepted electromagnetic theory, which, if generalized, challenged the notion that gravitational forces acted instantaneously.[32] Moreover, it appeared that at-

---

29. Whittaker had remarked at one point that "the fundamental branch of physical science is the theory of the ether" (Whittaker 1910, 365).

30. A comprehensive account of the state of stellar astronomy during the early years of the twentieth century is given by Eddington 1914. Eddington's book makes no reference to the ether, even though the book was written well before he became interested in Einstein's work on relativity.

31. Eddington 1914 reveals that the mathematics and physics of stellar astronomy consisted at this point mainly of spherical trigonometry, statistics, and the dynamical theory of gases.

32. Many physicists and astronomers believed that gravity was probably an electrical effect of some kind, so that even for those who accepted the ETM the idea that gravitational effects were subject to the Cambridge-school principle of relativity was perfectly reasonable.

tributing a finite rate of propagation to gravitational effects might lead to important and observationally verifiable corrections to Newtonian planetary theory. Eddington's work both at the RGO and during his first years as Plummian Professor in Cambridge was confined almost exclusively to stellar astronomy (rather than electromagnetic theory or cosmology), and it is in the context of this work (and with the understanding of the principle of relativity outlined above) that we must assess his changing attitude toward Einstein's general theory of relativity.

9.3. British Relativity and World War I

Historians of science have generally assumed that the widespread ignorance among British scientists regarding the development of Einstein's work during the period 1914–16 is largely attributable to the outbreak of hostilities between Britain and Germany in August 1914.[33] There is however very little evidence that Cambridge mathematicians and astronomers were taking a serious interest in the technical details of Einstein's work in the months that preceded the outbreak of World War I. When, for example, Britain's leading expert on the principle of relativity, Ebenezer Cunningham, published his 1914 textbook on the subject, he explicitly avoided discussion of Einstein's attempts to generalize the principle "in connection with a physical theory of gravitation" on the grounds that they were "highly speculative" (Cunningham 1914, vi). And, as we have just seen, British astronomers did not especially associate the application of the principle of relativity with the name of Einstein. The key issue for these men circa 1914 was not the physical status of Einstein's emergent field theory of gravitation (of which they were hardly aware) but the theoretical and observational implications of assuming that gravitation had to conform to the principle of relativity (the most important being that gravitational forces were propagated with the velocity of light). It is quite reasonable to suppose therefore that the outbreak of World War I actually had very little effect on British understanding of Einstein's new theory.

This supposition is further substantiated by Eddington's understanding of the significance of Einstein's work at the beginning of 1915. In February of that year Eddington contributed to the ongoing debate over the implications of attributing a finite velocity of propagation to gravitational effects in the form of a semi-popular article on "Gravitation" in the *Observatory* magazine

---

33. Earman and Glymour suggest that the general theory was little known in England before 1919 because "the war had interrupted receipt of German scientific journals" (1980a, 50).

(Eddington 1915). In this contribution to the regular column "Some Problems in Astronomy," Eddington expressed concern that no satisfactory physical theory of gravitation had yet been advanced and rejected all electrical and ethereal theories of gravitation as irrelevant to the observational and mathematical work of astronomers.[34] He nevertheless acknowledged that the ETM had cast doubt on the absolute validity of Newton's laws of motion. Astronomers now had to decide, he argued, whether it was the "transverse" or the "longitudinal" mass that should be associated with the gravitational mass of a planet.[35] Eddington also pointed out that since it had been shown that attributing a finite rate of propagation to gravitational effects was an "admissible hypothesis, we may inquire into other possible correspondences with electrical forces."[36] The most pressing correspondence to be explored in this respect was whether gravitation conformed to the "Principle of Relativity."

Eddington saw the consequences of applying the principle of relativity to gravitation as a direct extension of the role of that principle in electrical theory. Since in electrical science one could not tell whether the solar system was "at rest in the aether or rushing through it at many thousands of miles per second" (even though electromagnetic effects were known to be propagated with finite velocity), then, by analogy, the same might well be true for gravitational effects. It was also in this context that Eddington first made reference to Einstein's work on gravitation. Citing Einstein's 1911 paper, "On the Influence of Gravitation on the Propagation of Light," Eddington drew attention to Einstein's claim that if the principle of relativity applied to gravitation then "a wave of light will travel more slowly when it enters an intense gravitational field of force." It followed that a ray of light "seen close to the limb of the Sun would be apparently displaced 0.83 [seconds of arc] away from the Sun's limb" (1915, 97).

The nature of this reference to Einstein's work is of considerable relevance to our present concerns. It is clear that, even in 1915, Eddington did not associate Einstein's work with a fundamentally new theory of gravitation (nor was he apparently familiar with any of Einstein's papers after 1911). As

---

34. Eddington 1915, 94. Eddington was not against such theories in principle, but complained that they made no predictions that were testable by the astronomer.

35. Ibid. The electron would appear to have slightly different masses depending on whether it was being accelerated normally (longitudinally) or transversely to its direction of motion with respect to the observer.

36. Ibid., 96. Many astronomers believed Laplace to have shown conclusively that the rate of propagation of gravitational effects had to be instantaneous. Poincaré showed that this was not necessarily the case in 1905 (Whittaker 1951–53, 2:151).

far as Eddington was concerned, Einstein had shown only that if gravitational effects conformed to the principle of relativity, they must travel with finite velocity (probably the velocity of light) and that these two conditions might be accessible to empirical test. If the gravitational bending of light were detected, Eddington concluded, gravitation would have been "pulled down from its pedestal, and [would] cease to stand aloof from the other forces of nature" (1915, 98). He did not even mention that Einstein's prediction was based upon a new hypothesis concerning the physical nature of gravity (the "equivalence principle") and that without this hypothesis Einstein's prediction was meaningless. Given this state of affairs it seems unlikely that British astronomers would have taken much interest in Einstein's work during the period 1914–16 even if the war had not broken out. In fact a detailed summary of the physical principles, mathematical methods, and general strategy being employed by Einstein and Marcel Grossmann to build the theory *was* published in England in 1915 by the Dutch astronomer Adriaan Fokker but, as we shall shortly see, seems to have had little if any impact on Eddington and his colleagues.

Einstein published the completed general theory of relativity in late November 1915, but news of his great achievement arrived in Britain only in June 1916 in the form of a letter from de Sitter (working in the neutral Netherlands) to Eddington (then Secretary of the Royal Astronomical Society).[37] The letter is now lost, but it can easily be inferred from Eddington's reply that de Sitter gave some account of the astronomical consequences of Einstein's theory and offered to submit a substantial paper on the subject to the prestigious *Memoirs of the Royal Astronomical Society*. What seems to have prompted de Sitter to write to Eddington at this particular juncture was not Einstein's November publication announcing the completed field equations, but the publication by Einstein in mid-May 1916 of a more general summary of the new theory including some discussion of its cosmological consequences.[38] This was a topic that immediately attracted de Sitter's attention and one that he believed, quite correctly, would be of great interest to astronomers in the English-speaking world.[39]

In his reply to de Sitter, Eddington confirmed that he had to date heard only "vague rumours" about Einstein's "new work" (thereby indicating that

---

37. Although de Sitter's letter is lost, Eddington's reply, dated 11 June 1916, reveals that the former must have been written in late May or early June 1916.

38. Einstein, *Collected Papers*, vol. 6, doc. 30.

39. Had de Sitter merely been concerned to inform his British colleagues of Einstein's new theory, he would have contacted Eddington in 1915 when the field equations were first published.

he had not studied Fokker's 1915 paper), adding that no one in England had any knowledge of the "details of [Einstein's] paper."[40] Eddington also confirmed that he was "immensely interested" to learn more about Einstein's theory and suggested that, for speed of publication, de Sitter submit a paper to the *Monthly Notices* (rather than the *Memoirs*) of the Royal Astronomical Society.[41] In the event, de Sitter published a series of three expository papers in the *Monthly Notices* (and one less-technical account in the *Observatory* magazine) on Einstein's theory of gravitation and its astronomical consequences.[42] It was at this point that the state of hostilities between Britain and Germany became an important factor in the British reception of Einstein's work. Both de Sitter's letter to Eddington as well as his subsequent papers presented general relativity as a fundamentally astronomical theory. As in his 1911 paper on the role of the principle of relativity in astronomy, de Sitter was careful to avoid such contentious issues as the existence or nonexistence of the ether, focusing instead on the mathematical structure and observational consequences of general relativity. Moreover, having had their interest in Einstein's work piqued by de Sitter's initial paper, British astronomers, unable because of the war to obtain Einstein's original publications, were now heavily reliant on de Sitter's exegesis of Einstein's work.

9.4. Relativity in Leiden

> Einstein mentioned his plan to travel to Haarlem in order to hear your opinion and criticism of his theory, which would be very important to him.
>
> A. D. FOKKER TO H. A. LORENTZ, DECEMBER 1913[43]

Although it is the Cambridge reception of Einstein's work that is central to this chapter, it will be useful briefly to consider the transmission of the general theory from Berlin to Leiden. The mechanisms by which relativity was

---

40. The suggestion that de Sitter's initial letter included an offprint of "Einstein's paper" seems to originate in Plummer 1945, 117, and to have been "confirmed" verbally by S. Chandrasekhar. See Earman and Glymour 1980b, 183. Eddington's reply to de Sitter's first letter indicates that de Sitter had not enclosed any offprints and it is in fact unclear when Eddington first received any of Einstein's original publications.

41. Eddington to de Sitter, 11 June 1916 (WdS-ALO). Eddington pointed out that papers submitted to the *Memoirs* were required to undergo a lengthy process of refereeing and balloting, and that this, together with the difficulties of wartime production, could delay publication by months or even years.

42. De Sitter 1916a, 1916b, 1916c, and 1917. The papers in the *Monthly Notices* were completed respectively in August and October 1916 and July 1917.

43. Fokker to Lorentz, 4 December 1913, Einstein, *Collected Papers,* vol. 5, doc. 490.

established in Leiden not only played a significant role in shaping its subsequent Cambridge reception but strike an informative contrast with the Cambridge case. I have several times emphasized that the translation of theoretical technique from one site to another is neither a straightforward process nor one that is easily accomplished by written texts alone. In the study of the Cambridge reception of Maxwell's *Treatise* in chapter 6, we also saw that the development of an effective pedagogical account of a new theory was a complicated, protracted, and collective process. I shall shortly argue for the importance of de Sitter's work in Eddington's gradual mastery of the technicalities of general relativity, but this discussion will immediately raise the question of how de Sitter himself learned the theory so rapidly. In order to anticipate this difficulty I begin by explaining the resources through which de Sitter not only mastered general relativity but was able within a few months to make substantial contributions to its interpretation and development.

De Sitter was by no means a lone student of Einstein's work in Leiden circa 1916, but was part of what has been described as a "coherent and authoritative" college on the emergent relativistic theory of gravitation.[44] Two of de Sitter's colleagues, H. A. Lorentz and Paul Ehrenfest, were personally acquainted with Einstein, and both had long-standing interests in his work on relativity. Ehrenfest and Einstein were in fact close personal friends and regularly discussed the latest developments in physics both in conversation and through regular exchanges by letter.[45] Einstein first met Lorentz in 1911, and from 1913 the two men engaged in a "prolonged discourse" through correspondence and personal meetings on various aspects of the emergent general theory of relativity.[46] Several of Lorentz's research students also worked on problems associated with the general theory, some even joining the research program before Einstein had found the final form of his field equations. Adriaan Fokker, for example, collaborated with Einstein on gravitational theory in Zurich for a few months during the winter of 1913–14. It was following this close personal interaction with Einstein—which resulted in a joint publication—that Fokker wrote his 1915 overview of Einstein and Grossmann's work for the *Philosophical Magazine*.[47] Another of Lorentz's students, Johannes Droste, not only mastered the Einstein-Grossmann formulation of the field

---

44. Kox 1992, 39.
45. Ehrenfest's close relationship with Einstein is discussed in Klein 1970, chap. 12.
46. On the Lorentz-Einstein collaboration, see Illy 1989.
47. Sanchez-Ron 1992, 60. Einstein's collaboration with Fokker is discussed in Stachel 1989, 81–83. Fokker described his early interaction with Einstein in a letter to Lorentz in mid-December 1913. See Einstein, *Collected Papers*, vol. 5, doc. 490.

equations in 1915 but began important work on their solution. By May 1916, Droste had independently derived the first exact solution to the final form of Einstein's field equations.[48]

As the above remarks suggest, Leiden was probably the largest and best-informed site for the investigation, development, and criticism of the general theory of relativity in 1915.[49] Einstein himself visited the town on more than one occasion and appears to have derived considerable comfort from the support of his Dutch colleagues as he struggled to complete and then to interpret the new theory.[50] This rare enthusiasm for Einstein's work in Leiden also helps to explain both de Sitter's deep interest in general relativity and his rapid mastery of the theory in 1916. De Sitter had long been interested in new theories of gravitation, especially those that were capable of accounting for unexplained perturbations in the orbits of the planets. Surrounded as he was by colleagues who were not only following the development of Einstein's work but actively participating in the process (and even working and corresponding directly with Einstein), he would have been informed immediately of Einstein's success in mid-November 1915 in explaining the anomalous motion of Mercury's perihelion.[51] His interest in the theory having been piqued, moreover, de Sitter would have had a large pool of expertise to draw upon in mastering the new theory for himself.

The first installment of de Sitter's paper for the *Monthly Notices* reveals further that his pedagogical exposition of Einstein's work also owed a great deal to the collective expertise of his Leiden colleagues. He readily acknowledged that "many of the results" contained in the paper were "wholly or partially derived" from a lecture series on Einstein's work held by Lorentz, while others originated in personal discussions with Lorentz, Ehrenfest, and Droste (de Sitter 1916b, 707). While de Sitter was writing the paper during the summer of 1916, he also began his own researches into the theory's implications for planetary astronomy, the status of absolute rotation, and what boundary conditions were applicable at infinity. On the last two of these topics he began a regular correspondence with Einstein, the two men discussing the is-

---

48. Eisenstaedt 1989a, 216–18; Kox 1992, 41.

49. The only comparable site at this time was Göttingen. Thanks to the work of Minkowski, the mathematicians at Göttingen, especially David Hilbert, had long-standing interests in the development of the relativity principle. Einstein lectured there on the emergent general theory of relativity in the summer of 1915 and corresponded with Hilbert, who was also seeking the correct mathematical form of the field equations.

50. Klein 1970, 296.

51. Norton 1989b, 147–48.

sues in person when Einstein visited Leiden for two weeks in September–October 1916.[52] Further correspondence between the two men over the following year led Einstein to alter his interpretation of some aspects of the general theory, and de Sitter to establish an alternative cosmological model to the one proposed originally by Einstein.[53] The fruits of de Sitter's debate over cosmology with Einstein were also reported by de Sitter in the second and third installments of his paper for the *Monthly Notices*. It is important to keep in mind, therefore, that these installments were not simply expositions of the general theory of relativity, but to some extent original papers describing the current state of cosmological theory (according to the general theory of relativity) to an English-speaking audience.[54]

There are two important points to take from this brief study of the Leiden reception of Einstein's work. First, the translation of the theory from Berlin to Leiden was not accomplished simply by the publication of Einstein's 1915 and 1916 papers. It was, rather, an ongoing process enabled by regular correspondence, personal interaction, discussion, and collective interpretation. In this sense the Leiden reception of general relativity makes a sharp contrast with the Cambridge case, in which Eddington had few of the above resources to help him make sense of the new theory. Second, the production of de Sitter's pedagogical exegesis of Einstein's work—running ultimately to more than eighty pages of printed text—was similarly the combined product of de Sitter's outstanding command of dynamical astronomy, his long-standing interest in theories that attributed a finite rate of propagation to gravitational effects, his interest in the principle of relativity and its astronomical applications, Lorentz's lectures on the general theory of relativity, personal conversations with Lorentz, Ehrenfest, Droste, and others, correspondence and conversations with Einstein himself, and his own research on the cosmological consequences of Einstein's theory.[55] It is thus of some importance that the first account of Einstein's work to arrive in England was not only written by de Sitter but was composed in the unique environment of Leiden. The fact that several of de Sitter's colleagues were

---

52. Klein 1970, 302–4. When he returned to Berlin, Einstein wrote enthusiastically to a friend that the theory of relativity "has already come very much alive" in Leiden. Einstein to Besso, 31 October 1916, Einstein, *Collected Papers*, vol. 8, doc. 270.

53. For de Sitter's view on the importance of personal contact with Einstein, see de Sitter to Einstein, 1 November 1916, Einstein, *Collected Papers*, vol. 8, doc. 272. The technical details of the de Sitter-Einstein correspondence (1916–17) are discussed in Kersberg 1989.

54. Part 3 in particular constituted an announcement of the de Sitter universe.

55. On the importance of de Sitter's colleagues to the development of his research, see Eisenstaedt 1989a, 219–20, and Kerszberg 1989, 338.

teaching and researching the general theory made it easier for him to provide a coherent and authoritative account of the subject, while his own outstanding expertise in astronomy and cosmology suited him ideally to tailor his account to the interests of astronomers.

9.5. Eddington the Einsteinian

> Owing to historical tradition, there is an undue tendency to connect the principle of relativity with the electrical theory of light and matter, and it seems well to emphasize its independence.
>
> A. S. EDDINGTON, 1918 [56]

Returning now to Eddington's initial exchange of letters with de Sitter in the summer of 1916, we can begin to build an explanation of why and how he became the first person in Britain to master Einstein's work. Eddington's first letter to de Sitter confirms both the complete lack of awareness in Britain of the new theory and its cosmological consequences, and Eddington's own eagerness to learn more from de Sitter. The immediate effect of de Sitter's first letter appears to have been to prompt Eddington to offer a paper on Einstein's work to a session on "Gravitation" to be held at the British Association meeting in Newcastle in early September.[57] At the beginning of July 1916, Eddington informed de Sitter of the meeting and asked if he had any objection to him using the first installment of the proposed paper (which Eddington had yet to receive) as the basis of a presentation. Eddington's letter also confirms that English astronomers had still not seen "Einstein's paper," a remark which strongly implies that de Sitter had not enclosed any offprints of Einstein's work with his first letters. The earliest account of the new theory that Eddington actually studied was almost certainly the first installment of de Sitter's paper, which was completed sometime in August 1916.[58] The British Association session on gravitation was introduced by Cunningham, but it was Eddington's subsequent presentation (based on de Sitter's paper) that spelt out the astronomical consequences of the theory.[59] It is a measure of Cunningham and Eddington's ignorance of Einstein's work at this time that the official account of the session (published in 1917) made no

---

56. Eddington 1918b, v.

57. The session also contained papers on more mundane aspects of geodesy. See *Nature* 98 (1916): 120.

58. This is further supported by Eddington 1923, vi.

59. It is unclear whether Cunningham's introductory paper was part of the session as originally planned or added in the light of Eddington's proposed paper.

mention of their presentations, but referred the reader directly to de Sitter's first two papers in the *Monthly Notices*.[60]

It was over the autumn of 1916 that Eddington took his first steps towards mastering the new theory. A letter to de Sitter in mid-October indicates that Eddington had just read (for the first time) Einstein's paper of May 1916 and was reflecting upon the status of absolute rotation according to the general-relativistic account of gravitation.[61] Eddington followed de Sitter in expressing dislike for Einstein's practice of defining a body's rotation (and hence its gravitational field) in terms of boundary condition at infinity, claiming that this approach seemed to contradict Einstein's general rule that "observable phenomena" ought to be "entirely conditioned by other observable phenomena." The deferential tone of Eddington's letter indicates that he was very unsure of his ground at this stage, and he concluded by remarking that he would suspend his judgment on the matter until he had seen the second instalment of de Sitter's paper (in which de Sitter's differences with Einstein were explicitly addressed). Eddington's first exposition of the general theory (Eddington 1916), published at the end of December 1916, similarly displays a sound grasp both of the basic principles of the new theory and of the astronomically testable predictions that it made, but gives no indication that Eddington had begun to come to grips with the forbidding technicalities of Einstein's work. As Eddington himself noted some eighteen months later, "the main difficulty" to be overcome in mastering the general theory was that it required "a special mathematical calculus, which, though not difficult to understand, needs time and practice to use with facility" (Eddington 1918b, v).

It is worth pausing at this point briefly to consider the extent to which Eddington's attitude towards general relativity was shaped by the circumstances of his initial encounter with the theory. I argued above that as a member of the international astronomical community, Eddington adopted a far more pragmatic attitude towards the principle of relativity than did his Cambridge colleagues, whose primary interests lay in electromagnetic theory. The main consequence of the principle for astronomers was to require gravitational forces to travel with the velocity of light, a hypothesis that had already been explored in the hope of explaining the anomalous motion of some of the planets. Eddington's reception of the general theory in the form of letters and papers from de Sitter would have reinforced this view of relativity

---

60. *British Association Report* (1916, 364).
61. Eddington to de Sitter, 13 October 1916 (WdS-ALO). Eddington stated that his criticisms did not "detract in the least from the remarkable importance of the paper." The context implies that "the paper" referred to was Einstein's rather than de Sitter's.

in several ways. First, general relativity appeared from the outset as a fundamentally cosmological theory that was not only acceptable to astronomers but which also accounted for the previously unexplained motion of Mercury's perihelion and made other astronomically testable predictions. Second, de Sitter's high reputation as an astronomer together with his advocacy and technical mastery of Einstein's work would have reassured Eddington that the theory was technically sound and, far from representing a threat to astronomy, pointed to a new era in theoretical and observational cosmology. Third, de Sitter's exposition not only highlighted the cosmological consequences of general relativity but provided Eddington with a reliable and authoritative technical exegesis of the theory, in English, that culminated in a state-of-the-art discussion of the relative merits of the respective Einstein and de Sitter universes. A mathematical astronomer could hardly have wished for a more reliable, accessible, and original introduction to general relativity and its cosmological consequences.

These circumstances help to explain why Eddington, alone among his Cambridge colleagues, embraced the general theory and became its chief advocate in the English-speaking world. Far from regarding the theory as a threat to Cambridge views on geometry and electrodynamics, Eddington saw it as a new, physical explanation of gravitation that attributed gravitational effects with a finite rate of propagation, explained a puzzling anomaly in Mercury's motion, opened exciting avenues of research concerning the large-scale structure of the universe, and made further predictions that were testable by astronomers. It was for these reasons that Eddington was prepared from the start to entertain Einstein and de Sitter's assumption that the space of our universe was non-Euclidian. The depth and uniqueness of Eddington's early commitment to relativity is nicely illustrated in an exchange with James Jeans (3W 1898), a man five years Eddington's senior and, as we saw in chapter 7, an expert on Cambridge electromagnetic theory. In January 1917, Jeans published a competent overview of the relativistic theory of gravitation that concluded with the claim that Einstein's "crumpling up" of four-dimensional space should be considered "just as fictitious as the crumpling of ordinary space represented by the metric transformation $ds' = \mu ds$ in optical theory" (Jeans 1917, 58). According to Jeans, Einstein's use of non-Euclidian geometry was merely a convenient calculating device which correctly predicted such measurable phenomena as Mercury's motion around the sun. In this sense the metric transformations of general relativity were akin to those routinely employed by wranglers in optics and electrostatics (see chapter 8). These transformations were understood either as convenient fictions which helped to solve specific problems, or else as mathematical devices for generating new geometrical distributions of physical effects within Euclidian space. Jeans argued accordingly that Einstein was wrong to claim that space was actually non-Euclidian and therefore had provided no new *physical* explanation of gravitation (1917, 58).

Jeans's views may in two respects be taken as broadly representative of those held by Cambridge mathematical physicists trained in the late Victorian and Edwardian periods. First, as we saw in the previous two chapters, these men understood the principle of relativity to represent an analytical property of the equations of electromagnetism. They did not attribute the principle with axiomatic status and were therefore very suspicious of attempts to apply it to non-electromagnetic forces such as gravitation. Second, there was a long-standing antipathy in Cambridge to attributing equal status to Euclidian and non-Euclidian geometries. As Joan Richards (1979) has shown, the majority of Cambridge mathematicians interpreted the latter in terms of projective geometry and continued to believe that the space of our universe was necessarily Euclidian. These views were neatly illustrated with respect to Einstein's work by William Garnett (5W 1873) in 1918. Garnett lampooned Einstein's non-Euclidian universe by likening it (fig. 9.2) to the strange dream world depicted in Lewis Carroll's famous children's story *Through the Looking-glass and What Alice Found There*.[62] In Garnett's parody, Einstein's error in claiming that space was really non-Euclidian was analogous to confusing the distorted image of an object in a convex mirror with the actual shape of the object as seen with the naked eye.

The sharp difference of interpretation between Eddington's understanding of de Sitter's first two expository papers and that due to the majority of his Cambridge colleagues is captured in his response to Jeans in the pages of the *Observatory* magazine. Eddington argued that Jeans was wrong to portray general relativity as a mere computational device which employed an auxiliary geometrical space, and went on to give a "defence of the 'metaphysical' garb" of Einstein's work (Eddington 1917a, 94). It was not, he insisted, "the relativity-space—the 'shadow'" that had to be discarded, but rather the so-called "ordinary space" of Euclidian geometry. What Jeans did not appreciate, Eddington concluded, was that the "'the relativists' [were] not speaking according to their own conceptions" (94), but drawing upon geometrical techniques that were well known and considered acceptable to many Euro-

---

62. Carroll 1872; Garnett 1918. "Lewis Carroll" was the pen name of the Oxford humourist and mathematician Charles Dodgson.

FIGURE 9.2. William Garnett's comic sketch illustrating his claim that those who believed space to be non-Euclidian were confusing physical reality with a distorted image. "I will ask you to consider," Garnett wrote, "whether Alice through the Convex Looking-Glass, with her variable standards of length, mass and energy, is living in a wonderland more wonderful than that which Einstein and the other mathematicians of Relativity would have us believe that we inhabit" (Garnett 1918, 239). Most Cambridge-trained mathematicians would have agreed with Garnett that the space of our world was Euclidian. (By permission of the Syndics of Cambridge University Library.)

pean mathematicians.[63] Eddington was, nevertheless, careful to qualify his evident enthusiasm for Einstein's work by adding that he was "not necessarily defending the theory, but its mode of presentation" (94). As a devout Quaker who had declined military service in World War I on religious grounds, Eddington had to take care not to appear too strong an advocate of what was widely regarded as a German theory.[64]

Mere enthusiasm and qualified public support for Einstein's work would not however have been enough to establish Eddington as a leading exponent of the new theory. In order to offer an informed opinion on issues of the kind debated by Einstein and de Sitter, and to speak authoritatively on the theory to his most technically competent British colleagues, Eddington needed a sound grasp of the physical principles, mathematical methods, and

---

63. In other words, Einstein's non-Euclidian space was itself to be understood as physically real.
64. Eddington had already been criticized by Larmor for "viewing things from the German standpoint." See Eddington to Larmor, 7 June 1916 (JL-RS).

exemplary problem solutions of which the theory was composed. Acquiring such skills without the resources de Sitter had taken for granted in Leiden would be a daunting task had it not been for three factors militating in Eddington's favor. The first was his remarkable "capacity for rapid manipulation of mathematical tools." As we have seen, Eddington was the only man in the history of the Mathematical Tripos to become senior wrangler after just two years of undergraduate study, and, in his subsequent career, became famous for the speed with which he could solve well-formulated but technically complex problems in mathematical physics.[65] Second, Eddington was unusual among Cambridge applied mathematicians in having mastered the rudiments of differential geometry in Herman's lectures at Trinity College a decade earlier. Lastly, de Sitter's papers on the general theory were, as we have seen, intended to be introductory and expository, and were informed pedagogically by Lorentz's experience of teaching the new theory in Leiden.

Despite these advantages, Eddington still took a further eighteen months to master the most technical aspects of general relativity. A letter from Einstein to de Sitter in mid-January 1917 implies that Eddington had not only begun to study Einstein's 1914 paper "The Formal Foundation of the General Theory of Relativity"—a paper intended clearly to explain the mathematical methods and formal structure of the theory—but had detected some errors and inconsistencies in the work. The same letter also suggests that Einstein was now forwarding offprints of important publications to Eddington via de Sitter.[66] Progress nevertheless remained slow. Almost eight months later, when Eddington received the third installment of de Sitter's trilogy for the *Monthly Notices,* he responded with some general comments (of the kind he had made ten months earlier) but confessed that he still "could not venture to go into the analysis in detail." Eddington's ability to deal confidently with the mathematical machinery of general relativity continued to develop over the following year as he prepared an official *Report on the Relativity Theory of Gravitation* (1918b; hereafter *Report*) for the Physical Society of London. He nevertheless continued to rely on de Sitter as an author-

---

65. Douglas 1956, 109–10.
66. Einstein to de Sitter, 23 January 1917, Einstein, *Collected Papers,* vol. 8, doc. 290. Einstein's refers only to "our colleague" in this letter, but the context makes Eddington the most likely candidate. The paper concerned (*Collected Papers,* vol. 6, doc. 9) can be inferred from Einstein's section and page references. Einstein claimed in the opening paragraph that the theory had previously consisted of a "coloured mixture" of physical and mathematical postulates. It is unclear whether this work, published in November 1914, had been sent by de Sitter or had arrived in England just after the outbreak of the war. It should be noted that the version of the general theory given in this paper is not the same as the final one published in 1915.

ity on all aspects of the subject and to defer to his judgment on matters of technical detail. When Eddington completed the *Report* in April 1918, he wrote at once to de Sitter asking if he would be prepared to read the proof-sheets for "general criticism and detection of blunders."[67] Eddington had by this time acquired sufficient mastery of Einstein's field equations to derive some novel mathematical relationships which, although implicit in David Hilbert's version of general relativity, had never before been stated explicitly. Eddington still found it necessary to consult de Sitter to ensure that the results were correct and had not already appeared in print in Continental Europe.[68]

This was almost certainly the last occasion on which Eddington felt the need to rely on de Sitter's guidance in his writings on relativity. The *Report* was the first complete account of the general theory to appear in English and was praised by at least one reviewer as "one of the masterpieces of contemporary scientific literature."[69] Eddington began by tracing the technical development of the principle of relativity from Einstein's original publication of 1905 through Minkowski's reformulation in terms of four-dimensional space-time, to Einstein's introduction of the equivalence principle and expression of gravitational force in terms of the metric properties of space and time. Aware that many of his British readers would approach the theory from the perspective of the ETM, Eddington took care to emphasize that Einstein's work was in no way derivative of electromagnetic theory.[70] There then followed a detailed account of the mathematical theory of tensors, which prepared the reader for a chapter on Einstein's field equations and completed theory of gravitation. In the final four chapters, Eddington dealt with the "crucial" predictions made by the theory, the gravitational effects of a continuous distribution of matter, the "least action" formulation of the theory due to Hilbert, and, finally, the cosmological consequence of the theory, including Einstein and de Sitter's respective accounts of the structure of the universe. For many British mathematicians and physicists the *Report* (including a second edition published in 1920) represented the definitive English-language account of general relativity and further established Eddington's emergent reputation as the theory's master and champion in Britain.

---

67. Eddington to de Sitter, 16 August 1917 (WdS-ALO). A letter of 24 June 1918 confirms that de Sitter did check the proofs before the *Report* went to press.
68. Eddington to de Sitter, 28 April 1918 (WdS-ALO).
69. Quoted in Douglas 1956, 39.
70. Eddington emphasized, for example, that the change in mass of moving matter predicted by relativity was "in no way dependent upon the electrical theory of matter" (1918b, 11).

9.6. General Relativity versus the ETM

One reason Eddington's *Report* was keenly anticipated and well received in Britain in 1918 is that during the year preceding its publication Eddington had engaged in a partly public and partly private debate with Sir Oliver Lodge over the respective cosmological consequences of the ETM and the general theory of relativity. Lodge had made his reputation as young professor of physics at University College, Liverpool, during the 1880s and 1890s. An outstanding public lecturer and pioneering experimenter in wireless telegraphy, he had been made principal of Birmingham University in 1900 and was knighted in 1902. Lodge had also collaborated with FitzGerald and Larmor during his Liverpool days, and remained a stalwart supporter and popular-public champion of the ETM and the ether concept.[71] The exchanges between Lodge and Eddington from August 1917 to February 1918 are especially important because they not only helped to shape the attitude of the British scientific community towards the ETM and the general theory of relativity, but further underline a number of significant differences between the British reception of the special and general theories of relativity respectively.

We have seen that Einstein's 1905 paper on the electrodynamics of moving bodies initially aroused little interest among British physicists. Even after 1909, when it became clear that the Einsteinian interpretation of the relativity principle involved the outright rejection of the ether concept, there was still no heated controversy over the issue because it remained experimentally unresolvable. But the appearance of the general theory in 1915 destabilized this situation in three ways. First, by 1915 both Einstein himself and his controversial interpretation of the principle of relativity were well known to the European physics community, so that the announcement of a new relativistic theory of gravitation was not likely to pass largely unnoticed in Britain, as had his papers of 1905. Second, because the new theory was interpreted and popularized in Britain mainly in the context of astronomy, it was primarily perceived as a new theory of gravitation. And since gravitation had remained beyond the purview of the ETM, it was most unlikely that general relativity would be conflated with the ETM by Cambridge mathematicians. Finally, general relativity made some very specific predictions (the advance of Mercury's perihelion, the gravitational bending of light, and gravitational red shift) which were experimentally testable and did not appear as

---

71. Lodge's career is discussed in Hunt 1991a.

consequences of the ETM. From the British point of view, Einstein had unexpectedly stolen the initiative by taking the debate over the validity of the Einsteinian principle of relativity out of the realm of the electrodynamics of moving bodies and into astronomy. In this new context the ETM suddenly appeared vulnerable to experimental refutation.

In order to sustain the credibility of the ETM as the general theory of relativity became increasingly well known in Britain during 1916 and 1917, it would be necessary to show that the former theory was equally capable of accommodating the phenomena predicted and explained by Einstein. By far the most important of the these phenomena was Einstein's successful explanation of the motion of Mercury's perihelion. It was Lodge who, in the summer of 1917, attempted to defend the ETM by showing that it too had cosmological consequences which might be capable of explaining Mercury's anomalous motion. He pointed out that according to the ETM the electromagnetic mass of a planet varied quite considerably during its orbit around the sun, a variation which he explained in the following way. The total electromagnetic mass of a planet at any instant was composed of its electrostatic rest energy and its electromagnetic "extra mass" due to its motion through the ether. But during a complete revolution around the sun, the planet's extra mass would vary considerably due to the mean drift of the whole solar system through the ether. During one half of its orbit the sun's velocity through the ether would add to the electromagnetic mass of the planet, while during the other half of the orbit the solar velocity would have to be subtracted (Lodge 1917a, 82). Lodge then claimed that if the *gravitational mass* was a constant quantity associated only with the *electrostatic rest mass,* then "an increase in [electromagnetic] inertia, without corresponding increase of gravitational control, [could not] fail to have astronomical consequences" (82). He then showed not only that it was possible to explain the motions of the perihelia of Mercury and Mars in this way, but that the existence of such orbital perturbations also offered a way of determining the absolute velocity of the solar system through the ether (87).[72] Lastly, he addressed the question of the bending of a light beam as it passed by the sun, showing how what he called "temporary matter" could be associated with the electromagnetic momentum of the beam so that the bending became a predictable and quantifiable phenomenon without introducing the principle of relativity (93).[73]

72. Lodge thought this velocity might be about two and a half times the earth's orbital velocity.
73. Lodge was anticipating a positive detection of this effect and obtained a value of 0.74″ for the deflection (roughly half of that predicted by Einstein). Earman and Glymour (1980a, 57–58) claim that this so-called "Newtonian" value was invented by Eddington to ensure a crucial test

By the summer of 1917, Lodge was well aware that Eddington had become the most important supporter of general relativity in Britain, but since the latter was also one of Britain's foremost experts on mathematical and observational astronomy, Lodge sent him an advance copy of his paper asking for Eddington's opinion on its astronomical plausibility. Eddington replied that while he found the paper interesting, it seemed to him unlikely that it would be possible to "find any solar motion which will set Mercury right without upsetting something that is already accordant."[74] He went on to point out that although Lodge's ingenious theory could explain the perturbations in the orbits of Mercury and Mars, it also introduced new perturbations into the orbits of the other planets which were not observed. And despite the conciliatory tone of his correspondence with Lodge, Eddington clearly felt that Lodge's paper had to be answered publicly. He wrote a note for the *Philosophical Magazine* which, while acknowledging that Lodge's ideas would no doubt be "widely preferred" in Britain to those of Einstein, set out his own objections to Lodge's argument (Eddington 1917b, 163).

Eddington sent the paper to Lodge (then an editor of the *Philosophical Magazine*) inquiring in a note whether he would be prepared to "communicate" it to the journal. He added that he would be "extremely glad" if Lodge would accede to his request, as it would "remove any possibility of an appearance of hostility, which was, of course, far from my thoughts, for (although I am somewhat of an Einsteinian) I regarded the suggestion as particularly interesting."[75] Lodge agreed to communicate the paper to the *Philosophical Magazine*, but Eddington was not prepared to let matters rest having offered only qualitative objections to Lodge's arguments. He immediately wrote a second note—also communicated by Lodge—in which he attacked Lodge's arguments quantitatively by undermining his competence as a practitioner of the ETM (Eddington 1917c). Eddington argued that Lodge had not distinguished properly between the longitudinal and transverse masses of the planets and showed that according to his own "more rigorous calculation" the motion of the perihelion of Mercury was actually independent of the velocity of the solar system through the ether. Lodge could only reply in print by suggesting that the solar system might be in a state of acceleration (due to galactic rotation) so that Eddington's calculations were

---

between the Einsteinian and Newtonian values. Eddington certainly knew of Lodge's calculation, which assumed that Newton's gravitational law held good.

74. Eddington to Lodge, 2 August 1917 (OJL-UCL).
75. Eddington to Lodge, 6 August 1917 (OJL-UCL).

invalid, and by insisting that the success of his original calculation in explaining the perturbations in the orbits of Mercury and Mars was too good to be "accidental" (Lodge 1917b, 519). In private Lodge began an intense correspondence with Joseph Larmor in an effort to improve his own mathematical arguments.

Throughout the autumn of 1917, Lodge and Larmor worked hard to explain planetary motion using the ETM. At one point Larmor became mentally exhausted wrestling with Einstein's explanation of the advance of Mercury's perihelion, but Lodge, having advised Larmor to "put away [his] arithmetic," soon turned to him again for help with mathematics.[76] There turned out to be enormous flexibility in the assumptions that could be made in applying the ETM to planetary motion, so that the debate with Eddington might have gone on almost indefinitely. In practice, however, two factors hampered Larmor and Lodge in their struggle with Eddington. First, as we saw in chapter 8, by 1915 the brilliant young Cambridge mathematicians who had developed the ETM after 1900 had moved on to other projects. And since Lodge had only limited knowledge of higher mathematics, Larmor (now almost sixty years old) found himself working alone to build a consistent account of planetary motion based on the ETM.[77] Second, Larmor and Lodge were not competent practitioners of the general theory of relativity and had little grasp of the complicated mathematics of planetary motion. They were therefore unable to criticize Einstein's calculations, and were obliged to accept Eddington's assessment of the fine accuracy of their own predictions.[78] In the end, however, it was Eddington who brought the debate to a close through clever manipulation of Lodge's arguments.

In December 1917, Lodge wrote to Eddington announcing a somewhat surprising result he had come upon while trying to respond to Eddington's latest criticism. Lodge had originally associated the gravitational mass of each planet with its electrostatic rest mass because he assumed (as did Eddington) that if the gravitational and inertial properties of matter increased proportionally with velocity through the ether, the motions of the planets would remain unaffected overall. But it turned out that if one attributed gravitational effect to *all* electromagnetic mass, the perturbations originally calculated by Lodge would, on average, be doubled. A considerable disadvantage

---

76. Larmor to Lodge, 22 September 1917 (OJL-UCL) Ms. Add. 89/32.
77. Warwick 1989, 206–7.
78. Several of Einstein's calculations were also open to question at this time, as we shall shortly see.

to adopting this proposal was that the ETM then failed correctly to predict the advance of Mercury's perihelion, but Lodge was very satisfied to discover that the ETM appeared to have astronomical consequences, regardless of the detailed assumptions made in applying it to gravitational problems. In order to appreciate the significance of Eddington's reply to Lodge, it is necessary to keep in mind precisely what was at stake in their debate. Lodge's claim that the ETM had important astronomical consequences included the assumption that the Newtonian law of gravitation was to be retained. Indeed, Larmor and Lodge's calculations concerning the relative gravitational and inertial effects of electromagnetic mass were only meaningful if it was assumed that the gravitational constant was not itself a function of velocity through the ether.[79] Eddington, on the other hand, in advocating the general relativistic explanation of gravitation, was in favor of abandoning the Newtonian law of gravitation altogether. Thus it was not simply the ETM he was attacking in his debate with Lodge. Eddington believed that if the predictions made by Einstein's theory were verified, they would provide unequivocal evidence that Newton's law of gravitation was false.

In his reply to Lodge, Eddington began by welcoming the discovery that the ETM had astronomical consequences regardless of whether or not gravitational effect was ascribed to electromagnetic mass. This meant, he explained, that provided a significant fraction of the mass of ordinary matter was of electromagnetic origin (and that the solar system was not at rest in the ether), then *all* forms of the ETM definitely predicted large perturbations in the planetary orbits.[80] But since, Eddington continued, these large perturbations were not observed, it was proper to infer that the Newtonian law of gravitation did not hold when bodies moved relative to the ether. He concluded:

> In thinking over the matter in the last two months, I had felt that this discussion had produced valuable evidence *(evidence—*not *proof)* for the principle of relativity in its Minkowskian form, with or without the recent extensions by Einstein. We know that optical and electrical laws have entered into a strange conspiracy to prevent us determining our motion through the aether, by methods which at first sight seemed almost certain of success. In general-

---

79. If the gravitational constant also varied, it would not be possible to determine the earth's motion through the ether.

80. It must be remembered that most physicists, including Eddington, believed that a significant portion of the mass of matter was probably of electromagnetic origin, so that Eddington's argument carried considerable weight.

izing this, it has been assumed as an hypothesis (I believe without a vestige of proof) that gravitation conforms to the same principle. By working out your theory we obtain the first definite indication that gravitation has actually joined the conspiracy.[81]

Here Eddington was drawing a clever analogy between the origins of the principle of relativity in electromagnetic theory and its application to gravitation. The principle had been established in the former case in response to the failure by experimenters to measure effects that seemed to follow straightforwardly from established electromagnetic theory. Eddington now claimed that since the ETM clearly predicted major gravitational effects which astronomers had failed to detect, gravitation too must be subject to the principle. It then followed that the Newtonian law of gravitation would have to be modified or abandoned entirely. Eddington conceded that Lodge would probably consider this "the point of view of an incurable relativist," but he had managed to acknowledge the validity of Lodge's arguments while cleverly turning them to his own advantage.[82]

Eddington's positive remarks also offered Lodge a convenient way out of an increasingly embarrassing public exchange.[83] In order to continue the debate, Lodge and Larmor would have either to argue that the solar system was stationary in the ether, or to find a new version of the ETM that explained the advance of Mercury's perihelion without introducing unobserved perturbations in the orbits of the other planets. The first option would have been self-defeating, as it provided no account of Mercury's anomalous motion; the second was beyond Lodge's abilities. Having considered the matter for a few weeks, Lodge wisely acquiesced in Eddington's interpretation of his arguments. His final comments on the topic, published in February 1918, acknowledged the shortcomings of his own astronomical theories but ended with the optimistic remark that "we may perhaps discover something constructive" from the debate (Lodge 1918, 143). He then followed Eddington in suggesting that the non-existence of the astronomical effects predicted by

81. Eddington to Lodge, 8 December 1917 (OJL-UCL).
82. Ibid. Eddington's show of evenhandedness was something of a sham, but as a conscientious objector his position in Cambridge, even as Plumian Professor, was extremely delicate. Eddington later appealed to Lodge to try to dissuade the Cambridge Tribunal from forcing him to work as a farm laborer, "as indeed [had] happened at the same Tribunal to another relativist—Cunningham." Eddington to Lodge, 22 July 1918 (OJL-UCL).
83. An account of the debate given in *Nature* was throughly unsympathetic to Lodge, concluding that relativity was "left in a stronger position than ever" ("Our Astronomical Column," *Nature*, 13 September 1917, 33).

the ETM might "serve to strengthen our belief in some form of the Principle of Relativity," though he made no reference to Einstein's recent work. Thus was Lodge finally led to admit that physicists now had to "face the possibility that gravitation too obeys a compensating law, and declines to enable us to receive information about absolute motion of matter through aether" (144). These remarks by no means signaled Lodge's acceptance of Einstein's work, but they did represent an authoritative and public acknowledgment that the principle of relativity applied to gravitation, and that the ETM could not explain Mercury's anomalous motion.

9.7. Relativity Comes to Cambridge

> A slight man of average height, in academic gown, reserved almost to the point of shyness, [Eddington] rarely looks at his class. His keen eyes look at or through the side wall as he half turns from the blackboard and seems to think aloud the significance of the tensors which he has just written on the board. The mathematical theory of relativity is developed *ab initio* before our eyes and the symbols are made to live and take on meaning.[84]

The year 1918 marked an important watershed in Eddington's deepening association with the general theory of relativity. The successful conclusion of his public debate with Lodge was followed by the first of many popular lectures in which Eddington sought to captivate a general audience with the theory's strange and counterintuitive implications. Speaking on 1 February 1918 at the Royal Institution in London, he described a remarkable universe in which moving rods became shorter, time and space ceased to be absolute entities, and the earth was held in its orbit by "trying to find the shortest way through a space and time which have been tangled-up by an influence radiating from the sun" (Eddington 1918a, 34). The completion and publication of the *Report* for the Physical Society later in the year further established him in the scientific world not just as a brilliant astronomer and talented popularizer, but as Britain's leading authority on relativity and relativistic cosmology. The cessation of hostilities between Britain and Germany in November 1918 also enabled Eddington to commence concrete preparations for yet another relativity-related project. He and Sir Frank Dyson, the Astronomer Royal, intended to test Einstein's prediction that the path of a ray of light would be deflected when it passed close to a massive gravitating body.

---

84. This anonymous account of Eddington lecturing in 1922 is reproduced in Douglas 1956, 51.

Astronomers had known since 1911 that it might be possible to test the prediction by photographing the field of stars close to the sun's limb during a total solar eclipse. The effect of the sun's gravitation could then be calculated by comparing the photograph with one taken of the same star field when the sun was elsewhere in the sky. Sometime during 1917, Dyson decided to send two British expeditions to photograph an especially favorable eclipse due on 29 May 1919.[85] He invited Eddington, who, as I noted above, had participated in a similar a project in 1912, to lead the expedition bound for the island of Principe off the west coast of Africa.[86] On 6 November 1919, at a joint meeting of the Royal Society and the Royal Astronomical Society, Dyson and Eddington announced that the results obtained by the expeditions had verified Einstein's prediction.[87] At the end of the discussion that followed the announcement, Sir J. J. Thomson, President of the Royal Society and chairman of the joint meeting, summed up the proceedings by remarking that this was "the most important result obtained in connection with the theory of gravitation since Newton's day."[88] The unexpected demise of Newton's celebrated law of gravitation prompted sensational headlines in the world's press, which, in turn, generated widespread public interest in relativity, its little-known German author, and the English astronomer who had verified the theory.[89]

These events also altered the status of relativity within the scientific community. Einstein's theory was suddenly placed before a multinational audience of mathematicians, astronomers, and physicists, most of whom understood it to have been verified by the British eclipse expeditions. Yet few even among these professional scientists could claim either to have mastered the theory or to be prepared to expend the time and effort required to do so. As the brilliant German theoretical physicist Max Born recalled some years later, the general theory was so conceptually and mathematically difficult that it seemed "frightening" when first encountered in Einstein's papers. Born eventually came to grips with the essentials of the theory through a combination of paper-based study and "numerous discussions with Einstein" in

---

85. Earman and Glymour 1980a, 71. The eclipse was especially favorable because it occurred in a field of bright stars.

86. Eddington's participation was also a condition of his exemption from military service (Earman and Glymour 1980a; Chandrasekhar 1983, 25–26).

87. For an account of the eclipse expedition, Eddington's construction of the "crucial experiment," and the public announcement of the result, see Earman and Glymour 1980a.

88. Anon., "Joint Eclipse Meeting of the Royal Society and the Royal Astronomical Society," Observatory 42 (1919): 289–398, at 342.

89. For accounts of the world-wide response to these announcements, see Pais 1982, 303–12, Glick 1987, Kevles 1978, chap. 12, and Missner 1985, 267–91.

Berlin, but he resolved "never to attempt any work in this field."[90] Apart from its intrinsic technical difficulty, general relativity made few empirically testable predictions and was of little consequence to the development of other branches of theoretical physics. Born concluded that while Einstein's masterpiece represented "the most amazing combination of philosophical penetration, physical intuition and mathematical skill," it was best "enjoyed and admired from a distance."[91]

Born's comments point to a problem that severely limited the spread of general relativity for more than forty years after its initial inception. It is likely that, circa 1919, only a few dozen astronomers and physicists in the world had sufficient technical grasp of the theory either competently to criticize its content or to contribute to its development.[92] Virtually all of those who were competent practitioners had in any case been following Einstein's steps towards the final form of his field equations for some years, had long-standing interests in closely related aspects of astronomy, physics, and/or mathematics, and were associated either directly or indirectly with Einstein's intimate circle of colleagues.[93] For the majority of scientists there were several serious obstacles to mastering the theory. First, apart from what many people regarded as its intrinsic conceptual beauty and mathematical elegance, there was, as I have already noted, little professional incentive to spend the considerable time and effort required to master the theory's technical intricacies. Second, those who did seek such mastery would, like Born, have found it very difficult to make sense of Einstein's papers without help from someone who already understood their content. This point was neatly illustrated above by the respective cases of de Sitter and Eddington: where de Sitter was able within months to make original contributions to relativistic cosmology, Eddington took two years simply to develop a firm

---

90. For an excellent example of a physicist writing to Einstein asking him to clarify how the field equations were applied to a specific problem, see Reissner to Einstein, 22 June 1915, Einstein, *Collected Papers*, vol. 8, doc. 90.

91. Born's remarks are quoted in Eisenstaedt 1989b, 289.

92. It is difficult to put even an approximate figure on how many scientists in the world had a really sound technical grasp of general relativity circa 1919. If we assume that the number corresponds roughly to those who made significant research contributions to general relativity by 1919 then it is around twenty.

93. The wider reception of general relativity is beyond the scope of this study, but a brief survey of early contributors indicates that they either had the requisite interests in mathematics and astronomy, and/or were associated with major centers such as Leiden or Göttingen, where the development of Einstein's work had been closely followed. More peripheral contributors such as the Russian A. A. Friedmann and the American H. P. Robertson would perhaps make interesting case studies along the lines of this chapter.

grasp of the theory's conceptual foundations, mathematical structure, and canonical-problem solutions.

Finally, there were few pedagogical aids in the form of expository courses, introductory textbooks, graded exercises, and problem sets to help would-be students gradually and systematically develop the necessary concepts and skills. The rarity of authoritative material of this kind in 1919 is hardly surprising, as the theory not only drew upon a novel and very demanding mixture of physical, geometrical, and computational methods, but remained opaque in some of its most intriguing aspects. I mentioned above that de Sitter and Einstein disagreed over the boundary conditions to be applied at infinity, what physical cause was to be ascribed to gravity, and what model of the universe followed most naturally from the general theory. These issues raised serious questions concerning both the physical and mathematical interpretation of the general theory, but there was even uncertainty over how the theory's three main predictions were properly calculated from the field equations. Einstein had employed an indirect and not strictly valid route to calculating the deflection of a light ray due to a gravitational field, a procedure which, as Earman and Glymour have noted, "caused confusion among those less adept than he at getting the right answer" (1980a, 55). Similar difficulties attended the reliable derivation of the gravitational red shift, while the proper interpretation of important properties of the first complete solution to Einstein's field equations—the so-called Schwarzchild solution—remained undecided.[94] Einstein's published derivation of the motion of Mercury's perihelion was sound, but following the subtleties of his "almost precarious path of reasoning" required a very thorough understanding of the use of approximations in non-Euclidian geometry.[95] Any would-be teacher or textbook writer would need sufficient mastery of his subject to explain these difficulties and to develop proofs and derivations of his own.

With the approach of the academic year 1919–20, Eddington took his first steps towards addressing some of the above problems by offering an introductory course on relativity at Cambridge. In order to appreciate the

---

94. The derivation the red shift is discussed in Earman and Glymour 1980b. The debate over the meaning of the singularity in the Schwarzchild solution, which would lead eventually to the notion of "black holes," is discussed in Eisenstaedt 1989a.

95. Earman and Jannsen have shown that Einstein's first published derivation of Mercury's motion is very difficult to follow in mathematical detail without access to his earlier, unpublished, attempts to tackle the problem. The derivation became more straightforward once the Schwarzchild solution was found, but still caused considerable confusion among British mathematicians. See Earman, Janssen, and Norton, eds., 1993, 130, 155, 159.

ambitious and potentially hazardous nature of this enterprise, it is important to keep in mind that Eddington's experience as a teacher of relativity was at this time confined to the preparation of individual popular lectures and one technical overview in the form of the *Report*. Offering a term-long course in which the theory was explained from first principles to a class of trained mathematicians and physicists was a very different matter—something that Ebenezer Cunningham learned to his cost. Cunningham had followed Einstein's attempt to apply the principle of relativity to gravitation in the years immediately prior to World War I, and had done his best to keep up with developments while working as a farm laborer during the war. But when he tried to teach the theory in Cambridge in late April 1919 (some nine months before Eddington did so), he found that he so lacked the deep understanding of the subject necessary to satisfy his class that he had to abandon the course.[96]

The pedagogical difficulties confronting Eddington in 1919 were in some respects similar to those experienced by teachers of Maxwell's *Treatise* in Cambridge in the 1870s. As we saw in chapter 6, it took around seven years of combined teaching and research for men like Niven to clarify what could reliably be taught and examined as "Maxwell's theory." Einstein's relativistic theory of gravitation was at least as conceptually novel as Maxwell's field theory of electromagnetism, and, as we have just seen, there was similar uncertainty among its early interpreters as to which aspects were reliable, which required clarification, and which needed major revision. In the published *Report* of 1918, Eddington had generally followed the calculations and explanations of Einstein or de Sitter, but he would have to be able to justify these assertions if he were formally to defend the theory in a Cambridge classroom.[97] From a purely mathematical perspective, Einstein's theory would almost certainly have been more difficult to teach in Cambridge in the early 1920s than had Maxwell's in the 1870s. Niven could rely on Routh to prepare students in the necessary mathematical methods, and Routh and Maxwell shared a common background in Hopkins's undergraduate coaching. Some of the students attending Eddington's proposed lectures would have studied the mathematical methods upon which the general theory was based, but Eddington would have had to explain how their familiar application to problems in three-dimensional Euclidean space was generalized to treat problems

---

96. Cunningham 1969, 64; McCrea 1978, 118. Cunningham offered a course in the Easter term 1919 entitled the "The Principle of Relativity." The following term he changed this to "Electron Theory." See *Cambridge University Reporter*, 24 April 1919, 624, and 9 October 1919, 99.

97. The origins of Eddington's arguments in the *Report* are briefly discussed in Earman and Glymour 1980a, 56, and 1980b, 200; and Eisenstaedt 1989a, 220.

in four-dimensional, non-Euclidian, space and time (fig. 9.3).[98] In short, he would have to build the pedagogical apparatus through which students could be taught in about eight weeks what he had taken more than two years to master.

The first evidence of Eddington's intent to begin teaching relativity at Cambridge appeared in the *Cambridge University Reporter* in early October 1919. The list of lectures proposed by the Special Board of Physics and Chemistry for the forthcoming Michaelmas term (October to December), included a course entitled "Outlines of the Theory of Relativity" to be held by Eddington at a venue "to be announced."[99] This notice turned out to be erroneous as the course was never delivered, but it does provide a clue to Eddington's strategy for introducing the new subject to the university. It appears to have been some time around the beginning of the Michaelmas term that he decided that the data from the eclipse expedition supported Einstein's rather than Newton's theory of gravitation.[100] In anticipation of the November announcement of this important result, Eddington probably informed the university that he intended to offer a course on relativity in the Lent term (January to March) of 1920.[101] This course was duly delivered but, as things turned out, it was not Eddington's first public discussion of Einstein's theory in Cambridge.[102] On 24 November 1919, he and E. T. Cottingham addressed the Cambridge Philosophical Society on the "theory of relativity and the recent eclipse observations." The announcement of this talk, just a couple of weeks after the sensational press reporting of the London meeting, generated such interest that "hundreds were turned away unable to

---

98. Herman, for example, offered a two-term course on "Differential Geometry" for students of Part I of the Mathematical Tripos. *Cambridge University Reporter,* 9 October 1919, 99.

99. *Cambridge University Reporter,* 9 October 1919, 107.

100. Eddington made the final measurements from the photographic plates in late August 1919. See Dyson et al. 1920, 321. On 12 September at the British Association, Eddington was prepared to say only that the measured deflection lay between the Newtonian and Einsteinian values. By 22 October, at a meeting of the $\nabla^2 V$ Club in Cunningham's rooms in St. John's College, he announced that the deflection confirmed Einstein's theory. The $\nabla^2 V$ was an elite Cambridge club devoted to the "discussion of questions in mathematical physics." *British Association Report* (1919), 156–57; minutes of the $\nabla^2 V$ Club, 22 October 1919 (AHQP).

101. The most likely explanation for the erroneous announcement is a misunderstanding over the term in which the course was to be taught.

102. It should be noted that the announcement of the verification at the $\nabla^2 V$ Club on the 22 October was not strictly a "public" announcement, as membership of the club was restricted by internal election. Seventeen members were present on this occasion, including Cunningham, Harold Jeffreys, E. A. Milne, and L. A. Pars. The President remarked that this meeting was "historic." Minutes of the $\nabla^2 V$ Club, 22 October 1919 (AHQP).

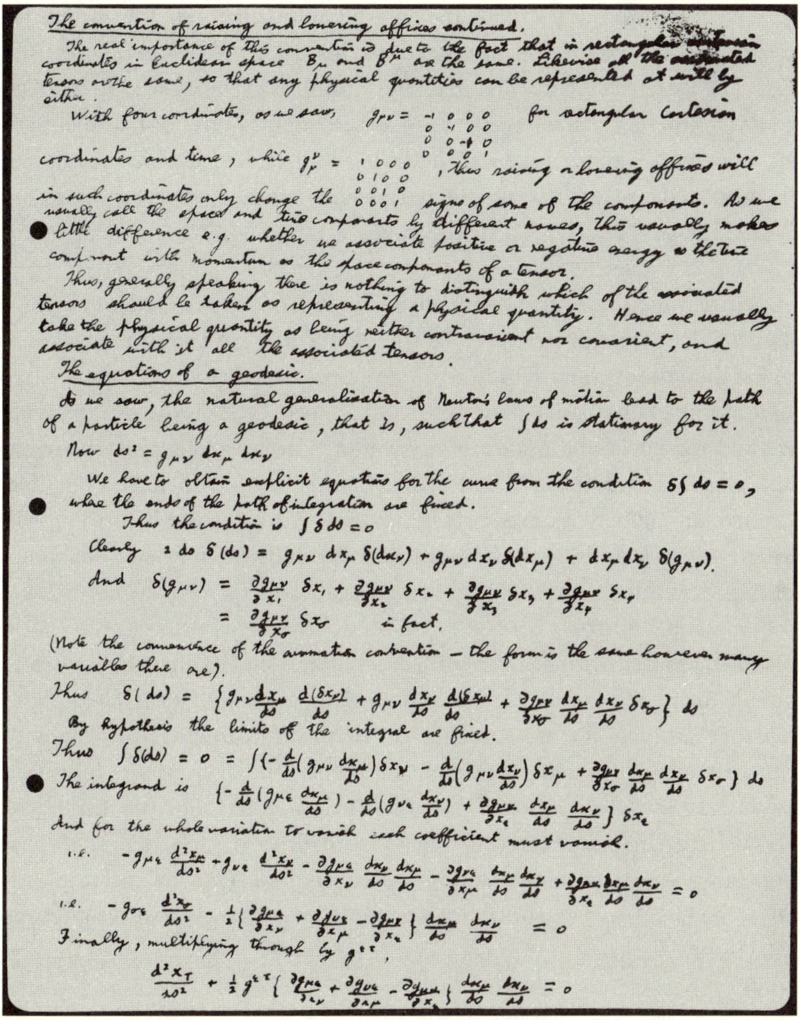

The convention of raising and lowering suffixes continued.

The real importance of this convention is due to the fact that in rectangular cartesian coordinates in Euclidean space B_μ and B^μ are the same. Likewise all the associated tensors are the same, so that any physical quantities can be represented at will by either.

With four coordinates, as we saw, $g_{\mu\nu} = \begin{matrix} -1 & 0 & 0 & 0 \\ 0 & -1 & 0 & 0 \\ 0 & 0 & -1 & 0 \\ 0 & 0 & 0 & 1 \end{matrix}$ for rectangular Cartesian coordinates and time, while $g^\nu_\mu = \begin{matrix} 1 & 0 & 0 & 0 \\ 0 & 1 & 0 & 0 \\ 0 & 0 & 1 & 0 \\ 0 & 0 & 0 & 1 \end{matrix}$, thus raising or lowering suffixes will in such coordinates only change the signs of some of the components. As we usually call the space and time components by different names, this usually makes little difference e.g. whether we associate positive or negative energy as the time component with momentum as the space components of a tensor.

Thus, generally speaking there is nothing to distinguish which of the associated tensors should be taken as representing a physical quantity. Hence we usually take the physical quantity as being neither contravariant nor covariant, and associate with it all the associated tensors.

The equations of a geodesic.

As we saw, the natural generalisation of Newton's laws of motion lead to the path of a particle being a geodesic, that is, such that $\int ds$ is stationary for it.

Now $ds^2 = g_{\mu\nu}\, dx_\mu\, dx_\nu$

We have to obtain explicit equations for the curve from the condition $\delta \int ds = 0$, where the ends of the path of integration are fixed.

Thus the condition is $\int \delta\, ds = 0$

Clearly $2\, ds\, \delta(ds) = g_{\mu\nu}\, dx_\mu\, \delta(dx_\nu) + g_{\mu\nu}\, dx_\nu\, \delta(dx_\mu) + dx_\mu\, dx_\nu\, \delta(g_{\mu\nu})$.

And $\delta(g_{\mu\nu}) = \frac{\partial g_{\mu\nu}}{\partial x_1}\delta x_1 + \frac{\partial g_{\mu\nu}}{\partial x_2}\delta x_2 + \frac{\partial g_{\mu\nu}}{\partial x_3}\delta x_3 + \frac{\partial g_{\mu\nu}}{\partial x_4}\delta x_4$

$= \frac{\partial g_{\mu\nu}}{\partial x_\sigma}\delta x_\sigma$ in fact.

(Note the convenience of the summation convention – the form is the same however many variables there are).

Thus $\delta(ds) = \left\{ g_{\mu\nu}\frac{dx_\mu}{ds}\frac{d(\delta x_\nu)}{ds} + g_{\mu\nu}\frac{dx_\nu}{ds}\frac{d(\delta x_\mu)}{ds} + \frac{\partial g_{\mu\nu}}{\partial x_\sigma}\frac{dx_\mu}{ds}\frac{dx_\nu}{ds}\delta x_\sigma \right\} ds$

By hypothesis the limits of the integral are fixed.

Thus $\int \delta(ds) = 0 = \int \left\{ -\frac{d}{ds}\left(g_{\mu\nu}\frac{dx_\mu}{ds}\right)\delta x_\nu - \frac{d}{ds}\left(g_{\mu\nu}\frac{dx_\nu}{ds}\right)\delta x_\mu + \frac{\partial g_{\mu\nu}}{\partial x_\sigma}\frac{dx_\mu}{ds}\frac{dx_\nu}{ds}\delta x_\sigma \right\} ds$

The integrand is $\left\{ -\frac{d}{ds}\left(g_{\mu\epsilon}\frac{dx_\mu}{ds}\right) - \frac{d}{ds}\left(g_{\nu\epsilon}\frac{dx_\nu}{ds}\right) + \frac{\partial g_{\mu\nu}}{\partial x_\epsilon}\frac{dx_\mu}{ds}\frac{dx_\nu}{ds} \right\} \delta x_\epsilon$

And for the whole variation to vanish each coefficient must vanish.

i.e. $-g_{\mu\epsilon}\frac{d^2x_\mu}{ds^2} - g_{\nu\epsilon}\frac{d^2x_\nu}{ds^2} - \frac{\partial g_{\mu\epsilon}}{\partial x_\nu}\frac{dx_\nu}{ds}\frac{dx_\mu}{ds} - \frac{\partial g_{\nu\epsilon}}{\partial x_\mu}\frac{dx_\mu}{ds}\frac{dx_\nu}{ds} + \frac{\partial g_{\mu\nu}}{\partial x_\epsilon}\frac{dx_\mu}{ds}\frac{dx_\nu}{ds} = 0$

i.e. $-g_{\sigma\epsilon}\frac{d^2x_\sigma}{ds^2} - \frac{1}{2}\left\{ \frac{\partial g_{\mu\epsilon}}{\partial x_\nu} + \frac{\partial g_{\nu\epsilon}}{\partial x_\mu} - \frac{\partial g_{\mu\nu}}{\partial x_\epsilon} \right\} \frac{dx_\mu}{ds}\frac{dx_\nu}{ds} = 0$

Finally, multiplying through by $g^{\epsilon\tau}$,

$$\frac{d^2x_\tau}{ds^2} + \frac{1}{2}g^{\epsilon\tau}\left\{ \frac{\partial g_{\mu\epsilon}}{\partial x_\nu} + \frac{\partial g_{\nu\epsilon}}{\partial x_\mu} - \frac{\partial g_{\mu\nu}}{\partial x_\epsilon} \right\} \frac{dx_\mu}{ds}\frac{dx_\nu}{ds} = 0$$

FIGURE 9.3. A page from the notes taken by L. H. Thomas at Eddington's 1923 lecture course on the "Mathematical Theory of Relativity." Eddington's lectures presented the theory of relativity not as a development of electron theory (as did Cunningham) but as a fundamentally new theory of space, time, and measurement. This sheet records Thomas's attempt to follow Eddington's explanation of how to find the path (a "geodesic") traced by a particle moving freely through space according to the general theory of relativity. Note how (top of the sheet) Eddington also taught his students the mathematical techniques and conventions they would need in order to follow Einstein's theory. For previous generations of Cambridge students, coaches like Routh and Webb had done the same thing in connection with Maxwell's electromagnetic field theory. Thomas Collection, LTP-NCSU. (By permission of North Carolina State University Libraries.)

get near the room."¹⁰³ Capitalizing on evident fascination in Cambridge for Einstein's work, Eddington announced his intention to hold an open lecture on relativity in the Great Hall of Trinity College. On the afternoon of 2 December 1919, he used an engaging mixture of humor and drama to enthral an "enormous" and "breathless" audience with an "intelligible account" of the new theory of gravitation.¹⁰⁴ These lectures stimulated further interest in Einstein's work, thereby setting the stage for Eddington's first formal course on relativity, due to begin just seven weeks later.

By the time the course commenced on 20 January 1920, it had been renamed "Principles of Relativity," a title that would have sounded more familiar than the original "Outlines of the Theory of Relativity" to the average Cambridge mathematician and physicist.¹⁰⁵ The fact that the twice-weekly lectures were held in the late afternoon at the Cavendish Laboratory further indicates that Eddington hoped to attract as large an audience as possible. Courses for mathematicians and physicists were usually held before lunch in order to leave the afternoon free for private study, laboratory work, or sporting activities, those specifically for mathematicians taking place in the Arts School or the lecturer's college.¹⁰⁶ Holding the course in the late afternoon meant that any member of the university interested in the subject would be able to attend. It is also clear that Eddington put a good deal of work into making this first formal presentation of the fundamental principles of relativity interesting and intelligible to a broad audience of physicists and mathematicians. No record of the course's content has survived, but we do know that it formed the basis of what is sometimes described as his "semi-popular" introduction to the subject, *Space, Time and Gravitation* (1920). Indeed, the fact that the manuscript was completed just seven weeks after the final lecture makes it almost certain that Eddington wrote much of the text while preparing and delivering the course.¹⁰⁷ Moreover, having recognized the book's origins in the Cavendish lectures, we can see that the epithet "semi-popular" is somewhat misleading. It is certainly the case that Eddington

---

103. *Proceedings of the Cambridge Philosophical Society* 20 (1920–21): 214; Eddington to Einstein, 1 December 1919, quoted in Earman and Glymour 1980b, 183. Cottingham had accompanied Eddington on the Principe expedition.

104. *Nature*, 11 December 1919, 385; Douglas 1956, 107–8.

105. The original title appeared again in the *Cambridge University Reporter*, 14 January 1920, 515, but had been altered to "Principles of Relativity" in the *Reporter*, 20 January 1920, 530.

106. McCrea 1987, 57.

107. Eddington's last lecture was on 14 March; the Preface of *Space, Time and Gravitation* is dated 1 May. L. A. Pars, who attended the course, confirms the close relationship between the course and the book (Douglas 1956, 108).

avoided the use of very specialized mathematical methods, but he did provide a clear, concise, and intellectually demanding account of the theory's conceptual foundations. The text is actually designed to appeal not so much to a general audience as to a class of trained physicists and mathematicians seeking an accessible introduction to Einstein's work. Those who followed the lectures or studied the book would certainly have been far better equipped to tackle the theory in its fully mathematical form.

In the final lecture of his Cavendish course, Eddington announced that he intended to deliver a series of "more mathematical" lectures on relativity in the forthcoming Easter term (April–June).[108] The fact that these lectures had not been announced in the *Reporter* at the beginning of the Autumn or Lent terms (as was usual), suggests that Eddington wanted to gauge the response to his introductory course at the Cavendish Laboratory before committing himself to a much more technical presentation of the subject. In any event, a notice by the Special Board of Mathematics on 22 April 1920 announced that he would shortly commence a thrice-weekly lecture series in the Arts School, entitled "Relativity Theory of Gravitation (Mathematical)."[109] This course, intended primarily for graduates and advanced undergraduates preparing for Part II of the Mathematical Tripos, was thereafter held annually by Eddington in the Easter term.[110] Like his introductory course the previous term, the substance of these mathematical lectures soon appeared in print. When a French edition of *Space, Time and Gravitation* was published in 1921, Eddington appended a supplement containing a more technical overview of the general theory.[111] By August 1922, these draft notes had been expanded to form the text of completely new book, *The Mathematical Theory of Relativity* (1923; hereafter *Theory of Relativity*). The *Theory of Relativity* was quickly acknowledged as an original and masterful exposition of the origins, emergence, and conceptual foundations of Einstein's theory, the mathematical apparatus upon which it relied, and its technical development until 1922.[112] It is a measure of the importance and rapid impact

---

108. Douglas 1956, 108.

109. *Cambridge University Reporter,* 22 April 1920, 823. These lectures were held on Tuesdays, Thursdays, and Saturdays at midday from 27 April.

110. During the latter 1930s and early 1940s Eddington sometimes taught the course over two or three terms. These changes can be traced in the annual list of lectures published in the *Reporter* in early October.

111. Eddington 1921a; Plummer 1945, 117. The supplementary new chapter was written in the eight months following the course (Eddington 1923, v).

112. Eddington took a further eighteen months to turn the supplementary chapter into a book (Eddington 1923, v).

of the work—which soon became a standard textbook in the field—that a second edition was called for in 1924 and translated into German and Russian. The new edition enabled Eddington not only to correct a number of errors in the original text, but to add some twenty-three pages of explanatory notes based on his own teaching experience and the "kindness of correspondents" in drawing his attention to remarks and derivations that required clarification.[113]

It was while preparing his first courses of lectures and drafting his two books on relativity that Eddington also made his earliest original contributions to the subject. In 1921 he generalized Hermann Weyl's attempt to include electromagnetic forces in Einstein's general relativistic scheme, and, in so doing, sought to shed new light "on the origin of the fundamental laws of physics" (Eddington 1921b, 105). In 1922 he tackled the problem of the production and propagation of gravitational waves. By analyzing the changing structure of space and time in the vicinity of a rapidly rotating material bar, Eddington showed, first, which of the several types of waves generated would actually convey energy away from the bar, and, second, that the propagated waves would travel with the velocity of light in all frames of reference.[114] There are two pedagogical senses in which Eddington's *Theory of Relativity* should also be considered an original text. First, the book contained his own original contributions as part of the general exposition of the theory. Second, this exposition necessarily included original points of clarification and interpretation. In Eddington's own words, the *Theory of Relativity* contained "numerous independent developments connected with the logical presentation of Einstein's theory as a whole" (Douglas 1956, 191).[115] The latter included alternative derivations of formulae as well as the simplification and physical interpretation of important results. Most important among these were improved derivations or interpretations of the red shift and light-bending effects, and a generalization and discussion of the meaning of the Schwarzchild solution to the field equations.[116]

---

113. Eddington, *Theory of Relativity*, 2d ed. (1924), 241. This edition was reprinted eight times to 1965. The *Physics Citation Index* (Small 1981) reveals the numerous occasions on which Eddington's book was cited in research publications on relativity through the 1920s.

114. Eddington 1922. Eddington's demonstration of the propagated nature of some gravitational effects drew on the same arguments in Rayleigh's *Theory of Sound* as had FitzGerald's demonstration of the propagated nature of electromagnetic effects (Eddington, *Theory of Relativity*, 129).

115. On the widespread use of the *Theory of Relativity* as a textbook, see Douglas 1956, 52–53, 191, Chandrasekhar 1983, 32, and Sanchez-Ron 1987b, 176, and 1992, 68–69.

116. These contributions are briefly discussed in Earman and Glymour 1980a, 56, and 1980b, 200–202, and Eisenstaedt 1989a, 221.

A prominent feature of Eddington's work on relativity at the beginning of the 1920s is the almost seamless relationship it displays between his courses, books, and research publications: the Cavendish lectures became *Space, Time and Gravitation*, while the mathematical course (repeated annually) formed the substance of the *Theory of Relativity;* the original research problems were not only formulated and solved while the latter book was being drafted but eventually incorporated within its pages.[117] These links help to explain both the quality of Eddington's formal expositions of the theory as well as the timing of his first research contributions. The preparation of the lectures and books required him to subject the foundations, formal structures, and predictions of the theory to severe scrutiny. In crafting a consistent exegesis of the entire subject, Eddington would have had to resolve apparent contradictions and paradoxes both to his own satisfaction and to that of his students. What emerged from this process were two outstanding expositions of relativity, both of which culminated in Eddington's own attempt, following Weyl, to find the most general foundations for a combined field theory of gravity and electromagnetism.[118]

The role of teaching in the development of Eddington's work is further highlighted by the sharp contrast that existed between the respective styles of his popular and academic lectures. Where the former were famous for being lucid, fluent, and entertaining, the latter were described by one student as "confused and hesitant." Another of Eddington's Cambridge students, Charles Goodwin, parodied the start of a typical lecture as follows: "Eddington gave a moan and then stopped for what seemed a very long time. He then moaned again and stopped again for a very long time. Then he shook his head vigorously and said: 'No! That's wrong.'"[119] These remarks clearly exaggerate Eddington's failings as a formal lecturer, since other accounts of his teaching (see epigraph to this section) are more flattering, but they do highlight the quite different purposes he ascribed to popular and academic lectures.[120] The former were mainly intended to convince a general audience that the theory was exciting and plausible. Eddington accordingly read them from a carefully prepared script in such a way as to make

---

117. The intimate relationship between the content of Eddington's lectures on the mathematical theory of relativity and that of the *Theory of Relativity* is illustrated in Thomas's lecture notes taken in the Easter Term of 1923. See AHQP.

118. Eddington described his generalization of Weyl's work in the *Theory of Relativity* as emerging from his attempt to present general relativity as a logical whole (Douglas 1956, 191).

119. Hoyle 1994, 146.

120. For another favorable account of Eddington's teaching, see Wali 1991, 82–83.

them appear spontaneous.[121] Academic lectures, by contrast, were intended to convey his own interpretation and technical mastery of relativity to a class of highly trained mathematicians. To this end he taught from a few sheets of notes, concentrating on the conceptual foundations and logical structure of the theory, and apparently reconstructed technical proofs and theorems, partly from memory, as and when required.[122]

There is in this respect a striking parallel with the teaching styles of Niven and Larmor discussed in earlier chapters. According to Fred Hoyle, who reported Goodwin's unflattering parody of Eddington's classroom oratory, the latter was in another respect a "very good lecturer." In particular Hoyle had in mind the fact that able students "remembered and thought a lot about the big issues [Eddington] raised, long after [they had] forgotten apparently much better presented lectures from others" (1994, 147). As with Niven and Larmor's approach to teaching electromagnetic theory, Eddington's focus on conceptual, foundational, and interpretational matters in relativity had pedagogical strengths as well as weaknesses. To those students who preferred to copy down pages of systematic deductions from well-established mathematical and physical principles, Eddington's halting monologue, sometimes illegible jottings on the blackboard, and occasional indecision on important issues no doubt seemed irritating and confusing. But for those seeking an account of current thinking on the status of general relativity as a physical theory as well as a sense of where it might be accessible to further clarification and development, his approach almost certainly provided a more powerful spur to original and critical thought than did courses in which the content was carefully contrived to appear closed and incontrovertible. The fact that Eddington as a teacher of relativity was able to range so effectively from the stepwise derivation and interpretation of specific formulae to reflection upon the meaning and limitations of the theory as a whole owed a great deal to his own unaided struggles to master general relativity and its cosmological consequences over the previous four years.

Recognition of Eddington's course on relativity as an official component of Part II of the Mathematical Tripos was marked in the mid 1920s by the appearance of relevant questions on the annual Tripos examination papers. These questions announced the effective acceptance of relativity theory in Cambridge and offer some insight into the kinds of technical expertise

121. McVittie 1987, 69.
122. Hoyle 1994, 146. Eddington sometimes brought a copy of the *Theory of Relativity* to his lectures to use instead of lecture notes (McCrea 1986, 276).

Eddington expected his advanced students to possess. For example, one question (fig. 9.4) set in 1927 could only have been answered correctly by someone with a sound mathematical and physical grasp of Einstein's field equations. The candidate was required to prove a fundamental property of the form of the equations applicable to empty space, to show how this property could be used to derive a more general form applicable to continuous matter, and to explain the important difference between the role of potential energy in Newtonian and Einsteinian mechanics.[123] A question set two years later (fig. 9.4) shows that Eddington also expected his students to understand the new meaning of the concepts of "space" and "time" peculiar to the general theory of relativity. In order to display this knowledge, they were required to derive the Lorentz transformations (see chapter 7) and the law for the composition of velocities (see chapter 8) as special cases of the general principles upon which the theory was founded (and to apply the results to solve a straightforward problem).[124] A final part of the question probed the student's knowledge of the physical significance of coordinate systems by inviting discussion of the validity of Einstein's prediction that light emitted from atoms in an intense gravitational field would have a different frequency from that emitted by similar atoms on earth.[125]

The Tripos questions set by Eddington through the 1920s and 1930s gradually built into an archive of canonical problems that served the various purposes discussed in chapter 3. They provided an informal syllabus of the material covered in Eddington's course, informed students of the level of technical expertise expected in the examination, and provided a collection of practical exercises through which the required skills could be developed. As in other areas of mathematical study in Cambridge over the previous century, past examination problems formed a key point of pedagogical interaction between lecturers, supervisors, and students.[126] In this sense, Eddington's questions reveal a major shift in the meaning of the "principle of relativity" in undergraduate mathematical studies. The rapid disappearance of references to "correlated" electrical systems "traveling through the aether"

---

123. Question 17F, Mathematical Tripos (Part II), Schedule B, 6 June 1927, afternoon paper. The solution to the questions is found in Eddington's *Theory of Relativity*, chap. 4, 115–19, 135, 148.

124. Question 19D, Mathematical Tripos (Part II), schedule B, 3 June 1929, afternoon paper. Eddington's derivation and discussion of these formulas appears in *Theory of Relativity*, 8–21, 91–93.

125. As I noted above, there had been considerable controversy regarding the reliability of this derivation. See Earman and Glymour 1980b.

126. Burcham 1987, 161. Exercises of this kind for today's students of general relativity are to be found in books such as Lightman et al. 1975.

> SCHEDULE B.
>
> MONDAY, *June* 6, 1927. 1½—4½.
>
> PAPER II.
>
> *Candidates are reminded that greater weight is attached to one or two complete answers than to a larger number of fragments.*
>
> 1 A. Explain what is meant by a transvectant $(f, \phi)^n$ of two binary forms f, ϕ, and prove that the system obtained by transvection from two finite and complete systems is itself finite and complete.
>
> Obtain the complete system for a binary quintic f, and express in terms of members of this system the transvectants $(H, f)^4$, $(H, f)^2$, where H is the Hessian of the quintic f.
>
> 17 F. Prove the fundamental property of Einstein's tensor $G_{\mu\nu}$, viz.
>
> $$(G_\mu^\nu)_\nu - \tfrac{1}{2}\partial G/\partial x_\mu = 0,$$
>
> and explain how this leads to the identification of the stress-energy tensor with $G_\mu^\nu - \tfrac{1}{2}g_\mu^\nu G$.
>
> Explain how the potential energy of matter in a gravitational field appears in Einstein's theory as an additional (non-tensor) quantity not included in the energy tensor.

> SCHEDULE B.
>
> MONDAY, *June* 3, 1929. 1½—4½.
>
> PAPER II.
>
> *More credit is given for complete answers than for a proportionate number of fragments.*
>
> 1 A. Shew that an integer n which is not of the form
> $$4^a(8^b+7) \quad (a=0,1,2,\ldots, b=0,1,2,\ldots)$$
> can be expressed as the sum of the squares of three integers.
>
> 19 D. Obtain the formulae of the Lorentz transformation from the fundamental principles of relativity theory, and deduce the law of composition of velocities. Apply the latter to Fizeau's experiment on the propagation of light in a stream of water.
>
> Examine critically the argument which predicts that the frequency of light from a solar source will differ from that from a corresponding terrestrial source in the ratio of $\sqrt{g_{44}}$ at the sun and the earth.

FIGURE 9.4. The presence of these questions on the Tripos examination papers of 1927 (question 17F) and 1929 (question 19D) mark Eddington's success in introducing the general theory of relativity to Cambridge in the decade after World War I. The training system that had lent conceptual credibility and problem-solving power to Larmor's electronic theory of matter in the first decade of the twentieth century now served a similar function for Einstein's new relativistic physics. Eddington's examination questions follow his lectures in treating relativity as a self-contained theory that could be developed in terms of its own principles and axioms. I have included question 1 on each paper to give a flavor of the other subjects covered. Cambridge University Examination Papers, Mathematical Tripos, Part II, Monday, 6 June 1927, 1.30–4.30, q. 17F; and Monday, 3 June 1929, 1.30–4.30, q. 19D. (By permission of the Syndics of Cambridge University Library.)

in favor of those to "the fundamental principles of relativity" signal Eddington's success not just in marginalizing critics like Larmor and Lodge, but in actively instilling new physical sensibilities and technical competencies in the post–World War I generation of Cambridge undergraduates. Unlike Cunningham, who continued to teach the principle of relativity in the context of electromagnetic theory (see chapter 8), Eddington introduced and taught general relativity as a fundamentally new theory of space, time, and gravitation. It is a mark of the profound difference between these respective approaches that when in 1923 the Special Board of Physics and Chemistry decided to begin teaching the special theory of relativity to physics undergraduates, they chose not to send students to Cunningham's lectures but to invite Eddington to offer an annual short course at the Cavendish Laboratory.[127]

9.8. Teaching and Researching Relativity in Cambridge

Eddington's early success at imparting enthusiasm and technical competence in general relativity is illustrated by the experiences of one of his first research students in this area, Leopold Pars. Having obtained first-class honors in Parts I and II of the Mathematical Tripos in 1917 and 1919 respectively, Pars attended Eddington's Trinity lecture on relativity in December 1919 as well as his first two courses on the subject in the Lent and Easter terms of 1920.[128] During the latter course, Pars was inspired to write a dissertation on the general theory and soon became personally acquainted with Eddington as a research supervisor. According to Pars, Eddington was "always ready to help and advise," and, in 1921, the former was awarded the first Smith's Prize for a dissertation entitled "On the General Theory of Relativity."[129] One of Eddington's most distinguished students from the mid 1930s, Fred Hoyle, recalled not only attending Eddington's lectures during the academic year 1935–36, but answering his examination questions on general relativity in Part II of that year's Mathematical Tripos.[130] Hoyle's early research focused on nuclear decay and astrophysics, but his training in general relativity stood him in excellent stead a decade later when he, Hermann

---

127. *Cambridge University Reporter*, 5 October 1923, 89. These short courses consisted of four or five lectures given on Tuesday and Thursday afternoons at 4:45. Eddington gave them throughout the mid 1920s.
128. Douglas 1956, 108.
129. Douglas 1956, 108; Barrow-Green 1999, 312.
130. Hoyle 1994, 152.

Bondi, and Tommy Gold began developing what would become known as the "steady-state" theory of the universe.[131]

It is difficult to make an accurate assessment of the total number of students trained by Eddington during the interwar period, but occasional references by his ex-pupils to the numbers attending his and other advanced classes indicate that it was around 150.[132] Although the majority of these students made no significant research contributions to the development of relativity theory, they did constitute a very significant fraction of the small number of relativity-literate mathematicians and physicists in the world. Those of Eddington's students who did undertake research in this area (approximately ten percent) can usefully be divided into three categories. First, there are those like Pars, W. R. Andress, R. W. Narliker, and G. N. Clark who learned relativity directly from Eddington's books and lectures and went on immediately to prepare research dissertations on the subject under his direction.[133] The most significant member of this group is Clark, who discovered important errors in complicated calculations on relativistic dynamics made by Tullio Levi-Civita and de Sitter. Clark subsequently published his work in the prestigious *Proceedings of the Royal Society* with Eddington as co-author.[134]

Second, there were men like Llewellyn H. Thomas, Paul Dirac, William McCrea, S. Chandrasekhar, and Fred Hoyle who studied relativity with Eddington but only later, and independently, put that knowledge to research use.[135] I have already noted Hoyle's work on steady-state cosmology in the 1940s and 1950s, but work by Thomas and Dirac points to the impact of Eddington's teaching even in the mid 1920s. Eddington's course impressed upon both men the restrictions placed on the expression of physical relationships by the invariance requirements of relativity, and taught them the mathematical techniques by which these restrictions could be imposed (fig. 9.3). Thus it was Thomas, rather than one of the Continental masters of quantum mechanics, who, in 1926 showed how a major inconsistency in the

---

131. Ibid., 400–401.

132. Eddington's classes in 1922 and 1923 contained around eight students (Douglas 1956, 51; interview with L. H. Thomas in Kuhn et al. 1967). This seems to have been typical, as Dirac's advanced lectures on quantum mechanics attracted a similar number of students circa 1930 (Wali 1991, 82).

133. Andress, Narliker, and Clark were awarded Rayleigh prizes at Cambridge in 1930, 1932, and 1938 respectively for dissertations involving the general theory of relativity (Barrow-Green 1999, 313–14).

134. Eddington and Clark 1938. Clark and Eddington's work is discussed in Douglas 1956, 111–12, and Damour and Schäfer 1992.

135. Sanchez-Ron 1987b, 181; McCrea 1987, 63; Wali 1991, 82. It is very likely that E. A. Milne learned relativity in the same way, but I have found no direct evidence to support this claim.

quantum-mechanical treatment of electron spin could be solved by deriving a particular expression relativistically. Indeed, Thomas initially tackled the problem by analogy with a similar one solved by Eddington in the *Theory of Relativity*.[136] Dirac turned the same technique into a research "game"—which he "indulged in at every opportunity"—according to which "whenever one saw a bit of physics expressed in nonrelativistic form, [one would] transcribe it to make it fit in with special relativity" (Kragh 1990, 11). Eddington took a personal interest in Dirac's work, helping him to master difficult passages in the *Theory of Relativity* as well as commenting on and communicating his research papers.[137] The game of seeking the form of fundamental equations that would render them consistent with the special theory of relativity became an extremely powerful research tool in Dirac's hands during the decade after 1926 as he laid the foundations of relativistic quantum mechanics.[138]

The final group of research students to benefit from Eddington's tutelage is that composed of men like Georges Lemaître and G. C. McVittie who studied relativity before arriving in Cambridge and were subsequently drawn to the university by Eddington's reputation as a master of the subject. Lemaître, for example, began his undergraduate career in 1911 as a student of engineering at the Catholic University of Louvain in Belgium. Following the disruption of his studies by military service in World War I, he turned to the study of mathematics and physics. It was after graduating in these subjects in 1920, and while preparing for ordination as a priest, that Lamaître began to study the theory of relativity. These studies were hampered by the fact that none of the physicists in Louvain were familiar with Einstein's work on gravitation, but Lemaître managed in 1923 to write three essays on the subject that were sufficiently impressive to win him funding to study abroad. Lemaître chose first to work at the Cambridge Observatory, where, under Eddington's guidance, he wrote a research paper on the long-standing problem of the motion of rigid bodies according to the principle of relativity. It was also through working with Eddington and studying the *Theory of Relativity* that Lemaître not only mastered relativistic cosmology but began investigating difficulties with de Sitter's solution to Einstein's field equations that would lead eventually to Lemaître's theory of the expanding universe.[139]

136. Sanchez-Ron 1987b, 181–82.
137. Kragh 1990, 12. Dirac found the *Theory of Relativity* "rather tough at first" and was pleased that Eddington was "actually present" to explain problems (Dirac 1977, 115).
138. This work is outlined in Kragh 1990, chap. 3.
139. Kragh 1987, 116–22; Godart 1992, 437–43.

McVittie was first introduced to the general theory in Edinburgh through a postgraduate lecture course held by Edmund Whittaker in 1927–28 on "field theories that attempted the unification of gravitation and electromagnetism." Despite the fact that Whittaker was also engaged in research in this area, McVittie decided to move to Cambridge where Eddington could act as his supervisor. Eddington suggested to McVittie that he try to find an exact solution to Einstein's most recent unified field theory of gravitation and electromagnetism—based on a Riemannian space that exhibited "distant parallelism"—and compare it with the corresponding solution given by Einstein's original field equations of 1916. The modifications to the theory of distant parallelism introduced by McVittie in the light of his work were published in part in the *Proceedings of the Royal Society* (communicated by Eddington) in 1929 and formed the core of a dissertation for which he was awarded a Cambridge Ph.D. in 1930.[140] The Ph.D. degree had been introduced to Cambridge in 1920 to enable research students to work for a definite qualification, thereby encouraging non-Cambridge graduates to enrol at the university.[141]

This brief survey of some of Eddington's more prominent students during the interwar period highlights his unique importance in establishing and preserving a recognizably Einsteinian version of relativity theory in Cambridge. A more detailed investigation would almost certainly enable us to find far more specific mathematical and theoretical connections between Eddington's own account of the theory and that subsequently developed by his students. I alluded above to connections of this kind in the cases of Thomas and Dirac, and at least one other historian (Sanchez-Ron 1992, 70) has suggested that Eddington's work in relativistic cosmology played a major role in shaping the respective researches of McVittie, McCrea, and Chandrasekhar. To pursue this line of inquiry further would require us to delve more deeply into the specific research projects pursued by Eddington and his most able students. As we saw in the cases of Thomson, Poynting, Larmor, and Cunningham, firmly establishing direct links between training and research is a highly technical matter and one that in the case of general relativity is beyond the scope of this study. It is in any case sufficient for my present purposes to have shown that Eddington's teaching and research was *the* major factor in making Cambridge an important center for relativity after

---

140. McVittie 1987, 69–70; Sanchez-Ron 1992, 75.
141. The introduction of the degree was nevertheless controversial and was not generally accepted until after World War II. See Wilson 1983, 417–20, and Hoyle 1994, 127.

1920. I shall accordingly conclude with some general remarks on the significance of the process by which Eddington learned the new theory and the importance of training in its establishment in Cambridge.

First let us briefly recap the largely fortuitous range of factors that prompted Eddington's initial interest in the theory and that enabled him to become its master. We have seen that he possessed outstanding ability in mathematical physics, had a knowledge of differential geometry that was rare among British astronomers, and approached the principle of relativity in terms of gravitational theory rather than electrodynamics. It is also the case that his refusal of military service enabled him to make a careful and protracted study of relativity at a time when most of his Cambridge peers were engaged in war-related projects. The fact that it was de Sitter who introduced Eddington to Einstein's work is yet another very significant factor. As we have seen, de Sitter's correspondence and papers quickly convinced Eddington that the general theory was interesting, credible, and, perhaps most important of all, of fundamental astronomical and cosmological significance. De Sitter's expository publications subsequently fueled Eddington's interest by setting out the basics of the theory in English and by developing its cosmological consequences in detail. Finally, de Sitter was able to help and support Eddington on technical matters as he grappled with the details of the theory and to supply him with original publications by Einstein and others that were not then available in England.[142]

In order to appreciate the relevance of these events to our understanding of the transmission of relativity theory from Europe to Cambridge, we must pause for a moment to consider some of the lessons learned in earlier chapters of this study. We have seen that mathematical theories are best propagated in local contexts though carefully contrived pedagogical regimes in which students are able to work under the direct supervision of a competent practitioner. It was for this reason, as I argued in chapter 5, that textbooks by Edward Routh and others were of limited use in training when abstracted from the Cambridge context. In the last two chapters I have extended this argument by suggesting that the transmission of novel theories is similarly reliant on networks of communication between skilled practitioners and not simply on the exchange of published research papers. This is not to say that mathematical physicists engaging a new theory are in an ex-

---

142. The best guide to these papers occurs in the bibliography to Eddington's *Report*, which was completed before the end of the war. This includes papers published during the war by Einstein, de Sitter, Lorentz, Droste, David Hilbert, Karl Schwartzchild, and Tullio Levi-Civita.

actly analogous position to bright young students learning the discipline for the first time, but the differences and similarities are sufficiently striking to warrant serious consideration.

The professional mathematical physicist has a much broader knowledge of the discipline, greater confidence in tackling technically forbidding work, and a better sense of how to find help in the published literature or from colleagues when stuck. Yet, as Cunningham's case nicely illustrates, without the guidance of a competent teacher the very theoretical skills possessed by an experienced physicist can lead him to reinterpret or reject important work couched in an unfamiliar style. Moreover, even physicists who are aware of some novel aspects of a new theory and have a strong desire to master its technicalities can find it very difficult to do so solely from printed sources. As we have seen, there were very few mature physicists who became competent practitioners of the general theory of relativity in the wake of its verification in 1919. Joseph Larmor's experience is an informative case in point. In 1917 the sixty-year-old Larmor was inspired by Einstein's work to restudy Riemann's papers on non-Euclidian geometry with "much more zest than in the old days," but, having exhausted himself struggling in vain to understand the general theory of relativity, he wrote despairingly to Oliver Lodge: "If only one were young, a life could last." Even in December 1919 Larmor remained unable to make the general theory yield the red shift and light deflection effects claimed by its supporters, and he eventually concluded, "As they [the relativists] all agree I must presume that simply I don't understand them." [143] There were in fact very few British physicists, Cunningham included, who, like Eddington, had mastered the general theory of relativity in a sufficiently Einsteinian sense to see the gravitational red shift as a necessary prediction of the field equations.[144]

Returning now to the difficulties associated with mastering a new theory from afar, we can see that they are comparable at least in kind with those experienced by the student trying to learn from a textbook without the aid of a teacher. In both cases it is skills that are acquired only with time and practice, as well as those that are tacitly assumed by the author, that are not effectively communicated in purely written form. On this showing, Eddington's easiest route to mastering the general theory of relativity—assuming that he already considered the exercise worthwhile—would have been to

---

143. Larmor to Lodge, 13 September 1917, and Larmor to Lodge, 10 December 1919 (OJL-UCL).

144. Earman and Glymour 1980b, 198. Eddington did offer reasons why this effect might not be measurable, but these were not based on the fundamentals of the theory.

travel to Leiden or to Berlin to meet and attend the lectures of Lorentz, de Sitter, or Einstein.[145] Given that this was impossible, his next best option was to receive an expository account of the theory, in English, written not only to appeal to an astronomer but from a site at which the theory was already being made accessible to students. This was precisely what happened in Eddington's case, but, as we have seen, he still required outstanding mathematical ability, a rare knowledge of differential geometry, and four years of hard work fully to master the theory. I emphasized this point above by contrasting Eddington's learning experience with that of de Sitter, and a similar point can be made in a purely Cambridge setting by comparing the former's route to the general theory with that of his first research student, L. A. Pars. Pars began leaning relativity from Eddington in 1920, yet, despite being a rather less able mathematical physicist than his teacher, he had within a year produced a sufficiently original dissertation on the subject to win the first Smith's Prize.

This last point brings us appropriately to Eddington's role as a teacher of relativity in Cambridge in the interwar period. By the end of World War I he had become the only mathematical physicist in Britain to have acquired something akin to an Einsteinian understanding of the general theory, but this personal mastery provided no guarantee that the theory would take root in undergraduate and graduate studies in his university. The hostility in Cambridge to many aspects of Einstein's work in fact made it a rather unlikely candidate prior to 1920 as a site for the establishment of a major center of teaching and research in relativity. Viewed in this light, Eddington's decision to begin teaching and examining the general theory in Cambridge in the early 1920s, and to write two major textbooks on the subject, were momentous ones both for the university and for the discipline. It is widely accepted that the announcement of the eclipse results in 1919 played a major role in the subsequent acceptance of relativity in Britain and elsewhere, but, as I noted in the introduction to this chapter, an important distinction must be drawn between convincing physicists that a new theory has been verified and actually persuading and enabling them to learn it and apply it in their own research. The vast majority of Cambridge mathematicians and physicists probably were convinced after 1919 that the theory of relativity was correct, but most believed it to be of little or no relevance to their own research and did not trouble to teach themselves its principles and methods.

145. Einstein held annual courses on the general theory, including the necessary mathematical methods, from the academic year 1916–17. Einstein, *Collected Papers,* vol. 3. Appendix B, and vol. 6, Appendix A.

Thus when the Cavendish-trained experimenter, Norman Campbell, appended a supplementary chapter on "Relativity" to his textbook *Modern Electrical Theory* in 1923, he lamented:

> It remains an indubitable fact that, in spite of the attempts to enlighten him, the average physicist ... is still ignorant of Einstein's work and not very much interested in it. Physicists of great ability, who would be ashamed to admit that any other branch of physics is beyond their powers, will confess cheerfully to a complete inability to understand relativity. (Campbell 1923, v)

Campbell's reference to attempts to enlighten physicists on relativity refer not to formal courses on the subject but to the textbooks published by himself, Eddington, and Cunningham, and to the flood of popular expositions that appeared in the early 1920s. Campbell had been trying to popularize the special theory of relativity to what he called "the man in the laboratory" (1923, v) since before World War I, and was therefore acutely aware that even four years after Eddington's successful eclipse expedition in 1919, very few British physicists had a working knowledge of the special, let alone the general theory.[146] Campbell's remarks are important because they confirm that even in 1923 very few British physicists were interested in mastering relativity, and that this was not due to the want of good introductory or advanced textbooks on the subject.[147] Thus we may safely conclude that the widespread acceptance of relativity does not provide an adequate explanation of how, when, or why it became an essential component of every aspiring physicist's undergraduate training. Most studies of the early reception of relativity focus almost exclusively on the factors that led to the theory being accepted as true by a handful of elite physicists around the world, but fail to address the question of its wider uptake within the physics community.[148]

In order to answer this question, at least for the Cambridge case, it is necessary not only to focus on training but to differentiate between the fates of the special and general theories respectively. First the special theory. We have just seen that British physicists remained largely ignorant of relativity in the early 1920s, a state of affairs that can be attributed to the practical irrelevance of the subject to the kind of experimental atomic physics being

---

146. Campbell's reasons for taking an interest in Einstein's work from 1909 are discussed in Warwick 1993.

147. Campbell referred to Eddington's *Space, Time and Gravitation* as a "masterly book, the finest model of 'semi popular' exposition our generation has produced" (1923, v).

148. For a recent survey of such studies see Brush 1999.

pursued at institutions such as the Cavendish Laboratory.[149] It was at exactly this time, however, that special relativity was becoming an important tool in work on the emergent quantum theory of the atom. To take a simple example, many of Europe's leading theoretical physicists routinely applied Einstein's expression $E = mc^2$ in atomic theory, an equation whose derivation required a basic knowledge of special relativity.[150] Those who wished to understand these developments, especially following the appearance of a radically new quantum mechanics in the mid 1920s, had to learn special relativity as an adjunct to their studies in atomic theory. By the late 1920s and early 1930s, even Ernest Rutherford, who had previously had little time for recent developments in theoretical physics, had become acutely aware that local expertise in quantum theory and special relativity were important even for a primarily experimental group like the one he directed at the Cavendish Laboratory.[151]

It was against this background that Eddington offered his annual short course on special relativity at the Cavendish from 1923. Unlike the general theory, special relativity could be learned fairly quickly from a competent teacher, required no advanced mathematical methods that were not already known to the average physicist, did not contain a technical component that was still open to debate, and could easily be applied to a wide range of simple problems. Questions on the special theory began to appear on examination papers for the "Physics" option of the Natural Sciences Tripos (taken by undergraduates at the Cavendish Laboratory) in the mid 1920s, and, by the early 1930s, the more ambitious students were required to derive the Lorentz transformations and show how they altered the classical conceptions of mass and momentum.[152] It nevertheless seems likely that many students at the Cavendish in the 1920s acquired only a tenuous grasp of the theory from Eddington's lectures, not least because it still remained of little practical importance to their experimental work. As a typical Cavendish experimenter from this period revealingly recalled, Eddington's course "made everything

---

149. For discussion of the separate cultures of theory and experiment in Cambridge in the 1920s, see Hughes 1998, 342–47.

150. In 1923, for example, Louis de Broglie had taken the important step of ascribing wave-like properties to subatomic particles in an attempt to resolve an inconsistency between special relativity and the old quantum theory (Wheaton 1983, 286–97).

151. The changing attitude at the Cavendish Laboratory towards the new quantum mechanics in this period is discussed in Hughes 1998.

152. See, for example, *Cambridge University Examination Papers,* Natural Sciences Part II, 1927 Physics (2) question 6; and 1932 Physics (1) question 2.

crystal clear to me at the time, but invariably I found myself floundering some weeks later."[153]

The one Cambridge physicist associated with the Cavendish Laboratory who certainly did have a sound grasp of relativity in the late 1920s was Neville Mott, but he had studied the Mathematical Tripos as an undergraduate and would have attended Eddington's Part II lectures on the general theory in 1926.[154] In fact all of the theoreticians working on relativistic quantum mechanics at Cambridge in the early 1930s had studied advanced mathematics at the university and almost certainly learned their relativity from Eddington's courses.[155] By the late 1930s, the special theory of relativity had become so important to students taking options in theoretical physics in the Mathematical Tripos that a new course on "Electrodynamics and Special Theory of Relativity" was put on in tandem with Eddington's lectures on the general theory.[156] This one-term course would have been far less demanding than Eddington's lectures, but would have served equally well as a preparation for Dirac's class on advanced quantum mechanics.[157] It is probably safe to say that from the mid 1930s at the latest, no student with a serious interest in quantum mechanics could complete his undergraduate studies satisfactorily without mastering the special theory of relativity.

The case of the general theory is rather different. We have already seen that there was little incentive for the vast majority of physicists to try to master this theory and that those who were tempted to explore its inner workings had to surmount formidable conceptual and technical difficulties. Unlike special relativity, the general theory was of little relevance to the development and application of mainstream quantum mechanics and did not therefore become a standard component in the physicist's training. Those who undertook research in this area prior to the 1960s were usually concerned with the mathematical formalism of the general theory itself and with its application to the large-scale structure of the universe.[158] This limited

153. Bruyne 1984, 87. It should be remembered that aspiring physicists in Cambridge studied the Natural Sciences Tripos (Physics option) at the Cavendish Laboratory and not the Mathematical Tripos.

154. Mott 1984, 125.

155. These were Dirac, Mott, H. R. Hulme, and H. M. Taylor. See Mott 1984, 128, and Jeffreys 1987, 37.

156. *Cambridge University Reporter*, 7 October 1939, 106. The new course was taught by F. C. Powell, who had taken Eddington's course in Part II of the Mathematical Tripos of 1928.

157. Dirac's lectures followed the contents of his 1930 book, *The Principles of Quantum Mechanics*, which included the relativistic quantum mechanics of the electron.

158. The few papers on quantum gravity from the interwar period are surveyed in Stachel 1995, 317–19.

utility of the general theory highlights two specific ways in which Eddington's Cambridge teaching of the subject was especially important to its preservation. First, without his books, courses, and willingness to help young research students, it is very unlikely that Cambridge would have become an important center for work on both the special and general theories of relativity in the 1920s and 1930s. In order to underline this point we can again draw an illuminating comparison with the case of Leiden. We have seen that circa 1916 there was not only enormous expertise and enthusiasm for general relativity in this town, but that it was also one of the first sites in the world at which the theory was taught to students. It is nevertheless the case that no appreciable research school in relativity studies emerged in Leiden in the post–World War I period, the reasons for this odd state of affairs being easy to find. De Sitter was based outside the town at the Observatory and so had little direct contact with students, while the ageing Lorentz, who had already retired, ceased to deliver his lectures on the general theory. The only physicist in Leiden who built a school in the interwar period was Lorentz's successor, Paul Ehrenfest, but he took no active role in the development of relativity after 1913, choosing instead to focus his research and teaching on statistical physics.[159] The Leiden case thus illustrates how quickly an established theoretical tradition can wane when it ceases actively to be promoted to new generations of students.

This brings us to the second sense in which recognition of the general theory's marginality to mainstream physics helps to place Eddington's teaching in proper perspective. It is clear with the benefit of hindsight that the period from the mid 1920s to the late 1950s constituted what has recently been dubbed the "low water mark" of the theory. During much of this period only a small number of theoreticians were actively engaged in its development, the representation of general relativity in the physics curriculum falling so low that by 1961 a contemporary observer suggested that it might disappear altogether within half a century.[160] By the early 1960s the theory was in fact in the early stages of a revival that would eventually make it far more widely practiced that it had ever been before.[161] This revival would surely have been much harder to accomplish had not the likes of Eddington helped to train a whole new generation of relativists during the first half of the so-called low-water period. When placed in this context, his accomplishment in training

159. Kox 1992, 45–46.
160. Kuhn 1977, 189. See also Eisenstaedt 1989b and Will 1988, 3–18.
161. Mercier 1992; Kaiser 1998; Stachel 1995, 297–98.

around 150 students, some ten percent of whom made original contributions to the theory, is very considerable indeed.[162]

The story of the Cambridge reception of general relativity provides a rare opportunity to study the transmission of a new and highly technical theory from one site to another in circumstances in which the exchange of personnel between the sites was impossible. This episode casts additional light on the concerns of earlier chapters of this study by making visible those difficulties inherent in learning a new theory that generally pass unnoticed and unrecorded in face-to-face encounters and formal pedagogical settings. It took Eddington around four years to gain sufficient mastery of general relativity not only to write two authoritative expositions of the subject but to begin teaching it to Cambridge students and to make his own original contributions; once he had incorporated his technical expertise in the pedagogical economy of the university, general relativity was passed on annually to classes of Cambridge students without special comment on its conceptual or technical difficulty. The success of Eddington's teaching over the subsequent two decades also highlights the importance of training in the establishment of new theories. British physicists might have been convinced circa 1920 that Einstein's predictions had been verified experimentally, but few of them had much technical grasp of relativity or accepted Einstein's interpretation of his theory. For members of an older generation of wranglers, such as Larmor and Shaw, the complicated mathematical machinery of general relativity awaited some kind of dynamical interpretation and did not imply that the space of our universe was really non-Euclidian. However, for the new generation trained by Eddington, issues of this kind simply did not arise. They were taught a broadly Einsteinian approach to general relativity from the start, thereby becoming part of a small but international community of practitioners who shared a common understanding of the theory's meaning as well as the theoretical problems it raised. In this chapter we have glimpsed something of the work and resources required to establish and sustain this community at just one important site.

162. I have not included here the students who attended his annual short course at the Cavendish Laboratory, as I have found no indication of the numbers of students who attended.

~ EPILOGUE ~

Training, Continuity, and Change

> Imagine that the natural sciences were to suffer the effects of a catastrophe.... Widespread riots occur, laboratories are burnt down, physicists are lynched, books and instruments are destroyed . . . a Know-Nothing political movement takes power and successfully abolishes science teaching in schools and universities, imprisoning and executing the remaining scientists. Later still there is a reaction against this destructive movement and enlightened people seek to revive science, although they have largely forgotten what it was . . . all they possess are fragments . . . parts of theories unrelated either to the other bits and pieces of theory which they possess or to experiment.
> ALASDAIR MACINTYRE, "A DISQUIETING SUGGESTION"[1]

What is the most disturbing aspect of MacIntyre's disquieting suggestion? He rightly reasoned that the specter of murdered scientists and burning laboratories would be shocking to a society that prized knowledge of the physical world above other forms. Yet, from a philosophical perspective, it is the thought that science might not be recoverable from written sources alone that is surely the most troubling. MacIntyre mused that a revivalist movement might have children "learn by heart the surviving portions of the periodic table and recite as incantations some of the theorems of Euclid" (1981, 1) with little or no real sense of their former meaning. The remaining textbooks and journal articles might even be made coherent within a new, pseudo-scientific tradition, but science as we know it would have been lost, perhaps irretrievably. This suggestion is disquieting because it challenges the assumption, deeply held in the Western metaphysical tradition, that scientific truth transcends culture. If something incommunicable through the written word is required to make science recoverable, then our best-verified theories and most secure forms of reasoning must be reliant to some extent on tacit skills and sensibilities carried within science's own culture. Once

---

1. MacIntyre 1981, 1.

that tradition is broken, even the most intelligent reader will fail to find scientific truths self-evident. Both the value and practical limitations of this conclusion are illustrated by the rise of natural knowledge in the West. Scholars in medieval and renaissance Europe took more than four centuries to assimilate and transcend the most advanced mathematical and astronomical works from the ancient Greek and Islamic worlds. That they did so at all was a function of the size and longevity of the scholarly community that developed within and around the cathedral schools and universities.[2]

These problems of recovery are much easier to comprehend when scientific knowledge is approached pedagogically. We are inclined to think of theories as collections of mathematically expressed propositions from which predictions and problem solutions can straightforwardly be derived. Yet this is a view of theory proper to those who are already its masters. Theories are not made from the top down, and they are learned exactly the other way around. Each generation of students learns to theorize by struggling with simple exercises and a very limited, sometimes fictional understanding of current physical theory, before moving on to more difficult techniques and problems over a period of years. The emergent sense of a theory's self-evidence and independent existence is as much the *outcome* as it is the *cause* of this learning process. Reviving modern mathematical physics would accordingly be more difficult than recovering classical Greek geometry. As we have seen, the community that made the former discipline was built on a highly specialized and protracted training processes. Those who would revive it would need not only to develop the innumerable skills required to make a living craft from a lifeless jumble of diagrams, symbols, and syntactical relationships, but also to find methods of preserving and propagating the bits of theory they managed painstakingly to reconstruct. Nor should they expect to recover in a few years or among a handful of practitioners a culture which was several centuries in the making and which drew upon a global community of contributors. The length and complexity of the training regimes explored in this study are a measure of the sheer volume of taken-for-granted skill that now constitutes the common language through which physicists routinely communicate; in MacIntyre's terms, it is a measure of what would need to be revived if mathematical physics as we know it were to be recovered.

Turning this pedagogical argument around, it is interesting to note that insofar as MacIntyre's scientific apocalypse is currently in prospect, it is far

---

2. Lindberg 1992, 364.

more likely to occur through physics failing to reproduce its own culture than through its destruction by anti-science movements. In Britain at least, the currently perceived threat to the elite technical disciplines comes not from lynch mobs but from an increasingly desperate shortage of qualified science and mathematics teachers and a shrinking number of able students prepared to submit themselves to a rigorous scientific education. As we saw in chapter 5, since the 1860s the top universities have come to rely on secondary education to provide a steady supply of highly trained and motivated students. If physicists want to pass on their unique culture to future generations, they may have to find new ways of recruiting and motivating a sufficiently large number of suitable students, and extend their undergraduate programs to include more elementary training. The way in which they do so will surely play a role in shaping their discipline's future form. These concerns with the historical development of mathematical knowledge raise issues that range beyond the Cambridge case study which I have used as an example. With this in mind I shall conclude by commenting briefly on the wider historical and explanatory scope of some of the arguments advanced over the previous chapters. My remarks are intended not so much to summarize or to defend these arguments, as to open them for further development and debate.

The first point returns us to my comment towards the end of chapter 1 that much of this study can be read as a historical ethnography concerned with the reproduction of technical culture. This remark was intended to highlight the relationship I sought to establish between the real-time productivity of new and local forms of pedagogical practice, and the preservation and expansion of mathematical forms of life over much longer periods of time. Couched in more historical terms, this relationship is one that enables us to view and to analyze the rise of modern mathematical physics in the *longue durée*. As I noted in chapter 1, the recent social and cultural turns in the history of science have tended to efface questions concerning the diachronic, internal development of technical disciplines, in favor of those seeking the place of the sciences within broader and synchronic cultural contexts. The latter approach has proved extremely fruitful in science studies but, as Yves Gingras has recently argued, it should also *enable* rather than *preclude* the development of new historical perspectives on the longer-term trajectory of the sciences. Gingras points to the relentless mathematization of physics that has taken place over the last four hundred years, and highlights the fundamental role that considerations of mathematical symmetry have come to play in the search for new physical theories. With this in mind, he calls for a history of the rise of *private science* to parallel and complement recent studies of the rise of *public science* in the eighteenth and nineteenth centuries.[3]

The present study provides one route to understanding how the emergent mathematical physics of the late seventeenth century first excluded the majority of scholars—by virtue of the special technical expertise it demanded—but then, from the mid-eighteenth century, underwent a rapid and long-term expansion both in student numbers and technical content. On my showing the "private" world in which physics was systematically mathematized and globalized was nurtured and sustained through new regimens of training that were implemented in the coaching room, student residence, and examination hall. In this sense my study is concerned less with the rise of specific physical theories and mathematical methods than with the emergence of a special and widely distributed cultural space in which it became possible to develop, preserve, and transcend ever more sophisticated forms of mathematical theorizing. I do not mean to imply that physical theories have developed in isolation from experiments over the past two hundred years, but only that the theorists's world has come to constitute a distinct form of life worthy of the same cultural analysis as the experimental workplace.

It is an obvious corollary of the above claims that many of the general characteristics I have attributed to mathematical study in Cambridge are equally applicable to other institutions of technical training in Europe, North America, and elsewhere in the nineteenth and twentieth centuries. If the story I have told is to provide a plausible explanation of how an esoteric form of life was expanded and transmitted to new sites, then the sites in question ought to share a good deal in common. This issue was touched upon at the end of chapter 3. There I noted that the new methods of technical training instituted in the early nineteenth century in the Ecole Polytechnique and Neumann's physics seminar at Königsberg mirrored some of those I have identified as important in Cambridge during roughly the same period. Whether similar or comparable methods of training were generally employed in institutions of higher education over the nineteenth century is a matter for further investigation, but I suggest that they were and that from the early twentieth century virtually all major mathematical and theoretical physicists were trained by some form of the methods described in the preceding chapters.

---

3. Gingras 2001, 11. Gingras uses "private" in the sense of "exclusive." On science and the public sphere, see Clark, Golinski, and Schaffer 1999, 23–26.

Some support for this suggestion is provided by a 1912 survey on the mathematical training offered to physicists in the major European countries and the United States.[4] Conducted by the Göttingen physicist and mathematician Carl Runge (on the behalf of the International Commission on the Teaching of Mathematics), the survey was intended to ascertain which mathematical topics were typically included in a physics education, and how and by whom physicists were taught. The survey's results are relevant because they show that a decade into the twentieth century the mathematical methods taught to physicists were "very much the same almost everywhere," and that there was considerable international concern for further standardizing and improving the physicist's mathematical knowledge. The main problems identified by the survey were that physicists were too often taught by pure mathematicians who did not explain the relevant applications of the methods they were teaching, that students were not given nearly enough graded exercises and problems through which to develop and apply their mathematical skills, and that teachers frequently failed to offer enough "personal intercourse" to students for the "discussion of individual difficulties."[5] Runge concluded that the mathematical instruction offered to physicists in some institutions was "very much in need of reform" if these shortcomings were to be surmounted. It is beyond the scope of this study to trace if and how this reform was accomplished, but it is striking that the methods of teaching implicitly recommended were almost exactly those pioneered and employed by Cambridge coaches and lecturers over the previous century. The important sentiment being expressed here is that insofar as undergraduate physicists were not already being trained by the kinds of method used in Cambridge and some other leading European universities, they should be brought into line with these methods as quickly as possible. The few ethnographic and historical studies of twentieth-century physics training that we currently have suggest that this goal was accomplished, at least in North America, by the middle decades of the century.[6]

The final two points I would like to raise concern the relationship between training and innovation in physical theory. Throughout this study I have emphasized the conservative nature of the kind of institutionalized training system that emerged in Victorian Cambridge. We have seen that

4. Runge 1913.
5. Ibid., 598, 601–2. Runge argued that "personal intercourse" between teacher and student should be "insisted upon."
6. See for example Traweek 1988, White 1991, and Kaiser 1998.

these systems are not only effective in preserving the local methods and ideals of physical theory, but are surprisingly resistant to alternative approaches from elsewhere. My first point is that this conservatism should not blind us to the productive power inherent in such regimes. It is probably true to say that individual regimes are unlikely to foster radically new theories without major input from external sources. Cambridge's two most innovative Victorian theorists, for example (William Thomson and James Clerk Maxwell), both left the university shortly after graduation and found important intellectual resources for their work elsewhere. It is nevertheless the case that a large number of important, if less radical, works were produced by Cambridge graduates building mainly on the tools of their undergraduate training. In addition to the work of the Cambridge Maxwellians (discussed in chapter 6), one has only to think of Lord Rayleigh's *Theory of Sound,* Horace Lamb's *Hydrodynamics,* George Darwin's foundational papers on geophysics, and A. S. Eddington and James Jeans's similarly important work on the constitution of stars. It is also extremely important to keep in mind that the skills of Cambridge-trained mathematicians were put to a wide range of more mundane but nonetheless vital applications in applied science and engineering. To take but one relatively well-researched example, the first and highly influential theory of aeronautical stability was developed by one of Routh's pupils, George Hartley Bryan (5W 1886), based on the former's dynamical theory of rigid bodies. The theory of flight and flight instrumentation was further developed by Cambridge mathematicians working in connection with the National Physical Laboratory and, during World War I, at the Royal Aircraft Factory at Farnborough. These men directly deployed the tools of their undergraduate training and even found occasional inspiration in specific Tripos problems they had tackled in their student days.[7]

This brings me to a final point concerning the relationship between training and radical innovation in physical theory. My approach to theorizing through the acquisition and use of technical skills has led me to focus on the *reception* of two highly innovative theories (electromagnetic field theory and the theory of relativity) while largely ignoring the origins of the theories themselves. I think my reasons for doing so are clear, but it is appropriate at this point to ask whether a pedagogical approach can tell us anything significant about the work of physicists like Maxwell and Einstein, who are noteworthy because they transcended and redefined the fundamental tools

---

7. Hashimoto 1990, chap. 1; Bradley 1994, 90.

of their discipline. Is there more to be said on this issue than that a training in physics played a necessary but entirely insufficient part in explaining their accomplishments? I believe there is more to be said, and to illustrate the points I want to make I shall return to the training and early research of Paul Dirac, a man who is not only regarded as Cambridge's greatest theoretical physicist of the twentieth century but one whose accomplishments are believed by some to rank alongside those of Maxwell and Einstein.[8]

As Peter Galison has recently noted (2000b, 145), Dirac stood throughout the twentieth century as the "theorist's theorist," an epithet that is easily understood by recalling the bare bones of his early scientific biography. Dirac arrived in Cambridge in 1923 (aged 21) with little knowledge of physics and none whatsoever of the emergent quantum theory of the atom. Within five years he had played a major role in founding quantum mechanics and published an equation describing the relativistic, quantum mechanical behavior of the electron. This equation revolutionized quantum physics and has since born Dirac's name. By the time he was thirty-one years old, Dirac had been elected a Fellow of the Royal Society, appointed to the Lucasian Chair of Mathematics at Cambridge, and awarded a Nobel Prize for his contributions to theoretical physics. Dirac's public persona and style of work also exemplified the image of the theoretician as an introspective and unworldly thinker; he generally worked alone and secretively, and adopted an infamously taciturn style in conversation. Even his Cambridge colleagues had little idea of precisely what problems he was currently attempting to solve.[9] Dirac's accomplishments, like those of his great predecessor in the Lucasian Chair, Isaac Newton, seemed to his contemporaries to come from nowhere but his solitary genius.

It is a striking feature of several recent studies of Dirac's life and work that they implicitly challenge this mythology by relating Dirac's early research to his training.[10] What they show is that the very features of Dirac's education that seem at first sight to make him an unlikely candidate for unlocking the mathematical mysteries of the quantum world turn out on closer inspection to provide a powerful means of explaining his success.[11] For example,

8. Kragh 1990, ix, 12.

9. Neville Mott recalled that "all of Dirac's discoveries just sort of fell on me and there they were. I never heard him talk about them, or he hadn't been in the [Cavendish Laboratory] talking about them. They just came out of the sky" (Kragh 1990, 57).

10. By far the most detailed study of Dirac's training and early research is Mehra and Rechenberg 1982, Part 1.

11. Dirac's training has attracted an unusual amount of interest among historians of physics. For example, Darrigol (1992) says nothing at all about Werner Heisenberg's education but

Dirac's undergraduate training consisted of two degrees taken at Bristol University, the first in electrical engineering, the second in applied mathematics. At the end of these studies his knowledge of contemporary physics was confined largely to a basic understanding of relativity, a subject he learned through a course of philosophical lectures on Einstein's work and through private study of Eddington's *Space, Time and Gravitation*. These biographical details made Dirac something of an outsider when he arrived in Cambridge. Virtually all of his contemporaries had been trained in the Mathematical Tripos and had learned some version of the electronic theory of matter as described in chapters 7 and 8. Dirac, by contrast, had learned electromagnetism in terms of practical circuit theory and Oliver Heaviside's operational calculus. Having learned relativity from Eddington's book, moreover, Dirac understood the theory not as an adjunct to the electrodynamics of moving bodies but as a self-contained theory of space and time that required other physical theories to conform to certain mathematical transformations. What emerges is that Dirac's Bristol experience equipped him with a view of physical theory and a range of mathematical skills that were different from those possessed by Cambridge-trained undergraduates. These skills, in operational calculus, projective geometry, and relativity, played a vital role in enabling him to develop his own version of quantum mechanics from the mid 1920s.

Dirac's first two years in Cambridge were also important in shaping his early research career. He had intended to concentrate on relativity, but under the direction of his supervisor, R. H. Fowler, his research interests changed. Fowler was the foremost expert in Britain on atomic theory, and one of the very few men in the country in touch with Continental attempts to develop a fully fledged quantum mechanics. Fowler taught Dirac the Bohr theory of the atom, impressed upon him the importance of Hamiltonian mechanics in quantum calculations, and gave Dirac an advance copy of Werner Heisenberg's now historic paper on the computational principles of the new quantum mechanics. Dirac meanwhile had deepened his knowledge of Hamiltonian mechanics and projective geometry by studying standard Cambridge treatises by E. T. Whittaker and H. F. Baker respectively. Having mastered Heisenberg's paper, Dirac found an important relationship between the new theory and Hamiltonian mechanics, tried to make the theory conform to the principle of relativity, and developed a consistent

---

considers Dirac's in some detail. The study most directly concerned to link Dirac's training and research is Galison 2000b.

operational algebra of the new quantum variables. It is worth noting too that Dirac's lack of knowledge in some areas of physics was almost as important as his expertise in others. On the one hand, he was able to study atomic theory from German textbooks and journals without the prejudices of a Cambridge-trained undergraduate; on the other, his distance from the German physics community meant he could study Heisenberg's theory in mainly mathematical terms. While those steeped in the old quantum theory were troubled by some of the mathematical properties of Heisenberg's theory, Dirac saw those same properties as the key to the theory's power and further development. Drawing upon the above mentioned treatises by Whittaker and Baker, Dirac was able to develop his quantum algebra just fast enough to stay ahead of his Continental competitors.[12]

In light of the present study we ought to be very surprised that someone with such a meager knowledge of physics was able after just two years of graduate study at Cambridge to intervene so incisively and productively in the development of a theory whose masters worked in Munich, Göttingen, and Copenhagen. Judging by the relativity example discussed in the previous chapters, we should expect Dirac to have taken several years to master the quantum theory of the atom, let alone to find himself in a position to develop it more rapidly than Heisenberg and his collaborators. What recent studies of Dirac's work indicate is that this intervention can be made comprehensible, perhaps even predictable, through a close study of his training. Even my thumbnail sketch of his early career indicates how Dirac's education individuated him from his Cambridge peers, none of whom made similar contributions to quantum physics. Indeed, we can see why it seemed to his contemporaries, knowing little of Continental atomic theory and less of Dirac's educational biography, that his brilliant papers from the mid 1920s "just came out of the sky." My sketch also points to a constructive account of how Dirac was able to build so rapidly on Heisenberg's work. As Olivier Darrigol has noted in this context, modern theorists "live in a world of highly developed theories . . . in order to obtain new theories, they extend, combine, or transpose available pieces of theory" (1992, xxii). This is surely correct, but if we are to offer an *explanation* of how novel theoretical work is accomplished, we need to know which pieces of theory a given theorist knew and why he or she chose to put them together in one way rather than another. Dirac would make an excellent case study in this respect, as we

12. My overview of Dirac's work is taken from Mehra and Rechenberg 1982, Part 1, Darrigol 1992, chaps. 11 and 12, and Kragh 1998, chap. 2.

already know a great deal about the subjects he studied, the supervision he received, and the books he read—even down to which textbook passages he drew upon while building his own theories.[13]

This is not the place to undertake such a study, but it is worth considering what kind of explanation of innovative work it might produce. One of Dirac's Cambridge contemporaries, W. M. McCrea, observed towards the end of his career that, thanks to a dismissive remark made by one of his lecturers, he had not studied Whittaker's textbook on analytical dynamics. He now realized that he had neglected a book "which Dirac, working just round the corner [at St. John's College], was finding to be so precisely what he required" (McCrea 1986, 276). McCrea was clearly reflecting on the way seemingly chance events in a scientist's education could strongly influence the relative success of his research career. Read in stronger terms the remark invites us to wonder whether any able mathematical physicist who had followed the same course of study as had Dirac could have made the same contributions to modern physics. An affirmative answer to this question points to an explanation of theoretical innovation that is the polar opposite to the "insights of genius" approach I criticized in chapter 1. Where the latter localizes innovation in special and inaccessible properties of the scientist's mind, the former makes him a vehicle for a collection of skills that are drawn from the wider community, and whose uniqueness derive from the scientist's particular course of training.

This model of innovative work makes strong connections between Dirac's contributions to quantum theory and, say, the emergence of the Poynting vector in the 1880s. In both cases seemingly sudden and intuitive theoretical insights turn out to emerge from a more mundane series of problems, skills, and interests which derive directly from the relevant theoretician's educational biography. There is moreover another common feature of these examples which might help us understand what kind of innovation is accessible to this sort of analysis. In both cases the key innovations in question were either anticipated or would very shortly have been made by other theoreticians.[14] This suggests that training is most likely to be of immediate relevance in situations where a well-formulated problem exists to which a theoretician happens to possess the tools required to effect a speedy and uncontroversial solution. This kind of approach may well have limita-

---

13. On Dirac's use of Baker's and Whittaker's treatises, see Mehra and Rechenberg 1982, 45–46, 67–68, 71–72, 104, 139, 150–52, 161–65, 168, 172–74, 178 185, 202–3, 215.

14. Poyning's work is discussed in chapter 6. On Dirac's competitors, see Kragh 1992, 20, 62.

tions, but it has the considerable advantage of making at least some novel work accessible to historical explanation. Pursuing it would at the very least show where those limits lie and at what point, if any, we are required to admit other explanatory devices in the form of resources extraneous to a scientist's training, innate mental qualities, or otherwise inexplicable leaps of intuition. For the present I leave this problem as an exercise to the reader.

APPENDIX A

Coaching Success, 1865–1909

The following table gives the names (where known) of the coaches of the "upper ten" wranglers for the period 1865–1909. I have not found systematic records for the years prior to 1865 (the system was discontinued in 1909). Pupils beyond the tenth wrangler are not included partly because the record is very patchy, but also because the vast majority of wranglers who undertook significant research were in the upper ten.

YEAR	WRANGLER	COLLEGE	COACH
1865	Strutt, J. W.	Trinity	Routh
	Marshall, A.	St. John's	Routh
	Taylor, H. M.	Trinity	Routh
	Mitchell, C. T.	Caius	Routh
	Ashton, J.	Sidney	Routh
	Wood, A.	Johns	Routh
	Cumming, L.	Trinity	Routh
	Watson, H.C.	Trinity	Routh
	Blanch, J.	St. John's	Routh
	Caldwell, R. T.	Corpus	Routh
1866	Morton, R.	Peterhouse	Routh
	Aldis, T. S.	Trinity	Aldis
	Niven, W. D.	Trinity	Routh
	Stuart, J.	Trinity	Routh

YEAR	WRANGLER	COLLEGE	COACH
	Hill, E.	St. John's	Besant
	Pirie, G.	Queens	Routh
	Toller, T. N.	Christ's	Aldis
	Gross, E. J.	Caius	Routh
	Dick, G. R.	Caius	Routh
	Osborn, T. G.	Trinity Hall	Routh
1867	Niven, C.	Trinity	Routh
	Clifford, W. K.	Trinity	Frost
	Lambert, C. J.	Pembroke	Aldis/Routh
1868	Moulton, J. F.	St. John's	Routh
	Darwin, G. H.	Trinity	Routh
	Smith, C.	Sidney	Routh
	Christie, W. H. M.	Trinity	Routh
1869	Hartog, N. E.	Trinity	Routh
	Elliot, J.	St. John's	Besant
	Moir, A.	Christ's	Routh
	Waymouth, S.	Queens	Routh
	Wright, R. T.	Christ's	Routh
1870	Pendlebury, R.	St. John's	Routh
	Greenhill, A. G.	St. John's	Besant
	Levett, E. L.	St. John's	Routh
	Hunter, H. St. J. A.	Jesus	Frost
	Stephen, W. D.	Corpus	Besant
	Haslam, W. H.	St. John's	Besant
	Henderson, G.	Pembroke	Routh
	Blaikie, J. A.	Caius	Frost
1871	Hopkinson, J.	Trinity	Routh
	Glaisher, J. W. L.	Trinity	Routh
	Spence, W. M.	Pembroke	Aldis
	Hart, H.	Trinity	Routh
	Temperley, E.	Queens	Routh
	Appleton, R.	Trinity	Besant
	Malcolm, W.	Christ's	Routh
	Carver, T. G.	St. John's	Routh
	Genese, R. W.	St. John's	Routh

APPENDIX A

YEAR	WRANGLER	COLLEGE	COACH
	Dey, A.	Peterhouse	Routh
1872	Webb, R. R.	St. John's	Routh
	Lamb, H.	Trinity	Routh
	Lock, J. B.	Caius	Routh
	Richardson, J. G.	Trinity	Frost
	Rives, G. L.	Trinity	Frost
	Cook, C. H. H.	St. John's	Routh/Besant
	Taylor, W. W.	Trinity	Routh
	Warren, J. W.	Caius	Routh
	Boughey, A. H. F.	Trinity	Dale
1873	Harding, T. O.	Trinity	Routh
	Nanson, E. J.	Trinity	Routh
	Gurney, T. T.	St. John's	Besant
	Prior, C. H.	Caius	Routh
	Garnett, W.	St. John's	Besant
	Terry, T. W.	Trinity	Routh/Dale
	Hicks, W.	St. John's	Routh
	Richie, W. I.	Trinity	Frost
	Lock, G. H.	Clare	Routh
	Gregory, P. S.	King's	Dale
1874	Calliphronas, G. C.	Caius	Routh
	Ball, W. W. R.	Trinity	Routh
	Harris, J. R.	Clare	Routh
	Craik, A.	Emmanuel	Routh
	Dickson, J. D. H.	Peterhouse	Routh
	Stuart, G. H.	Emmanuel	Routh
	Clarke, H. L.	St. John's	Routh
	Butcher, J. G.	Trinity	Routh
	Cox, J.	Trinity	Routh
	Elliott, T. A.	St. John's	Routh
1875	Lord, J. W.	Trinity	Routh
	Burnside, W.	Pembroke	Besant
	Chrystal, G.	Peterhouse	Routh
	Scott, R. F.	St. John's	Routh
	Griffiths, G. J.	Christ's	Frost
	Body, C. W. E.	St. John's	Routh

COACHING SUCCESS

YEAR	WRANGLER	COLLEGE	COACH
	Lewis, T. C.	Trinity	Routh
	Marshall, J. W.	Peterhouse	Routh
	Wilson, J.	Christ's	Routh
	Sharpe, J. W.	Caius	Routh
1876	Ward, J. T.	St. John's	Routh
	Mollison, W. L.	Clare	Routh
	Poynting, J. H.	Trinity	Routh
	Trimmer, F. G.	Trinity	Routh
	Glazebrook, R. T.	Trinity	Dale
	Hargreaves, R.	St. John's	Besant
	Bishop, W. C.	Emmanuel	Routh/Webb
	McCann, H. W.	Trinity	Routh
	Sunderland, A. C.	Trinity	Routh
	Findlay, C. F.	Trinity Hall	Routh
1877	MacAlister, D.	St. John's	Routh
	Gibbons, F. B. De M.	Caius	Routh
	Rowe, R. C.	Trinity	Routh
	Smith, J. P.	Trinity	Routh
	Coates, C. V.	Trinity	Greenhill/Routh
	Knight, S.	Trinity	Routh
	Wilson, S. R.	Sidney	Wilson, J.
	Greaves, J.	Christ's	Besant
	Walters, F. B.	Queens	Frost
	Milton, H.	Caius	Besant
1878	Hobson, E. W.	Christ's	Routh
	Steggall, J. E. A.	Trinity	Routh
	Graham, C.	Caius	Routh
	Edwards, J.	Sidney	Routh
	Pinsent, H. C.	St. John's	Dale
	Macaulay, W. H.	King's	Routh
	Adair, J. F.	Pembroke	Besant
	Sargant, E. B.	Trinity	Routh
	Allcock, C. H.	Emmanuel	——
	Martin, J. A.	Sidney	Routh
1879	Allen, A. J. C.	Peterhouse	Routh
	Walker, G. F.	Queens	Routh

YEAR	WRANGLER	COLLEGE	COACH
	Pearson, C.	King's	Routh
	Gunston, W. H.	St. John's	Routh
	Hill, M. J. M.	Peterhouse	Routh
	Wallis, A. J.	Trinity	Routh
	Green, A.	Christ's	Besant
	Bell, W. G.	Trinity Hall	Routh
	Pilkington, J. H.	Pembroke	Besant
	Wood, J.	Queens	Routh
1880	Larmor, J.	St. John's	Routh
	Thomson, J. J.	Trinity	Routh
	Allcock, W. B.	Emmanuel	Routh
	Cox, H.	Trinity	Routh
	Mackenzie, H. W. G.	Emmanuel	Webb
	McIntosh, A.	Queens	Routh
	Welsford, J. W. W.	Caius	Routh
	Johnson, G. W.	Trinity	Frost
	Maclean, A. J.	King's	Routh
	Harrison, J.	King's	Dale
1881	Forsyth, A. R.	Trinity	Routh
	Heath, R. S.	Trinity	Routh
	Steinthal, A. E.	Trinity	Routh
	Dodds, J. M.	Peterhouse	Routh
	Jones, G. M. E.	Pembroke	Routh
	Pollock, C. A. E.	Trinity	Routh
	Fountain, E. O.	Pembroke	Routh
	Stokes, W. F.	Sidney	Besant
	Leahy, A. H.	Pembroke	Routh
	Hopkinson, E.	Emmanuel	Routh
1882 January	Herman, R. A.	Trinity	Routh
	Yeo, J. S.	St. John's	Webb
	Loney, S. L.	Sidney	Routh
	Brill, J.	St. John's	Routh
	Randell, J. H.	Pembroke	Routh
	Robson, H. C.	Sidney	Webb
	Parker, J.	St. John's	Routh
	Harker, A.	St. John's	Webb
	Littlewood, E. T.	Peterhouse	Routh

YEAR	WRANGLER	COLLEGE	COACH
	Hensley, W. S.	Christ's	Webb
1882	Welsh, W.	Jesus	Routh
June	Turner, H. H.	Trinity	Routh
1883	Mathews, G. B.	St. John's	Besant
	Gallop, E. G.	Trinity	Routh
	Lachlan, R.	Trinity	Routh
1884	Sheppard, W. F.	Trinity	Routh
	Workman, W. P.	Trinity	Routh
	Bragg, W. H.	Trinity	Routh
	Young, W. H.	Peterhouse	Routh
1885	Berry, A.	King's	Routh
	Love, A. E. H.	St. John's	Webb
	Richmond, H. W.	King's	Routh
1886	Dixon, A. C.	Trinity	Webb
	Fletcher, W. C.	St. John's	Routh/Webb
	Coates, W. M.	Queens	Routh
	Hill, F. W.	St. John's	Routh
	Bryan, G. H.	Peterhouse	Routh
1887	Baker, H. F.	St. John's	Routh
	Flux, A. W.	St. John's	Besant
	Iles, J. C.	Trinity	Webb
	Michell, J. H.	Trinity	——
	Peace, J. B.	Emmanuel	Routh/Webb
1888	Orr, W. M.	St. John's	Routh
	Brunyate, W. E.	Trinity	Webb
	Sampson, R. A.	St. John's	Webb
	Valentine-Richards, A. V	Christ's	——
	Guest, J. J.	Trinity	Webb
1889	Walker, G. T.	Trinity	Webb
	Dyson, F. W.	Trinity	Webb
	Gaul, P. C.	Trinity	Webb
	Macdonald, H. M.	Clare	——
	Lay, C. J.	Catherine's	Besant

YEAR	WRANGLER	COLLEGE	COACH
	Cook, A. G.	St. John's	Webb
	Ramsey, A. S.	Magdalene	Gunston/Besant
	Geake, C.	Clare	Webb
	Jackson, C. S.	Trinity	Webb
	Todhunter, R.	Clare	Webb
1890	Bennett, G. T.	St. John's	Webb
	Segar, H. W.	Trinity	Webb
	Brand, A.	Pembroke	Webb
	Vaughan, A.	Trinity	Webb
	Crawford, L.	King's	Webb
	Reeves, J. H.	St. John's	Webb
	Troup, J. M.	Pembroke	Webb
	Alexander, J. J.	St. John's	Webb
	Dobbs, W. J.	St. John's	Webb/Love
1891	Goowillie, J.	Corpus	Webb
	Mair, D. B.	Christ's	Webb
	Mayall, R. H. D.	Sidney	Webb
	Stamp, A. E.	Trinity	Webb
	Daniels, A. E.	Peterhouse	Webb
	Selby, F. J.	Trinity	Webb
	Abbott, R. C.	Trinity	——
	Charles, A. W.	Trinity Hall	——
	Cullis, C. E.	Caius	Webb/Gunson
1892	Cowell, P. H.	Trinity	Webb
	Sharpe, F. R.	Christ's	Webb
	Hough, S. S.	St. John's	Webb
	Munro, A.	Queens	Webb
	Pocklington, H. C.	St. John's	Webb
	Adamson, H. A.	Trinity	——
	Chevalier, R. C.	St. John's	——
	Morton, W. B.	St. John's	Webb
1893	Manley, G. T.	Christ's	Hobson
	Hurst, G. H. J.	King's	Webb
	Sanger, C. P.	Trinity	Webb
	Gaul, C. A.	Trinity	Webb
	Peel, T.	Magdalene	Webb
	Dale, J. B.	St. John's	Webb

YEAR	WRANGLER	COLLEGE	COACH
	Caunt, S. W.	Catherine's	—
	Russell, B. A. W.	Trinity	Webb
	Atkins, H. E.	Peterhouse	Webb
1894	Adie, W. S.	Trinity	Webb
	Sedgwick, W. F.	Trinity	Webb
	Philip, W. E.	Clare	Webb
	Carslaw, H. S.	Emmanuel	Webb
	Lawrence, F. W.	Trinity	Webb
	Leathem, J. G.	St. John's	Webb
	Browne, A. B.	Christ's	Webb
	Osborn, G. F. A.	Trinity	Webb
	Campbell, A. V. G.	Trinity	Webb
	Jacob, J.	Caius	Coates
1895	Bromwich, T. J. I'A.	St. John's	Webb
	Grace, J. H.	Peterhouse	Webb
	Whittaker, E. T.	Trinity	Hobson/Webb
	Godfrey, C.	Trinity	Webb/Herman/Love
	Wilson, G. H. A.	Clare	Webb
	Howard, F. A.	Pembroke	—
	Western, A. W.	Trinity	—
	Carter, F. W.	St. John's	Baker
	Rumsey, C. A.	Trinity	Webb/Coates
	Smallpiece, A. J.	St. John's	Webb
1896	Fraser, W. G.	Queens	Webb
	Barnes, E. W.	Trinity	Hobson/Webb
	Carson, G. E. St. L.	Trinity	Hobson/Webb
	Wilkinson, A. C. L.	Trinity	Hobson
	Edwardes, F. E.	St. John's	Webb
	Houston, W. A.	St. John's	Webb
	Cook, S. S.	St. John's	Webb
	Turner, E. G.	St. John's	Webb
	Radford, E. M.	Trinity	Webb
	Allen, H. S.	Trinity	Webb

YEAR	WRANGLER	COLLEGE	COACH
1897	Austin, W. H.	Trinity	Webb
	Whipple, F. J. W.	Trinity	——
	Frankland, F. W. B.	Clare	Hobson/Macdonald
	Ezechiel, P. H.	Trinity	Herman
	Walker, G. W.	Trinity	Love
	Darnley, E. R.	Trinity	Love
	Gibberd, W. W.	Trinity	——
	Channon, F. G.	Corpus	Webb
	Thompson, T. P.	Christ's	Hobson/Manley
	Blandford, J. H.	St. John's	Webb
1898	Hudson, R. W. H. T.	St. John's	Webb
	Cameron, J. F.	Caius	Webb
	Jeans, J. H.	Trinity	Webb/Herman/Walker
	Hardy, G. H.	Trinity	Webb/Herman/Love
	Moulton, H. F.	King's	Berry/Richmond
	Siddons, A. W.	Jesus	Coates
	Watkin, E. L.	St. John's	Webb
	Pococck, A. H.	Trinity	Herman
	Lenfesty, S. De J.	Peterhouse	Coates
	Lloyd, L. S.	Christ's	Herman
1899	Birtwistle, G.	Pembroke	Webb
	Paranjpye, R. P.	St. John's	Webb
	McLaren, S. B.	Trinity	Webb
	Bevan, P. V.	Trinity	Webb
	Eckhardt, J. C.	St. John's	Webb
	Bell, G. M.	Trinity	Coates
	Bell, J. G.	Trinity	Herman
	Hartley, W. E.	Trinity	Robson
	Valentine, G. D.	Trinity	Herman
	McAlpin, M. C.	Trinity	Herman
1900	Wright, J. E.	Trinity	Herman
	Aldis, A. C. W.	Trinity Hall	Hobson
	Bottomley, W. C.	Trinity	Hobson
	Balak, R.	St. John's	Baker

COACHING SUCCESS 521

YEAR	WRANGLER	COLLEGE	COACH
	Chadwick, J.	Pembroke	Baker/Webb
	Boutflower, C.	Trinity	Coates
	Marples, P. M.	Jesus	Coates
	Hackforth, E.	Trinity	Coates
	Casson, R.	St. John's	Webb/Baker
	MacBean, W. R.	Caius	Webb
1901	Brown, A.	Caius	Webb/Bromwich
	Knapman, H.	Emmanuel	Bennett
	Brown, H. A.	Caius	Webb/Herman
	Jackson, W. H.	Clare	Wilson, G. H. A.
	Thompson, A. P.	Pembroke	Baker
	Cama, B. N.	St. John's	Webb/Coates
	Cama, C. N.	St. John's	Webb/Herman
	Gharpurey, H. G.	St. John's	Webb/Hobson
	Kidner, A. R.	St. John's	Webb/Bromwich
	Strachan, J.	Clare	Wilson, G. H. A.
1902	Cunningham, E.	St. John's	Webb
	Slator, F.	St. John's	Webb/Brom./ Bak./Hobs.
	Webb, H. A.	Trinity	Hobson
	Wood, P. W.	Emmanuel	Allcock
	Hughes, W. R.	Jesus	Herman
	Goddard, J.	St. John's	Baker/Herman
	Picken, D. K.	Jesus	Herman
	Thomas, J. L.	Trinity	Herman
	Dixon, A. L.	Sidney	Mayall
	Garnett, W. H. S.	Trinity	Hobson
1903	Bateman, H.	Trinity	Gunston/Herman
	Marrack, P. E.	Trinity	Herman
	Barnes, J. S.	Trinity	Hobson
	Gold, E.	St. John's	Baker
	Hills, G. F. S.	Trinity	Herman
	Phillips, S. H.	St. John's	Baker
	Durell, C. V.	Clare	Wison, G. H. A.
	Gabbatt, J. P.	Peterhouse	Hobson
	Tanner, E. L.	Clare	Grace
	Irwin, D.	Sidney	Coates

YEAR	WRANGLER	COLLEGE	COACH
1904	Eddington, A. S.	Trinity	Herman
	Blanco-White, G. R.	Trinity	Herman/Hobson
	Stratton, F. J. M.	Caius	Gallop/Cameron/Grace
	Beckett, J. N.	St. John's	Bromwich/Hobson/Grace
	Grantham, S. G.	Trinity	Herman
	Starte, O. H. B.	Clare	Grace
	Bennett, T. L.	Trinity	Herman
	Ross, E. B.	Trinity	Hobson/Grace
	Small, W. J. T.	Caius	Grace
	Foyster, H. T.	Trinity	Herman
1905	Littlewood, J. E.	Trinity	Herman
	Mercer, J.	Trinity	Hardy/Jeans
	Smith, H.	Trinity Hall	Coates
	Trimble, C. J. A.	Trinity	Herman
	Booth, L. B.	Christ's	Hobson/Herman
	Priestley, H. J.	Jesus	Coates
	Hassé, H. R.	St. John's	Young/Herman
	Forsyth, J. P.	Caius	Gallop/Cameron/Grace
	Page, W. M.	King's	Leathem
	Fraser, P.	Queens	Munro
1906	Rajan, A. T.	Trinity	Herman
	Sewell, C. J. T.	Trinity	Herman
	Harrison, W. J.	Clare	Wilson/Grace
	Bell, B. H.	Trinity	Herman
	Milne, W. P.	Clare	Grace
	Dé, B.	St. John's	———
	Scholfield, W. F.	Trinity	Herman
	Titterington, E. J. G.	St. John's	Grace
	Watson, H. A.	Queens	Grace
	Eason, E. K.	Jesus	Welsh/Coates
1907	Watson, G. N.	Trinity	Herman
	Turnbull, H. W.	Trinity	Herman
	Hill, A. V.	Trinity	Herman

YEAR	WRANGLER	COLLEGE	COACH
	Nayler, W. A	Trinity	Herman
	Wilton, J. R.	Trinity	Herman
	Baynes, D. L. H.	Clare	Grace
	Norton, H. T. J.	Trinity	Herman
	Shaw, H. K.	Trinity	Herman
	Stewart, F. W.	Trinity	Herman
1908	Brodetsky, S.	Trinity	Herman
	Ibbotson, A. W.	Pembroke	Grace/Bromwich
	Minson, H.	Christ's	Herman
	Knox-Shaw, J.	Sidney	Herman
	Barnes, G. G.	St. John's	Herman
	Moody, R. H.	Emmanuel	Bennett/Allcock
	MacRobert, T. M.	Trinity	Herman
	Dunkley, H. F.	St. John's	Grace
	Leak, H.	Caius	Grace
	Bishop, T. B. W.	Emmanuel	Bennett/Allcock
1909	Daniell, P. J.	Trinity	Herman
	Neville, E. H.	Trinity	Herman
	Mordell, L. J.	St. John's	Bromwich
	Berwick, W. E. H.	Clare	———
	Darwin, C. G.	Trinity	Herman
	Livens, G. H.	Jesus	Grace/Welsh
	Thompson, A. W. H.	Trinity	Herman
	Waterfall, C. F.	Queens'	Bromwich
	Rau, B. N.	Trinity	Bromwich
	Bonhote, E. F.	Clare	Wilson/Grace

The major sources for the information above are *The Cambridge Chronicle* and *The Cambridge Express*. These weekly newspapers published the annual results of the Mathematical Tripos and gave short biographical accounts of each wrangler. The biographies appear more consistently for the high wranglers and generally include the names of their coaches. The other important sources are: *The Cambridge Independent Press*, which frequently gives biographies of the high wranglers; and *The Graphic*, which frequently published likenesses of the first three wranglers together with a short biography of each man. Other journals that sometimes provide similar information are *The Times, The Daily Graphic, The Cambridge Graphic, The Illustrated London News,* and *The Cambridge Review*. All other information has been obtained from biographies, reminiscences, and obituary notices.

APPENDIX B

Coaching Lineage, 1865–1909

The table below ranks the coaching successes of the private tutors who appeared twice or more in the previous table. The name of each coach is followed by the year in which he took the Tripos, his position in the Tripos (TP), his coach, and the number of (known) students he coached to the upper ten. The numbers given for Frost, Besant and Routh underestimate their success, as they were coaching before 1865.

NAME	YEAR	TP	COACH	NO OF STUDENTS
Routh	1854	1	Hopkins	133
Webb	1872	1	Routh	100
Herman	1882	1	Routh	48
Besant	1850	1	Parkinson	21
Hobson	1878	1	Routh	18
Grace	1895	2	Webb	16
Coates	1886	3	Routh	13
Baker	1887	1	Routh	9
Frost	1839	2	——	8
Bromwich	1895	1	Webb	6
Dale	1862	3	——	6
Love	1885	2	Webb	5
Wilson	1895	5	Webb	5
Aldis	1861	1	Routh	4
Bennett	1890	1	Webb	4

NAME	YEAR	TP	COACH	NO OF STUDENTS
Allcock	1880	3	Routh	3
Gunston	1879	4	Routh	3
Cameron	1898	2	Webb	2
Gallop	1883	2	Routh	2
Welsh	1882	1	Routh	2
Young	1884	4	Routh	2

Bibliography

Archival Sources

AHQP	Archive for History of Quantum Physics. Microfilm collection held at various archives including the library of the National Museum of Science and Industry, London.
CC-CL	Cambridgeshire Collection, Cambridgeshire Libraries.
CP-UCL	Carl Pearson Collection, University College London.
EJR-PCC	Edward Routh Papers, Peterhouse College, Cambridge.
JAF-CUL	J. A. Fleming (Add 8083), Cambridge University Library.
JAF-UCL	J. A. Fleming Collection (MS Add 122/34), University College London.
JCM-CUL	James Clerk Maxwell Collection (Mss Add 7655), Cambridge University Library.
JDF-UAL	J. D. Forbes Papers, University of St. Andrews Library.
JL-SJCC	Joseph Larmor Collection, St, John's College, Cambridge.
JTW-SJCC	Joseph Timmis Ward Diary (W2), St. John's College (Library), Cambridge.
LHT-NCSU	L. H. Thomas Collection, North Carolina State University Libraries
OB-KCC	Oscar Browning Collection, King's College, Cambridge.
OJH-IEE	Oliver Heaviside Collection, Institute of Electrical Engineers, London.
OJL-UCL	O. J. Lodge Collection (MS Add 89/65), University College London.
RGO-CUL	Royal Greenwich Observatory, Cambridge University Library.
RLE-CUL	Leslie Ellis Collection, Cambridge University Library.
RLE-TCC	Leslie Ellis Collection (Add Ms c.67.82), Trinity College, Cambridge.
RRW-SJCC	Robert Rumsey Webb Papers, St. John's College (Library), Cambridge.
SP-SJCC	Stephen Parkinson Papers, St. John's College, Cambridge.
WdS-ALO	Willem de Sitter correspondence, Box 31, Archive of Leiden Observatory.
WHP-TCC	William Whewell Papers, Wren Library, Trinity College, Cambridge.
WMH-SJCC	W. M. Hicks Collection, St, John's College, Cambridge.
WNS-CUL	William Napier Shaw Collection (Mss Add 8124), Cambridge University Library.

Books and Articles

Airy, G. B. 1826. *Mathematical Tracts on Physical Astronomy, the Figure of the Earth, Precession and Nutation, and the Calculus of Variations.* 2d ed., 1831. Cambridge: J. Deighton and Sons.

———. 1896. *Autobiography of Sir George Biddell Airy.* Edited by W. Airy. Cambridge: Cambridge University Press.

Alder, K. 1999. "French Engineers Become Professionals; or, How Meritocracy Made Knowledge Objective." In Clark et al., eds., 1999, 94–125.

Anon. 1800. "Some Account of Dr Waring the Late Celebrated Mathematician." *The Monthly Magazine; or, British Register* 9:46–49.

Anon. 1810. *Cambridge Problems.* Cambridge: J. Deighton.

Anon. 1820. *Cambridge Problems.* Cambridge: J. Deighton.

Anon. 1824. *Gradus ad Cantabrigiam.* Published under the pseudonym "Brace of Cantabs." London: John Hearne.

Anon. 1908a. "The Mathematical Threory of Electricity and Magnetism." *Philosophical Magazine* 16:830–31.

Anon. 1908b. "The Mathematical Threory of Electricity and Magnetism." *Electrician* 61:800.

Anon. 1912. "Mathematical Theory of Electricity and Magnetism (2nd edition)." *Electrician* 69:497–98.

Anon. 1919. "The Theory of Electricity." *Philosophical Magazine* 38:199–200.

Appleyard, E. S. 1828. *Letters from Cambridge.* London: Richardson.

Atkinson, S. 1825. "Struggles of a Poor Student Through Cambridge." *The London Magazine,* April, 491–510.

Babbage, C. 1864. *Passages from the Live of a Philosopher.* London: Longmans, Green and Co.

Baker, H. F. 1922. *Principles of Geometry,* vol. 1, *Foundations.* Cambridge: Cambridge University Press.

Baker, H. F., and J. E. Littlewood. 1936–38. "Francis Sowerby Macaulay 1862–1937," *Obituary Notices of Fellows of the Royal Society* 2:357–61.

Barnard, H. 1862. "Military Schools." *American Journal of Education* 12:53–130.

Barnes, B., and S. Shapin, eds. 1979. *Natural Order: Historical Studies of Scientific Culture.* Beverly Hills and London: Sage.

Barrow, I. 1655. *Euclidis Elementorum Libri XV.* Cambridge.

Barrow-Green, J. 1999. "'A Correction to the Spirit of Too Exclusively Pure Mathematics': Robert Smith (1689–1768) and his Prizes at Cambridge University." *Annals of Science* 56:271–316.

Bateman, H. 1909. "The Conformal Transformations of a Space of Four Dimensions and their applications to Geometrical Optics." *Proceedings of the London Mathematical Society* 7:70–89.

———. 1910. "The Transformation of the Electrodynamical Equations." *Proceedings of the London Mathematical Society* 8:223–64.

Becher, H. 1980a. "William Whewell and Cambridge Mathematics." *Historical Studies in Physical Sciences* 11:1–48.

———. 1980b. "Woodhouse, Babbage, Peacock, and Modern Algebra." *Historia Mathematica* 7:389–400.

———. 1984. "The Social Origins and Post-Graduate Careers of a Cambridge Intellectual Elite, 1830–1860." *Victorian Studies* 28:97–127.

———. 1991. "William Whewell's Odyssey: From Mathematics to Moral Philosphy." In Fisch and Schaffer, eds., 1991, 1–29.

———. 1995. "Radicals, Whigs and Conservatives: The Middle and Lower Classes in the Analytical Revolution at Cambridge in the Age of Aristocracy." *British Journal for the History of Science* 28:405–26.

Bechler, Z., ed. 1982. *Contemporary Newtonian Research.* London: D. Reidel.
Bede, C. 1853. *The Adventures of Mr Verdant Green, an Oxford Freshman.* London: James Blackwood.
Bellone, E. 1980. *A World on Paper: Studies in the Second Scientific Revolution.* Translated by Mirella and Riccardo Giacconi (Italian ed., 1976). Cambridge, Mass.: The MIT Press.
Bennett, J. A. 1986. "The Mechanic's Philosophy and the Mechanical Philosophy." *History of Science* 24:1–28.
Bentley, M. 1996. *Politics Without Democracy 1815–1914.* London: Fontana Press.
Berry, A. 1910. "Review of Hardy (1908)." *Mathematical Gazette* 5:303–5.
Bertoloni Meli, D. 1993. *Equivalence and Priority: Newton versus Leibniz.* Oxford: Clarendon Press.
Biot, J. B. 1816. *Traité de physique.* 4 vols. Paris.
Bloor, D. 1976. *Knowledge and Social Imagery.* Chicago: University of Chicago Press.
Boase, F. 1908. *Modern English Biography.* 6 vols. Netherton and Worth: Truro.
Born, M. 1909. "Die Theorie des starren Elektrons in der Kinematik des Relativitäts-Prinzipes." *Annalen der Physik,* 30:1–56.
Bos, H. J. M. 1980. "Mathematics and Rational Mechanics." In Rousseau and Porter, eds., 1980, chap. 8.
Bradley, J. K. 1994. "The History and Development of Aircraft Instruments, 1909–1919." Ph.D. diss., London University.
Bristed, C. 1852. *Five Years in an English University.* 2 vols. New York: Putnam and Co.
Brock, W. H., and M. H. Price. 1980. "Squared Paper in the Nineteenth Century: Instrument of Science and Engineering and Symbol of Reform in Education." *Educational Studies in Mathematics* 11:365–81. Dordrecht: D. Reidel Publishing.
Brockliss, L. 1987. *French Higher Education in the Seventeenth and Eighteenth Centuries: A Cultural History.* Oxford: Clarendon Press.
Brown, L., et al. 1995. *Twentieth Century Physics.* Vol. 1. Bristol: IOP Publishing Ltd.
Bruyne, de, B. 1984. "A Personal View of the Cavendish." In Hendry, ed., 1984, 81–84.
Brush, S. G. 1999. "Why was Relativity Accepted?" *Perspectives in Physics* 1:184–214.
Bucherer, A. H. 1907. "On the New Principle of Relativity in Electromagnetism." *Philosophical Magazine* 13:413–20.
———. 1908. "On the Principle of Relativity and on the Electromagnetic Mass of the Electron: A Reply to Mr. Cunningham." *Philosophical Magazine* 15:316–18.
Buchwald, J. Z. 1985. *From Maxwell to Microphysics: Aspects of Electromagnetic Theory in the Last Quarter of the Nineteenth Century.* Chicago: University of Chicago Press.
———. 1994. *The Creation of Scientific Effects: Heinrich Hertz and Electric Waves.* Chicago: University of Chicago Press.
Buchwald, J. Z., ed. 1995. *Scientific Practice: Theories and Stories of Doing Physics.* Chicago: University of Chicago Press.
Burcham, W. E. 1987. "Some Thoughts on Physics Courses in Cambridge: 1931–37." In Williamson, ed., 1987, chap. 18.
Bury, J. P. T. 1967. *Romilly's Cambridge Diary 1832–42.* Cambridge: Cambridge University Press.
Bury, M. E., and J. D. Pickles. 1994. *Romilly's Cambridge Diary 1842–1847.* Cambridge: Cambridgeshire Records Society.
Cajori, F. 1993. *A History of Mathematical Notations.* 2 vols. in one. New York: Dover Publications. Original 2 vols. published respectively in 1928 and 1929.
Campbell, L., and W. Garnett. 1882. *The Life of James Clerk Maxwell.* London: Macmillan and Co.
Campbell, N. R. 1923. *Modern Electrical Theory: Supplementary Chapter 16, Relativity.* Cambridge: Cambridge University Press.
Campion, W., and W. Walton. 1857. *Solutions of the Problems and Riders Proposed in the Senate-House Examinations for 1857.* Cambridge: Macmillan and Co.

Cannell, D. M. 1993. *George Green: A Biographical Memoir.* London: Athlone Press.
Cantor, G. N., and M. J. S. Hodge, eds. 1981. *Conceptions of Ether.* Cambridge: Cambridge University Press.
Carr, J. 1821. *The First Three Sections of Newton's Principia; with Copious Notes and Illustrations.* London: Baldwin, Cradock and Joy.
Carroll, Lewis (C. L. Dodgson). 1871. *Through the Looking-glass and What Alice Found There.* London: Macmillan and Co.
Cassidy, D. 1986. "Understanding the History of Special Relativity." *Historical Studies in the Physical and Biological Sciences* 16:177–95.
Caswall, E. 1836. *Pluck Examination Papers for Candidates at Oxford and Cambridge.* Oxford. Published under the pseudonym Scriblerus Redevivus.
Cayley, A. 1872. "Solutions of a Smith's Prize Paper for 1871." *Messenger of Mathematics* 1:37–47.
Chandrasekhar, S. 1983. *Eddington: The Most Distinguished Astrophysicist of his Time.* Cambridge: Cambridge University Press.
Chree, C. 1908. "Mathematical Aspects of Electricity and Magnetism." *Nature* 78:537–38.
Chrystal, G. 1882. "Clerk Maxwell's 'Electricity and Magnetism.'" *Nature* 12 (January): 237–40.
Clark, W., J. Golinski, and S. Schaffer, eds. 1999. *The Sciences in Enlightened Europe.* Chicago and London: University of Chicago Press.
Clifford, W. K. 1879. *Lectures and Essays.* Edited by L. Stephen and F. Pollock. 2 vols. London: Macmillan and Co.
Coddington, H. 1823. *An Elementary Treatise on Optics.* Cambridge: J. Deighton and Sons.
Collins, H. M. 1985. *Changing Order: Replication and Induction in Scientific Practice.* London, Beverly Hills, New Delhi: SAGE Publications.
Coombe, J. A. 1841. *Solutions of the Cambridge Problems for the Years 1840, 1841.* Cambridge: John W. Parker.
Cozzens, S., and R. Bud, eds. 1992. *Invisible Connexions.* Bellingham: SPIE.
Crosland, M., and C. Smith. 1976. "The Transmission of Physics from France to Britain: 1800–1840." *Annals of Science* 33:1–61.
Cross, J. J. 1985. "Integral Theorems in Cambridge Mathematical Physics, 1830–55." In Harman, ed., 1985, chap. 5.
Crowther, J. G. 1952. *British Scientists of the Twentieth Century.* London: Routledge and Kegan Paul.
Cunningham, E. 1907a. "On the Electromagnetic Mass of a Moving Electron." *Philosophical Magazine* 14:538–47.
———. 1907b. "The Structure of the Ether." *Nature* 76:222.
———. 1908. "On the Principle of Relativity on the Electromagnetic Mass of the Electron: A Reply to Dr, A, H. Bucherer." *Philosophical Magazine* 16:423–28.
———. 1910. "The Principle of Relativity in Electrodynamics and an Extension Thereof." *Proceedings of the London Mathematical Society* 8:77–98.
———. 1912. "The Application of the Mathematical Theory of Relativity to the Electron Theory of Matter." *Proceedings of the London Mathematical Society* 10:116–27.
———. 1914. *The Principle of Relativity.* Cambridge: Cambridge University Press.
———. 1915. *Relativity and the Electron Theory.* London: Longmans, Green and Co.
———. 1969. "Ebenezer." Unpublished autobiographical typescript dictated in 1969. Copy held in the Library of St. John's College, Cambridge.
Curthoys, M. C., and H. S. Jones. 1995. "Oxford Athleticism, 1850–1914: A Reappraisal." *History of Education* 24:305–17.
Cushing, J. T. 1981. "Electromagnetic Mass, Relativity, and the Kaufmann Experiments." *American Journal of Physics* 49:1133–46.
Damour, T., and G. Schäfer. 1992. "Levi-Civita and the General-Relativistic Problem of Motion." In Eisenstaedt and Kox, eds., 1992, 393–99.

BIBLIOGRAPHY 531

Darrigol, O. 1992. *From c-Numbers to q-Numbers: The Classical Analogy in the History of Quantum Theory*. Berkeley, Los Angeles, and Oxford: University of California Press.

———. 1996. "The Electrodynamic Origins of Relativity Theory." *Historical Studies in the Physical and Biological Sciences* 26:241–312.

Darwin, G. H. 1907–16. *Scientific Papers by G. H. Darwin*. 5 vols. Cambridge: Cambridge University Press.

Dealtry, W. 1810. *The Principles of Fluxions*, Cambridge: J. Deighton.

Dear, P. R., ed. 1991. *The Literary Structure of Scientific Argument: Historical Studies*. Philadelphia: University of Pennsylvania Press.

———. 1995. *Discipline and Experience: The Mathematical Way in the Scientific Revolution*. Chicago: University of Chicago Press.

DeMorgan, A. 1832. "Wood's Algebra." *Quarterly Journal of Education* 3:276–85.

———. 1835a. "Peacock's Algebra." *Quarterly Journal of Education* 9:293–311.

———. 1835b. "Ecole Polytechnique." *Quarterly Journal of Education* 10:330–40.

———. 1872. *Budget of Paradoxes*. London: Longmans, Green and Co.

DeMorgan, S. E. 1882. *Memoir of Augustus De Morgan*. London: Longmans, Green and Co.

Dening, G. 1995. *The Death of William Gooch: A History's Anthropology*. Honolulu: University of Hawai'i Press.

Desmond, A., and J. Moore. 1991. *Darwin*. London: Michael Joseph.

de Sitter, W. 1911. "On the Bearing of the Principle of Relativity on Gravitational Astronomy." *Monthly Notices of the Royal Astronomical Society* 71:388–415.

———. 1916a. "Space, Tme, and Gravitation." *Observatory* 39:412–19.

———. 1916b. "On Einstein's Theory of Relativity and its Astronomical Consequences. First Paper." *Monthly Notices of the Royal Astronomical Society* 76:699–728.

———. 1916c. "On Einstein's Theory of Relativity and its Astronomical Consequences. Second Paper." *Monthly Notices of the Royal Astronomical Society* 77:155–84.

———. 1917. "On Einstein's Theory of Relativity and its Astronomical Consequences. Third Paper." *Monthly Notices of the Royal Astronomical Society* 78:3–28.

Dickens, H. 1934. *The Recollections of Sir Henry Dickens*. London: William Heinemann Ltd.

Dirac, P. A. M. 1930. *The Principles of Quantum Mechanics*. Oxford: Clarendon Press.

———. 1977. "Recollections of an Exciting Era." In Weiner, ed., 1977, 109–46.

Doncel, M. G. 1987. *Symmetries in Physics (1600–1980)*. Barcelona: Servei de Publicacions.

Douglas, A. V. 1956. *The Life of Arthur Stanley Eddington*. Edinburgh: Thomas Nelson and Sons Ltd.

Dreyfus, H. L., and P. Rabinow. 1983. *Michel Foucault: Beyond Structualism and Hermeneutics*. 2d ed. Chicago: University of Chicago Press.

Dykes Spicer, A. 1907. *The Paper Trade*. London: Methuen.

Dyson, F., et al. 1920. "A Determination of the Deflection of Light by the Sun's Gravitational Field, from Observations made of the Total Eclipse May 29, 1919." *Philosophical Transactions of the Royal Society* 220:291–33.

Earman, J., and C. Glymour. 1980a. "Relativity and Eclipses: The British Eclipse Expeditions of 1919 and their Predecessors." *Historical Studies in Physical Sciences* 11:49–85.

———. 1980b. "The Gravitational Red Shift as a Test of General Relativity: History and Analysis." *Studies in the History and Philosophy of Science* 11:175–214.

Earman, J., M. Janssen, and J. Norton, eds., 1993. *The Attraction of Gravitation: New Studies in the History of General Relativity*. Boston, Basel, and Berlin: Birkhäuser.

Eddington, A. S. 1914. *Stellar Movements and the Structure of the Universe*. London: Macmillan and Co.

———. 1915. "Gravitation." *Observatory* 338:93–98.

———. 1916. "Gravity and the Principle of Relativity." *Nature* 98:328–30.

———. 1917a. "Einstein's Theory of Gravitation." *Observatory* 40:93–95.

———. 1917b. "Astronomical Consequences of the Electrical Theory of Matter: Note on Sir Oliver Lodge's Suggestion." *Philosophical Magazine* 34:163–67.

———. 1917c. "Astronomical Consequences of the Electrical Theory of Matter: Note on Sir Oliver Lodge's Suggestion II." *Philosophical Magazine* 34:321–27.

———. 1918a. "Gravitation and the Principle of Relativity." *Nature* 101:15–17, 34–36.

———. 1918b. *Report on the Relativity Theory of Gravitation*. London: Fleetway Press Ltd.

———. 1920. *Space, Time and Gravitation*. Cambridge: Cambridge University Press.

———. 1921a. *Espace, temps et gravitation*. Paris: Hermann.

———. 1921b. "A Generalization of Weyl's Theory of the Electromagnetic and Gravitational Fields." *Proceedings of the Royal Society* 99:104–22.

———. 1922. "The Propagation of Gravitational Waves." *Proceedings of the Royal Society* 102:268–82.

———. 1923. *The Mathematical Theory of Relativity*. Cambridge: Cambridge University Press.

———. 1942–44. "Joseph Larmor 1857–1942." *Obituary Notices of Fellows of the Royal Society* 4:197–207.

Eddington, A. S., and G. L. Clark. 1938. "The Problem of N Bodies in General Relativity Theory." *Proceedings of the Royal Society* 166:465–75.

Edgerton, D. 1999. "From Innovation to Use: Ten (Eclectic) Theses on the History of Technology." *History and Technology* 16:1–26.

Ehrenfest, P. 1909. "Gleichförmige Rotation starrer Körper und Relativitätstheorie." *Physikalishe Zeitschrift* 10:918.

Einstein, A. 1922. *Sidelights on Relativity*. London: Methuen.

———. 1989–. *The Collected Papers of Albert Einstein*. 8– vols. Princeton: Princeton University Press.

Eisenstaedt, J. 1989a. "The Early Interpretation of the Schwazchild Solution." In Howard and Stachel, eds., 1989, 213–33.

———. 1989b. "The Low Water Mark of General Relativity, 1925–1955." In Howard and Stachel, eds., 1989, 277–92.

Eisenstaedt, J., and A. J. Kox, eds. 1992. *Studies in the History of General Relativity*. Boston, Basel, and Berlin: Birkhäuser.

Ellis, L., and M. O'Brian. 1844. *The Senate-House Problems for 1844 with Solutions*. Cambridge: Cambridge University Press.

Ellis, R. L. 1863. *The Mathematical and Other Writings of Robert Leslie Ellis*. Edited by W. Walton. Cambridge: Deighton, Bell, and Co.

Emerson, R. W. 1956. *English Traits*. Edited by H. M. Jones. Cambridge, Mass.: Harvard University Press.

Enros, P. C. 1983. "The Analytical Society (1812–1813): Precursor of the Renewal of Cambridge Mathematics." *Historia Mathematica* 10:24–47.

Erdélyi, E. 1948. "Harry Bateman." *Royal Society Obituary Notices* 5:591–618.

Evans, J. H. 1834. *The First Three Sections of Newton's Pricipia*. Cambridge: Pitt Press.

Evans, E. J. 1996. *The Forging of the Modern State: Early Industrial Britain 1783–1870*. 2d ed. London and New York: Longman.

Faith, N. 1990. *The World the Railways Made*. London: Bodley Head.

Falconer, I. 1987. "Corpuscles, Electrons and Cathode Rays: J. J. Thomson and the 'Discovery of the Electron.'" *British Journal for the History of Science* 20:241–76.

Feingold, M. 1993. "Newton, Leibniz, and Barrow too: An Attempt at a Reinterpretation." *Isis* 84:310–38.

———. 1997. "The Mathematical Sciences and New Philosophies." In Tyacke, ed., 1997, chap. 6.

Ferrers, N. M., and J. Stuart Jackson. 1851. *Solutions of the Cambridge Senate-House Problems for the Years 1848–51.* Cambridge: Macmillan and Co.

Fisch, M., and S. Schaffer, eds. 1991. *William Whewell: A Composite Portrait.* Oxford: Clarendon Press.

FitzGerald, G. F. 1902. *The Scientific Writings of the George Francis FitzGerald.* Edited by Joseph Larmor. London: Longmans, Green and Co.

Fleming, J. A. 1934. *Memoirs of a Scientific Life.* London and Edinburgh: Marshall, Morgan and Scott.

Fokker, A. D. 1915. "A Summary of Einstein and Grossmann's Theory of Gravitation." *Philosophical Magazine* 29:77–96.

Forman, P. 1971. "Weimar Culture, Causality, and Quantum Theory, 1918–1927: Adaptation by German Physicists and Mathematicians to a Hostile Intellectual Environment." *Historical Studies in the Physical Sciences* 3:1–115.

Forsyth, A. R. 1885. *A Treatise on Differential Equations.* London: Macmillan and Co.

———. 1898. "On the Transformations of Coordinates which Lead to New Solutions of Laplace's Equation." *Proceedings of the London Mathematical Society* 29:165–206.

———. 1918. *Solutions of the Examples in A Treatise on Differential Equations.* London: Macmillan and Co.

———. 1929–30. "James Whitbread Lee Glaisher 1848–1928." *Proceedings of the Royal Society* 126:i–xi.

———. 1935. "Old Tripos Days At Cambridge." *Mathematical Gazette* 19:162–79.

Foucault, M. 1977. *Discipline and Punish: The Birth of the Prison.* Translated by A. Sheridan. New York: Penguin Books.

Frend, W. 1787. *Considerations on the Oathes Required at the Time of Taking Degrees.* London: J. Deighton.

Frost, P. 1880. "On the Potential of the Electricity on Two Charged Spherical Conductors Placed at a Given Distance." *Quarterly Journal of Pure and Applied Mathematics* 17:164–68.

Galison, P. 1997. *Image and Logic: A Material Culture of Microphysics.* Chicago: University of Chicago Press.

———. 1988. "History, Philosophy, and the Central Metaphor." *Science in Context* 2:197–212.

———. 2000a. "Einstein's Clocks: The Place of Time." *Critical Inquiry* 20:355–89.

———. 2000b. "The Suppressed Drawing: Paul Dirac's Hidden Geometry." *Representations* 72:145–66.

Galison, P., and A. C. Warwick, eds. 1998. *Cultures of Theory,* special issue of *Studies in the History and Philosophy of Modern Physics,* 29B, no. 3.

Gallop, E. G. 1886. "The Distribution of Electricity on the Circular Disc and Spherical Bowl." *Quarterly Journal of Pure and Applied Mathematics* 21:229–56.

Galton, F. 1869. *Hereditary Genius: An Inquiry into its Laws and Consequences.* London: Macmillan and Co.

———. 1908. *Memories of My Life.* 2d ed. London: Methuen and Co.

Garber, E. 1999. *The Language of Physics: The Calculus and the Development of Theoretical Physics in Europe, 1750–1914.* Boston, Basel, and Berlin: Birkhäuser.

Garnett, W. 1918. "Alice Through the (Convex) Looking Glass." *Mathematical Gazette* 9:237–41, 249–52, 293–98.

Gascoigne, J. 1984. "Mathematics and Meritocracy: The Emergence of the Cambridge Mathematical Tripos." *Social Studies of Science* 14:547–84.

———. 1988. "From Bentley to the Victorians: The Rise and Fall of British Newtonian Natural Theology." *Science in Context* 2:219–56.

———. 1989. *Cambridge in the Age of the Enlightenment: Science, Religion, and Politics from the Restoration to the French Revolution.* Cambridge: Cambridge University Press.

Geison, G. L. 1978. *Michael Foster an the Cambridge School of Physiology.* Princeton: Princeton University Press.

Geison, G.L., and F. L. Holmes, eds. 1993. "Research Schools: Historical Reappraisals." Volume 8 of *Osiris: A Research Journal Devoted to the History of Science and its Cultural Influences.*

Gingras, Y. 2001. "What did Mathematics do for Physics?" *History of Science* 39:383–416.

Glaisher, J. W. L. 1871. "Advertisement." *Messenger of Mathematics* 1:iii–iv.

———, ed. 1879. *Solutions of the Cambridge Senate-House Problems and Riders for the Year 1878.* London: Macmillan and Co.

———. 1886–87. "The Mathematical Tripos." *Proceedings of the London Mathematical Society* 18:4–38.

Glazebrook, R. T. 1896. *James Clerk Maxwell and Modern Physics.* London: Cassell and Company, Ltd.

———. 1926. "The Cavendish Laboratory: 1876–1900." *Nature* 118 (Supplement, 18 December): 52–58.

———. 1928. "H. A. Lorentz." *Nature* 121:287–88.

———. 1931. "Early Days at the Cavendish Laboratory." In Thomson et al., 1931, 130–41.

———. 1932–35. "Sir Horace Lamb 1849–1934." *Obituary Notices of Fellows of the Royal Society* 1:376–92.

Glick, T. F., ed. 1987. *The Comparative Reception of Relativity.* Dordrecht: Reidel.

Godart, O. 1992. "Contributions of Lemaître to General Relativity (1922–1934)." In Eisenstaedt and Kox, eds., 1992, 437–52.

Goldberg, S. 1970. "In Defense of Ether: The British Response to Einstein's Special Theory of Relativity, 1905–1911." *Historical Studies in the Physical Sciences* 2:89–124.

———. 1970–71. "The Abraham Theory of the Electron: The Symbiosis of Experiment and Theory." *Archive for History of Exact Sciences* 7:7–25.

———. 1976. "Max Planck's Philosophy of Nature and his Elaboration of the Special Theory of Relativity." *Historical Studies in the Physical Sciences* 7:125–60.

———. 1984. *Understanding Relativity: Origin and Impact of a Scientific Revolution.* Oxford: Clarendon Press.

Goldman, L., ed. 1989. *The Blind Victorian: Henry Fawcett and British Liberalism.* Cambridge: Cambridge University Press.

Goldstein, J. 1984. "Foucault among the Sociologists: The 'Disciplines' and the History of the Professions." *History and Theory* 23:170–92.

Golinski, J. 1998. *Making Natural Knowledge: Constructivism and the History of Science.* Cambridge: Cambridge University Press.

Gooch, R. 1836. *Facetiae Cantabrigienses.* 3d ed. Published under the pseudonym Socius. London: William Cole.

Gooday, G. 1990. "Precision Measurement and the Genesis of Physics Teaching Laboratories in Victorian Britain." *British Journal for the History of Science* 23:25–51.

Gooding, D., T. Pinch, and S. Schaffer. 1989. *The Uses of Experiment: Studies in the Natural Sciences.* Cambridge: Cambridge University Press.

Goody, J. 1977. *The Domestication of the Savage Mind.* Cambridge: Cambridge University Press.

Gould, P. 1993. "'A Thing Most Inexpedient and Immodest': Women and the Culture of University Physics in Late Nineteenth Century Cambridge." Undergraduate diss., Whipple Museum Library, Cambridge.

Grattan-Guinness, I. 1970. *The Development of the Foundations of Analysis from Euler to Riemann.* Cambridge, Mass.: MIT Press.

———. 1972a. "A Mathematical Union: William Henry Young and Grace Chisholm Young." *Annals of Science* 29:105–86.

———. 1972b. "University Mathematics at the Turn of the Century: Unpublished Recollections of W. H. Young." *Annals of Science* 28:369–84.

———. 1985. "Mathematics and Mathematical Physics from Cambridge, 1815–1840: A Survey of the Achievements and of the French Influences." In Harman, ed., 1985, 84–111.

———. 1990. *Convolutions in French Mathematics, 1800–1840*. 3 vols. Basel and Boston: Birkhauser.

Greenberg, J. 1986. "Mathematical Physics in Eighteenth-Century France." *Isis* 77:59–78.

———. 1995. *Theories of the Shape of the Earth*. Cambridge: Cambridge University Press.

Greenhill, A.G. 1876. *Solutions of the Cambridge Senate-House Problems and Riders for the Year 1875*. London: Macmillan and Co.

Gregory, D. F. 1841. *Examples in the Process of the Differential and Integral Calculus*. Cambridge: J. and J. J. Deighton.

Griffen, N., and A. C. Lewis, 1990. "Betrand's Russell's Mathematical Education." *Notes and Records of the Royal Society* 44:51–71.

Groenewegen, P. 1995. *A Soaring Eagle: Alfred Marshall 1824–1924*. Aldershot: Edward Elgar.

Guicciardini, N. 1989. *The Development of the Newtonian Calculus in Britain 1700–1800*. Cambridge: Cambridge University Press.

———. 1999. *Reading the Principia: The Debate on Newton's Mathematical Methods for Natural Philosophy from 1687 to 1763*. Cambridge: Cambridge University Press.

Gunning, H. 1854. *Reminiscences*. 2 vols. London: George Bell.

Haley, B. 1978. *The Healthy Body and Victorian Culture*. Cambridge, Mass.: Harvard University Press.

Halladay, E. 1990. *Rowing in England: A Social History*. Manchester: Manchester University Press.

Hamilton, H. P. 1826. *Principles of Analytical Geometry*. Cambridge: J. Deighton and Sons.

Hankins, T. L. 1970. *Jean d'Alembert: Science and the Enlightenment*. Oxford: Clarendon Press.

Hanson, H. 1983. *The Coaching Life*. Manchester: Manchester University Press.

Hardy, G. H. 1967. *A Mathematician's Apology*. Cambridge: Cambridge University Press.

Hargreaves, R. 1908. "Integral Forms and their Connection with Physical Equations." *Transactions of the Cambridge Philosophical Society* 21:107–22.

Harman, P. M. 1990–95a. "Introduction." In Maxwell 1990–95, 1:1–32.

———. 1990–95b. "Introduction." In Maxwell 1990–95, 2:1–37.

Harman, P. M., ed. 1985. *Wranglers and Physicists: Studies on Cambridge Physics in the Nineteenth Century*. Manchester: Manchester University Press.

Harrison, H. M. 1994. *Voyager in Time and Space: The life of John Couch Adams, Cambridge Astronomer*. Sussex: The Book Guild.

Harwood, J. 1993. *Styles of Scientific Thought: The German Genetics Community 1900–1933*. Chicago and London: University of Chicago Press.

Hashimoto, T. 1990. "Theory, Experiment, and Design Practice: The Formation of Aeronautical Research, 1909–1930." Ph.D. diss., The Johns Hopkins University.

Hassé, H. R. 1910. "The Equations of Electrodynamics and the Null Influence of the Earth's Motion on Optical and Electrical Phenomena." *Proceedings of the London Mathematical Society* 8:178–94.

———. 1913. "The Equations of the Theory of Electrons Transformed Relative to a System in Accelerated Motion." *Proceeding of the London Mathematical Society* 12:181–206.

———. 1951. "My Fifty Years of Mathematics." *Mathematical Gazette* 35:153–64.

Heilbron, J. L. 1986. *The Dilemmas of an Upright Man: Max Planck as Spokesman for German Science*. Berkeley and Los Angeles: University of California Press.

———. 1993. "A Mathematicians' Mutiny, with Morals." In Horwich, ed., 1993, 81–129.

Hemming, G.W. 1899. *Billiards Mathematically Treated*. London: Macmillan and Co.

Hendry, J., ed. 1984. *Cambridge Physics in the Thirties.* Bristol: Adam Hilger.
Henrici, O. 1873. "Introductory Address." *Nature* 8:492.
———. 1911. "Professor Olaus Henrici." *The Central* (Imperial College, London) 8:67–80.
Hicks, W. M. 1878. "On the Velocity and Electric Potential between Parallel Plates." *Quarterly Journal of Pure and Applied Mathematics* 15:274–315.
Hill, M. J. M. 1879. "The Steady Motion of Electricity in Spherical Current Sheets." *Quarterly Journal of Pure and Applied Mathematics* 16:306–23.
———. 1880. "Solutions by Means of Elliptic Functions to some Problems in the Conduction of Electricity and Heat in Plane Figures." *Quarterly Journal of Pure and Applied Mathematics* 17:284–92.
Hilton, B. 1989. "Manliness, Masculinity and the mid-Victorian Temperament." In Goldman, ed., 1989, 60–70.
Hobson, E. W., and A. E. H. Love, eds. 1913. *Proceedings of the Fifth International Congress of Mathematicians.* 2 vols. Cambridge: Cambridge University Press.
Hodgkin, L. 1981. "Mathematics and Revolution from Lacroix to Cauchy." In Mehrtens, Bos, and Schneider, eds., 1981, 50–71.
Holton, G. 1973. *Thematic Origins of Scientific Thought: Kepler to Einstein.* Cambridge, Mass: Harvard University Press.
Hopkins, W. 1841. *Remarks on Certain Proposed Regulations Respecting the Studies of the University.* Cambridge: J. and J. J. Deighton.
———. 1853. "Presidential Address." *British Association Report,* xli–lvii.
———. 1854. *Remarks on the Mathematical Teaching of the University of Cambridge.* Privately Printed. Copy in Cambridge University Library.
Hopkinson, John. 1901. *Original Papers by the Late John Hopkinson.* Edited by B. Hopkinson. 2 vols. Cambridge: Cambridge University Press.
Horwich, P., ed. 1993. *World Changes: Thomas Kuhn and the Nature of Science.* Cambridge Massachusetts: MIT Press.
Howard, D., and J. Stachel, eds. 1989. *Einstein and the History of General Relativity.* Boston, Basel, and Berlin: Birkhäuser.
Howson, G. 1982. *A History of Mathematics Education in England.* Cambridge: Cambridge University Press.
Hoyle, F. 1994. *Home is Where the Wind Blows: Chapters from a Cosmologist's Life.* Mill Valley, Calif.: University Science Books.
Huber, V. A. 1843. *The English Universities.* 2 vols. An abridged translation edited by F. W. Newman. London: William Pickering.
Hughes, J. 1998. "'Modernists with a Vengance': Changing Cultures of Theory in Nuclear Science, 1920–1930." In Galison and Warwick, eds., 1998, 339–67.
Hunt, B. J. 1986. "Experimenting on the Ether: Oliver J. Lodge and the Great Whirling Machine." *Historical Studies in Physical Sciences* 16:111–34.
———. 1988. "The Origins of the FitzGerald Contraction." *British Journal for the History of Science* 21:67–76.
———. 1991a. *The Maxwellians.* Ithaca and London: Cornell University Press.
———. 1991b. "Rigorous Discipline: Oliver Heaviside Versus the Mathematicians." In Dear, ed., 1991, 72–95.
Hymers. J. 1830. *Analytical Geometry of Three Dimensions.* Cambridge: J. and J. J. Deighton.
Iliffe, R. 1998. "Isaac Newton: Lucatello Professor of Mathematics." In Lawrence and Shapin, eds., 1998, 121–55.
———. 2003. "Butter for Parsnips: Authorship, Audience and the Incomprehensibility of the *Principia.*" In *Scientific Authorship,* ed. M. Biagioli and P. Galison. London: Routledge.

Illy, J. 1981. "Revolutions in a Revolution." *Studies in History and Philosophy of Science* 12:175–210.

———. 1989. "Einstein teaches Lorentz, Lorentz teaches Einstein: Their Collaboration in General Relativity, 1913–1920." *Archive for History of Exact Sciences* 39:247–89.

Jameson, F. J. 1851. *The Principles of the Solution of Senate-House "Riders" (1848–1851)*. Cambridge: Macmillan and Co.

Jeans, J. 1908. *The Mathematical Theory of Electricity and Magnetism* (2d ed., 1911; 3d ed., 1915). Cambridge: Cambridge University Press.

———. 1917. "Einstein's Theory of Gravitation." *Observatory* 40:57–58.

Jeffreys, B. S. 1987. "A Cambridge Research Student in the 1920s." In Williamson, ed., 1987, 32–43.

Johnston, S. 1991. "Mathematical Practitioners and Instruments in Elizabethan England." *Annals of Science* 48:319–44.

Jungnickel, C., and R. McCormmach. 1986. *Intellectual Mastery of Nature: Theoretical Physics from Ohm to Einstein*. 2 vols. Chicago: University of Chicago Press.

Kaiser, D. 1998. "A ψ is just a ψ?: Pedagogy, Practice and the Reconstruction of General Relativity, 1942–1975." In Galison and Warwick, eds., 1998, 321–38.

Kaye, G. W. C., and T. H. Laby. 1936. *Physical and Chemical Constants*. 8th ed. London: Longmans and Co.

Kerszberg, P. 1989. "The Einstein-de Sitter Controversy of 1916–1917 and the Rise of Relativistic Cosmology." In Howard and Stachel, eds., 1989, 325–66.

Kevles, D. J. 1978. *The Physicists: The History of a Scientific Community in Modern America*. Cambridge, Mass: Harvard University Press.

Klein, M. J. 1970. *Paul Ehrenfest: The Making of a Theoretical Physicist*. Amsterdam: North Holland Publishing Company.

Kox, A. J. 1992. "General Relativity in the Netherlands, 1915–1920." In Eisenstaedt and Kox, eds., 1992, 39–56.

Kragh, H. 1987. "The Beginning of the World: Georges Lemaître and the Expanding Universe." *Centaurus* 32:114–39.

———. 1990. *Dirac: A Scientific Biography*. Cambridge: Cambridge University Press.

Kuhn, T. S. 1970. *The Structure of Scientific Revolutions*. 2d ed. (1st ed., 1962). Chicago: University of Chicago Press.

———. 1976. "Mathematical versus Experimental Traditions in the Development of Physical Science." Reprinted in Kuhn 1977, 31–65.

———. 1977. *The Essential Tension: Selected Readings in Scientific Tradition and Change*. Chicago: University of Chicago Press.

Kuhn, T. S., et al. 1967. *Sources for History of Quantum Physics: An Inventory and Report*. Philadelphia: The American Philosophical Society.

Kushner, D. 1993. "Sir George Darwin and a British School of Geophysics." In Geison and Holmes, eds., 1993, 196–223.

Lacroix, S.F. 1816. *Elementary Treatise on the Differential and Integral Calculus*. Translated by George Peacock, John Herschel, and Charles Babbage. Cambridge: J. Deighton and Sons.

Lagrange, J. L. 1788. *Méchanique analytique*. Paris: Mme Ve Courcier.

Lamb, H. 1883. "On Electrical Motions in a Spherical Conductor." *Philosophical Transactions of the Royal Society* 174:519–49.

———. 1884. "On the Induction of Electric Currents in Cylindrical and Spherical Conductors." *Proceedings of the London Mathematical Society* 15:139–49.

———. 1895. *Hydrodynamics*. Cambridge: Cambridge University Press.

———. 1931. "Clerk Maxwell as Lecturer." In Thomson et al., 1931, 142–46.

Laplace, P. S. 1798–1827. *Traité de mécanique céleste*. 5 vols. Paris.

Larmor, J. 1900. *Aether and Matter*. Cambridge: Cambridge University Press.

———. 1910–11. "Edward Routh." *Proceedings of the Royal Society* 84:xii–xvi.
———. "Sir William Davidson Niven." *Proceedings of the London Mathematical Society* 16: xxviii–xliii.
———. 1929. *Mathematical and Physical Papers.* 2 vols. Cambridge: Cambridge University Press.
———. 1936–39. "A. A. Robb." *Obituary Notices of Fellows of the Royal Society* 2:315–21.
Latour, B. 1987. *Science in Action: How to Follow Scientists and Engineers through Society.* Milton Keynes: Open University Press.
Lattis, J. 1994. *Between Copernicus and Galileo: Cristoph Clavius and the Collapse of Ptolemaic Cosmology.* Chicago and London: University of Chicago Press.
Laue, M. von. 1911. *Das Relativitätsprinzip.* Braunschweig: Friedr. Vieweg und Sohn.
Lawrence, C., and S. Shapin, eds. 1998. *Science Incarnate: Historical Embodiments of Natural Knowledge.* Chicago: University of Chicago Press.
Leathem, J. G. 1906. *Examples in the Mathematical Theory of Electricity and Magnetism.* London: Edward Arnold.
Lightman, A. P., et al. 1975. *Problem Book in Relativity and Gravitation.* Princeton: Princeton University Press.
Lindberg, D. C. 1992. *The Beginnings of Western Science.* Chicago and London: University of Chicago Press.
Lindsay, R. B. 1945. "Historical Introduction." In Dover edition of Rayleigh 1877–78, vol. 1.
Littlewood, J. E. 1986. *Littlewood's Miscellany.* Edited by B. Bollobas. Cambridge: Cambridge University Press.
Livens, G. H. 1911. "The Initial Accelerated Motion of a Perfectly Conducting Electrified Sphere." *Philosophical Magazine* 21:640–48.
———. 1918. *The Theory of Electricity.* Cambridge: Cambridge University Press.
Lodge, O. J. 1917a. "Astronomical Consequences of the Electrical Theory of Matter." *Philosophical Magazine* 34:81–94.
———. 1917b. "Astronomical Consequences of the Electrical Theory of Matter. Supplementary Note." *Philosophical Magazine* 34:517–21.
———. 1918. "Continued Discussion of the Astronomical Consequences of the Electrical Theory of Matter." *Philosophical Magazine* 35:141–56.
Lorentz, H. A. 1934–39. *Collected Papers.* 9 vols. The Hague: Martinus Nijhoff.
Love, A. E. H. 1892. *A Treatise on the Mathematical Theory of Elasticity.* Cambridge: Cambridge University Press.
MacAlister, E. F. B. 1935. *Sir Donald MacAlister of Tarbert.* London: Macmillan and Co.
Macdonald, H. M. 1923. "Charles Niven, 1845–1923." *Proceedings of the Royal Society* 104: xxvii–xxviii.
MacIntyre, A. 1981. *After Virtue: A Study in Moral Philosophy.* London: Duckworth; Notre Dame, Ind.: University of Notre Dame Press.
MacLeod, R., ed. 1982. *Days of Judgement: Science, Examination and the Organisation of Knowledge in Late Victorian England.* Driffield: Nafferton.
Magnello, E. 1998. "Pearson, Karl." In *The Encyclopedia of Biostastics,* edited by P. Armitage and T. Colton, 3308–15. Chichester: John Wiley.
Mangan, J. A. 1981. *Athleticism in the Victorian and Edwardian Public School: The Emergence and Consolidation of an Educational Ideology.* Cambridge: Cambridge University Press.
Mangan, J. A., and J. Walvin. 1987. *Manliness and Morality: Middle-class masculinity in Britain and America 1800–1940.* Manchester: Manchester University Press.
Mansfield, N. 1993. "Grads and Snobs: John Brown, Town and Gown in Early Nineteenth-century Cambridge." *History Workshop Journal,* Issue 35, 184–98.
Marcus, G. E., and M. J. Fisher. 1986. *Anthropology as Cultural Critique: An Experimental Moment in the Human Sciences.* Chicago: University of Chicago Press.

Marner, J. 1994. "Coaching Wranglers: E. J. Routh and the Pursuit of Mathematical Physics in Victorian Cambridge." MSc diss., Imperial College, London.
Maxwell, J. C. 1873. *A Treatise on Electricity and Magnetism*. 2 vols. 2d ed. 1881, 3d ed. 1891. Oxford: Clarendon Press.
——. 1881. *An Elementary Treatise on Electricity*. Edited by William Garnett. Oxford: Clarendon Press.
——. 1890. *Scientific Papers of James Clerk Maxwell*. Edited by W. D. Niven. 2 vols. Cambridge: Cambridge University Press.
——. 1990–95. *The Scientific Letters and Papers of James Clerk Maxwell*. Edited by P. M. Harman. 2 vols. Cambridge: Cambridge University Press.
Mayo, C. H. P. 1923. "A Great Schoolmaster." *Mathematical Gazette* 11:325–29.
McCauley, R. N. 1998. "Comparing the Cognitive Foundations of Religion and Science." Report 37, Emory Cognition Project, Emory University (Atlanta, Georgia).
McCormmach, R. 1966–67. "The Atomic Theory of John William Nicholson." *Archive for History of Exact Sciences* 3:160–84.
——. 1969. "Editor's Introduction." *Historical Studies in the Physical Sciences* 1:vii-xi.
——. 1970. "H. A. Lorentz and the Electromagnetic View of Nature." *Isis* 61:459–97.
McCrea, W. 1978. "Ebenezer Cunningham." *Bulletin of the London Mathematica Society* 10:116–26.
——. 1986. "Cambridge Physics 1925–1929." *Interdisciplinary Science Reviews* 11:269–84.
——. 1987. "Cambridge 1923–26: Undergraduate Mathematics." In Williamson, ed., 1987, chap. 5.
McLaren, S.B. 1913. "A Theory of Gravity." *Philosophical Magazine* 26:636–73.
——. 1925. *Scientific Papers*. Cambridge: Cambridge University Press.
McVittie, G. C. 1987. "An Anglo-Scottish University Education." In Williamson, ed., 1987, chap. 6.
Medawar, P. 1996. "Is the Scientific Paper a Fraud?" In *The Strange Case of the Spotted Mice*. Oxford: Oxford University Press.
Mehra, J., and H. Rechenberg. 1982. *The Historical Development of Quantum Theory*. Vol. 4. New York, Heidelberg, and Berlin: Springer-Verlag.
Mehrtens, H., H. Bos, and I. Schneider, eds. 1981. *Social History of Nineteenth Century Mathematics*. Boston, Basel, and Stuttgart: Birkhäuser.
Mercier, A. 1992. "General Relativity at the Turning Point of Its Renewal." In Eisenstaedt and Kox, eds., 1992, 109–21.
Miller A. I. 1981. *Albert Einstein's Special Theory of Relativity: Emergence (1905) and Early Interpretation (1905–1911)*. London: Addison Wesley.
Miller, D. P. "Between Hostile Camps: Sir Humphrey Davy's Presidency of the Royal Society of London, 1820–27." *British Journal for the History of Science* 16:1–47.
Miller, E. 1961. *Portrait of a College: A History of the College of St. John the Evangelist Cambridge*. Cambridge: Cambridge University Press.
Miller, W. H. 1831. *The Elements of Hydrostatics and Hydrodynamics*. Cambridge: J. and J. J. Deighton.
Milne, E. A. 1945–48. "James Jeans." *Obituary Notices of Fellows of the Royal Society* 5:573–89.
——. 1952. *Sir James Jeans: A Biography*. Cambridge: Cambridge University Press.
Minkowski, H. 1908. "Die Grundleichungen für die elektromagnetischen Vorgänge in bewegten Körpern." *Göttinger Nachrichten*, 53–111.
——. 1909. "Raum und Zeit." *Physikalische Zeitschrift* 20:104–11.
——. 1915. "Das Relativitätsprinzip." *Annalen der Physik* 47:927–38.
Missner, M. 1985. "Why Einstein became Famous in America." *Social Studies of Science* 15:267–91.
Moody, T. W. 1959. *Queen's, Belfast, 1845–1949: The History of a University*. 2 vols. Belfast: Queen's University Belfast.

Moggridge, D. E. 1992. *Maynard Keynes: An Economist's Biography.* London: Routledge
Moorehead, C. 1992. *Bertrand Russell: A Life.* London: Sinclair Stevenson.
Morgan, H. A. 1858. *A Collection of Problems and Examples in Mathematics Selected by H. A. Morgan.* Cambridge: Macmillan and Co.
———. 1871. *The Mathematical Tripos: An Enquiry into its Influence on a Liberal Education.* London, Oxford, and Cambridge: Rivingtons.
———. 1898. "Rev. Dr. Frost, FRS." *Cambridge Review* 19:405.
Mott, N. 1984. "Theory and Experiment at the Cavendish circa 1932." In Hendry, ed., 1984, 125–32.
Moulton, H. 1923. *Life of Lord Moulton.* London: Nisbet and Co.
Nahin, P. J. 1988. *Oliver Heaviside: Sage in Solitude.* New York: IEEE Press.
Neville, E. H. 1928. "Robert Alfred Herman," *The Cambridge Review*, 10 February, 237–39.
Newall, H. F. 1910. "1885–1894." In Thomson et al., 1910, 102–58.
Nicholson, J. W. 1910a. "The Accelerated Motion of an Electrified Sphere." *Philosophical Magazine* 20: 610–18.
———. 1910b. "The Accelerated Motion of a Dielectric Sphere." *Philosophical Magazine* 20:828–35.
———. 1912. "On Uniform Rotation, the Principle of Relativity, and the Michelson-Morley Experiment." *Philosophical Magazine* 24:820–27.
———. 1915. "Electromagnetic Inertia and Atomic Weight." *Proceedings of the Physical Society* 27:217–29.
———. 1917. "S. B. Maclaren." *Proceedings of the London Mathematical Society* 16:xxxiii–xxxvii.
Nind, W. 1854. *Sonnets of Cambridge Life.* Cambridge: Macmillan.
Niven, C. 1881. "On the Induction of Electric Currents in Infinite Plates and Spherical Shells." *Philosophical Transactions of the Royal Society* 172:307–53.
Niven, W. D. 1877. "On the Theory of Electric Images, and its Application to the Case of Two Charged Spherical Conductors." *Proceedings of the London Mathematical Society* 8:64–83.
———. 1880–81a. "On the Electrical Capacity of a Conductor Bounded by Two Spherical Surfaces Cutting at any Angle." *Proceedings of the London Mathematical Society* 12:27–36.
———. 1880–81b. "Applications of Lamé's Coordinates to Determine the Distributions of Electricity on a Conductor in the Form of an Ellipsoid Placed in a Field of Electric Force." *Messenger of Mathematics* 10:119–21.
———. 1890. "Preface." In Maxwell 1890, 1:ix–xxix.
Noakes, R. 1992. "Poynting's Use of Acoustics: The Manufacture of Electromagnetic Energy Flow." Undergraduate diss., Department of History and Philosophy of Science, Cambridge University.
Norton J. 1989a. "What Was Einstein's Principle of Equivalence?" In Howard and Stachel, eds., 1989, 5–47.
———. 1989b. "How Einstein Found his Field Equations, 1912–1915." In Howard and Stachel, eds., 1989, 101–59.
———. 1993. "General Covariance and the Foundations of General Relativity: Eight Decades of Dispute." *Reports on Progress in Physics* 56:791–858.
O'Brian, M. 1844. *A Treatise on Plane Co-ordinate Geometry,* Part I. Cambridge: Deighton.
Olesko, K. M. 1991. *Physics as a Calling: Discipline and Practice in the Koenigsberg Seminar for Physics.* Ithaca: Cornell University Press.
———. 1993. "Tacit Knowledge and School Formation." In Geison and Holmes, eds. 1993, 16–29.
Owens, L. 1985. "Pure and Sound Government: Laboratories, Playing Fields, and Gymnasia in the Nineteenth-Century Search for Order." *ISIS* 76:182–94.
Pais, A. 1982. *'Subtle is the Lord . . .': The Science and the Life of Albert Einstein.* Oxford: Oxford University Press.

Panteki, M. 1987. "William Wallace and the Introduction of Continental Calculus to Britain: A Letter to George Peacock." *Historia Mathematica* 14:119–32.

Parry, J. 1993. *The Rise and Fall of Liberal Government in Victorian Britain.* New Haven and London: Yale University Press.

Peacock, G. 1820. *Examples Illustrative of the Use of the Differential and Integral Calculus.* Cambridge: J. Deighton and Sons.

———. 1841. *Observations on the Statutes of the University of Cambridge.* London and Cambridge: J. and J. J. Deighton.

Pearson, E. S. 1938. *Karl Pearson: An Appreciation of Some Aspects of his Life and Work.* Cambridge: Cambridge University Press.

Pearson, K. 1906–7. "Walter Frank Raphael Weldon, 1860–1906." *Biometrika* 5:1–52.

———. 1914. *The Life, Letters and Labours of Francis Galton.* Cambridge: Cambridge University Press.

———. 1936. "Old Tripos Days at Cambridge, As Seen from Another Viewpoint." *Mathematical Gazette* 20:27–36.

Pickstone, J. V. 1993. "Ways of Knowing: Towards a Sociology of Science, Technology and Medicine." *British Journal for the History of Science* 26:433–58.

Plummer, H. C. 1910. "On the Theory of Aberration and the Principle of Relativity." *Monthly Notices of the Royal Astronomical Society* 70:252–66.

———. 1945. "Arthur Stanley Eddington 1882–1944." *Obituary Notices of Fellows of the Royal Society* 5:113–25.

Poisson, S. D. 1811. *Traité de méchanique.* Paris.

Polanyi, M. 1958. *Personal Knowledge: Towards a Post-Critical Philosophy.* Chicago: University of Chicago Press.

Poynting, J. H. 1920. *Collected Scientific Papers.* Cambridge: Cambridge University Press.

Prichard, A. 1886. *Annals of Our School Life.* Oxford: Printed for Private Distribution by Horace Hart.

———. 1897. *Charles Prichard: Memoirs of his Life.* London: Seeley and Company.

Pryme G. 1870. *Autobiographical Recollections.* Edited by A. Bayne. Cambridge: Deighton, Bell and Co.

Pupin, M. 1923. *From Immigrant to Inventor.* New York: Charles Scribner's Sons.

Pyenson, L. 1985. *The Young Einstein: The Advent of Relativity.* Bristol and Boston: Adam Hilger.

Rabinow, P., ed. 1984. *The Foucault Reader.* New York: Pantheon Books.

Ravetz, J. R. 1996. *Scientific Knowledge and its Social Problems.* Reissue with new introduction by the author. New Brunswick and London: Transaction Publishers.

Rayleigh, Lord. 1877–78. *The Theory of Sound.* 2 vols. London: Macmillan. Second, enlarged, editions of vols. 1 and 2 were published in 1894 and 1896 respectively. A Dover edition with an historical introduction by Robert Bruce Lindsay was published in 1945.

———. 1899–1920. *Scientific Papers.* 6 vols. Cambridge: Cambridge University Press.

———. 1942. *The Life of Sir J. J. Thomson.* Cambridge: Cambridge University Press.

Richards, J. L. 1979. "The Reception of a Mathematical Theory: Non-Euclidean Geometry in England, 1868-1883." In Barnes and Shapin, eds., 1979, 143–66.

———. 1986. "Projective Geometry and Mathematical Progress in Mid-Victorian Britain." *Studies in History and Philosophy of Science* 17:297–325.

———. 1991. "Rigor and Clarity: Foundations of Mathematics in France and England, 1800–1840." *Science in Context* 4:297–319.

Roget, J. L. 1851. *Cambridge Customs and Costumes.* Cambridge: Macmillan.

Rothblatt, S. 1968. *The Revolution of the Dons: Cambridge Society in Victorian England.* London: Faber and Faber.

———. 1974. The Student Sub-culture and the Examination System in Early 19th Century Oxbridge." In Stone, ed., 1974, 247–303.

———. 1982. "Failure in Early Nineteenth-Century Oxford and Cambridge." *History of Education* 11:1–21.

Rouse, J. 1987. *Knowledge and Power: Toward a Political Philosophy of Science*. Ithaca and London: Cornell University Press.

Rouse Ball, W. W. 1889. *A History of the Study of Mathemtics at Cambridge*. Cambridge: Cambridge University Press.

———. 1899. *Notes on the History of Trinity College Cambridge*. London: Macmillan.

———. 1907. "Edward John Routh." *The Cambridge Review*, 13 June, 480–81.

———. 1908. *A History of the First Trinity Boat Club*. Cambridge: Bowes and Bowes.

———. 1912. "The Cambridge School of Mathematics." *Mathematical Gazette* 6:311–23.

Rousseau, G. S., and R. Porter, eds. *The Ferment of Knowledge: Studies in the Historiography of Eighteenth Century Science*. Cambridge: Cambridge University Press.

Routh, E. J. 1868. *An Elementary Treatise on the Dynamics of a System of Rigid Bodies*. 2d ed. London and Cambridge: Macmillan and Co.

———. 1877. *A Treatise on the Stability of a Given State of Motion, Particularly Steady Motion*. London: Macmillan and Co.

———. 1889. "Obituary Notice of Thomas Gaskin." *Proceedings of the Royal Society* 46:i–iii.

———. 1891. *A Treatise on Analytical Statics*. Cambridge: Cambridge University Press.

———. 1892. *The Advanced Part of a Treatise on the Dynamics of a System of Rigid Bodies*. London: Macmillan and Co.

———. 1898a. *A Treatise on Dynamics of a Particle*. 2 vols. Cambridge: Cambridge University Press.

———. 1898b. *Die Dynamik der Systemme starrer Körper*. Translated by F. Klein. 2 vols. Leipzig: A. Schepp.

Routh, E. J., and H. W. Watson. 1860. *Cambridge Senate-House Problems and Riders*. Cambridge: Macmillan and Co.

Rowe, D. 1989. "Klein, Hilbert, and the Göttingen Mathematical Tradition." *Osiris* 5:186–213.

Runge, C. 1913. "The Mathematical Training of the Physicist in the University." In Hobson and Love, eds., 1913, 2:598–607.

Sanchez-Ron, J. M. 1987a. "The Reception of Special Relativity in Great Britain." In Glick, ed., 1987, 27–58.

———. 1987b. "The Role Played by Symmetries in the Introduction of Relativity in Great Britain." In Doncel, ed., 1987, 165–84.

———. 1992. "The Reception of General Relativity Among British Physicists and Mathematicians (1915–1930)." In Eisenstaedt and Kox, eds., 1992, 57–88.

Satthianadhan, S. 1890. *Four Years in an English University*. Madras: Srinivasa.

Schaffer, S. J. 1992. "A Manufactory of Ohms: Late Victorian Metrology and its Instrumentation." In Cozzens and Bud, eds., 23–56.

———. 1995. "Accurate Measurement is a English Science." In Wise, ed., 1995, chap. 6.

Schott, G.A. 1907. "On the Radiation from Moving Systems of Electrons, and on the Spectrum of Canal Rays." *Philosophical Magazine* 13:657–87.

Schuster, A. 1910. "The Clerk Maxwell Period." In Thomson et al., 1910, 14–39.

Searby, P. 1997. *A History of the University of Cambridge*, vol. 3, *1750–1870*. Cambridge: Cambridge University Press.

Seeley, R. 1868a. "A Plea for More Universities." *The London Student*, April, 9.

———. 1868b. "Recreation." *The London Student*, July, 205–8.

Servos, J. W. 1986. "Mathematics and the Physical Sciences in America, 1880–1930." *Isis* 77:611–29.

Shapin, S. 1994. "'The Mind Is Its Own Place': Science and Solitude in Seventeenth Century England." *Science in Context* 4:191–218.

———. 1994. *A Social History of Truth: Civility and Science in Seventeenth Century England.* Chicago and London: University of Chicago Press.

Sherman, W. H. 1995. *John Dee: The Politics of Reading and Writing in the English Renaissance.* Amherst: University of Massachusetts Press.

Siegel, D. M. 1985. "Mechanical Image and Reality in Maxwell's Electromagnetic Theory." In Harman, ed., 1985, chap. 7.

———. 1991. *Innovation in Maxwell's Electromagnetic Theory.* Cambridge: Cambridge University Press.

Simpson, T. K. 1966. "Maxwell and the Direct Experimental Test of His Electromagnetic Theory." *Isis* 57:411–32.

Slesser, G. M. 1858."On the Motion of a Body Referred to Moving Axes." *Quarterly Journal of Pure and Applied Mathematics* 2:341–53.

Small, H. 1981. *Physics Citation Index 1920–1921.* Philadelphia: Institute for Scientific Information.

Smedley, F. E. 1850. *Frank Fairlegh; or, Scenes from the Life of a Private Pupil.* London: Blackfriars Publishing Co.

Smith, S. 1810. "Remarks of the System of Education in Public Schools." *Edinburgh Review* 16:326–34.

Smith, C., and M. Norton Wise. 1989. *Energy and Empire: A Biographical Study of Lord Kelvin.* Cambridge University Press.

Snobelen, S. 2000. "William Whiston: Natural Philosopher, Prophet, Primitive Christian." Ph.D. diss., Cambridge University.

Soddy, F. 1907. "Is the Electronic Theory of Matter Legitimate?" *Nature* 76:25–26.

Spencer Jones, H. 1951. "George Henry Livens." *Monthly Notices of the Royal Astronomical Society* 3:159–60.

Stachel, J. 1986. "Eddington and Einstein" In Ullmann-Margalit, ed., 1986, 225–50.

———. "Einstein's Search for General Covarience, 1912–1915." In Howard and Stachel, eds., 63–100.

———. 1995. "History of Relativity." In Brown et al., eds., 1995, 249–356.

Staley, R., ed. 1994. *The Physics of Empire: Public Lectures.* Cambridge: Whipple Museum of the History of Science.

Stephen, L. 1885. *Life of Henry Fawcett.* London: Smith, Elder and Co.

Stewart, L. 1992. *The Rise of Public Science: Rhetoric, Technology and Natural Philosophy in Newtonian Britain, 1660–1750.* Cambridge: Cambridge University Press.

Stewart, L., and P. Weindling. 1995. "Philosophical Threads: Natural Philosophy and Public Experiment among the Weavers of Spitalfields." *British Journal for the History of Science* 28:37–62.

Stokes, G. G. 1907. *Memoirs and Scientific Correspondence of Sir George Stokes.* Vol. 1. Cambridge: Cambridge University Press.

Stone, L., ed. 1974. *The University in Society.* Vol 1. Princeton University Press.

Strutt, R .J. 1924. *The Life of John William Strutt, Third Baron Rayleigh.* London: Edward Arnold and Co.

Stuart, J. 1912. *Reminiscences.* London: Cassell and Company.

Sutherland, J. 1995. *Victorian Fiction: Writers, Publishers, Readers.* London: Macmillan.

Sviedrys, R. 1970. "The Rise of Physical Science in Victorian Cambridge." *Historical Studies in Physical Sciences* 2:127–51.

———. 1976. "The Rise of Physics Laboratories in Britain." *Historical Studies in Physical Sciences* 7:405–36.

Swenson, L. S. 1972. *The Ethereal Ether: A History of the Michelson-Morley Aether-Drift Experiment, 1880–1930.* Austin and London: University of Texas Press.

Tait, P. G. 1872. "Artificial Selection." *Macmillan's Magazine* 25:416–21.

Tanner, J. R. 1917. *The Historical Register of the University of Cambridge.* Cambridge: Cambridge University Press.
Temple, G. 1956. "Edmund Taylor Whittaker." *Biographical Memoirs of Fellows of the Royal Society* 2:299–325.
Terrall, M. 1999. "Metaphysics, Mathematics, and the Gendering of Science in Eighteen-Century France." In Clark et al., eds., 1999, 246–71.
Thackeray, W. M. 1967. *Vanity Fair.* First published 1847. Edited with an introduction by Gilbert Phelps. London: Pan Books.
———. 1994. *The History of Pendennis.* First two-volume edition 1849–50. Edited with an Introduction by J. Sutherland. Oxford: Oxford University Press.
Thompson, S. P. 1910. *The Life of William Thomson.* 2 vols. London: Macmillan and Company.
Thomson, J. J. 1880. "On Maxwell's Theory of Light." *Philosophical Magazine* 9:284–91.
———. 1881a. "On the Electric and Magnetic Effects Produced by the Motion of Electrified Bodies." *Philosophical Magazine* 11:229–49.
———. 1881b. "On some Electromagnetic Experiments with Open Circuits." *Philosophical Magazine* 12:49–60.
———. 1884a. "On the Determination of the Number of Electrostatic Units in the Electromagnetic Unit of Electricity." *Philosophical Transactions of the Royal Society* 174:707–21.
———. 1884b. "On Electrical Oscillations and the Effects Produced by the Motion of an Electrified Sphere." *Proceedings of the London Mathematical Society* 15:197–218.
———. 1885. "Note on the Rotation of the Plane of Polarization of Light by a Moving Medium." *Proceedings of the Cambridge Philosophical Society* 77:250–54.
———. 1886. "Report on Electrical Theories." *British Association Report* (Aberdeen, 1885): 97–155. London: John Murray.
———. 1888. *Application of Dynamics to Physics and Chemistry.* London: Macmillan.
———. 1893. *Notes on Recent Researches in Electricity and Magnetism.* Oxford: Clarendon Press.
———. 1912. "The Functions of Lectures and Text-books in Science Teaching." *Nature* 88: 399–400.
———. 1936. *Recollections and Reflections.* London: G. Bell and Sons Ltd.
Thomson, J. J., et al. 1910. *A History of the Cavendish Laboratory 1871–1910.* London, Longmans, Green and Co.
———. 1931. *James Clerk Maxwell: A Commemoration Volume 1831–1931.* Cambridge: Cambridge University Press.
Thomson, W., and P. G. Tait. 1867. *Treatise on Natural Philosophy.* Cambridge: Cambridge University Press.
Todhunter, I. 1852. *A Treatise on Differential Calculus.* London: Macmillan.
———. 1872. "Audi Aliam Partem." *Macmillan's Magazine* 26:60–69.
———. 1873. *The Conflict of Studies and Other Essays on Subjects Connected with Education.* London: Macmillan and Co.
Topper, D. R. 1971. "Commitment to Mechanism: J. J. Thomson, the Early Years." *Archive for History of Exact Sciences* 7:393–410.
Traweek, S. 1988. *Beamtimes and Lifetimes: The World of High Energy Physicists.* Cambridge, Mass.: Harvard University Press.
Trusdell, C. 1984. *An Idiot's Fugitive Essays on Science.* New York, Berlin, and Heidelberg: Springer-Verlag.
Turner, A. J. 1973. "Mathematical Instruments in the Education of Gentlemen." *Annals of Science* 30:51–88.
Turner, H. H. 1907–8. "Edward John Routh." *Monthly Notices of the Royal Astronomical Society* 68:239–41.
Tyacke, N., ed. 1997. *The History of the University of Oxford.* Volume 4. Oxford: Clarendon Press.

Ullmann-Margalit, E., ed. 1986. *The Prism of Science.* Dordrecht, Reidel.
Van der Waerden, B. L. 1954. *Science Awakening.* English translation by A. Dresden. Groningen, Holland: P. Noordhoff Ltd.
Venn, J. A. 1913. *Early Collegiate Life.* Cambridge: W. Heffer and Sons Ltd.
———. 1940–54. *Alumni Cantabrigienses,* Part II (1752–1900). 6 vols. Cambridge: Cambridge University Press.
Vince, S. 1797, 1799, 1808. *A Complete System of Astronomy.* 3 vols. Cambridge: Cambridge University Press.
Vint, J. 1956. "Henry Roland Hassé." *Journal of the London Mathematical Society* 31:252–55.
Wali, K. C. 1991. *Chandra: A Biography of S.Chandrasekhar.* Chicago and London: University of Chicago Press.
Walker, G. W. 1910. "The Initial Accelerated Motion of Electrical Systems of Finite Extent, and the Reaction Produced by the Resulting Radiation." *Philosophical Transactions Royal Society* A 210:145–97.
Walton, W. 1842. *A Collection of Problems in Illustration of the Principles of Theoretical Mechanics.* Cambridge: W. P. Grant.
———. 1863. *The Mathematical and Other Writings of Robert Leslie Ellis.* Cambridge: Deighton, Bell, and Co.
Walton, W., and W. Mackenzie. 1854. *Solutions of the Problems and Riders Proposed in the Senate-House Examinations for 1854.* Cambridge: Macmillan and Co.
Walton, W., et al. 1864. *Solutions of Problems and Riders Proposed in the Senate-House Examination for 1864.* Cambridge: Macmillan and Co.
Warwick, A. C. 1989. "The Electrodynamics of Moving Bodies and the Principle of Relativity in British Physics 1894–1919." Ph.D. diss., Cambridge University.
———. 1991. "On the Role of the FitzGerald-Lorentz Contraction Hypothesis in the Development of Joseph Larmor's Electronic Theory of Matter." *Archive for History of Exact Sciences* 43:29–91.
———. 1992. "Cambridge Mathematics and Cavendish Physics: Cunningham, Campbell and Einstein's Relativity 1905–1911," Part I: "The Uses of Theory." *Studies in the History and Philosophy of Science* 23:625–56.
———. 1993. "Cambridge Mathematics and Cavendish Physics: Cunningham, Campbell and Einstein's Relativity 1905–1911," Part II: "Comparing Traditions in Cambridge Physics." *Studies in the History and Philosophy of Science* 24:1–25.
———. 1994. "The Worlds of Cambridge Physics." In Staley, ed., 1994, 57–86.
———. 1995a. "The Sturdy Protestants of Science: Larmor, Trouton and the Earth's Motion Through the Ether." In Buchwald, ed., 1995, 300–43.
———. 1995b. "The Laboratory of Theory, or What's Exact about the Exact Sciences?" In Wise, ed., 1995, chap. 12.
Watson, F. 1909. *The Beginning of the Teaching of Modern Subjects in England.* London: Pitman and Sons Ltd.
Weinberg, S. 2000. "Could We Live without Quarks?" *Times Literary Supplement,* 18 February, 8.
Weiner, C., ed. 1977. *History of Twentieth Century Science.* New York and London: Academic Press.
Westfall, R. S. 1980. "Newton's Marvellous Years of Discovery and Their Aftermath: Myth versus Manuscript." *Isis* 71:109–21.
Wheaton, B. R. 1983. *The Tiger and the Shark.* Cambridge: Cambridge University Press.
Whewell, W. 1819. *An Elementary Treatise on Mechanics.* Cambridge: J. Deighton.
———. 1823. *A Treatise on Dynamics.* Cambridge: J. Deighton and Sons.
———. 1835. *Thoughts on the Study of Mathematics as Part of a Liberal Education.* Cambridge: Cambridge University Press.
———. 1837. *On the Principles of English University Education.* Cambridge: J. and J. J. Deighton.

———. 1845. *Of a Liberal Education in General; and with Particular Reference to the Leading Studies of the University of Cambridge*. Vol. 1. London: John W. Parker.

———. 1850. *Of a Liberal Education in General; and with Particular Reference to the Leading Studies of the University of Cambridge*. Vol. 2. London: John W. Parker.

Whipple, F. J. W. 1899. "The Stability of the Motion of a Bicycle." *Quarterly Journal of Pure and Applied Mathematics* 30:312–48.

White, P. 1991. *The Idea Factory: Learning to Think at MIT*. Dutton: Penguin, USA.

Whiteside, D. T. 1982. "Newton the Mathematician." In Bechler, ed., 1982, 102–27.

Whittaker, E. T. 1902. *A Course of Modern Analysis*. Cambridge: Cambridge University Press.

Whittaker, E.T. 1910a. *A History of Theories of the Aether and Electricity*. London: Longmans, Green and Co.

———. 1910b. "Recent Researches on Space, Time and Force." *Monthly Notices of the Royal Astronomical Society* 70:363–66.

———. 1942–44. "Andrew Russell Forsyth 1858–1942." *Obituary Notices of Fellows of the Royal Society* 4:209–27.

———. 1951–53. *A History of the Theories of the Aether and Electricity*. 2 vols. London: Thomas Nelson.

Wiener, N. 1953. *Ex-Prodigy: My Childhood and Youth*. New York: Simon and Schuster.

———. 1956. *I Am a Mathematician: The Later Life of a Prodigy*. Cambridge, Mass. : The MIT Press.

Will, C. M. 1988. *Was Einstein Right? Putting General Relativity to the Test*. Oxford: Oxford University Press.

Williams, R. 1958. *Culture and Society, 1780–1950*. London, Chatto and Windus.

Williams, P. 1991. "Passing on the Torch: Whewell's Philosophy and the Principles of English Education." In Fisch and Shaffer, eds., 1991, 117–47.

Williamson, R., ed. 1987. *The Making of Physicists*. Bristol: Adam Hilger.

Wilson, A. T., and J. S. Wilson. 1932. *James M. Wilson: An Autobiography, 1836–1931*. London: Sidgwick and Jackson.

Wilson, D. 1983. *Rutherford: Simple Genius*. London: Hodder and Stoughton.

Wilson, D. B. 1982. "Experimentalists among the Mathematicians: Physics in the Cambridge Natural Sciences Tripos, 1851–1900." *Historical Studies in Physical Sciences* 12:325–71.

———. 1985. "The Educational Matrix: Physics Education at Early-Victorian, Cambridge, Edinburgh and Glasgow Universities." In Harman, ed. 1985, 12–48.

———. 1987. *Kelvin and Stokes: A Comparative Study in Victorian Physics*. Bristol: Adam Hilger.

Wilson, W. 1956. "J. W. Nicholson." *Biographical Memoirs of Fellows of the Royal Society* 2:209–14.

Winstanley, D. A. 1935. *Unreformed Cambridge*. Cambridge: Cambridge University Press.

Winter, A. 1998. "A Calculus of Suffering: Ada Lovelace and the Bodily Constraints on Women's Knowledge in Early Victorian England." In Lawrence and Shapin, eds., 1998, 202–39.

Wise, M. N., ed. 1995. *The Values of Precision*. Princeton: Princeton University Press.

Wolstenholme, J. 1867. *Mathematical Problems*. London: Macmillan and Co.

———. 1878. *Mathematical Problems*. 2d, enlarged ed. London: Macmillan and Co.

Wood, J. 1790–99. *Principles of Mathematics and Natural Philosophy*. 4 vols. Cambridge: J. Deighton and Co.

Woodhouse, R. 1803. *The Principles of Analytical Calculation*. Cambridge: J. Deighton.

———. 1818. *An Elementary Treatise on Astronomy*. Vol. 2, *Containing Physical Astronomy*. Cambridge: J. Smith.

———. 1821, 1823. *A Treatise on Astronomy Theoretical and Practical*. 2 vols. Cambridge: J. Deighton and Sons and G. W. and B. Whittaker.

Wordsworth, C. 1874. *Social Life at the English Universities in the Eighteenth Century*. Cambridge: Deighton, Bell and Co.

———. 1877. *Scholae Academicae: Some Account of the Studies at the English Universities in the Eighteenth Century.* Cambridge: Cambridge University Press.
Wright, J. M. F. 1825. *Solutions of the Cambridge Problems for 1800–1820.* 2 vols. London: Black and Armstrong.
———. 1827. *Alma Mater; or Seven Years at the University of Cambridge.* 2 vols. London: Black, Young, and Young.
———. 1828. *A Commentary on Newton's Principia: With a Supplementary Volume.* London: Black, Young and Young.
———. 1829a. *Self-Instructions in Pure Arithmetic.* Cambridge: W. P. Grant and E. J. Armstrong.
———. 1829b. *Self-Examination in Euclid.* Cambridge: W. P. Grant.
———. 1830. *The Principia of Newton: With Notes, Examples, and Deductions; Containing all the is Read at the University of Cambridge.* Cambridge: W. P. Grant; London: Whittaker and Co.
———. 1830–31. *The Private Tutor and Cambridge Mathematical Repository.* Cambridge: W. P. Grant.
———. 1831. *Hints and Answers: Being a Key to a Collection of Cambridge Mathematical Examination Papers, as Proposed at the Several Colleges.* Cambridge: W. P. Grant.
Wynne, B. 1982. "Natural Knowledge and Social Context: Cambridge Physicists and the Luminiferous Ether." In *Science in Context,* ed. B. Barnes and D. Edge, 212–31. Milton Keynes: Open University Press.
Yavetz, I. 1995. *From Obscurity to Enigma: The Work of Oliver Heaviside, 1872–1889.* Basel: Birkhäuser.
Yeo, R. 1988. "Genius, Method, and Morality: Images of Newton in Britain, 1760–1860." *Science in Context* 2:257–84.

Index

Abbott, R. C., 518
Abraham, Max, 375, 384, 411–12
Acts (disputations), 53–54; characteristics of, 122–23; depiction of, *123*; discontinuation of, 131; preparation for, 59–60, 130, 134; and research, 138; Senate House examination contrasted with, 129; Senate House examination eclipsing, 55, 64, 131, 201; Whewell on, 128
Adair, J. F., 515
Adams, John Couch, 3, 86n.106, 136–37, 275, 275nn. 88, 89
Adams, William Grylls, 7n.13, 23
Adamson, H. A., 518
Adie, W. S., 519
advanced lecture courses, 271, 284
aether. *See* ether
Aether and Matter (Larmor): Cunningham and, 410, 425–26; electronic theory of matter in, 3, 376, 390; and Larmor's electrodynamics in Tripos questions, 377; and Lorentz transformation, 372, 374, 411, 413; transformation technique in, 392, 401, 415, 428
Airey, J. A. L., 257
Airy, George: and Adams's research on Uranus, 275n.89; and Atkinson, 76; as Lucasian professor, 75, 80n.94, 99n.141; *Mathematical Tracts*, 74, *75*, 77, 97n.134, 136; and Peacock, 73–74, 75, 78, 99; physics training at Cambridge, 3; on Rayleigh's *Theory of Sound*, 324; on reforming Cambridge mathematics teaching, 264; Routh marries daughter of, 233n.8; scribbling paper of, 135–36; as senior wrangler and Smith's prizeman, 75; on teaching electromagnetism to undergraduates, 291; William Thomson's criticism of, 108n.164; undergraduate experience of, 72–76; walking by, 192; writing out mathematical knowledge, 132
Aldis, A. C. W., 520
Aldis, T. S., 512
Aldis, William Steadman, 512, 513, 524
Alexander, J. J., 518
algebraic geometry, 21–22, 28, 30n.59
Allcock, C. H., 515
Allcock, W. B., *316*, 516, 521, 523, 525
Allen, A. J. C., 515
Allen, H. S., 519
Alpine Club, 216n.86
analysis: calculus of variations, 29, 34, 75, 82, 181; development of methods of, 29, 66; differential equations, 29, 145, 254, 411; methods of transmitting, 33; in reformed Cambridge curriculum of 1840s, 101. *See also* calculus; Continental analysis
Analytical Society, 67–68, 75n, 95
Andress, W. R., 490, 490n.133
Appleton, R., 513

archive of examination questions, 155, 161, 487
Ashton, J., 512
astronomy, 27, 28, 59, 87, 240, 267, 452–54, 463–64
Astronomy (Vince), 74
athleticism: anti-intellectual, 221–22, 224; benefits of, 224; and mathematics, 176–226; scholarly culture usually contrasted with, 178; as widespread student activity, 223. *See also* physical exercise
Atkins, H. E., 519
Atkinson, G. B., 281
Atkinson, Solomon, 76, 77–78, 131–32, 135
atom: Bohr theory of, 508; quantum theory of, 497, 507; Rutherford's model of, 394. *See also* electron
Austin, W. H., 519

Babbage, Charles, 67–68, 68n.56, 76, 77, 145, 146
Baker, H. F., 508, 509, 517, 524
Balak, R., 520
Ball, W. W. R., 514
Banks, Sir Joseph, 36, 37
Barker, Thomas, 335n.92
Barnes, E. W., 519
Barnes, G. G., 523
Barnes, J. S., 521
Barnwell, 214, 214n.79
Barrow, Isaac, 32
Basel University, 32–33
Bateman, Harry: Cambridge career of, 417; and Cunningham on new theorem of relativity, 402–3, 416–24, 433; emigrates to United States, 435; and general theory of relativity, 446; at Liverpool University, 417, 434n.75; on Maxwell's equations applied to ponderable matter, 429, 430; photograph of, *417*; relation to relativity theory, 432–33; as senior wrangler of 1903, 417, 521
Baynes, D. L. H., 522
Becher, H., 67, 78, 107n.160
Beckett, J. N., 521
Bede, Cuthbert, 92, 92n.122
Bell, B. H., 522
Bell, G. M., 520
Bell, J. G., 520
Bell, W. G., 516
Bellone, Enrico, 168–69, 168n
Bennett, G. T., 518, 521, 523, 524
Bennett, T. L., 521
Bentley, A. T., 335n.92
Bentley, Edward, 92n.122
Bentley, Richard, 13, 14, 14n.31, 17, 130
Berlin, 176
Bernoulli, Daniel, 66
Bernoulli, Jacob, 32–33

Bernoulli, Johann, 29, 32–33
Berry, A., 517
Berwick, W. E. H., 523
Besant, William: blackboard introduced by, 234n.13; in lineage of Cambridge coaches, 231, 231n.4; ranking as coach, 524; retirement of, 280; as sportsman, 200; success as coach, 238; teaching methods of, 236–37, 236n.18; "upper ten" wranglers coached by, 513–18; and Webb, 247
Bestelmeyer, Adolf, 384
Bevan, P. V., 520
Biot, Jean Baptiste, 72
Birmingham Philosophical Society, 346
Birtwistle, G., 520
Bishop, T. B. W., 523
Bishop, W. C., 515
blackboards, 93, 134, 136, 234, 234n.13
Blaikie, J. A., 513
Blanch, J., 512
Blanco-White, G. R., 521
Blandford, J. H., 520
Bloor, David, 396, 397
Board of Mathematical Studies: on bookwork-and-rider questions, 163; establishment of, 102; on improvement in physical subjects, 323; on intercollegiate lectures, 269; Mathematical Tripos split by, 267; and Public Lecturers, 105; and reintroduction of electromagnetism into curriculum, 287; strengthening mathematical physics in 1860s, 291–92; and syllabus for Mathematical Tripos, 162
boat race, Oxford versus Cambridge, 213
Body, C. W. E., 514
Bondi, Hermann, 489–90
Bonhote, E. F., 523
bookwork-and-rider questions, 163
bookwork questions, 139–40, 152, 163
Booth, L. B., 522
Born, Max, 431n.69, 476–77
Bottomley, W. C., 520
Boughey, A. H. F., 514
Boutflower, C., 520
Boyle, Robert, 30, 37
Bragg, W. H., 3, 517
Brand, A., 518
Brass, John, 69–70, 75n
Brill, J., 516
Bristed, Charles: on division of mental labor in Britain, 189; on the ideal wrangler, 215; mental breakdown of, 184; on physical exercise, 191, 193–94; on pressures of the Tripos, 182, 183–84, 186, 190; on private tutors, 87–89, 241; on prostitution in Cambridge, 214; on written examinations, 126–27, 128–29

INDEX   551

Bristol College, 87
Britain: centers of mathematical learning in, 171; mixed mathematics in, 34–37. *See also* Cambridge University; Oxford University; public schools; Royal Society
British Association for the Advancement of Science, 178, 286, 294, 462
Brockliss, Laurence, 33n.67
Brodetsky, S., 522
Bromhead, Edward, 171n.125
Bromwich, T. J. I'A., 519, 521, 522, 523, 524
brothels, 214
Brown, Alexander, 245, 521
Brown, H. A., 521
Brown, John, 69, 69n.59, 71–72, 75n
Browne, A. B., 519
Browning, Oscar, 176, 199
Brunyate, W. E., 517
Bryan, George Hartley, 506, 517
Bucherer, Alfred, 384, 385, 386, 400, 413, 413n
Buchwald, J. Z., 279
Burcham, Thomas, 208
Burnside, W., 514
Butcher, J. D., 514

calculus: Bernoullis teaching, 32–33; in development of mathematical physics, 28–29; development tied to physical problems, 29; d-notation in, 34, 51, 67, 68, 73, 76, 77, 79, 95; elliptic integrals, 86; Lacroix's textbook on, 66, 68, 71, 72, 73, 77, 145; Leibniz in development of, 14n.35, 29, 32, 34, 66, 68; Newton in development of, 14, 14n.35, 29, 34, 35, 66, 118; Peacock's book of worked examples of, 145; Poynting's knowledge of, 21; in reformed Cambridge curriculum of 1840s, 101; Webb's teaching of, 249. *See also* fluxions
calculus of variations, 29, 34, 75, 82, 181
Caldwell, R. T., 512
Calliphronas, G. C., 514
Cama, B. N., 521
Cama, C. N., 521
Cambridge Mathematical Journal (CMJ), 101, 157
Cambridge Philosophical Society, 480
Cambridge University: admissions increasing in early 1800s, 77n.85; athleticism and mathematics at, 176–226; breadth of mathematical studies in 1870s, 323–24; as center of British mathematics, 170–71, 182; conservative pedagogy at, 67, 356, 358, 438, 439, 505–6; Continental analysis coming to, 67–84; Corpus Christi College, 198, 317; Ecole Polytechnique compared with, 174; escalating mathematical competence of freshmen, 260, 263–64; first-generation Cambridge Maxwellians, 325–33, 353; general theory of relativity at, 443–500; Georgian, 39–40, 46–47, 52–84; King's College, 270; Larmor's electronic theory of matter taught at, 376–90; liberal education ideal at, 53, 127, 169; localization of British electromagnetic theory at, 356, 375; and mathematical knowledge generated elsewhere, 358; mathematical physicists trained at, 2–3; mathematics and natural science emphasized at, 57, 180, 182; mathematics honor students, 1825–1900, 3n.4; mixed mathematics at, 37–38, 49–113; on non-Euclidean geometry, 465; Oxford and Cambridge boat race, 213; Parliamentary Commission of 1850s, 103–4, 110; pedagogical geography of mid-Victorian, 290; Peterhouse College, 232, 233n.8, 266, 290; Ph.D. degree introduced at, 492; private tutors at, 37–38, 40, 49–52, 58–66; professorships of mathematical subjects in eighteenth century, 60n.26; Queens College, 131, 135; relativity theory at, 43–44, 358, 359–60, 362, 398, 399–500; research school in electromagnetism, 286, 351, 352–56, 388–89; research schools lacking before 1870, 270; research style at, 179, 222; second-generation Cambridge Maxwellians, 333–52, 353, 354; special theory of relativity at, 362, 399–442; structure of, 53; three major sites of physics teaching at, 289; training system at mid-Victorian, 26; translating the mathematical culture of, 252–64; undergraduate instruction in eighteenth century, 59; university-wide syllabus emerging in, 64. *See also* Cavendish Laboratory; lectures; private tutors; public teaching; St. John's College; Senate House examination (Mathematical Tripos); Trinity College
Cameron, J. F., 520, 521, 522, 525
Campbell, A. V. G., 519
Campbell, L., 351
Campbell, Norman, 496, 496n.147
Campion, W., 167n.118
Cantor, G. N., 359n.3
Carslaw, H. S., 519
Carson, G. E. St. L., 519
Carter, F. W., 519
Carver, T. G., 513
Casson, R., 520
Caunt, S. W., 519
Cavendish Laboratory: building of, 265; and Cambridge research school in electromagnetism, 351n.127; Eddington's "Principles of Relativity" course at, 482–83; Eddington's special relativity course at, 497–98; "electron" as used in, 394, 394n.76; Maxwell as director of, 287, 303, 352; Maxwell not encouraging elec-

Cavendish Laboratory (*continued*)
tromagnetic research at, 304, 305–6; and Maxwell's view of research, 114; Niven working at, 319; as site of physics teaching at Cambridge, 289, *290*; J. J. Thomson as director of, 343; J. J. Thomson's early work at, 335–36, 343
Cayley, Arthur: in Alpine Club, 216n.86; copybook style of teaching of, 136, 136n.45; and Hopkins, 87, 136n.45; as senior wrangler, *207*; Smith's Prize questions published by, 156n.88; solution to Routh exam question published by, 158; William Thomson's criticism of, 264n.72
celestial mechanics, 34, 82, 101, 240
Chadwick, J., 520
Chandrasekhar, S., 490, 492
Channon, F. G., 520
charge, electrical, 296, 307, 326–27, 332
Charles, A. W., 518
Chevalier, R. C., 518
Chree, Charles, 299, 306, 317, 382
Christie, W. H. M., 513
Chrystal, George, 295n.19, 301, 302, 317, 322, 322n.63, 514
Clairaut, Alexis Claude, 66
Clapham Grammar School, 256n.54
Clark, G. N., 490, 490n.133
Clarke, Charles, 233
Clarke, H. L., 514
classical physics, 3, 397n
Classical Tripos, 76–77, 77n.82, 180
Clavius, Christoph, 33, 33n.65
Clifford, W. K., 3, 197, 237, 238, 266n.76, 513
coaches (tutors). *See* private tutors
coaches (vehicles), 89–90, *91*, 103
Coates, C. V., 515
Coates, W. M., 517, 520, 524
Coddington, Henry, 77
Coggin, Henry, *209*
college examinations, 126
college fellows: appointment based on original research, 270; college lectures by, 59; intercollegiate lectures by, 229; new breed in mid-nineteenth century, 215; and private tuition, 65, 113; religious tests abolished for, 270–71; top wranglers as candidates, 56, 56n.17
college lecturers, 50n.3; collapse of influence of, 94, 102, 127; college examinations and, 126; and demise of coaching system, 280; lectures of, 59; methods of, 134; as moderators or examiners, 61; Pollock and, 63; private tutors and, 50, 60, 65, 96, 98, 98n.136, 113, 149; and reforms of 1868, 265–66; Whewell on private tuition and, 103, 109
Common University Fund, 269

competition: among Routh's students, 238; athleticism fueling, 224; in Cambridge style of research, 179, 222; cumulative, competitive learning, 78; Galton on emphasis on, 183; manliness of competitive mathematics, 223, 224; in team sports, 179, 182, 191, 212, 213, 215, 225, 226; by women, 218. *See also* competitive examinations
competitive examinations: concern about excessive emphasis on, 94, 96; "fights," 88, 236; Hopkins on, 104; and meritocracy, 225; and natural selection, 110–12; Senate House examination as increasingly competitive, 56–57, 58, 182; undergraduate studies changed by, 179–80; Whewell on, 127–28. *See also* written examinations
conduction, electrical, 296, 327–29, 333–34, 348
"Conformal Representation and the Transformation of Laplace's Equation" (Cunningham), 421
conformal transformations, 418, 421–23
conservation of energy, 325
Continental analysis: academies and, 39–40; British mathematicians and, 34, 66–67; private tutors and, 50–51, 65–84; and problem-solving tradition, 144–51; and Senate House examination, 67, 70, 97; and teaching regime at Ecole Polytechnique, 174; Whewell on, 95, 102–3, 146, 181
continuum mechanics, 427
Cook, A. G., 517
Cook, C. H. H., 514
Cook, S. S., 519
Coombe, John, 161
copy-books, 136, 137
"corkscrew," the, 197, 197n.45
Corpus Christi College, 198, 317
Cottingham, E. T., 480
Cowell, P. H., 518
Cox, H., *316*, 516
Cox, J., 514
Craig, John, 13, 13n.29
Craik, A., 514
crammers, 258
Crawford, L., 518
cricket, 178, 195, 198, 213, 219, 222
Cullis, C. E., 518
culture: cultural turn in history of science, 47; Hopkins on, 107; scientific truths seen as transcending, 501–2; translating the Cambridge mathematical culture, 252–64; Whewell on, 106–7. *See also* material culture
Cumming, L., 512
Cunningham, Ebenezer: and Bateman on new theorem of relativity, 402–3, 416–24, 433; at

British Association session on gravitation, 462; "Conformal Representation and the Transformation of Laplace's Equation," 421; as conscientious objector in World War I, 435–36; and decline of electronic theory of matter, 434–36, 437, 439, 440–42; Eddington contrasted with, 448, 489; and Einstein's "On the Electrodynamics of Moving Bodies," 411, 414–16, 426, 433; "Electron Theory" course of, 437, *438*, 441–42, 479n.96; on the ether, 424–28; and general theory of relativity, 446; lectureship at St. John's College, 434–35, 436, 437; at Liverpool University, 411, 417, 434n.75; on Maxwell's equations applied to ponderable matter, 429; photograph of, *410*; on principle of relativity, 411–14, 416; *The Principle of Relativity*, 435, 455; relation to relativity theory, 409, 432–33; *Relativity and the Electron Theory*, 435, 437, 437n.83; relativity theory lectures of, 479, 479n.96; as senior wrangler of 1902, 409, 521; training in Cambridge, 409–14
current electricity, 303, 307, 322
Curthoys, M. C., 225

Daldy, F. F., *316*
Dale, J. B., 518
Dale, Thomas, 310–11, 514, 515, 516, 524
D'Alembert, Jean LeRond, 33, 33n.66, 35, 66, 148, 173
D'Alembert's Principle, 29, 145
Daniell, P. J., 523
Daniels, A. E., 518
Darnley, E. R., 520
Darrigol, Olivier, 360, 409n.23, 430n.65, 507n.11, 509
Darwin, C. G., *203*, 523
Darwin, Charles, 110
Darwin, George H.: Cambridge training and geophysics of, 506; from Clapham Grammar School, 256n.54; health and excellent spirits before sitting the Tripos, 190n; on Maxwell's lecture on Rayleigh, 324n.67; on Maxwell's shortcomings as analyst, 137, 297; physics training at Cambridge, 3; on problem solving and research, 137, 138, 273–74, 275, 278–79; as second wrangler of 1868, 513
Dawson, John, 62–63, 62n.35, 63n.39, 65
Dé, B., 522
Dealtry, William, 63n.41, 71n.66, 73n.73
Dear, Peter, 28
degrees, conferring of, 207–12
$\nabla^2 V$ Club, 480n.100
DeMorgan, Augustus, 142, 149, 150, 151–54, 218, 231
Dening, Greg, 46–47

Descartes, René, 17n.42, 28
de Sitter, Willem: in bibliography of Eddington's *Report*, 493n; on British papers on relativity, 453–54; Clark discovering errors in, 490; correspondence with Eddington, 457–58, 462, 463, 493; correspondence with Einstein, 460–61, 467, 478; in Eddington's mastery of general relativity, 459, 464, 468; and failure to develop research school in Leiden, 499; learning general theory of relativity, 459, 460–62, 477
Dey, A., 514
Dick, G. R., 513
Dickens, Charles, 91, 237
Dickens, Henry, 237–38
Dickson, J. D. H., 514
differential equations, 29, 145, 254, 411
differential geometry, 451, 467, 495
Dirac, Paul: at Cambridge, 507, 508–9; Cunningham declining to supervise, 437n.85; in Powell's special relativity course, 498; *The Principles of Quantum Mechanics*, 498n.157; studying with Eddington, 490, 490n.132, 492; as "theorist's theorist," 507; training and early research of, 507–10; transcribing problems into relativistic form, 491
Dirichlet, Gustav, 34n.69
displacement currents, 331, 332, 336, 338
disputations. *See* Acts (disputations)
distant parallelism, 492
Dixon, A. C., 517
Dixon, A. L., 521
Dixon, James, 242
Dobbs, W. J., 518
Dodds, J. M., 516
Douglas, A. V., 484
Dreyfus, H. L., 4n.8
Droste, Johannes, 459–60, 493n
Dublin, 171, 171n.125
Dunkley, H. F., 523
Durell, C. V., 521
"Dynamical Theory of the Electric and Luminiferous Medium" (Larmor), 368, 377, 394, 401
"Dynamical Theory of the Electromagnetic Field, A" (Maxwell), 325
Dynamics of a System of Rigid Bodies (Routh), 253, 258
Dyson, Sir Frank, 475–76, 517

Earman, J., 448n, 455n, 470n.73, 478, 478n.95
Eason, E. K., 522
Eckhardt, J. C., 520
Ecole Polytechnique (Paris), 174, 504
Eddington, A. S.: as astronomer, 452, 463–64; Cambridge Philosophical Society address of, 480; Cambridge style of research of, 275,

Eddington, A. S. (*continued*)
275n.88; Cambridge training and astronomical work of, 506; correspondence with de Sitter, 457–58, 462, 463; debate with Lodge over electronic theory of matter, 469–75; de Sitter's influence on, 459, 464, 468, 493; differential geometry known by, 451, 467, 495; early research in relativity, 484; general theory of relativity supported by, 437, 444; "Gravitation," 455–56; "Gravitation and the Principle of Relativity," 463; on Jeans on non-Euclidean geometry, 465; Leiden group contrasted with, 461; masters general theory of relativity, 462–68, 477–78, 493, 494–95; *The Mathematical Theory of Relativity*, 483–84, 484n.113, 485, 491; mathematical training of, 449–51; photograph of, *450;* as Plumian Professor of Astronomy, 452, 455; "Principles of Relativity" course, 482–83; as rapid problem-solver, 467; on relativity and gravitation, 455–57; relativity theory introduced to Cambridge by, 44, 478–89, 495; "Relativity Theory of Gravitation (Mathematical)" course, 483, 485; *Report on the Relativity Theory of Gravitation*, 467–69, 475, 479, 493n; at Royal Greenwich Observatory, 451–52, 455; as senior wrangler of 1904, 450, 521; *Space, Time, and Gravitation*, 482–83, 485, 496n.147, 508; special relativity course of, 497–98; *Stellar Movements and the Structure of the Universe*, 454n.30; students of, 489–93; teaching style of, 485–86; testing predictions of general relativity, 447, 452, 475–76, 480, 482; at Trinity College, 449–51, 452
Edinburgh University, 326
Edwardes, F. E., 519
Edwards, J., 515
E = mc², 497
Ehrenfest, Paul, 430, 430n.67, 431, 459, 460, 461, 499
Einstein, Albert, 43–44; annual courses on general relativity, 495n.145; in bibliography of Eddington's *Report*, 493n; correspondence with Searle, 399–402; and de Sitter, 460–61, 467, 478; and Ehrenfest, 459; on the ether as superfluous, 424–25, 426–27, 427n.59; "The Formal Foundation of the General Theory of Relativity," 467, 467n.66; Jeans referring to, 384, 385–86, 407; in Leiden, 460, 461; and Lorentz, 459; on Lorentz transformations, 372, 372n.29; on Maxwell's equations applied to ponderable matter, 429, 430; "On the Influence of Gravitation on the Propagation of Light," 456; in Plummer's account of relativity, 453; relativity an axiom for, 381, 387, 401, 403, 432, 433, 439, 441, 443; unified field theory of, 492. *See also* "On the Electrodynamics of Moving Bodies" (Einstein); relativity theory
elastic solid theory, 240
electrical images, method of, 326
electricity: charge, 296, 307, 326–27, 332; conduction, 296, 327–29, 333–34, 348; imponderable-fluid theory of, 294, 296, 333; polarization, 370; problems of teaching, 265; in Tripos of 1880s, 267
electromagnetic theory of light: Cambridge Maxwellians and, 329; in Maxwell's *Treatise*, 296, 331, 360; and Poynting's paper on energy flow, 347; Routh ignoring, 307; J. J. Thomson's work on, 336–38; in Tripos questions, 303
electromagnetism: action-at-a-distance theories of, 295, 307, 329, 333; Cambridge research school on, 286, 351, 352–56, 388–89; Dirac studying, 508; Electromagnetic View of Nature, 369, 437; experimental detection of, 331; first-generation Cambridge Maxwellians on, 325–33, 353; internationalization of theory of, 375–76; as mathematical discipline at Cambridge, 377; Maxwell's dynamical theory of electromagnetic field, 279, 295, 298; Maxwell's lectures on, 287, 299, 300, 304, 305, 305n.38; potential concept in, 278; Poynting's paper on energy flow, 2, 4–9, 18–26, 344–52; reintroduced into Senate House examination, 287, 291, 292, 325, 331–32, 354; relativity theory and Cambridge, 42–43; Routh teaching, 306–9; second-generation Cambridge Maxwellians on, 333–52, 353, 354. *See also* electricity; field theory of electromagnetism; light; magnetism
electron: acceptance of, 376; Cavendish Laboratory on, 394, 394n.76; Dirac on quantum mechanical behavior of, 507; in Larmor's electrodynamics, 42, 368, 369, 370, 371; Lorentz's deformable, 412; mass of, 376, 384, 385, 386, 395; positive, 369, 376, 377
electronic theory of matter (ETM): Cambridge pedagogy and establishment of, 376–90; cosmological consequences of, 469–75; Cunningham and Bateman on, 403, 423–24; Cunningham on relativity and, 414, 416; development of, 367–76; Eddington on Newtonian mechanics and, 456; general theory of relativity and, 434, 446–47, 468, 469–75; in Jeans's *Mathematical Theory*, 381–86; in Larmor's lectures on electromagnetic theory, 386–87; Larmor's students applying, 390–95; as little applicable to atomic physics, 437; Lorentz-FitzGerald contraction in, 361–62, 370–71, 373–74; Lorentz transformations in, 372–74, 372n.29, 375, 381, 411, 412, 413; and

INDEX   555

Mathematical Tripos questions, 377–81, *378*, 434; Nicholson on relativity as derivable from relativity, 403, 428–32; relativity theory discrediting, 43, 395; training and decline of, 432–42; training and persistence of, 396–98
electrostatic rest mass, 470, 472–73
electrostatics, 300, 307, *308*, 317, 326
Elementary Treatise on Mechanics, An (Whewell), 73
Elements (Euclid): and disputations, 140n.52; in Georgian Cambridge teaching, 59, 94; the new analysis contrasted with, 30; Newton's copy of, *31;* Pollock teaching, 63; private tutors and, 94; problem sheets and, 139; textbooks compared with, 141; and Whewell's *Dynamics*, 148
Elliot, J., 513
Elliott, T. A., 514
elliptic integrals, 86
Ellis, Leslie, 86–87, 99, 157, 162
Ellis, Robert, 167n.116, 184–85, 190, 191, 207–8
Emerson, Ralph Waldo, 189, 194, 225
Emerson, William, 119n.9
end-of-chapter exercises, 38, 41, 117, 168, 172, 382
engineering, 506
Enros, P. C., 67
equivalence principle, 444–45, 446, 468
error, 22–23
Essay (Green), 108
ether: in astronomy, 454, 454n.30; Cambridge physicists and, 43; Cunningham on, 424–28; Einstein abandoning as superfluous, 424–25, 426–27, 427n.59, 469; Larmor on, 366–67; in Maxwell's dynamical theory of electromagnetic field, 295–96, 360, 368; Michelson-Morley experiment and, 359, 360–61, 371, 374, 375
ETM. *See* electronic theory of matter
Euclidean geometry, 28, 95, 107, 180
Euclid's *Elements*. *See Elements* (Euclid)
Euler, Leonhard, 29, 34, 35, 66, 148, 173
Evans, J. H., 142n.57
Everett, J. D., 363, 363n.11
evolutionary theory, 110, 111–12
examination questions, 138–44; archive of, 155, 161, 487; bookwork questions, 139–40, 152, 163; as exemplars, 41; as exercises in textbooks, 117, 167, 171; pedagogical use of past, 161–62; problem questions, 139, 141–44; publications of collections of problems, 143–44, 143n.64, 153n.85; publishing solutions to past, 149–50, 160–63, 166–67; and research, 117, 155–60, 272; in Routh's teaching, 235 36; two types of, 138–39
examinations: college, 126; oral and written cultures of, 118–29. *See also* competitive examinations; examination questions; oral examinations; written examinations
examination scripts, 18–21, 23, 24, 25, 163, 163n.110
examiners, 55, 76, 78n.87
examples: examination questions as textbook, 117, 167; the "examples age," 221; learning through, 41, 160; worked examples in textbooks, 145, 161
exercise, physical. *See* physical exercise
exercises: end-of-chapter exercises, 38, 41, 117, 168, 172, 382; in physics training, 505; in science teaching, 502
experimental physics: Maxwell on material culture of, 114–15; J. J. Thomson studying, 335; wranglers knowing little of, 377. *See also* Professor of Experimental Physics
experimental science: Maxwell on material culture of, 114–15; as public in contrast to mathematics, 30; as site-specific, 11, 15; social and cultural aspects of, 10–11; theoretical work compared with, 12–18; as work, 16. *See also* experimental physics
Ezechiel, P. H., 520

Faraday, Michael, 286
Fawcett, Henry, 187, 188, 216, 218, 281
Fawcett, Millicent, 281
Fawcett, Philippa, 218, 281–82
Feingold, Mordechai, 15, 30
fellows, college. *See* college fellows
Fermat, Pierre, 28
Ferrers, Norman, 160, 162, 266n.76, 326n.75
field theory of electromagnetism: Cambridge Maxwellians and, 329, 332–33; Cambridge research school on, 352–56; and Larmor's electrodynamics, 42; in Maxwell's *Treatise*, 42, 287, 293–94; pedagogical difficulties compared with general relativity, 479; Poynting vector and, 2, 4, 6, 7, 8; Routh not mentioning, 307, 316. *See also* Maxwell's field equations
"fights" (examinations), 88, 236
Findlay, C. F., 515
FitzGerald, George Francis: Cambridge research school compared with, 288, 354, 355–56, 375; on electromagnetic wave propagation, 340, 341–42, 342n.112, 350; Larmor inspired by, 367–68; on Larmor's electronic theory of matter, 368; Lodge collaborating with, 469; Maxwell's *Treatise* studied by, 288, 343; and J. J. Thomson's work on moving charged conductor, 338–39. *See also* Lorentz-Fitzgerald contraction hypothesis
Fleming, J. A., 305n.38

Fletcher, W. C., 517
flight, 506
Flux, A. W., 517
fluxions: in British mixed mathematics, 34; in examination questions, 120, 140; in Newton's *Principia*, 14; private tutors' prepared manuscripts on, 144; public interest in, 35; in Smith's Prize examinations, 120; treatises on, 118, 119, 142; Wright studying, 71
Foggo, David, 89n.113
Fokker, Adriaan, 457, 458, 459
Fontenelle, Bernard, 32, 33n.64
football (soccer), 198, 213, 219
Forbes, J. D., 137n, 184, 326
"Formal Foundation of the General Theory of Relativity, The" (Einstein), 467, 467n.66
Forsyth, A. R., 254, 276, 319n.59, 320, 320n, 516
Forsyth, J. P., 522
Foucault, Michel, 3, 4, 223, 229
Fountain, E. O., 516
Fourier, Joseph, 30, 34n.69
Fowler, R. H., 508
Foyster, H. T., 522
Frank Fairlegh (Smedley), 92
Frankland, F. W. B., 519
Fraser, P., 522
Fraser, W. G., 519
free trade, and evolution of mathematicians, 109–13
Frend, William, 62
Fresnel, Augustin, 30
Friedmann, A. A., 477n.93
Frost, Percival, 200, 231, 237–38, 266n.76, 513–16, 524
funking fits, 187

Gabbatt, J. P., 521
Gabled, Christophe, 33, 33n.67
Galileo, 27, 28
Galison, Peter, 507
Gallop, E. G., 517, 521, 522, 525
Galton, Francis, 85, 135, 183, 184, 185, 190, 191
Garber, E., 29n.55
Garnett, W. H. S., 521
Garnett, William, 163–66, *165*, 172, 351, 465, *466*, 514
Gaskin, Thomas, 156, 168
Gaul, C. A., 518
Gaul, P. C., 517
Geake, C., 518
Geison, G. L., 11n.22
general theory of relativity: and astronomy, 452–54, 463–64; Cambridge response to, 443–500; Eddington bringing to Cambridge, 44, 478–89, 495; Eddington's students and, 489–93; Eddington supporting, 437, 444; Eddington testing predictions of, 447, 452, 475–76, 480, 482; electronic theory of matter and, 434, 446–47, 468, 469–75; equivalence principle, 444–45, 446, 468; general acceptance of, 447, 495; growing British awareness of, 437; in Leiden, 458–62, 499; "low water mark" of, 499; mathematical difficulty of, 446, 463, 476–77; obstacles to mastery of, 477–78; origins of, 444–45; predictions of, 445–46, 469–70, 477; publication of, 457; reception contrasted with that of special theory, 498–99; revival of, 499; in Tripos questions, 486–89, *488;* World War I and British knowledge of, 455–58
Genese, R. W., 514
Géométrie descriptive (Monge), 70
geometry: algebraic, 21–22, 28, 30n.59; differential, 451, 467, 495; Euclidean, 28, 95, 107, 180; non-Euclidean, 446, 464–65, *466*, 494, 500; physico-mathematics contrasted with, 28
Germany: Berlin, 176; British and German students compared, 176–78; British textbooks translated into German, 253, 253n.47; Cambridge and German approaches to relativity, 429–30, 441; Göttingen, 252, 460n.49, 477n.93
Gharpurey, H. G., 521
Gibberd, W. W., 520
Gibbons, F. B. De M., 515
Gifford, A. C., *199*
Gingras, Yves, 503–4
Glaisher, J. W. L., 167n.118, 270, 270nn. 83, 84, 340n.107, *341*, 513
Glazebrook, Richard T.: Cambridge pedagogy shaping work of, 355; in Cambridge research school in electromagnetism, 351, 355; as fifth wrangler of 1876, 515; on Hall effect, 353–54; on Maxwell's *Treatise*, 296–97; and Poynting, 350–51; on Rayleigh's question in Tripos of 1876, 309–11; as second-generation Cambridge Maxwellian, 333; on study of optical effects at Cavendish Laboratory, 305n.41; work owing nothing to personal relationship with Maxwell, 287–88
Glymour, C., 448n, 455n, 470n.73, 478
Goddard, J., 521
Godfrey, C., 519
Gold, E., 521
Gold, Tommy, 490
Goldberg, Stanley, 405, 406, 407, 409, 409n.23
Goldstein, J., 4n.8
golf, 219, 222
Goodenough, Samuel, 36
Goodwin, Charles, 485, 486
Goodwin, Harvey, 158
Goody, Jack, 16–17, 18

INDEX 557

Goowillie, J., 518
Göttingen, 252, 460n.49, 477n.93
Grace, J. H., 519, 521, 522, 523, 524
Graham, C., 515
grammar schools, mathematical teaching in, 254–64
Granchester grind, 192, 192n.36, 193
Grantham, S. G., 521
gravitation: in general theory of relativity, 444–45, 455–57, 468, 469, 474; Newton's theory of, 12, 28, 473, 474, 476
"Gravitation" (British Association session), 462
"Gravitation" (Eddington), 455–56
gravitational mass, 470, 472–73
gravitational red shift, 444, 445, 469, 478, 484, 494
"Gravitation and the Principle of Relativity" (Eddington), 463
Great Divide, 9–10, 16–17, 25
Greaves, J., 515
Green, A., 516
Green, George, 108, 171n.125
Greenhill, Alfred George, 167n.118, 323, 513
Green's functions, 298
Green's theorem, 311
Gregory, David, 13n.27
Gregory, Duncan, 101, 157
Gregory, P. S., 514
Griffiths, G. J., 514
Gross, E. J., 513
Grossmann, Marcel, 446, 457, 459
Guest, J. J., 517
Guicciardini, N., 13n.28
Gunston, W. H., 516, 521, 525
Gurney, Theodore T., 198, 198n, 514
Gwatkin, Richard, 68, 72, 76, 78, 79, 86, 145

Hackforth, E., 520
Hall effect, 353, 353n
Halley, Edmond, 31–32
Hamilton, Henry Parr, 77
Hamilton, William Rowan, 171n.125
Hamiltonian mechanics, 508
Harding, Thomas O., 198, 198n, 300, 514
Hardy, G. H.: Cambridge attitude toward mathematics of, 219, 258, 276n; as fourth wrangler of 1898, 520; and Herman's differential geometry lectures, 451; as pure mathematician, 221, 276n, 434; sitting Tripos after six terms, 260; as sportsman, 222, 224
Hargreaves, Richard, 423n.49, 515
Harker, A., 516
Harris, J. R., 514
Harrison, J., 516
Harrison, John, 255n.54
Harrison, W. J., 522

Hart, H., 513
Hartley, W. E., 520
Hartog, N. E., 513
Harvard University, 225
Harwood, John, 44
Haslam, W. H., 513
Hassé, H. R.: on "examples age," 221; on geometry at Cambridge, 421; on legitimate use of principle of relativity, 432n.70; at Liverpool University, 434n.75; on Maxwell's equations applied to ponderable matter, 429, 430, 430n.65; on reforms of 1909, 285; as seventh wrangler of 1905, 522; "theory of relativity" used by, 406n.17; during World War I, 436
Heath, R. S., 320, 516
Heaviside, Oliver: Cambridge research school compared with, 288, 354, 355–56, 355n.136, 375; Dirac studying work of, 508; Maxwell's *Treatise* studied by, 288, 343, 354; and Poynting's work on energy flow, 7; Searle's electrodynamics deriving from, 401
Heisenberg, Werner, 508, 509
Helmholtz, Hermann von, 222
Hemming, George, 214, 219, 256n.54
Henderson, G., 513
Hensley, W. S., 516
Herman, Robert A.: and Bateman, 417; coaching room of, *290;* and Eddington, 451, 467; and electronic theory of matter, 388; and Jeans, 381; photograph of, *283;* ranking as coach, 524; as senior wrangler of 1882, 516; teaching methods of, 282–83; "upper ten" wranglers coached by, 519–23
Hermann, Jacob, 29
Herschel, John, 67–68, 68n.56, 77, 145
Hertz, Heinrich, 331
Hicks, W. M., 331n.88, 514
Hilary, Henry, 257
Hilbert, David, 460n.49, 468, 493n
Hill, A. V., 522
Hill, E., 513
Hill, F. W., 517
Hill, M. J. M., 516
Hills, G. F. S., 521
Hints and Answers (Wright), 149
Hobbes, Thomas, 30n.59, 31
Hobson, E. W., 281, 282, 515, 524
Hodge, M. J. S., 359n.3
Hodgkin, T. E., 220
Hogg, R. W., *199*
Holmes, F. L., 11n.22
Holthouse, Cuthbert, *203, 210*
Hopkins, William, 80–81; and Airey, 257; assessing incoming students, 85–86, 190; and Cayley, 87, 136n.45, *207;* on culture, 107; elite

Hopkins, William (*continued*)
   mathematicians studying with, 87; and Leslie Ellis, 87, 87n.109; and Robert Ellis, 185, 208; and Fawcett, 187; and Galton, 183; infinitesimal impulses method of, 157–58, 238n.23; in lineage of Cambridge coaches, 231; on making mathematics relevant to commerce, 107, 264; on mathematics and moral virtue, 106, 181; as mathematics coach, 52, 80, 84–86, 106; and Maxwell, 137n, 186, 296; new generation of coaches surpassing, 231; Peacock on, 98; physics training at Cambridge, 3; portrait of, *81*; on private tuition, 84, 84n.99, 100–101, 103–5, 108, 110; as researcher, 107–8, 112; and Routh, 231, 233, 235, 524; rowing recommended by, 195; and Stokes, 87, 108; teaching methods of, 85, 99, 109; and William Thomson, 87, 108, 135; as Tory, 106, 107; on undergraduate education and research, 52, 97, 99–101, 102; and university lecturers, 269
Hopkinson, E., 516
Hopkinson, John, 187, 216–17, 217n, 513
Horologium Oscillatorium (Huygens), 13, 13n.30
Hough, S. S., 518
Houston, W. A., 519
Howard, F. A., 519
Hoyle, Fred, 486
Huber, V. A., 174n.133
Hudson, Hilda, *417*
Hudson, R. W. H. T., 520
Hughes, W. R., 521
Hunt, B. J., 279, 359n.3, 361
Hunter, H. St. J. A., 513
Hurst, G. H. J., 518
Hustler, James, 69n.59, 73, 74
Huygens, Christiaan, 13, 14, 14n.35, 28
hydrodynamics, 240, 267, 326, 353–54
Hydrodynamics (Lamb), 253n.47, 506
hydrostatics, 58, 59, 75, 87, 240
Hymers, John, 86, 231

Ibbotson, A. W., 522
Iles, J. C., 517
incommensurability, 43–44
industrialization, 179, 189, 225
infinitesimal impulses, 157–58, 238n.23
innovation, training and, 505–11
insomnia, 187–88
instruments, mathematical, 133
intercollegiate lectures: Board of Mathematical Studies and, 269; and demise of coaching system, 264, 265–66, 271, 280, 282; establishment of, 229, 264, 266–67; as forums for research, 271; Niven's lectures on Maxwell's *Treatise*, 317; personal interaction with pupils, 271–72; as site of physics teaching at Cambridge, 289; for year 1880–81, *268*
International Commission on the Teaching of Mathematics, 505
inversion, 418–19, *419, 420,* 421, *422*
Irwin, D., 521

Jackson, C. S., 518
Jackson, Henry, 216
Jackson, J. Stuart, 162
Jackson, W. H., 521
Jacob, J., 519
Jameson, Francis, 163
Janssen, M., 478n.95
Jeans, James H.: Cambridge training and astronomical work of, 506; on Lorentz-FitzGerald contraction, 357; on non-Euclidean geometry, 464–65; physics training at Cambridge, 3; as second wrangler of 1898, 520; sitting Tripos after six terms, 260. See also *Mathematical Theory of Electricity and Magnetism*
Jebb, John, 118n.4
Jerrard, J. H., 87
Johnson, G. W., 516
Jones, G. M. E., 516
Jones, H. S., 225
Joseland, Henry, 257
junior optimes, 55, 56, 208, 210

Kaufmann, Walter, 384, 385
Kelvin, Lord. See Thomson, William
Keynes, John Maynard, 220
Kidner, A. R., 521
King Edward VI Grammar School (Birmingham), 256, 257
King's College, 270
Klein, Felix, 252–53, 252n.46, 263, 275
Knapman, H., 521
Knight, S., 515
knowledge: fragility of text-based transmission of, 44, 477–78, 494–95; learning and knowing, 1–2, 3, 24. See also learning; research; tacit knowledge
Knox-Shaw, J., 522
Kragh, H., 491
Kuhn, Thomas, 3–4, 4n.10, 43, 172–73, 174, 229–30

Lachlan, R., 517
Lacroix, S. F., 66, 68, 71, 72, 73, 77, 78n.88, 145
Ladies' Diary (magazine), 35, 119
Lady Margaret boat club, 194
Lagrange, Joseph Louis, 29, 35, 40, 66, 70, 72, 146, 148
Lagrangian dynamics, 278, 336

INDEX 559

Lagrangian function, modified, 246, 336
Lamb, Horace: Eddington as pupil of, 449; as first-generation Cambridge Maxwellian, 325, 326; as focusing on specific problems, 330–31; *Hydrodynamics*, 253n.47, 506; Klein commissioning work by, 252n.46; lectures on hydrodynamics, 326, 353–54; as mathematician and teacher, 326n.75; on Maxwell as analyst, 297n.22; W. D. Niven as link between Charles Niven and, 332; papers on flow of currents, 328–29, 330; physics training at Cambridge, 3; as second wrangler of 1872, 514
Lambert, C. J., 513
Landen, John, 35
Laplace, Pierre-Simon: leading role in academic life of, 35; mathematical training of, 33; on propagation of gravitation, 456n.36; *Traité de mécanique céleste*, 29, 66, 70, 71, 72, 146; in Whewell's *Treatise*, 148; work environment of, 40
Laplace's equation: Bateman on, 422; Cunningham's "Conformal Representation and the Transformation of Laplace's Equation," 421; in Jeans's *Mathematical Theory*, 383, *419, 420*; and Maxwell's method of electrical images, 326; in Rayleigh's Tripos question of 1876, 313; in Webb's coaching, 249, 251, 278
Larmor, Joseph: on absolute frame of reference, 366–67, 373; Cunningham taught by, 409, 410; "Dynamical Theory of the Electric and Luminiferous Medium," 368, 377, 394, 401; Eddington compared with, 486; electrodynamics of, 42–43, 360, 361–62; electronic theory of matter developed by, 367–76; on the ether and magnetic effects, 424, 424nn. 54, 55; on foundations of Poynting's theorem, 347n.122; intellectual communication by, 356; lectures on electromagnetic theory, 386–87, 435; Lodge collaborating with, 469; mixed mathematics defended by, 220–21; on Niven, 319, 328; paper on electromagnetic induction of 1883, 364–66; physics training at Cambridge, 3; and relativity theory, 44, 472, 474, 494, 500; on Routh, 235, 246, *316*; as second-generation Cambridge Maxwellian, 333, 363; as senior wrangler in 1880, 364, *365*, 516; "The Theory of the Aberration of Light," 371–72; training and early career of, 363–67; Tripos question set by, 377–79, *378;* withdraws from electrodynamic research, 376. See also *Aether and Matter*
Latour, Bruno, 7n.15, 45
Laub, Jacob, 429, 430
Laue, Max von, 435, 436, 440
Lawrence, F. W., 519

Lay, C. J., 517
Leahy, A. H., 516
Leak, H., 523
learning: cumulative, competitive, 78; implicating the body in, 223–24; industrializing, 179, 189, 225; and knowing, 1–2, 3, 24; in physics, 2–18; teachers in, 228; of theories, 502. See also training
Least Action, Principle of, 29, 278, 367, 368, 369
Leathem, J. G., 519
lectures: advanced lecture courses, 271, 284; in eighteenth-century Cambridge education, 59; new university lectureships in mathematics, 229; in preparation for disputations, 134; private tuition displacing, 79, 80, 102, 103, 113; professorial, 60, 127, 267; Public Lecturers created, 104, 104n.152, 105, 113; Roget cartoon on, *93;* tutorial relationship contrasted with, 88; university lectures, 229, 264, 282, 343. See also college lecturers; intercollegiate lectures
Leeming, William, 257
Leibniz, Gottfried, 14n.32, 29, 34, 66, 68
Leiden: general relativity's transmission to, 458–62, 477n.93, 499; research school failing to develop at, 499
Lemaître, Georges, 491
Lenfesty, S. De J., 520
Levett, E. L., 513
Levett, Rawden, 255–56, 257
Levi-Civita, Tullio, 490, 493n
Lewis, T. C., 515
liberal education: Cambridge ideal of, 53, 127; mathematics and, 178–82; sport in, 179; versus technical subjects, 169; Whewell defending, 94, 95, 96, 102, 103, 106, 109, 180, 224–25; written examinations and, 127, 129
Liberals, 110
light: convection, 370; general relativity and bending of, 444, 445, 447, 452, 457, 469, 470, 475–76, 478, 484; optics, 58, 59, 75, 87, 101, 240, 267, 288; velocity of, 294, 348, 349, 405, 414, 441. See also electromagnetic theory of light
Lindsay, R. B., 277
Littlewood, E. T., 516
Littlewood, J. E., 219–20, 221, 222, 224, 259–60, 434, 451, 522
Livens, G. H.: on dynamics of charged sphere, 390–93; as fourth wrangler of 1909, 523; in mathematics education, 437; and relativity, 395; at Sheffield University, 434n.75; *The Theory of Electricity*, 387–88, 387n.61, 389
Liverpool University, 411, 417, 434n.75
Lloyd, L. S., 520
Lock, G. H., 514

Lock, J. B., 514
Lodge, Oliver: debate with Eddington over electronic theory of matter, 469–75; on electromagnetic wave propagation, 340, 341; on ether and magnetic effects, 425n.54; intellectual communication by, 356
Loney, S. L., 516
Long Vacation, 266, 269
Lord, J. W., 514
Lorentz, H. A.: in bibliography of Eddington's *Report*, 493n; and Einstein, 459; and failure to develop research school in Leiden, 499; Jeans referring to, 384, 386, 407; Larmor distinguishes himself from, 375; lectures on general relativity, 460, 461; in Plummer's account of relativity, 453; wranglers conflating Larmor and, 380–81. *See also* Lorentz-FitzGerald contraction hypothesis; Lorentz transformations
Lorentz-FitzGerald contraction hypothesis: at Cambridge, 43, 359n.3, 360, 361–62; Jeans on, 357, 385, 386; in Larmor's electronic theory of matter, 361–62, 370–71, 373–74; in Larmor's Tripos question, 378; for preserving ether-based electrodynamics, 359–60, 361
Lorentz transformations: in Cambridge versus German approaches to relativity, 429–30, 441; Cunningham on Maxwell's equations and, 411, 412–13, 414, 415, 422–23, 429; introduced by Einstein and Lorentz, 372n.29; in Larmor's electrodynamics, 372–74, 375, 411; Minkowski on, 423n.49; in Tripos questions, 487, 497; in wranglers' understanding of relativity, 381
Love, A. E. H., 252n.46, 253n.47, 517, 520, 524

MacAlister, Donald, 189, 204–5, 263, 515
Macaulay, F. S., 258–60, 259
Macaulay, W. H., 515
MacBean, W. R., 520
MacCullagh, James, 171n.125, 367–68
Macdonald, H. M., 377, 378, 517
MacIntyre, Alasdair, 501, 502–3
Mackenzie, H. W. G., 516
Mackenzie, W., 167n.118
Maclaurin, Colin, 148
Maclean, A. J., 516
MacRobert, T. M., 523
magnetism: electrons explaining magnetization, 370; problems of teaching, 265; in Tripos of 1880s, 267
magnetostatics, 307, 317
Mair, D. B., 518
Malcolm, W., 513
Manchester Grammar School, 257
Mangan, J. A., 212, 221
Manley, G. T., 518
manliness, 212–18; in Cambridge style of research, 179; competitive mathematics as mark of, 223, 224; mid-Victorian ideal giving way to neo-Spartan version of, 221–22
Marples, P. M., 520
Marrack, P. E., 417, 521
Mars, 471, 472
Marshall, Alfred, 256, 257, 512
Marshall, J., 316
Marshall, J. W., 515
Martin, J. A., 515
Maser, Hermann, 254
Mason College (Birmingham), 345–46
masturbation, 214
material culture: of experimental science, 11; of mathematical physics, 16, 17, 41; Maxwell on, 114–15; of mixed mathematics, 114–75
mathematical instruments, 133
mathematical physics: athleticism and, 176–226; calls for more emphasis on, 264–65; Cambridge University as training ground in, 2–3; emergence of, 27–30, 503–4; as global enterprise, 10; Great Divide separating experts from lay people, 9–10; invention of mathematical methods of, 28–29; material culture of, 16, 17, 41; muscularity and, 221; quantum mechanics, 491, 497, 498, 507, 508–9; recoverability of, 502; survey of 1912 on mathematical training of physicists, 505; technical unity imparted by common methods, 278; translating the Cambridge culture of, 252–64; way of knowing in, 4, 18, 24–25, 39; writing a pedagogical history of, 1–48. *See also* electromagnetism; mixed mathematics; Newtonian mechanics; relativity theory
Mathematical Problems (Wolstenholme), 261–63, 262
mathematical societies, 35, 119
Mathematical Theory of Electricity and Magnetism (Jeans), 381–86; and Cambridge school of electromagnetic theory, 388, 407; inversion technique in, 418, 419, 420; on Lorentz and Einstein, 384, 386, 407; solution of Laplace's equation in terms of spherical harmonic, 383, 419, 420
Mathematical Theory of Relativity, The (Eddington), 483–84, 484n.113, 485, 491
Mathematical Tracts (Airy), 74, 75, 77, 97n.134, 136
Mathematical Tripos. *See* Senate House examination (Mathematical Tripos)
mathematics: athleticism and, 176–226; breadth of Cambridge studies in 1870s, 323–24; Cambridge emphasizing, 57, 180, 182; certainty of, 30; as esoteric, 30; fragmenting in late nine-

INDEX 561

teenth century, 220–21; free trade and evolution of mathematicians, 109–13; in hierarchy of knowledge, 28; and liberal education, 178–82; moral dimension of, 106, 181; paper-based learning in, 115–17, 129–38; pedagogical devices of, 41; physico-mathematics, 28, 30, 171n.28; public spectacles pertaining to Cambridge, 201–12; reformed Cambridge curriculum of 1840s, 101–2, 109; Royal Society's equivocal view of, 36–37; seen as pinnacle of intellectual achievement, 205; shortage of teachers in, 503; training in before twentieth century, 27–40; translating the Cambridge culture of, 252–64; university teaching in eighteenth century, 33, 39. *See also* analysis; geometry; mathematical physics; mixed mathematics; pure mathematics

Mathews, G. B., 238, 517

Maxwell, James Clerk: as additional examiner for Tripos, 300n.28, 302; on British Association Committee on electrical standards, 286, 294; and Cambridge research school in electromagnetic theory, 352–53, 357–58; dynamical theory of electromagnetic field, 279, 295, 298; "A Dynamical Theory of the Electromagnetic Field," 325; examination question of inspiring research, 158–59; examination scripts used for drafts, 163n.110; exercise taken by, 196–97; field theory of electromagnetism, 2, 4, 6, 7, 8, 42, 287, 293–94, 307; illness of 1853, 200; intellectual resources from outside Cambridge, 506; lectures on electromagnetism, 287, 299, 300, 304, 305, 305n.38; on material culture of experimental physics, 114–15; mental state during Tripos, 188–89; and Niven, 319, 332; physics training at Cambridge, 3; on pressures of Cambridge, 185–86; "A Problem in Dynamics," 155, 164n.112; as Professor of Experimental Physics, 265, 287, 300, 304; and Rayleigh's acoustics, 324n.67; in reform of physics teaching at Cambridge, 291, 292; Routh beating in Tripos of 1854, 231, 231n.7; and Scottish natural philosophy, 326; Searle maintaining traditions of, 401; shortcomings as analyst, 137, 137n, 296–97; as student of Hopkins, 87; as teacher of electromagnetic theory, 303–6; theory and experimentation combined by, 264; at Trinity College, 326, 332; Tripos question of 1873, 163–66, 172; "A Vision," 1; writing on loose sheets of paper, 137. *See also* Maxwell's field equations; *Treatise on Electricity and Magnetism*

Maxwell's field equations: and Cambridge problem-solving approach to research, 329–30; Cunningham on Lorentz transformation and, 411, 412–13, 414, 415, 422–23, 429; and Glazebrook and Poynting's work on optical refraction, 288; and Larmor's electrodynamics, 370, 373, 380; in Niven's lectures, 322; in Poynting's work on energy flow, 348; in J. J. Thomson's work on moving charged body, 339; and Tripos questions of mid 1870s, 303; in wranglers' understanding of relativity, 381, 392, 401

Mayall, R. H. D., 518

McAlpin, M. C., 520

McCann, H. W., 515

McCormmach, R., 11n.21, 369

McCrea, William M., 409n.23, 490, 492, 510

McIntosh, A., *316*, 516

McLaren, S. B., 432n.70, 434n.75, 437, 520

McVittie, G. C., 491, 492

Mécanique analytique (Lagrange), 29, 70, 72

Medawar, P., 7n.14

Mercer, J., 522

Merchant Taylors' School, 257, 260, 381

Mercury: electronic theory of matter on orbit of, 470, 471–72, 473, 474, 475; general relativity on orbit of, 445, 445n.9, 447, 453, 460, 464, 469, 470, 478, 478n.95

meritocracy, 103, 180, 225

Michell, J. H., 517

Michelson, Albert, 360

Michelson-Morley experiment, 360–61; in crisis in physics circa 1900, 359–60; Jeans on, 385; Larmor's electronic theory of matter and, 371, 374, 375, 378, 395; Nicholson on, 431–32, 432n.71

Miller, R. K., 326n.75

Milne, E. A., 490n.135

Milne, W. P., 522

Milner, Isaac, 131

Milton, H., 515

Minkowski, Hermann, 417n.40, 423n.49, 429, 430, 433, 453, 460n.49, 468

Minor Scholarships, 256, 261

Minson, H., 522

Mitchell, C. T., 512

mixed mathematics: in Cambridge University, 37–38, 49–113; development in Britain, 34–37; in early seventeenth century, 27; in late eighteenth century, 37, 58; material culture and practice-ladenness of, 114–75; pure mathematics contrasted with, 220–21; university teaching methods as unsuited to, 30

moderators: for disputations, 54, 122, *123;* for Senate House examinations, 54, 55, 61, 76, 78; for written examinations, 123, *124*

Modern Electrical Theory (Campbell), 496

modified Lagrangian function, 246, 336

Moir, A., 513

Mollison, W. L., 515
Monge, Gaspard, 70
Moody, R. H., 523
Mordell, L. J., 523
Morgan, Henry A., 238, 256
Morley, Edward, 360
Morton, E. J. C., *316*
Morton, R., 512
Morton, W. B., 518
Mott, Neville, 498, 507n.9
Moulton, H. F., 520
Moulton, J. Fletcher, 236, 513
mountaineering, 216, 216n.86
moving axes, method of, 159–60, 160n.103, 238
Munro, A., 518
muscular Christianity, 213, 214, 215

Nanson, Edward, 198, 198n, 300–301, 302, 322, 328, 514
Narliker, R. W., 490, 490n.133
National Physical Laboratory, 506
Natural Science Tripos, 218, 260n.67, 497
natural selection, 110, 111–12
Nature (journal), 474n.83
Nayler, W. A., 522
Neumann, Franz, 174, 174n.133, 504
Neville, E. H., 252, *283*, 451, 523
Newton, Isaac: annotated copy of Euclid's *Elements*, *31*; in Cambridge undergraduate studies of eighteenth century, 57; in development of calculus, 14, 14n.35, 29, 34, 35, 66, 67, 118; educational biography of, 15–16, 16n.38, 17, 17n.42; eighteenth-century interest in, 35; in emergence of mathematical physics, 28, 29; mathematics mastered by, 32; *Optics*, 57; Whewell's *Treatise on Dynamics* and, 145–49; written examinations and study of, 118–19. *See also* Newtonian mechanics; *Principia Mathematica*
Newtonian mechanics: Cambridge teaching of in eighteenth century, 118, 119; Eddington on electronic theory of matter and, 456; liberal education versus, 169; theory of universal gravitation, 12, 28, 473, 474, 476; Whewell on, 95, 180
Nicholson, J. W.: atomic theory of, 395n.78; on axiomatic view of relativity, 432, 443; on four-dimensional representation of electromagnetism, 423n.50; and general theory of relativity, 446; on mass as electrodynamic, 394–95; on Michelson-Morley experiment, 431–32, 432n.71; at Queen's College, Belfast, 434n.75; relation to relativity theory, 432–33; on relativity as derivable from electronic theory of matter, 403, 428–32; on Walker on dynamics

of charged sphere, 392, 394; in World War I, 436
Nicol, William, 326
Nind, William, 201, 206n.63
Niven, Charles: in Cambridge research school in electromagnetism, 351; on field theory, 332; as first-generation Cambridge Maxwellian, 325; as focusing on specific problems, 330–31; Larmor influenced by, 364–65; W. D. Niven as link between Lamb and, 332; paper on flow of currents, 327–28, 330, 336; and Poynting vector, 348; and Scottish natural philosophy, 325–26; as senior wrangler of 1867, 513
Niven, W. D.: in Cambridge research school in electromagnetism, 351; Eddington compared with, 486; as first-generation Cambridge Maxwellian, 325, 353; and Glazebrook, 353; and Larmor, 364, 365, 387; lecturing on Maxwell's *Treatise*, 298, 317–25, 332, 344; as link between Charles Niven and Lamb, 332; paper on electrical charge, 326–27; photograph of, *318;* Poynting studying with, 24, 350, 351; and Pupin, 315; relationship with Maxwell, 319, 332; Routh as teacher of, 327, 512; and Scottish natural philosophy, 325–26; and second-generation Cambridge Maxwellians, 333, 334, 353; Thomson studying with, 24, 335, 336, 344, 351; at Trinity College, 290, 291, 298, 317
Nixon, Randal, 363
non-Euclidean geometry, 446, 464–65, 466, 494, 500
"non-reading" men, 55
normal science, 172–73, 174
Norton, H. T. J., 522
notes, 134, 135

O'Brian, Matthew, 162
Ohm, Georg, 30, 33
"On the Electrodynamics of Moving Bodies" (Einstein), 404–6; British reception of, 404, 406, 469; Cunningham and, 411, 414–16, 426, 433; title determining readership of, 407
"On the General Theory of Relativity" (Pars), 489
"On the Influence of Gravitation on the Propagation of Light" (Einstein), 456
optics, 58, 59, 75, 87, 101, 240, 267, 288
Optics (Newton), 57
oral examinations: and liberal education, 127; on Newton's *Principia*, 130, 139–40; seen as unfair, 127; as time consuming, 122; Whewell on, 128; written examinations contrasted with, 118–19. *See also* Acts (disputations)
order of merit, 54–55; abolition of, 202, 284; fel-

lows appointed based on research rather than, 270; Galton on, 191; grades of, 56; losing its elite status, 218; and moral dimension of mathematics, 181; printing of, 55; private tutors as undermining, 61; reading of, 201–7, *204, 212;* and split of Senate House examination into two parts, 267, 269; the *Times* publishing, 202, 202n.54, *203*. *See also* wranglers

originality, 154, 163, 166, 272

Orr, W. M., 517

Osborn, G. F. A., 519

Osborn, T. G., 513

overwork, 200

Owens, Larry, 225

Owens College (Manchester): Eddington studying at, 449; Poynting at, 345; J. J. Thomson studying at, 334, 335, 336, 335n.92, 344

Oxford, Cambridge, and Dublin Mathematical Messenger (journal), 171, 171n.28, 273, 274–75

Oxford University: as center of mathematics, 171; liberal education ideal at, 169; Oxford and Cambridge boat race, 213; private tutors as "coaches" at, 89, 89n.115; sport and academic achievement at, 222n.95, 225; written examinations employed by, 57

Page, W. M., 522

Paley, Raymond, 222–23, 224

paper and pen: copy-books, 136, 137; Maxwell on, 114–15; paper-based methods and practice-ladenness of theory, 168–75; price of paper, 132, 132n; private tutors and paper-based learning, 135; quill pens, 132, 133, 135, 136; research and paper-based training, 137–38; shift from oral discourse to, 115, 129–38; in theoretical work, 16, 17, 41; at Trinity College, 131–32; written assessment and use of, 116–17. *See also* written examinations

paradigms, 172–73, 174

Paranjpye, R. P., 520

Parker, J., 516

Parkinson, Stephen, 231, 236, 238, 247n.37

Parliamentary Commission, 103–4, 110, 269

Pars, Leopold, 489, 490, 495

partial differentials, 29, 34, 71, 71n.66

Peace, J. B., 517

Peacock, George: and Airy, 73–74, 75, 78, 99; Board of Mathematical Studies proposed by, 102; and Continental analysis, 67–68, 70, 72, 72n.72, 146; on culture, 106; and Ellis, 86; as Lowndean Professor, 98; as moderator for Senate House examination, 67, 72, 78; and Parliamentary Commission of 1850s, 104; on private tutors, 97, 98–99, 98n.136, 104, 108, 109; translation of Lacroix's calculus textbook, 68, 71, 145; on undergraduate education and research, 51, 52, 94, 97–99, 108, 109; walking encouraged by, 192; as Whig, 108; worked calculus examples of, 145

Pearson, Karl: British and German students compared by, 176–78; on Cambridge examination system, 183, 186; on Frost, 237; on mind and body combined, 177–78, 226; on research and problem solving, 276; Routh as coach of, 233, 233n.10, 237, 515

pedagogy. *See* training

Peel, Robert, 105, 106, 106n.157, 107n.159, 110

Peel, T., 518

pen and paper. *See* paper and pen

Pendennis (Thackeray), 49, 92, 206

Pendlebury, R., 513

Perrine, C. D., 452

Peterhouse College, 232, 233n.8, 266, *290*

Philip, W. E., 519

Phillips, S. H., 521

philomaths, 35, 119

Philosophical Magazine, 471

Philosophical Transactions, 2, 6, 7, 36, 347

Physical Astronomy (Woodhouse), 74

physical exercise, 191–200; balancing intellectual endeavor and, 218–19; benefits attributed to, 199–200, 215; hard study and, 181–82, 192, 223; mountaineering, 216, 216n.86; peer pressure to participate, 198, 200; Roget's sketch pun on, *193;* as secret of success in mental advancement, 217; tennis and golf, 219, 222; walking, 192–94, 198, 219, 222, 223; by women, 218. *See also* rowing; team games

Physical Society of London, 468, 475

physico-mathematics, 28, 30, 171n.28

physics: calculus' development tied to physical problems, 29; classical, 3, 397n; and mixed mathematics in early seventeenth century, 27; optics, 58, 59, 75, 87, 101, 240, 267, 288; three major sites of physics teaching at Cambridge, 289. *See also* mathematical physics

Picken, D. K., 521

Pickstone, J. V., 4n.10

Pilkington, J. H., 516

Pinsent, H. C., 515

Pirie, G., 513

Pitt, George, 311, *312*

Planck, Max, 453

Plummer, H. C., 452–53, 452n.26, 454

Pocklington, H. C., 518

Pocock, A. H., 520

Poincaré, Henri, 372n.29, 376, 381n.51, 453, 456n.36

Poisson, Siméon Denis, 73, 148

polarization, electric, 370

"poll men," 55, 55n.11

Pollock, C. A. E., 516
Pollock, Frederick, 63, 63nn. 40, 41, 76
positive electrons, 369, 376, 377
potential theory, 294, 298, 300, 301, 303, 307, 313, 325, 451
Powell, F. C., 498n.156
Poynting, J. H.: Cambridge pedagogy shaping work of, 334, 344–45, 351–52, 355; on Cambridge research school, 286, 351, 355; Larmor's electronic theory of matter transcending work of, 375; lecture on university training, 351; at Mason College, Birmingham, 345–46; as oarsman, 26; at Owens College, 345; paper on energy flow, 2, 4–9, 5, 18–26, 344–52, 355; physics training at Cambridge, 3; questions on electrical conduction, 334; research in electromagnetic theory, 288, 291; as second-generation Cambridge Maxwellian, 42, 333, 353; studying with Niven, 24, 350; studying with Routh, 24, 25–26, 515; as teacher, 346; at Trinity College, 345, 350; in Tripos of 1876, 1–2, 19, 20, 21–25, 159, 349–50; on velocity of light, 348, 349; on velocity of sound waves, 346
Poynting vector, 2, 4, 5, 7, 8–9, 348–49, 354, 510
practice: associated with skills, tools, and location, 12; paper-based methods and practice-ladenness of theory, 168–75; teacher as final arbiter of proper, 230. *See also* experimental science
Prichard, Charles, 80n.95, 81n, 89, 134–35, 256n.54
Priestley, H. J., 522
Principia Mathematica (Newton): Airy studying, 73; and bookwork questions, 139; at Bristol College, 87; in British mixed mathematics of eighteenth century, 34; in Cambridge undergraduate studies of eighteenth century, 57, 58, 59, 66, 94; contemporaries having trouble comprehending, 13–15; European mathematicians in reception of, 17–18; in Kuhn's account of normal science, 172–73; mathematics and mechanics unified in, 28; in oral examinations, 130, 139–40; private tutors and teaching of, 79, 144; reworked in Leibnizian language, 29; and Smith's Prize, 120; unpublished mathematical techniques in, 14, 14n.33; Whiston teaching, 32n.60; Wright's edition of, 82; Wright studying, 70, 71
Principle of Relativity, The (Cunningham), 435, 455
Principles of Analytical Calculation, The (Woodhouse), 67, 145
Principles of Quantum Mechanics, The (Dirac), 498n.157
Prior, C. H., 514
private science, 504

Private Tutor and Cambridge Mathematical Repository, The (journal), 82, 83
private tutors: in Cambridge mathematics instruction, 37–38, 40, 49–52, 58–66; as "coaches," 40, 49, 52, 89–94, 89n.113, 103; coaching success, 1865–1909, 512–25; and Continental analysis, 50–51, 65–84; craft of, 88–89; demise of coaching system, 280–85; fees of, 69, 69n.61, 80n.91, 84n.101; high point of coaching, 239; Hopkins on, 84, 84n.99, 100–101, 103–5, 108, 110; integrating into Cambridge teaching system, 112–13; intercollegiate lectures and, 264, 265–66, 271, 280; intermediate steps provided by, 153; mastering electromagnetic theory, 305; mastering new topics, 149; and Maxwell's *Treatise*, 306–17; as moderators, 61; natural selection and, 112; as necessity for honors, 76n.80, 78; as new approach to university education, 59; new developments in 1830s, 84–94; organizational economy of, 241; and paper-based learning, 135; Peacock on, 97, 98–99, 98n.136, 104, 108, 109; personal interaction with pupils, 70, 227–28, 241–45; prepared manuscripts of, 144; as "pupilizing," 69; regulation of, 61–62, 63, 79n.90; relationship with pupils, 88, 97, 98; relaxation of rules on, 50, 51, 62, 63–64, 65, 70, 71, 79–80; Routh and art of good coaching, 229–47; as site of physics teaching at Cambridge, 289; as sportsmen, 200; Tait on, 111, 112; two lines of descent in, 231; Victorian era attacks on, 94–109; Whewell on, 61, 62n.32, 96–97, 102–3, 109; for women, 281–82; written examinations and emergence of, 50, 60, 111, 118, 170
"Problem in Dynamics, A" (Maxwell), 155, 164n.112
problem questions, 139, 141–44
problems, examination. *See* examination questions
problem solving: archive of Tripos problems for improving, 144; Continental analysis and tradition of, 144–51; DeMorgan on Cambridge emphasis on, 152–54; in examinations versus research, 228–29; "fights" for practice in, 88, 236; as model for research, 272–80, 329–30; originality in, 154, 163, 166; paper-based, 117; personal interaction in learning, 70, 89; the "problem age," 221; Routh's approach to teaching, 243–46; sportsmanlike, 222–23, 224; teachers in, 230; written examinations and, 141
problems sheets, 122, 138, 139
Professor of Experimental Physics: Maxwell as, 265, 266, 287, 300, 304; J. J. Thomson as, 343

INDEX 565

prostitutes, 214, 214n.79
provincial universities, 255
Pryme, George, 62–63, 62n.38, 90
public schools: competitive games at, 191, 212–13, 222; Merchant Taylors' School, 257, 260, 381; St. Paul's School, 219–20, 258–60, 263; sending their most able students to Cambridge, 228; Tonbridge School, 257; Winchester College, 219, 258, 260; wranglers teaching mathematics at, 219, 254–64
public science, 504
public teaching, 50n.3; private tutors and, 50, 102, 113, 127; Public Lecturers created, 104, 104n.152, 105, 113; reform of, 264–72; Whewell and, 94–97, 109
Pupin, Michael, 198–99, 243, 244, 315–16
pure mathematics: in mathematical training of physicists, 505; and reform of Cambridge training system, 276, 284, 285, 434; in syllabus of 1880s, 267; William Thomson railing against Cayley's, 264n.72; and Webb's teaching methods, 249; younger generation of mathematicians favoring, 221

quantum mechanics, 491, 497, 498, 507, 508–9
Quarterly Journal of Applied Mathematics, 159, 160
Queens College, 131, 135
questionists, 55
questions, examination. *See* examination questions
quill pens, 132, 133, 135, 136

Rabinow, P., 4n.8
Radford, E. M., 519
railway travel, 102–3
Rajan, A. T., 522
Ramsey, A. S., 518
Randell, J. H., 516
rationality, 396
Rau, B. N., 523
Ravetz, J. R., 16n.39
Rayleigh, Lord (J. W. Strutt): in Cambridge research school in electromagnetism, 351; examination question of inspiring research by, 158–59; examination scripts used for drafts, 163n.110; physics training at Cambridge, 3; and Poynting's paper on energy flow, 2, 347, 349–50; preparation in mathematics of, 256–57; on problem-solving approach to mathematics, 276–77; and Pupin, 316; questions for Tripos of 1876, 1–2, 18–26, 159, 309–11, 313, 315, 349–50; as relaxed while taking Tripos, 187, 216; Routh coaching, 231n.5, 234n.11, 242, 245–46, 512; Searle maintaining traditions of,

401; Stuart urging to read Maxwell, 325; on Thomson and Forsyth, 320n; J. J. Thomson working under, 335, 336, 343. *See also Theory of Sound, The*
reading men, 55, 182, 221, 224, 224n
ready-made science, 7
red shift, gravitational, 444, 445, 469, 478, 484, 494
Reeves, J. H., 518
regimen of weekly study, 243–44
Relativitätsprinzip, Das (von Laue), 435, 436, 440
Relativity and the Electron Theory (Cunningham), 435, 437, 437n.83
relativity theory: as axiom, 381, 387, 401, 403, 431, 432, 433, 439, 441, 443, 446, 465; Bucherer's experimental proof of, 400; Cambridge response to, 43–44, 358, 359–60, 398, 399–500; Cunningham and Bateman's new theorem of relativity, 402–3, 416–24, 433; Cunningham's relation to, 409, 432–33; German textbooks on, 440–41; and incommensurability, 43–44; in Livens's *Theory of Electricity*, 387–88, 389; and supposed crisis in physics, 359; "theory of relativity" used, 406, 406n.17, 430; wranglers' understanding, 381, 388. *See also* general theory of relativity; special theory of relativity
religious tests, 270–71
"Report on Electrical Theories" (Thomson), 341n, 343
Report on the Relativity Theory of Gravitation (Eddington), 467–69, 475, 479, 493n
research: associated with postgraduate studies, 26; Cambridge debate over undergraduate education and, 51–52, 94–101; competition, fair play, and manliness in Cambridge style of, 179, 222; in examination questions, 117, 155–60, 272; examination scripts compared with, 23, 24, 25; intercollegiate lectures as forums for, 271; as open-ended, 279; and paper-based training, 137–38; problem solving as model for, 272–80, 329–30; reforms of 1870s and 1880s and, 270; Routh's teaching shaping, 246–47; and setting Tripos problems, 274; training and, 109, 228–29, 272–80, 288, 492. *See also* research schools
research schools: on electromagnetic theory in Cambridge, 287, 288, 352–56, 388–89; failing to develop in Leiden, 499; Maxwell's *Treatise* as focus of, 46, 352–53; Poynting on, 286, 351; reforms of 1870s and 1880s for facilitating, 270
Reynolds, Osborne, 3, 335n.92
Richards, Joan, 465
Richardson, George, 258, 260
Richardson, J. G., 514
Richie, W. I., 514
Richmond, H. W., 517

riders, 163
Rives, G. L., 514
Roberts, Samuel, 260
Robertson, H. P., 477n.93
Robson, H. C., 258, 516
Rogers, Mr. (Thomas), 73, 73n.74, 132
Roget, J. L., *93*, *193*, *196*
Rose, P. J. G., *248*
Ross, E. B., 522
rotating-wheel paradox, 430–31
Rothblatt, Sheldon, 49–50, 191n.33
Rouse, J., 4n.8
Rouse Ball, W. W.: on Cambridge school of mathematics, 388–89; on Euclid in oral examinations, 140n.52; on examination problems, 141; on private tuition, 50, 61, 61n.28, 62, 62n.38, 63, 63nn. 41, 43, 79, 241; on results of abandoning old Tripos, 285; on textbooks, 142n.61; on written examinations, 114, 115, 125, 126
Routh, Edward, 229–47; on abandoning order of merit, 284; Adams Prize-winning essay of, 246, 336; on applying potential theory to electrostatics, 300; and Bryan, 506; and Chrystal, 301; competition among students of, 238; *Dynamics of a System of Rigid Bodies*, 253, 258; Hopkins as coach of, 231, 233, 235, 524; humor of, 241; interaction with his pupils, 236; Klein on, 253; and Larmor, 364; lecture notes of, 234n.12, 240, *308*; marriage of, 233n.8; as mathematics coach, 25, 38, 41, 229–47; Maxwell's *Treatise* taught by, 306–9, 311, 315–17, 320–21, 323, 333, 334, 358; mixed mathematics defended by, 220; moving axes method, 159–60, 160n.103, 238; new techniques taught by, 238; and Niven, 327; performative dimension of teaching of, 242–45; Peterhouse fellowship and lectureship of, 232–33, 233n.8; physics training at Cambridge, 3; portrait of, *232;* Poynting studying with, 24, 25–26, 345; and Pupin, 198, 199, 243, 315–16; and Rayleigh, 231n.5, 234n.11, 242, 245–46; on Rayleigh's *Theory of Sound*, 324; reading of order of merit, 204–5; as researcher, 112, 246; research shaped by teaching of, 246–47; retirement of, 280; and second-generation Cambridge Maxwellians, 333, 353; as "senior wrangler maker," 231, 233, 236; as senior wrangler of 1854, 231; success as coach, 231, 233, 238, 524; team for Tripos of 1880, *316;* textbooks of, 260, 261, 493; Thomson studying with, 24, 335; Todhunter as coach of, 231; training methods of, 233–36, 239–40; transformation techniques used by, 373; translating mathematical culture of, 253, 255, 258, 260, 261; on Tripos problems, 161, 162, 168; "upper ten" wranglers coached by, 512–17; walking by, 200; and Webb, 247; and women students, 281–82, 282n.99
Rowe, R. C., 515
rowing: balancing physical and intellectual endeavor, 219; becoming means of exercise, 194–95; implicating the body in learning process, 223–24; intercollegiate, 194, *195*, 213; Oxford versus Cambridge boat race, 213; Pupin on, 199; reading versus rowing men, 224, 224n; Roget's "The Boats," *196;* seen as means of transport, 192; Trinity College first boat for Lent Term races of 1891, *220;* by wranglers, 198
Royal Aircraft Factory, 506
Royal Astronomical Society, 476
Royal Institution, 475
Royal Society: announcement of confirmation of general relativity, 476; equivocal view of mathematics, 36–37; *Philosophical Transactions*, 2, 6, 7, 36, 347; referees of Poynting's paper on energy flow, 8–9, 23
rugby, *199*, 213
Rumsey, C. A., 519
Runge, Carl, 227, 505
Russell, Bertrand, 220, *220*, 221, 258, 519
Rutherford, Ernest, 394, 497

St. John's College: Besant at, 231n.4, 236; coaches as sportsmen, 200; coaching tradition in, 231n.4; college examinations at, 126; Cunningham at, 409, 434–35, 436, 437; Dirac at, 510; as disproportionately powerful, 53; fellows appointed based on research, 270; Gwatkin at, 68, 72, 76, 86; and intercollegiate lectures, 266; Lady Margaret boat club, 194; Larmor at, 364; in pedagogical geography of mid-Victorian Cambridge, *290;* Prichard on lectures at, 134–35; private tuition paid for by, 80n.95, 86; rugby team of 1881, *199;* as seeking to dominate Senate House examination, 64; Webb at, 247
St. Paul's School, 219–20, 258–60, 263
Sampson, R. A., 517
Sanchez-Ron, J. M., 417n.38, 492
Sandars, Thomas, 213
Sanger, C. P., 518
Sargant, E. B., 515
Satthianadhan, Samuel, 198, 212, 217
Saunderson, Nicholas, 34–35, 35n.73
Scholfield, W. F., 522
schools of research. *See* research schools
Schott, G. A., 391, 415n.35
Schuster, Arthur, 335, 335n.92, 345, 449
Schwartzchild, Karl, 493n

INDEX

Schwartzchild solution, 478, 484
science: astronomy, 27, 28, 59, 87, 240, 267, 452–54, 463–64; emergence in the West, 502; laboratory revolution in, 115; normal science, 172–73, 174; private versus public, 504; ready-made science, 7; recoverability from written sources, 501–2; shortage of teachers in, 503; social and cultural studies not applied to mathematical work in, 11–12; sociology of, 44–45; tacit knowledge in, 45–46; truths of seen as transcending culture, 501–2. *See also* experimental science; physics; theory
Scott, Charlotte, 281
Scott, R. F., 189, 514
Searle, G. F. C., 399–402, 407
Sedgwick, W. F., 519
Seeley, Robert, 216
Segar, H. W., 518
Selby, F. J., 518
Senate House: in pedagogical geography of mid-Victorian Cambridge, *290*. *See also* Senate House examination (Mathematical Tripos)
Senate House examination (Mathematical Tripos), 52–58; abandoning the old Mathematical Tripos, 284–85, 284n.106; archive of questions from, 155, 161, 487; becomes Mathematical Tripos, 77; changes in late eighteenth century, 56–57; college examinations and, 126; as common standard for college teaching, 64; conformal transformation in questions, 421, 422; and Continental analysis, 67, 70, 97; disputation contrasted with, 129; electromagnetic theory reintroduced into, 287, 291, 292, 325, 331–32, 354; examiners, 55, 76, 78n.87; formal marking of, 84; general relativity in questions of, 486–89, *488;* honors students increasing, 77; as increasingly competitive, 56–57, 58, 182; interior of Senate House, *54;* Jeans's *Mathematical Theory* and, 381–86; Larmor's electronic theory of matter and questions in, 377–81, *378,* 434; leadership and excelling in, 181; losing its elite status, 218; and manliness, 212–18; mathematics and natural science emphasized in, 56, 58, 64; Maxwell on, 114; Maxwell's *Treatise* and problems in, 302–3, 309–16; moderators, 54, 55, 61, 76, 78; new days added in 1830s, 84, 94n; new methods and results in, 161; Part I, 267, 269, 283, 434; Part II, 267, 269, 284, 434; Part III, 267, 269; Peacock on changing, 98; pedagogical use of past problems, 161–62; and Poynting's paper on energy flow, 1–2, *19,* 20, 21–25, 159, 349–50; preparatory examinations for, 58; pressures of, 182–91; printing of questions for, 77; private teaching and, 61, 62, 63–64, 76n.80, 78–79; publications of collections of problems, 143–44, 143n.64, 153n.85; publishing solutions to past problems, 149–50, 160–63, 166–67; reforms of 1848, 102, 162–63; reforms of 1868, 264–65; reforms of 1909, 284, 434; research and setting Tripos problems, 274; research problems in, 117, 155–60, 272; scoring more than full marks, 167, 167n.116; sexual abstinence in preparation for, 214; sitting at end of just six terms, 260, 260n.67; split into two parts, 267, 269; syllabus for 1873–82, *239,* 240; tacit knowledge and, 151–55; as test of rational mind under duress, 216; textbooks and increasing difficulty of, 141–44, 150–51; J. J. Thomson's electromagnetic research in questions, 339–40, *341;* for training of mathematicians, 111, 112; Whewell on, 96; women sitting, 281–82; as written, 56, 58, 60, 64–65, 118, 119, 125, 130. *See also* order of merit
senior optimes, 55
senior wranglers, 56; announcement of, 202, 204–5; Alexander Brown in his study, *245;* Dawson in training of, 62; degree conferred upon, 207–8, *207;* Routh as "senior wrangler maker," 231, 233, 236
Sewell, C. J. T., 522
sexual abstinence, 214–15, 223
Shapin, Steven, 12
Sharpe, F. R., 518
Sharpe, J. W., 515
Shaw, H. K., 522
Shaw, William, 157, 500
Sheppard, W. F., 517
Siddons, A. W., 520
Simpson, Thomas, 35, 35n.76, 148
slates, 133, 133n.36
Slator, Frank, *248,* 521
Slesser, George, 159–60, 160n.103
Small, W. H. T., 522
Smallpiece, A. J., 519
Smedley, F. E., 92
Smith, Archibald, 157, 184
Smith, C., 513
Smith, H., 522
Smith, J. P., 515
Smith, Sydney, 191–92, 225
Smith's Prizes, 56, 56n.15, 119–20, 120n.15, *121,* 156, 231, 270
Snow, C. P., 219
social Darwinism, 110
societies, mathematical, 35, 119
sociology of scientific knowledge, 44–45

Soddy, F., 385n
Space, Time, and Gravitation (Eddington), 482–83, 485, 496n.147, 508
specialization, undergraduate, 284–85
special theory of relativity: Cambridge response to, 362, 399–442, 496–98; Campbell attempting to popularize, 496; in Cunningham's "Electron Theory" course, 437; Eddington's course on, 497–98; $E = mc^2$, 497; electromagnetic view of nature supplanted by, 437; a few Cambridge physicists acknowledging, 402, 406, 408; general theory supplanting, 445, 445n.8; and Michelson-Morley experiment, 359, 360–61; reception studies and, 404–9. *See also* "On the Electrodynamics of Moving Bodies" (Einstein)
Spence, W. M., 513
Spitalfields Mathematical Society, 35, 37n.82
sports. *See* team games
Stachel, J., 452n.25
Stamp, A. E., 518
Star Coach, 90, 91
Starte, O. H. B., 521
steady-state cosmology, 490
Steele, William, 158, 231, 232
Steggall, J. E. A., 515
Steinthal, A. E., 320, 516
Stellar Movements and the Structure of the Universe (Eddington), 454n.30
Stephen, Leslie, 187, 215, 216n.86
Stephen, W. D., 513
Stewart, Balfour, 335, 345
Stewart, F. W., 522
stoicism, 190–91
Stokes, George: Hopkins and, 108; physics training at Cambridge, 3; preparation in mathematics of, 87, 87n.110; Pupin and, 316; Searle maintaining traditions of, 401; Stokes's theorem as examination question, 156; walking habits of, 192–93
Stokes, W. F., 516
Stokes's theorem, 156
Strachan, J., 521
Stratton, F. J. M., 521
Strutt, J. W. *See* Rayleigh, Lord
Stuart, G. H., 514
Stuart, James: electro- and magnetostatics taught by, 317; and first-generation Cambridge Maxwellians, 325–26; and intercollegiate lectures, 266–67, 266n.75; lectures in electricity and magnetism, 332; potential theory taught by, 300, 300n.29, 325; and Scottish natural philosophy, 325–26; as third wrangler in 1866, 318, 513
Student's Guide, 260, 280

Sunderland, A. C., 515
supervisors, 285, 304

tacit knowledge: Cambridge mathematics and, 151–55; of the reader, 439; in social approach to scientific knowledge, 45–46
Tait, Peter Guthrie: as additional examiner in 1876, 302; debate with Todhunter over written examinations, 110–12; Hopkins as coach of, 87; on Maxwell exercising, 196; on Maxwell's *Treatise*, 292, 299n.26; physics training at Cambridge, 3; *Treatise on Natural Philosophy*, 221, 253n.47, 292, 294, 297n.23, 324, 325, 345; and Tripos problems as research topics, 158
Tanner, E. L., 521
Tanner, J. R., 3n.4
Taylor, H. M., 297n.23, 512
Taylor, W. W., 514
teaching: attempts to reform, 229; shortage of science and mathematics teachers, 503; teacher as final arbiter of proper practice, 230; textbooks and private study as ancillary to, 227–28; in theory transmission, 493–500. *See also* lectures; private tutors; public teaching; textbooks
team games: at American universities, 225; at Cambridge, 213; as competitive, 179, 182; cricket, 178, 195, 198, 213, 219, 222; football (soccer), 198, 213, 219; in modern higher education, 226; at public schools, 191, 212; rugby, 199, 213; and sexual abstinence, 214–15
telegraphy, 298, 307, 309
Temperley, Ernest, 281, 513
tennis, 219, 222
Terry, T. W., 514
textbooks: and analysis, 67, 68; as ancillary to teaching, 227–28; Atkinson on, 132; Eddington's *The Mathematical Theory of Relativity* as, 484; before the 1830s, 141–42; end-of-chapter exercises, 38, 41, 117, 168, 172, 382; examination questions as exercises in, 117, 167, 171; and increasing difficulty of Tripos questions, 141–44, 150–51; Jeans's *Mathematical Theory* as, 381–86; Kuhn on normal science and, 172, 173, 230; in material culture of mathematics, 41; Maxwell's *Treatise* as, 42, 287, 288, 289, 293–98, 352; Rayleigh's *Theory of Sound* as, 324; on relativity, 440–41, 478, 496; Routh's use of, 234–35; tacit knowledge and Cambridge, 151–55; and translating mathematical culture, 253–54, 260, 261–63; Whewell's *Treatise on Dynamics* as, 145–49; worked examples in, 161
Thackeray, William Makepeace, 49n.1, 91–92, 206
theory: associated with isolated genius, 12, 15, 16,

INDEX 569

23; distinct culture of, 504; in-the-learning rather than in-the-making, 25; paper-based methods and practice-ladenness of, 168–75; pedagogy and change in, 357–63; as site specific, 12–18, 403, 439, 440–41, 447, 493–94; teaching in transmission of, 493–500; top-down view of, 502; training and origins of, 506–11; uncertainty in, 23; written examinations as revolution in, 115

Theory of Electricity, The (Livens), 387–88, 387n.61, 389, 437

Theory of Sound, The (Rayleigh), 277; Eddington's work on gravity and, 484n.113; in electromagnetic theory, 323–25; and first-generation Cambridge Maxwellians, 353; FitzGerald influenced by, 340; Lamb drawing on, 330; as model of clarity, 297n.23; and Poynting's paper on energy flow, 344, 345, 346, 347, 349, 350, 352; Rayleigh's training and, 506; translation into German, 253n.47

"Theory of the Aberration of Light, The" (Larmor), 371–72

thermodynamics, 267, 278

Thomas, J. L., 521

Thomas, Llewellyn H., *438*, *481*, 485n.117, 490, 492

Thompson, A. P., 521

Thompson, A. W. H., 523

Thompson, Frederick, 257

Thompson, T. P., 520

Thomson, J. J.: on Adams's lectures, 136–37; on application of Lagrangian dynamics, 336; Cambridge pedagogy shaping work of, 334, 338, 339, 355; in Cambridge research school in electromagnetism, 351, 355; as Cambridge's leading exponent of Maxwell's theory, 334, 343; on Cayley's teaching, 136; on confirmation of general relativity, 476; on convection of light moving through transparent medium, 337, 339; determination of ratio "v," 335, 336; early work in electromagnetic theory, 334–44, 353; on electromagnetic theory of light, 336–38; electron identified by, 376, 394; and Forsyth, 320n; on genteel sports of 1880s, 221; as ideal type of high wrangler, 334–35; insomnia while taking Tripos, 187; intellectual communication by, 356; Larmor's electronic theory of matter transcending work of, 375; on Maxwell's *Treatise*, 321; on moving charged conductor, 338–39, 368–69; on Niven, 318n, 319–20; on personal interaction with pupils, 271–72; physics training at Cambridge, 3, 23–24; and Poynting, 23, 350, 351; on Poynting vector, 9; on problem solving versus research, 273, 274; as Professor of Experimental Physics, 343; on propagation of electromagnetic waves, 340–43; "Report on Electrical Theories," 341n, 343; and Routh, 24, 200, 235, 236, 261, *316*, 335; on rowing, 219; Searle's electrodynamics deriving from, 401; as second-generation Cambridge Maxwellian, 42, 333, 353; as second wrangler of 1880, 516; studying with Niven, 24, 335, 336, 344, 351; at Trinity College, 336, 338, 343

Thomson, William: on Airy, 108n.164; on Cayley, 264n.72; collected papers on electrostatics and magnetism, 304n.36; copy-book of, 136, 137, 222; and error in Maxwell's *Treatise*, 301; examination question inspiring research of, 158; as having left Cambridge, 291, 506; and Hopkins, 87, 108, 135, 190; on paper and pen at Cambridge, 135; physics training at Cambridge, 3; as referee for Poynting's paper on heat flow, 7n.13, 23; rowing by, 195, 219; theory and experimentation combined by, 264; *Treatise on Natural Philosophy*, 221, 253n.47, 292, 294, 297n.23, 324, 325, 345; undergraduate experience of, 185; work seen as kind of race by, 222

Times (newspaper), 202, 202n.54, 203

Titterington, E. J. G., 522

Todhunter, Isaac: on benefits of examination system, 217; on Cambridge training producing specialists, 264–65; debate with Tait over written examinations, 110–12; Hopkins as coach of, 87; on new results in Tripos papers, 157; on practice in problem solving, 230; as private tutor, 231; selecting his textbook examples from Tripos papers, 167

Todhunter, R., 518

Toller, T. N., 513

Tonbridge School, 257

Toplis, John, 171n.125

Toryism, 106, 107

training: associated with undergraduate studies, 26; athleticism and mathematics in Cambridge, 176–226; Cambridge tutors' methods, 38, 40; conservative pedagogy at Cambridge, 67, 356, 358, 438, 439, 505–6; convergence of assessment and, 117; Hopkins's teaching methods, 85; and innovation, 505–11; in mathematics before twentieth century, 27–40; mid-Victorian Cambridge system of, 26; as neglected in history of physics, 3–4; and persistence of false theories, 396–98; process versus content of, 38–39; recent origins of modern system of, 26; and research, 109, 228–29, 272–80, 288, 492; Roget's "The Boats" punning on, *196*, and theory change, 357–63; in way of knowing, 4, 39. *See also* teaching

Traité de mécanique céleste (Laplace), 29, 66, 70, 71, 72, 146
Traité de physique (Biot), 72
Traweek, S., 225–26
Treatise on Differential Equations (Forsyth), 254
Treatise on Dynamics (Whewell), 145–49
Treatise on Electricity and Magnetism (Maxwell), 286–356; broad audience intended for, 288–89; and Cambridge research school in Cavendish Laboratory, 46, 352–53; on current-sheets, 330; dynamical theory of electromagnetic field, 295, 298; electrical images method in, 326; errors in, 137, 296, 297, 297n.23, 301, 302, 321; first-generation Cambridge Maxwellians, 325–33, 353; four major sections of, 293; interpretation as collective product, 42, 299–306; mapping techniques in, 373; Maxwell's personal knowledge in, 289, 299; metrological dimension of, 294–95; Niven lecturing on, 317–25; physical analogies in, 321; Poynting's paper on energy flow citing, 8; private tutors and, 306–17; Routh teaching, 306–9, 311, 315–17, 320–21, 323; second edition of, 302, 319, 322; second-generation Cambridge Maxwellians, 333–52, 353, 354; as textbook, 42, 287, 288, 289, 293–98, 352; three aspects of, 298; translation into German, 253n.47; and Tripos problems, 302–3, 309–16
Treatise on Natural Philosophy (Thomson and Tait), 221, 253n.47, 292, 294, 297n.23, 324, 325, 345
Trimble, C. J. A., 522
Trimmer, F. G., 515
Trinity College: boat club formed at, 194; and Cambridge research school in electromagnetism, 351; college examinations at, 126; as disproportionately powerful, 53; Eddington at, 449–51, 452; Eddington lecture on relativity at, 482; fellows appointed based on research, 270; fellowship examinations as written, 130; first boat for Lent Term races of 1891, *220;* first-generation Cambridge Maxwellians at, 325, 326, 328, 331–32; Great Hall as examination room, *124,* 128; Herman at, 282; and intercollegiate lectures, 266; Jeans at, 381; Maxwell at, 326, 332; Niven at, *290, 291,* 298, 317; paper and pen in mathematics at, 131, 133–34, 135; Peacock at, 68, 72; in pedagogical geography of mid-Victorian Cambridge, *290;* Poynting at, 345, 350; and private tuition, 135n.43; as seeking to dominate Senate House examination, 64; Thackeray depicting, 92; J. J. Thomson at, 336, 338, 343; Whewell as Master of, 105

Tripos, Mathematical. *See* Senate House examination (Mathematical Tripos)
Trotter, Coutts, 266n.76
Troup, J. M., 518
Turnbull, H. W., 522
Turner, E. G., 519
Turner, H. H., 235, 241, 247, 517
tutors, private. *See* private tutors

uncertainty, 22–23
unified field theories, 492
Universities of Oxford and Cambridge Act (1877), 269
University College London, 231, 276
university lectures, 229, 264, 282, 343

Valentine, G. D., 520
Valentine-Richards, A. V., 517
Van der Waerden, B. L., 443n
Vanity Fair (Thackeray), 91
Varignon, Pierre, 29
Vaughan, A., 518
Verdant Green (Bede), 92, 92n.122
Vièta, François, 28
Vince, Samuel, 74
Virtual Velocities, Principle of, 29
"Vision, A" (Maxwell), 1

Walker, G. T., 260n.68, 517
Walker, G. W., 391–92, 394, 520
Walker, George F., 281, 515
walking, 192–94, 198, 219, 222, 223
Wallis, A. J., 516
Walters, F. B., 515
Walton, William, 161–62, 167n.118, 190
Ward, Joseph T., 210n.68, 211n.69, 243–44, 306, 311, 313, *314,* 515
Waring, Edward: Continental analysis contrasted with, 34; difficulty understanding Euler, 173n.130; as examiner, 40; as Lucasian Professor, 35n.76; mathematical education of, 35; paper in *Philosophical Transactions,* 36; as Smith's Prize examiner, 120n.15, *121*
Warren, J. W., 514
Waterfall, C. F., 523
Watkin, E. L., 520
Watson, G. N., 522
Watson, H. A., 522
Watson, H. C., 512
Watson, Henry W., 162, 216n.86
Watt, J. C., *316*
wave theory, 240
Waymouth, S., 513
Webb, H. A., 521
Webb, Robert R., 247–52; coaching notes of,

247–52, *250;* coaching rooms of, *290;* and Cunningham, 409; and demise of coaching system, 282; and electronic theory of matter, 388; and Jeans, 381; and Laplace's equation, 278, *314;* photograph of, *248;* ranking as coach, 524; at reading of order of merit in 1905, *204;* retirement of, 282; as senior wrangler of 1872, 514; as sportsman, 200; teaching methods of, 247; "upper ten" wranglers coached by, 515–21
Weber, Wilhelm, 33
Weinberg, Steven, 10
Weldon, W. F. R., 276
Welsford, J. W. W., *316,* 516
Welsh, W., 517, 522, 523, 525
Western, A. W., 519
Westfall, R. S., 16n.38, 17n.42
Weyl, Hermann, 484
Wheatstone's Bridge, 298, 303, 307, 309, 310, 311, 312
Whewell, William, 95; on competitive examinations, 127–28; as conservative Anglican, 95, 106; and Continental analysis, 95, 102–3, 146, 181; on culture, 106–7; *An Elementary Treatise on Mechanics,* 73; at Ellis's degree ceremony, 208; on examiners, 78n.86; large classes supported by, 106; on liberal education, 94, 95, 96, 102, 103, 106, 109, 180, 224–25; as Master of Trinity, 105; on mathematics curriculum, 101–2, 109; on oral disputation, 128; and Parliamentary Commission of 1850s, 103; on private tuition, 61, 62n.32, 96–97, 102–3, 109; on reasoning as practical art, 227; textbooks of, 77, 95; *Treatise on Dynamics,* 145–49; on undergraduate education and research, 51, 52, 94–97, 99, 101–4, 105–6, 109
Whigs, 103, 110
Whipple, F. J. W., 519
Whiston, William, 32n.60
Whittaker, Edmund T.: analysis textbook of, 221; on conductors in electromagnetic research, 328; Dirac studying work of, 508, 509, 510; Klein commissions work by, 252n.46; physics training at Cambridge, 3; on propagation of gravitational effects, 451n.22; on relativity, 453, 453n.27, 454, 454n.29; as third wrangler of 1895, 519; on unification of gravitation and electromagnetism, 492
Wiener, Norbert, 222–23, 224
Wilkinson, A. C. L., 519
Wilson, G. H. A., 519, 521, 522, 523, 524
Wilson, J., 515
Wilson, James, 159, 188, 238, 255n.54
Wilson, S. R., 515
Wilton, J. R., 522

Winchester College, 219, 258, 260
Wolstenholme, Joseph, 156, 238, 238n.22, 261–63
women: and physical capacity to study mathematics, 217–18; prostitutes, 214, 214n.79; in publication of order of merit, *203;* sitting the Mathematical Tripos, 281–82
Wood, A., 512
Wood, J., 516
Wood, P. W., 521
Woodcock, T., *316*
"wooden spoon," 56, 56n.14, 208–11, *209, 210,* 210n.68
Woodhouse, Robert, 67, 74, 145
Wordsworth, Christopher, 61, 192
Workman, W. P., 517
wranglers, 55; appointed to senior positions at Cambridge, 358; balancing physical exercise and intellectual endeavor, 218–19; becoming moderators, 78; Bristed on the ideal, 215; class background of, 107n.160; college fellows as, 60; as mathematicians, 112; private tuition by, 69; public schools in wrangler-making process, 228; reading of order of merit, 201–7; sexual abstinence for, 214; and Smith's Prizes, 56; and split of Senate House examination into two parts, 267; sporting activities of, 197–98; stresses in becoming, 182–91; teaching in public and grammar schools, 254–64; J. J. Thomson as ideal type of, 334–35; "upper ten," 1865–1909, 512–23; viewing their training vocationally, 218. *See also* senior wranglers
wrangler's walk, 192, 192n.36
Wren's Coaching Establishment, 93, 93n.128, 258
Wright, J. E., 520
Wright, J. M. F.: collections of Tripos solutions of, 149, 160; edition of Newton's *Principia* of, 82; *Hints and Answers,* 149; as mathematics teacher, 81–82; on paper and pen at Trinity, 131, 133–34; pass degree of, 72n.71; on past Tripos examination papers, 143n.64; on pressures of the Tripos, 186–87; *The Private Tutor and Cambridge Mathematical Repository,* 82, 83; on prostitution, 214n.79; reminiscences of Cambridge life, 89; undergraduate experience of, 69–72; on the "wooden spoon," 210n.68
Wright, R. T., 513
Wrigley, P. T., *316*
writing: blackboards, *93,* 134, 136, 234, 234n.13; notes, 134, 135; slates, 133, 133n.36. *See also* paper and pen; written examinations
written examinations: as altering skills and competencies, 115–16, 139–41; becoming common in British education, 258; earliest firsthand account of, 124–25; and liberal education, 127, 129; and meritocracy, 225; moderators in, 123,

written examinations (*continued*)
124; new cultural values associated with, 129; oral examinations contrasted with, 118–19; Oxford and Cambridge adopting, 57; and paper-based techniques of calculation, 116–17; as pervading British life, 110; pressures and anxieties of, 127–28; private tutor's emergence and, 50, 60, 111, 118, 170; ranking students made easier by, 123–24; Rouse Ball on, 114; Senate House examination becomes, 56, 58, 60, 64–65, 118, 119, 125, 130; as separating examination from adjudication, 123, 125; for Smith's Prize, 119–20, *121;* Tait-Todhunter debate over, 110–12; Trinity College fellowship examinations as, 130. *See also* competitive examinations

x-rays, 390–91

Yale University, 225
Yeo, J. S., 516
Young, W. H., 517, 522, 525